The Planet Venus

Yale Planetary Exploration Series

The Planet Venus

Mikhail Ya. Marov and David H. Grinspoon

with translations by Tobias Owen, Natasha Levchenko, and Ronald Mastaler

Yale University Press   New Haven & London

Published with the generous assistance of Martin Marietta
Astronautics Group.

Designed by Sonia L. Scanlon.
Set in Bulmer type by G & S Typesetters, Austin, Texas.
Printed in the United States of America by Thomson-Shore, Inc.,
Dexter, Michigan.

Library of Congress Cataloging-in-Publication Data
Marov, Mikhail I͡Akovlevich.
    The planet Venus / Mikhail Ya. Marov and David H.
  Grinspoon ; with translations by Tobias Owen, Natasha
  Levchenko and Ronald Mastaler.
        p.   cm. — (Yale planetary exploration series)
    Includes bibliographical references and index.
    ISBN 0-300-04975-7 (alk. paper)
    1. Venus (Planet)   I. Grinspoon, David Harry.   II. Title.
  III. Series.
  QB621.M37   1998
  523.42—dc21   98-10158
                    CIP

A catalogue record for this book is available from the British
Library.

10 9 8 7 6 5 4 3 2 1

Frontispiece: Sandro Botticelli, The Birth of Venus *(1483–1484).*
Courtesy Scala/Art Resource, N.Y.

To the three women who gave me this world, shared my life, and
provided its continuation: my mother, Mariya; my wife, Nataliya;
and my daughter, Irina.

—Mikhail Marov

CONTENTS

This book has a rather curious history. Preceding it are two editions of the book *Space Research*, first published in a Russian conference proceedings volume in 1977 and co-written by M. V. Keldysh, the past president of the USSR Academy of Sciences and leader of the Soviet space research program since its inception, and M. Ya. Marov, his longtime adviser and assistant in the program. Keldysh and Marov began to collaborate on a revision, but before it was finished, in 1978, Keldysh passed away. Marov completed the revision, and the book was published in 1981 by Nauka and was dedicated to Keldysh's memory on the occasion of what would have been his seventieth birthday.

Not long thereafter, Ed Tripp of Yale University Press invited Marov to translate *Space Research* into English, with further updating as appropriate. This project, alas, did not come to fruition owing to objections from some authorities of the former USSR Academy of Sciences. Tripp, at the suggestion of Toby Owen, then encouraged Marov to write his own book on Venus, which he had studied since the 1960s. Marov's goal was to develop a comprehensive book, combining historical background with the current state of knowledge and including helpful comparisons with Earth and the other inner planets. The first version of the manuscript was completed by Marov in 1988, soon after the Vega-1 and Vega-2 landers jettisoned from spacecraft targeted to rendezvous with Halley's comet. This event coincided with the end of the multiyear Soviet exploration of Venus and took place just prior to the start of the Magellan mission.

In 1991, Toby Owen and Natasha Levchenko translated Chapter 1. In 1992, David Grinspoon and Ronald Mastaler agreed to participate in developing the final manuscript — Mastaler as translator of Marov's original Russian manuscript and Grinspoon as editor of the translated text. Later, however, Marov and Grinspoon agreed that the book would benefit if Grinspoon participated as a co-author. Both authors significantly updated the original manuscript. The book now includes new data from the Magellan and Galileo spacecraft, and from the final reentry phase of Pioneer-Venus.

We wanted this book to be as comprehensive as possible. It covers the ancients' views on Venus and contemporary concepts involving Venus as a specific body undergoing planetary evolution. We have used a comparative planetology approach, for it is impossible to understand the natural conditions and mechanisms operating on Venus without considering other celestial bodies, in particular the other inner planets. We focus first on Venus, relative to its closest neighbors. This tactic helps us to understand why Venus evolved in such a peculiar way and eventually became a unique planet with slow retrograde rotation, impressive geology, no magnetic field, bizarre atmospheric dynamics, and an extremely hostile climate.

We start (Chapter 1) by discussing Venus as a planet in our solar system — examining its astronomical characteristics, mechanical and optical properties, and energy balance — and we end (Chapter 9) by summarizing our ideas about planetary evolution. We have dared to speculate on the limited experimental data available.

Chapter 2 is a historical review of how Venus has been studied through the ages, from the earliest observations with the naked eye to early telescopic observations to the contemporary sophisticated observatories and interplanetary missions. We include some illustrations and photographs from the former Soviet space program that have never before been published. In Chapter 3, we analyze the various surface landforms and volcano-tectonic history of Venus; this analysis is based mainly on the radar images of the Magellan, Venera-15, and Venera-16 spacecraft. This chapter illustrates the enormous progress that has been made in deciphering features on a planet obscured for centuries by its thick atmosphere and clouds. Chapter 4 focuses on the interior structure and thermal history of Venus; although the latest missions have provided much new data, many questions remain about its planetary geology and mechanisms responsible for the formation of diverse landforms.

Chapters 5–8 focus on the peculiar atmosphere of Venus. We consider the composition and structure of the lower and middle atmosphere; the structure and composition of the clouds; the thermal regime and planetary dynamics; and, finally, the physics of the upper atmosphere, problems of aeronomy, and interactions with the solar wind plasma. Because all these problems (and the problem of the atmospheric-surface interaction) are topically interrelated, we focus on how to link them by using evolutionary concepts (instead of exclusively analyzing data from available measurements) and by comparing and modeling the data. Marov and co-workers obtained many important experimental and theoretical results, including the first in situ measurements of the Venusian atmosphere and clouds; measurements of incident solar radiation, its spectral composition, and peculiar patterns of atmospheric dynamics; computer simulations of the radiative heat transfer in the lower atmosphere; and modeling of the structure and composition of the upper atmosphere. Grinspoon contributed to the detailed study of spectral patterns of the emitted radiation and gas isotopic ratios and to the development of contemporary views on the planet's evolution. Because our interests are not confined to the study of Venus, we have looked at the problem of planetary atmospheres in general. In a concluding chapter, we attempt to identify the most important problems for further studies of Venus, particularly with respect to future space missions.

We hope this book will be of interest to both specialists in planetary science and non-specialists. We believe that everyone interested in studying our solar system, and in particular the history of planetary space exploration involving the numerous Soviet and American missions to Venus, will find many new facts and concepts within these pages. We hope our use of a comparative approach will give the reader a better understanding not just of our closest planetary neighbor but also of Earth.

We are aware that some of our ideas are subject to debate. And, of course, the concepts we now consider fresh and innovative will age and evolve. As Vladimir Nabokov said in "The Art of Literature and Common Sense," "What we perceive as the present is the bright crest of an ever-growing past and what we call the future is a looming abstraction ever coming into concrete appearance."

A C K N O W L E D G M E N T S

Many people contributed to this project; it is impossible to list everyone. We are grateful to Yale University Press for proposing that we write this book—to Ed Tripp, who initiated the process, and to his successor, Jean Thomson Black, who has been tireless in her efforts to see it through to publication. We thank Toby Owen and Natasha Levchenko, who translated Chapter 1, and Ronald Mastaler, who translated the remainder. Our sincere gratitude goes to Noel Hinners of Martin Marietta (now Lockheed Martin), who worked closely with Marov on the implementation of the first Soviet-American agreement on cooperation in space in the early 1970s, and who helped obtain a generous grant from Martin Marietta to subsidize the book's publication.

We acknowledge the contributions of our closest colleagues with whom we have collaborated over the years. Marov wishes to note especially his collaborations with V. S. Avduevsky, M. K. Rozhdestvensky, V. V. Kerzhanovich, A. D. Kuzmin, V. N. Lebedev, V. P. Shari, A. P. Galtsev, V. E. Lystsev, A. T. Basilevsky, O. N. Rzhiga, V. P. Volkov, A. V. Kolesnichenko, and O. P. Krasitsky; David Grinspoon wishes to acknowledge J. S. Lewis, J. W. Head, M. Bullock, J. Pollack, and K. Zahnle for their input on various research problems involving Venus. We are grateful to NPO Lavochkin and NASA for permission to reproduce the photographs of space vehicles and pictures of space experiments, including the results of the Venera, Mariner, Pioneer-Venus, and Magellan missions. A. Dollfus provided several historical pictures of Venus; S. S. Limaye kindly sent composite images from Pioneer-Venus depicting cloud structures and atmospheric circulation patterns; D. Campbell provided images of Venus from the Arecibo observations. We have had the pleasure of discussing the many problems of Venus exploration with C. Sagan, J. Pollack, T. Owen, E. L. Akim, A. Kliore, H. Masursky, L. Esposito, Yu. A. Surkov, L. V. Ksanfomality, V. N. Zharkov, and S. Saunders. The Vernadsky-Brown microsymposia, initiated and headed for many years by V. L. Barsukov and J. Head, provided excellent opportunities for informal discussions and for establishing friendly relationships between planetary scientists of the former Soviet Union or Russia and the United States. We thank V. A. Krasnopolsky and A. T. Basilevsky for thorough readings of Chapters 8 and 3, respectively, and for valuable comments.

Marov wishes to acknowledge the assistance of N. D. Rozman, I. A. Belousova, and L. D. Lomakina in preparing the original Russian version of the manuscript.

The Planet Venus

Venus as a Planet of the Solar System

1

Visibility and Motion

Venus stands out among the celestial luminaries because of its great brightness. On the magnitude scale used in astronomy,[1] the maximum brightness of Venus is $M_V = -4.6$, whereas Jupiter does not exceed -2.7, Mars -2.0, and the brightest star, Sirius, is only -1.5. The Moon is the only heavenly body in our nighttime skies with which Venus cannot compete, as the brightness of the full moon is -12.7.

Venus is an inner (or inferior) planet. Its orbit is located inside the orbit of Earth at a distance from the Sun of 108.2 million km, or 0.723 AU. This orbit determines the peculiarities of the visible motions of Venus relative to the Sun. Fig. 1.1 illustrates the four extreme configurations of this motion. When the planet is in one of the two positions on the line connecting it with Earth and the Sun, it is in conjunction; inferior conjunction, V_1, is when the distance between Earth and Venus is at a minimum, and superior conjunction, V_3, is when the Earth-Venus distance is at a maximum. V_2 and V_4 correspond to the two maximum angular distances from Venus to the Sun, as seen from Earth. In astronomy, the apparent angular distance from any planet to the Sun is called its elongation. In the case of Venus, V_2 and V_4 (angles V_2-E-S and V_4-E-S) are respectively the maximum eastern and western elongations, reaching average values of 46°. When the effects of the inclination of the orbit plane to the ecliptic and the eccentricities of both orbits (Earth and Venus) are taken into account, the elongations may reach 52° (see, e.g., Sharonov, 1958, 1965).

The orbit of Venus is nearly circular. Its eccentricity is 0.0068—the lowest of any planet in the solar system and less than half that of Earth's orbit. The orbital velocity of Venus is 35.03 km/sec, compared to 29.79 km/sec for Earth. The period of revolution around the Sun (the planet's sidereal period) is 224.7 Earth days. The synodic period (the time between two identical configurations, such as two inferior conjunctions) is much longer; it averages 584 days. The difference in these figures is explained by the difference in the angular velocities of Earth and Venus: Earth covers on average $59'8.5''$ in 24 hours, whereas Venus traverses $1°36'8''$. This means that if we start timing from the moment Venus is at inferior conjunction, it will come back to V_1 in one sidereal period, whereas Earth will not manage to return to E in this same period of time. Thus, some additional time is needed for Venus to return to the Earth-Sun line. The relation between synodic S and sidereal P periods of an inferior planet can be expressed by a simple relationship called the synodic equation: $1/S = 1/P - 1/E$, where E = 365.26 days, that is, the sidereal period of Earth. As P = 0.6152 E, then S = 1.5987 E, or 583.92 days. This is 8/5 of an Earth year. Because such a simple fractional relationship exists between the synodic period of Venus and Earth's year, the conditions for the observation of this planet almost repeat every eight years (Table 1.1).

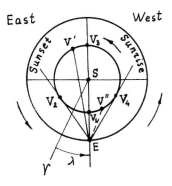

Fig. 1.1. The basic configuration of Venus (its position relative to Earth and the Sun). E is Earth, S is the Sun, $V_{1,2,3,4}$ is Venus in various configurations, and γ is the direction to the point of the vernal equinox.

Table 1.1   Conjunctions of Venus

Year	Inferior conjunction	Superior conjunction
1989	—	April 4
1990	January 19	October 31
1991	August 23	—
1992	—	June 13
1993	April 1	—
1994	November 3	January 17
1995	—	August 21
1996	June 11	—
1997	—	April 2
1998	January 16	October 28
1999	August 20	—
2000	—	June 10
2001	March 29	—
2002	October 31	January 14
2003	—	August 18
2004	June 8	—
2005	—	March 30

Because of the difference in the orbital angular velocities of Earth and Venus, the simple geometry of the motion of Venus relative to the Sun, obvious from Fig. 1.1, becomes rather complicated when the planet is observed from Earth. By drawing the motion of Venus during a year on a map of the stars, one finds a loop characteristic of the other planets as well. In the period from January to May 1989, for example, Venus appears briefly stationary as its motion changes from eastward to westward against the background stars (Fig. 1.2). Notice that the best conditions for observing Venus are determined by the time intervals from its rising and setting to sunrise and sunset. These intervals, in turn, depend on the elongation angle and on the declination of Venus and the Sun, that is, on the angular distance of these objects from the celestial equator. For an observer at temperate latitudes in the northern hemisphere of Earth, when the declination of Venus is greater than that of the Sun (it is located to the north of the Sun), the planet is visible in the sky after sunset or before sunrise, brightly shining in the dark sky. If the declination of Venus is less than that of the Sun (the planet is to the south of the Sun), it is visible only for a short period of time and in the twilight sky. Thus, the visibility of Venus is seasonal. The best times for observations are evenings in spring (eastern elongation) and in autumn before dawn (western elongation).

The phases of Venus are similar to those of the Moon.[2] At inferior conjunction (V_1 in Fig. 1.1), the planet turns its unlit side to us, and the phase angle $Q = 180°$. Near superior conjunction, when the planet is behind the Sun, $Q = 0°$, and one can see the planet's illu-

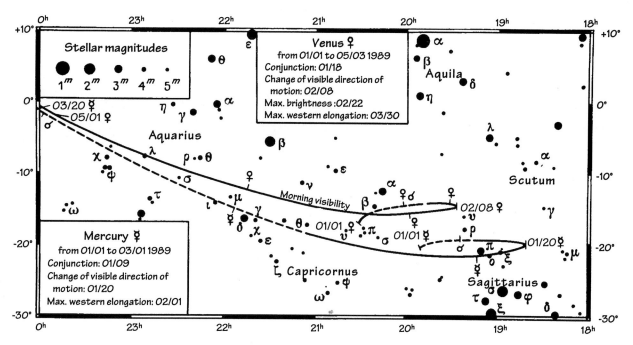

Fig. 1.2. The visible path of Venus on a map of the sky from 1 January through 3 May 1989. For comparison, a visible path of Mercury is also shown. A dashed line indicates that the planet cannot be observed in the northern hemisphere at middle latitudes.

minated disk. In intermediate phases $0° < Q < 180°$, Venus appears partially illuminated to an observer on Earth. In a telescope, the planet changes from a narrow sickle to a nearly full disk (Fig. 1.3). The visible angular diameter of the planet between inferior and superior conjunctions changes by a factor of 6.5, that is, from 9.9″ to 64.0″ (the Moon has an apparent diameter of 30′).

The varying brightness of Venus as it moves around the Sun, and through our skies, is the result of its simultaneously changing phase and apparent size. At inferior conjunction the apparent area is greatest, but the sunlit area is at a minimum. As Venus emerges from inferior conjunction, becoming visible in the morning sky, the disk becomes smaller but more fully illuminated. These competing effects result in Venus' reaching maximum brightness 36 days before and after inferior conjunction, halfway between inferior conjunction and greatest elongation, which occurs 72 days before and after inferior conjunction. In 1991, for example, greatest eastern elongation was on 13 June ($M_V = -4.3$), whereas greatest eastern brilliancy was on 17 July ($M_V = -4.5$). In the west, greatest brilliancy was 28 September ($M_V = -4.6$), and greatest elongation was 2 November ($M_V = -4.4$). For a period of about 8 days around inferior conjunction, Venus cannot be seen. Following this, Venus reappears in the morning for 263 days. Then, around superior conjunction, Venus is again lost in the glow of the Sun for about 50 days. The planet then emerges for 263 days of visibility as an "evening star" before disappearing in the evening twilight, as inferior conjunction approaches again. The time difference between the rising and setting of Venus and the Sun may reach almost four hours.

Observations of Venus during its inferior conjunctions determined that its orbit plane is inclined to the ecliptic by an angle i = 3° 23′ 39″. This explains why the disk of the planet is not projected onto the Sun's disk each time Venus passes through inferior conjunction. A transit of Venus across the Sun's disk can occur only when the planet is at in-

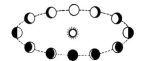

Fig. 1.3. The phases of Venus.

Fig. 1.4. Transit of Venus across the Sun's disk. The dates of the transits are indicated at the ends of the appropriate segments (in the format month/day/year).

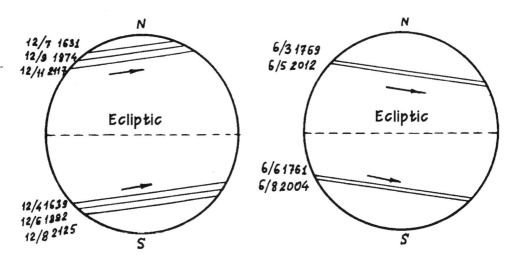

ferior conjunction at one of the points in its orbit where the orbit plane crosses the plane of Earth's orbit. Because of the necessary mutual orbital configurations, such events are comparatively rare (Fig. 1.4), repeating in cycles whose intervals are 8, 105.5, 8, and 121.5 years. The next transits will occur on 8 June 2004 and 5 June 2012. Observations of a transit of Venus in the seventeenth century established the presence of an atmosphere on Venus.

The atmosphere considerably influences the contours of the visible disk. Refraction shifts the position of the terminator[3] by producing a penumbral zone, where one can observe a gradual decrease of illumination from the day hemisphere to the night hemisphere. One can also see the effect of the atmosphere in the elongation of the horns of the crescent when Venus is near inferior conjunction (Fig. 1.5). Because of refraction, the lighted arc of the limb appears to be greater than 180°, and each horn seems to be shifted toward the dark half of the limb (Sharonov, 1953a). In the extreme case, when the crescent is very thin, a complete ring of light is observed surrounding the planet (Fig. 1.6). In addition to their intrinsic beauty, these phenomena can provide useful insights about some properties of the atmosphere, including the existence of high-altitude clouds.

Another unusual feature of Venus is its very slow rotation (which occurs in the retrograde direction, opposite that of Earth). One complete rotation takes 243.01 Earth days. This peculiarity was not known definitely until recently, and its discovery has a long history (see Chapter 2). This slow rotation is connected to the properties of the shape and interior structure of Venus, which probably go back to the earliest stages of its evolution. The parameters of the orbital and rotational motions of Venus are given in Tables 1.2 and 1.3.

Geometrical and Mechanical Characteristics

In our solar system Venus is the planet most similar to Earth in terms of size, mass, and average density, with Earth just slightly greater than Venus in each of these characteristics (see Table 1.3). Let us consider some of these data in more detail.

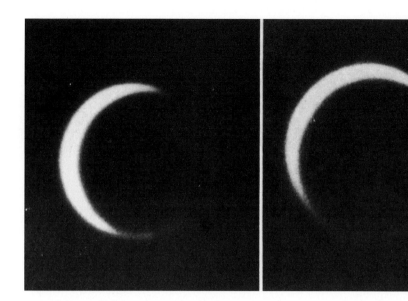

Fig. 1.5. The crepuscular arc around Venus. The elongation of the horns is clearly visible, almost encompassing the disk (photographs courtesy of V. M. Slafer, Lowell Observatory).

The diameter of Venus is about 650 km less than that of the Earth. The radius of its solid surface determined by ground-based radar measurements and by spacecraft is $R_{eq} =$ 6051.5 km. From the new Magellan altimetry data it is defined as $R_{eq} = 6051.84$ km (Ford and Pettengil, 1992). Because of the difficulty in determining this value, the optical radius that corresponds to the size of the visible disk has until recently been used to calculate ephemerides. The mean Venus radius measured from the external boundary of the cloud layer using today's data is $R_{opt} = 6119$ km.

It is important to stress that for Venus the value of the equatorial radius R_{eq} is essentially equal to the polar radius R_{pol}, while for the Earth $R_{eq} > R_{pol}$ by 22 km. The Earth's geometrical oblateness can be represented by $e = (R_{eq} - R_{pol})/R_{eq} = 1/298.25$.[4] For Venus, $e = 0$. This is directly related to Venus' anomalously low rotational velocity, resulting in its rotational period of 243.01 terrestrial days.[5] Only Mercury comes close to Venus in this parameter, as its rotational period is 58.6 Earth days. Consequently the oblateness of these planets is 2–3 orders of magnitude less than the oblateness of the rapidly rotating Earth and Mars. In contrast, Jupiter and Saturn, with rotational periods of only about ten hours, have e values 20–30 times higher than Earth's.

In contrast to rapidly rotating bodies whose configurations are approximately ellipsoidal (like Earth), Venus is well represented as a sphere with an average radius $R_{\female} = $ 6051.84 km. Altitude must be measured from an agreed-upon level. This may be taken as the mean radius of 6051.84 km or as the median radius (half the surface area below this level) of 6051.6 km. For surface maps, 6050.0 was often used as the zero reference level. The surface of Venus has also been represented with a model of its topographic decomposition, based on spacecraft measurements.

Another oddity of the rotation of Venus is its retrograde direction: opposite to the direction of motion along its orbit, that is, clockwise as viewed from above the North Pole, in contrast to all other planets, except Uranus and Pluto. But these two planets have their rotational axes almost in the ecliptic plane, whereas Venus' rotational axis is almost perpen-

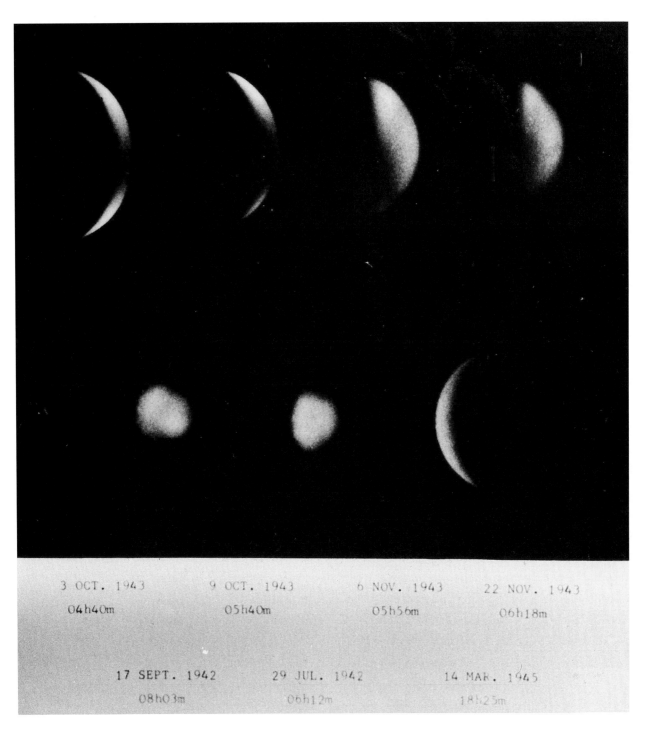

3 OCT. 1943   9 OCT. 1943   6 NOV. 1943   22 NOV. 1943

04h40m   05h40m   05h56m   06h18m

17 SEPT. 1942   29 JUL. 1942   14 MAR. 1945

08h03m   06h12m   18h25m

Fig. 1.6. Variation in the illumination of Venus' disk, as a function of phase (composite photographs, in yellow light, were obtained by H. Camichel with the 38 cm refractor in 1942–1943; the contrast is enhanced) (courtesy of A. Dollfus).

dicular to the ecliptic plane. Venus has a rotational inclination of 177.3°. In other words, its deviation from the normal to the ecliptic plane does not exceed 3°. This results in the absence of seasonal changes on the planet.

Slow retrograde rotation also explains the unusual duration of solar days on the planet. The solar day is determined by the synodic equation, except that the right-hand side will have a plus sign instead of a minus sign, taking into account the reverse rotation of Venus. Now it will look like 1/G = 1/P + 1/T. Here T = 243.01 days and P = 224.70

Table 1.2   Orbital elements for Venus and Earth (for the 1950.0 epoch)

	Mean distance from the Sun		Rotational period (mean Earth days)	
	Millions of km	A.U.	Sidereal	Synodic
Venus	108.1	0.7223	224.701	583.924
Earth	149.5	1.0000	365.256	—

	Eccentricity	Inclination to the plane of the ecliptic	Longitude of the ascending node	Longitude of perihelion
Venus	0.0068	3° 23′ 39″	76° 13′ 39″	130° 52′ 03″
Earth	0.0167	—	—	102° 04′ 50″

	Mean angular velocity along the orbit per day	Mean linear velocity along the orbit (km/sec)
Venus	1° 36′ 08″	35.03
Earth	0° 59′ 08″	29.79

days. So G = 116.69 days. Thus, on Venus the Sun rises in the west and sets in the east 116.69 Earth days later. One Venusian year consists of two (to be more precise, 1.925) solar days.

Venus has no moons. This is why estimates of its mass, based only on the characteristics of its orbital motion, were not of high accuracy until recently. The value for the mass of the planet has been revised from spacecraft tracking, first from Mariner-5 (Anderson et al., 1968) and then from Mariner-10 (Howard et al., 1974). An estimate is given in Table 1.3. When the mass is known, one can easily determine the gravitational acceleration at the planet's surface $g = 887.4$ cm/s. And finally, the average density of Venus determined as the ratio of its mass and volume is 5.25 g/cm^3. Taking into account the mass of the atmosphere does not significantly change these values.

The study of spacecraft orbits (Akim et al., 1978; Williams et al., 1983; Mottinger et al., 1985; Konopliv and Sjorgen, 1994) allowed a detailed description of the planet's gravitational field and placed constraints on models suggested by other disciplines. The best global gravity data set was obtained from the post-aerobraking phase of the Magellan mission, which provided high-resolution gravity mapping of Venus and significantly increased the ability to correctly model its interior. Correlation of gravity and altimetry data was used for the development of various spherical harmonic models. A peculiar feature, assumed in the earlier models, that the axis of the largest moment of inertia deviates from the planet's rotational axis by several degrees, has not been confirmed. However, the disk of Venus shows considerable anomalies at the equator, including marked bulges that prevent its figure from being described by a simple rotational ellipsoid or a three-axis ellipsoid. The figure of Venus shows a larger deviation from hydrostatic equilibrium than any of the other

Table 1.3   Geometrical and mechanical characteristics of Venus and Earth

	Equatorial radius (km)	Period of rotation (days)		Area of surface (relative to Earth)	Direction of rotation
		Sidereal	Solar		
Venus	6051.5[a]	243.01	116.69	0.95	retrograde
Earth	6378	1.00	—	1.00	prograde

	Volume (relative to Earth)	Mean density (g/cm^3)	Mass (relative to Earth)	Gravitational acceleration at the surface (cm/sec^2)
Venus	0.92	5.25	0.815	887.4
Earth	1.00	5.52	1.000	980.6

	Escape velocity (km/sec)
Venus	10.4
Earth	11.2

[a] Or 6051.84, according to Magellan altimetry data (Ford and Pettengill, 1992).

terrestrial planets. The new gravity data will help, in particular, to resolve the problems of mantle convection and lithospheric compensation mechanism and to constrain scenarios of planetary formation and thermal history.

Of great interest in motions of Venus is the presence of a peculiar orbital-rotational synchronization, suggestive of resonance. The planet's sidereal rotation period, 243.01 days, is surprisingly close to the period of resonant rotation relative to Earth, 243.16 terrestrial days. Thus at each inferior conjunction nearly the same side of Venus faces the Earth. The possibility that this is simply a coincidence cannot be denied, because an exact theory has not yet been developed to explain this phenomenon (Beletsky and Trushin, 1974; Ward and De Campli, 1979; Shapiro et al., 1979). The shift of the sub-Earth point during successive conjunctions is 0.52' to the west; that is, an exact repetition of a configuration should occur in 11,000 years.

When considering the possible orbital-rotational resonance of Venus, one should keep in mind some other interesting examples of relationships between the periods of rotation and revolution of other planets and their satellites. That the same side of the Moon always faces Earth is an obvious example of a 1:1 resonance between periods of rotation and revolution. Similar resonances have been found in the motions of other satellites of planets. The synchronization of revolution and rotation periods for Mercury (in resonance 3:2) is well known.

The synchronization of rotation and revolution of celestial bodies can be well described by the Cassini laws. These were established empirically for the revolution of the Moon by this outstanding French astronomer in the middle of the seventeenth century, and they were later generalized for a wide range of orbital and rotational motions. Planetary exploration has revealed that comparable relationships, or commensurabilities, exist for a few tens of pairs of bodies in the solar system. This tendency is observed in particular between mean motions of Jupiter and Saturn; the Galilean satellites of Jupiter called Io, Europa, and Ganymede; the four largest satellites of Uranus, and so on. An interesting feature is the commensurability in the motions of Neptune and Pluto: these planets cannot come closer than 18 AU to one another even though Pluto at perihelion is closer to the Sun than Neptune. Similar dynamics are characteristic of the orbits of asteroids.

Unfortunately, there is no clear explanation for orbital resonances among the inner planets of the solar system. Indeed, tidal interactions between celestial bodies (in particular, the Earth and Venus) can often be neglected if one takes into account that the effect of tidal forces diminishes with the cube of the distance between the interacting bodies; this effect decreases much more steeply than that of the gravitational force, which falls off as an inverse square. Nevertheless, the most convincing contemporary explanation for the synchronizations, and for the small value of eccentricity and orbital inclination, is derived from the mechanism of tidal friction (Fig. 1.7).

This type of resonance is clearly manifested in the motion of Mercury. It has been found that in the case of a planet with a non-isotropic moment of inertia, a stable mode is possible when the rotational period is 2/3 of the orbital period. For Mercury, with a sidereal orbital period of 88 days, the period of rotation is 58.6461 days. This is an interesting variant of a resonance, with the tidal interaction of the planet with the Sun leading to the transfer of its angular momentum, which results in the decrease of the rotational velocity and "capture" in the existing resonance mode. Detailed theoretical discussion of this interesting effect can be found in the works of Colombo and Shapiro (1965), Goldreich and Peale (1967), Beletsky (1975), and Beletsky and Khentov (1995).

It should be stressed that in order for Mercury to have a $3:2$ spin-orbit resonance, there must be a very slight compression of the inertia ellipsoid in the equatorial plane, that is, $(B - A)/C \sim 10^{-5}$, where A and B are principal moments of inertia about two orthogonal axes in the equatorial plane and C is the principal moment of inertia about the rotation axis. The deviation from a strictly concentric mass distribution in the vicinity of this plane is possibly conditioned by gravitational anomalies similar to the regions on the Moon with increased mass concentration (mascons). The largest of the hypothetical mascons on Mercury is associated with the huge basin Caloris (1300 km in diameter), which always faces the Sun when Mercury is at perihelion.

One may assume that the near-resonant rotation of Venus, with respect to Earth, is caused by Earth's gravitational attraction on the non-symmetric figure of Venus. However, to obtain the stable resonance state, Earth's stabilizing moment $T_\oplus$ must be greater than the tidal moment of Venus in the gravitational field of the Sun T_* ($T_\oplus > T_*$). This means that the asymmetry of the figure of Venus should be great enough to provide the difference in the moments of inertia relative to the equatorial axes A and B. In addition, the rela-

Fig. 1.7. The formation of a tidal bulge on a rotating planet, and its lag, owing to the motion of a satellite.

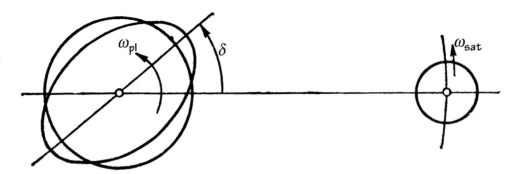

tive width of the resonance zone for Venus appears to be very small, considerably smaller than that for the Moon or Mercury. Using generalized Cassini laws, Beletsky and Trushin (1974) showed that in order to preserve the motion of a resonance zone in the presence of periodic solar perturbations, the following condition should be fulfilled for the main moments of inertia: $(B - A)/C \le 2.5 \times 10^{-5}$. The analogous ratio for the Moon is 5×10^{-4}. The given condition is critical, and it means that the phenomenon of a Venus resonance with Earth is in doubt.

Atmospheric tides may be considered an additional factor contributing to the synchronization of the rotation of Venus with Earth. Gold and Soter (1969, 1971, 1979) and Dobrovolskis and Ingersoll (1980) considered the possible role of daily pressure variations in the thick atmosphere of Venus due to solar heating. Indeed, pressure and mass are at a minimum in the afternoon atmosphere and reach maximum values in the morning. Such mass distribution is responsible for the moment T'_* in the gravitational field of the Sun that is proportional to the planetary period of revolution and has the opposite sign of T_*, causing an acceleration of the planet's rotation. This diminishes the slowing effect of the Sun's tidal element and facilitates the synchronization of Venus with the Earth. In this case the revised synchronization condition is $T_\oplus > T_* - T'_*$. Nevertheless, Bursa and Sima (1985) doubt that Earth controls the angular velocity of the rotation of Venus, on the basis of the analysis of the parameters of Venus' figure. Most important, the large bulges in the equatorial region of Venus do not coincide with the sub-Earth direction when Venus is at inferior conjunction.

An analysis of tidal interactions can shed light on the rotational peculiarities of some planets, and in particular it has led to one possible explanation of the retrograde rotation of Venus. According to a theory of planetary and satellite system formation developed by Eneev and Kozlov (1981a, 1981b), early tidal evolution of rotational motions of planets occurred much faster than such evolution in the contemporary epoch. Eneev and Kozlov found an intrinsic relationship between the process of formation, the mass, and the character of rotation of a planet, and they proposed a rather simple explanation of the occurrence of prograde or retrograde rotation as the result of direct and frontal impacts of bodies at the final accretionary stage.

Following this cosmogonic concept, Beletsky (1975) showed that "protoVenus," possessing retrograde rotation since its beginning, could be captured in the above-mentioned resonance with the Earth for the first 10^7 to 10^8 years after formation. Over the course of subsequent evolution, the tidal forces acted in such a direction as to "overturn" Venus from retrograde to regular prograde rotation; however, the deviation of the rotational axis from the original position did not exceed 2°. Thus although Venus has been locked in resonance-type rotation, it has not been strong enough to fundamentally alter its rotational direction, in contrast to the Moon or Mercury.

Energy and Reflectivity

The thermal balance of the planet is determined by the quantity of solar radiant energy received minus the energy reflected into space. Thus, the balance depends on the distance from the Sun a and on the planet's integral Bond albedo A,[6] as the inner sources of heat for all terrestrial planets can be neglected (their contribution does not exceed fractions of a percent). We define the solar constant E_s as the flux of solar radiation incident along the normal to a unit area on Earth's surface in the absence of the atmosphere. This quantity has been evaluated as $E_s = 1.917$ cal/cm^2 min or 1373 W/m^2 (Eddy 1977). These three values $(a, A,$ and $E_s)$, plus the Stefan-Boltzmann constant $\sigma = 5.67 \times 10^{-12}$ W cm^{-2} degree^{-4}, can be used to define the effective temperature T_E, an important parameter that serves as a measure of the energy received by the planet:

$$T_E = [E_s (1 - A)/4\sigma a^2]^{1/4} \qquad (1.1)$$

Here a is normalized to 1 AU, and the 4 in the denominator reflects the fact that the energy flux is incident on a disk but is radiated from a sphere.

The physical meaning of A and T_E may be understood by comparing the thermal flux from the planet with the radiation of a blackbody at the same distance from the Sun. A blackbody has an emissivity of 1, which means that $A = 0$ and the whole surface has the same temperature, called the equilibrium temperature. Obviously, $T_{eq} = (E_s/4\sigma a^2)^{1/4} = 328$ K at the orbit of Venus. In reality Venus is well known for its brilliance in our twilight skies, caused by the reflection of sunlight from the thick, uniform cloud cover in its atmosphere. Thus, for Venus the visible albedo is closer to 1 than 0, and thus $T_E < T_{eq}$.

The determinations of the integral spherical (often called bolometric) albedo of Venus, which were carried out for many years using the methods of ground-based and space photometry and radiometry, led to somewhat different values. This issue has been discussed in detail by Ksanfomality (1985). The mean value proposed by him, based on the most accurate measurements, is $A = 0.766 \pm 0.018$, in accord with the earlier finding of Irvine (1968). In this case we have $T_E = 228$ K. Now, taking into account the physical meaning of these variables, one should understand the effective temperature of Venus as the temperature when the energy radiated by a blackbody with the same surface area as Venus equals the energy radiated by the planet in the whole spectral range of electromagnetic radiation.

In spite of the planet's proximity to the Sun (and thus greater value of the solar flux

Fig. 1.8. The dependence of the monochromatic Bond albedo A for Venus, versus wavelength (normalized to the integral Bond albedo $A = 0.76$), in the near-ultraviolet, visible spectral region ("a") and in the near-infrared ("b") spectral region. Wavelength intervals identified with the presence of specific absorbers are shown. The dots indicate the dependence of maximum contrast, between the dark and light indices of UV absorption, on wavelength (according to Moroz et al., 1985a).

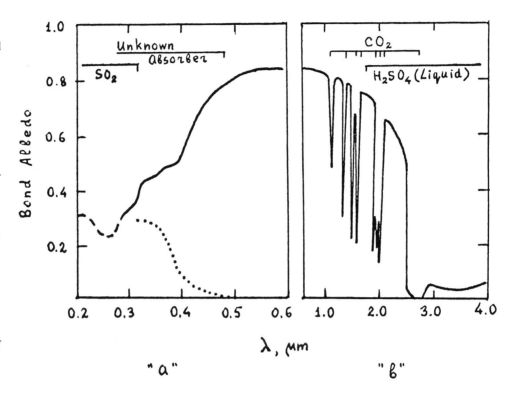

$E_s/(0.7223)^2 = 2620 \text{ W/m}^2$), the thermal flux radiated by Venus into space is no more than 70% of Earth's. This results from the difference in bolometric albedo, whereas the difference in surface area is less than 9%. The total solar radiation absorbed by Venus is 7.2×10^{10} MW, or slightly more than 600 W/m^2. Only 18 W/m^2, or 3% of it, reaches the surface of Venus. About 70% of the solar energy is absorbed in the altitude range 60–80 km, and the rest is absorbed in the cloud layer and underlying atmosphere (Kuzmin and Marov, 1974; Moroz et al., 1985a).

The data on the monochromatic Bond albedo A_λ reveal in which spectral intervals more energy is absorbed or reflected by the planet (see Fig. 1.8). Some spectral peculiarities are identified with the presence of atmospheric or cloud components responsible for the absorption at an altitude higher than 60 km. They are shown in the figure alongside the spectral dependence of the contrast of ultraviolet features on the disk of Venus—one of its persistent mysteries.

The brightness of Venus in the night sky changes dramatically as it goes around the Sun, depending on the phase angle Q. The dependence of the magnitude on Q (and thus the properties of the solar scattering for the observer at different angles) carries important information about the scattering media, in particular about the cloud layer and the aerosol particles in it (see Fig. 1.9). In spite of some spread in the data, one can derive certain limitations on the averaged optical properties of clouds on the planet. The brightness temperature T_B usually does not coincide with the effective temperature. Brightness temperature, sometimes also called color temperature, is the temperature of a blackbody that radiates the same amount of energy at the given wavelength (or in the narrow wavelength range) as the planet does. Venus has $T_B > T_E$ for almost the entire hemisphere and $T_B < T_E$ at the very

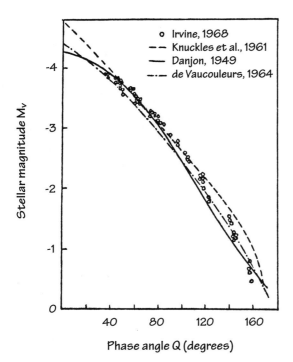

Fig. 1.9. Visual reflection phase curves as a function of Venus' stellar magnitude M_v ($\lambda = 5500\overset{\circ}{A}$) versus phase angle Q, based on photometric observations of Venus from Earth, performed by a series of researchers (according to Irvine, 1968).

edge of the disk, at the horizon. This is explained by the fact that brightness temperature depends on the spectral radiation properties and on the concentration of the cloud aerosols as well as on the temperature gradient in the vicinity of the radiating layer. The values of T_B for Venus (during observations at nadir) for several wavelength ranges in the far-infrared spectrum region are given in Table 1.4, and the dependence on the wavelength is shown in Fig. 1.10. In most of the spectral region of 7 to 100 μm, the average value of brightness temperature is in the interval $T_B = 220-250$ K, although Taylor et al. (1980, 1983) claim lower values around 14–15 μm (see Table 1.4). This value increases drastically, however, in the radio wavelength range, and it reaches the maximum value $T_B \approx 700$ K in the cm portion of the spectrum at $\lambda = 6-13$ cm. The results of the measurements of $T_B(\lambda)$, and of $A(\lambda)$, allow us to identify the presence of atmospheric components responsible for the respective attenuation of the radiation (see Fig. 1.10).

In conclusion, let us talk about one more important characteristic of the planet—its magnetic field. Direct measurements beginning with the spacecraft Venera-4, Mariner-5, and Mariner-10 led to the generally accepted conclusion that Venus has no intrinsic magnetic field (Dolginov et al., 1968, 1969, 1978; Bridge, 1967; Ness et al., 1974; Russell, 1976; Russell et al., 1980); however, the established upper limit just reached 30–100 gammas, or $(3 - 10) \times 10^{-4}$ **E**. The most complete observations, carried out by Pioneer-Venus (Russell, 1979a, 1979b), allowed the investigators to considerably lower this limit to 2 nT, or $(2 \times 10^{-5}$ **E**), 5×10^{-5} of Earth's value at the equator (0.31 **E**). Higher values registered earlier (up to 50–80 nT) were probably caused by effects induced by the interaction of the planet's ionosphere with the solar wind plasma. This may be attributed to occasional phenomena in limited regions of the radial field apparently extending through the base of the ionosphere. Thus, Venus does not have its own magnetic field (the upper limit

Table 1.4 The brightness temperature of Venus

Wavelength range, microns	T_B, K (nadir)	T_B, K (average over the disk)	Reference
8–13	240	225	Sinton and Strong, 1960; Sinton, 1963a
8–14	230–238	220–225	Sagan and Pollack, 1969; Diuer et al., 1976; Kunde, 1977; Chase et al., 1963, 1974; Logan, 1974
7.4–30	237–244		Ksanfomality, 1980
11.3–11.6	238–244		Taylor et al., 1980, 1983
14.6–14.8	170–185		Taylor et al., 1980, 1983
6–12	225–250		Ertel et al., 1984
18–31	240–265		Ertel et al., 1984
45–110	245		Ward et al., 1977
31–38	244		Kurtin et al., 1979
47–67	255		Kurtin et al., 1979
71–94	238		Kurtin et al., 1979
114–196	282		Kurtin et al., 1979
40–400	250		see Whitecomb, 1979
1000	300		see Whitecomb, 1979
2000–4000	350		see Whitecomb, 1979
Wavelength range, cm			
0.8–0.84		410–460	see Kuzmin and Marov, 1974, and references there
1.25–1.35		440–530	"
1.9–2.1		495–525	"
2.7–3.3		575–610	"
6.0		630–730	"
9.3–14.3		580–710	"
21		530–670	"
49.7		~510	"
70		~500	"

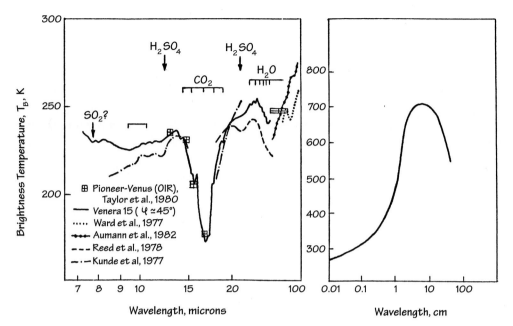

Fig. 1.10. Dependence of Venus' brightness temperature T_B versus wavelength λ, in the far-infrared range (left) and in the radiowave range (right). The results of measurements by various authors, and the identification of spectral properties of the radiation with atmospheric components, are shown (according to Kliore et al., 1985).

for magnetic moment does not exceed 3×10^{21} Gauss-cm^3). This imposes certain constraints on models of its internal structure.

The brief description of Venus in this chapter serves as an introduction that allows us to gain insight into specific features of this fascinating planet. We shall address these features in the following chapters, focusing on the contemporary ideas that were brought about by the impressive multiyear program of ground-based and space exploration of Venus. A historical review of the main issues concerning Venus, beginning in ancient times, is a valuable start to our exploration, and the next chapter is dedicated to this review.

2 Venus was the first luminary that humans learned to recognize among the thousands of stars strewn across the night sky. Prehistoric people noted the unique behavior of this "star," whose appearances were limited to the evening or pre-dawn hours (in contrast to Mars, Jupiter, and Saturn, which are also easily seen with the unaided eye). It even remained easily discernible in the evening or morning twilight. For this reason, Venus came to be called the "morning star" or "evening star." The ancient Greeks believed that these were two different objects, which they gave the names Vesper (evening) and Phosphor (morning). This planet eventually received the name of the goddess of sensual love and beauty in ancient mythology (known as Aphrodite by the Greeks and as Venus by the Romans).

The attempts to find an explanation for the whimsical visible motions of Venus and other "wandering stars" (see Fig. 1.2) date back to the cosmology of Aristotle (during the fourth century B.C.), which was based on the even earlier planetary theory of Eudoxus. The observed motions of the planets were explained as the uniform, misaligned rotation of concentric hollow spheres, to whose outer surfaces the planets were fastened, with Earth located at the center. The laws of planetary motion formulated in the second century B.C. by Claudius Ptolemaeus (known as Ptolemy) were considerably more thorough. The fundamentals of this theory, progressive for the time, were laid down by Ptolemy's predecessor Apollonius, who substituted the rotating planetary spheres with Aristotelian circles. In essence, the laws of Ptolemy, which predicted the positions of the planets based on extremely complex geometrical constructs, provided support for many centuries of astronomical observations and systematized the whole body of knowledge of that period.

Not until fifteen centuries later, when the heliocentric system of Nicholas Copernicus supplanted the Ptolemean geocentric system, was the true basis laid for a natural science of the solar system. Direct observations by Galileo Galilei confirmed the conclusions of Copernicus that the planets are spherical, dark opaque bodies, visible in the sky owing to the reflection of solar rays by one (the day) hemisphere, while the other (night) hemisphere remains dark. The beginning of the study of Venus' nature is attributed to Galileo, who first established the phases of the planet, analogous to the well-known phases of the moon (see Fig. 1.3). The remarkable results of these observations naturally led astronomers to surmise that the planets might have atmospheres and that life might exist on them. Such a possibility was seriously considered by Giordano Bruno.

That Venus does indeed have an atmosphere was first determined by the Russian scholar Mikhail V. Lomonosov at the Saint Petersburg Observatory in 1761. Lomonosov detected the refraction of solar rays while observing the transit of the planet across the disk of the Sun (see Lomonosov, 1955). Only the presence of refraction in a sufficiently thick at-

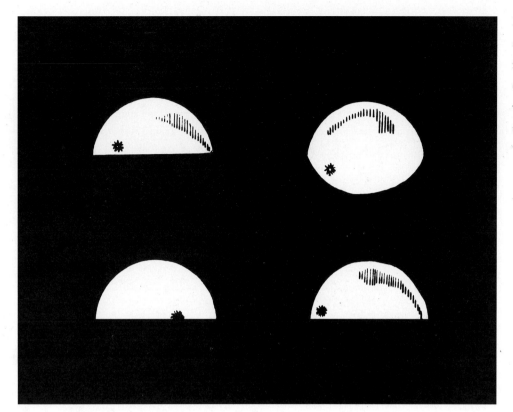

Fig. 2.1. Dark and light spots on the disk of Venus from sketches of J. Cassini (1666–1667, Italy). Observations were conducted in the clear air over Bologna.

mosphere could explain the appearance of a light ("fire") ring around the night disk of Venus during the initial phase of transit, on the side opposite from the direction of motion. Lomonosov described this phenomenon, which carries his name, as the appearance "of a hair-thin luminescence," which encircled a portion of the planet's disk that had not yet contacted the solar disk. "This bears witness to nothing less than the refraction of solar rays in the Venusian atmosphere," he wrote. Approximately 30 years later, the German astronomer I. Schroter and the English astronomer W. Herschel discovered the crepuscular phenomena on Venus and came to the correct conclusion that they result from the scattering of solar rays in the upper portion of the planet's atmosphere. This finding confirmed the discovery of Lomonosov.

Observations and Evaluation of Color

Attempts to view the surface of Venus through telescopes turned out to be unsuccessful. Indeed, its uniform disk shows almost no visible details or formations characteristic of other planets. Nevertheless, the existence of extremely weakly delineated bands, spots, and small differences in the brightness of individual regions was noted as early as the seventeenth century by a series of well-known investigators, including Fontana, Cassini, and Biancini. In 1645, Francesco Fontana, a lawyer and a lover of astronomy, was the first to detect a dark area almost in the center of the Venusian disk. Jean Dominique Cassini reported the existence of dark and light spots twenty years later while observing Venus in 1666–1667 in

Fig. 2.2. The first map of Venus, produced from observations of F. Biancini (1726). The bright and dark regions were interpreted as seas and continents, respectively. The observations were carried out on a 2.5″ refractor telescope with a focal length of 66″ and 100× magnification. Letters and numbers on the map served to identify other details on the disk.

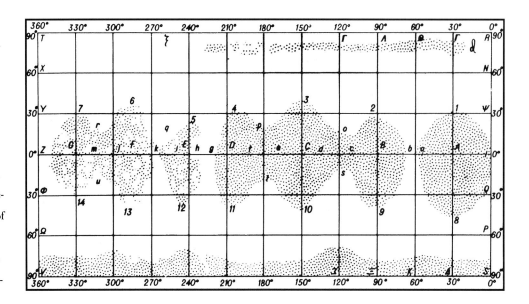

Bologna, Italy (Fig. 2.1). In 1726, F. Biancini, from the results of his observations, constructed the first map of Venus, on which he drew relatively darker and lighter regions, which he interpreted as "seas" and "continents" (Fig. 2.2). Subsequently, sketched onto the planet's disk were numerous other details, whose relative positions and configurations, and especially the explanations of their nature, supported a most diverse range of views about the planet.

Later, extensive series of visual and, subsequently, photographic studies of Venus, conducted on large instruments by E. Barnard, E. Antoniadi, P. Lowell, and other outstanding astronomers, and also studies conducted by amateur astronomers, made it possible at the beginning of the twentieth century to confidently specify certain regularly appearing, real features on the planet's disk. These studies disregarded haphazard, doubtful observational results. The doubt is introduced as a consequence of various types of distortion, primarily due to poor atmospheric conditions on Earth or the low quality of the equipment. Not infrequently these pseudoresults are brought about by psychophysiological effects, either preconceptions or the imagination of the observer. Among the imaginary features are, for example, star-shaped bright points or a system of bands on the disk of Venus, reminiscent of the "canals" of Mars. A detailed discussion of these data is contained in Sharonov's (1965) monograph.

Among the real properties of the highly variable and blurred details of Venus' visible disk is a darkening toward the terminator (the boundary between day and night), within which alternating light and dark zones or hazy arcs can sometimes be detected. Curves close to the terminator and individual rounded spots of diffuse outline, which differ slightly in tone from the surrounding background, were less reliably identified. These shapes, from the results of observations by Dollfus (1953) using a 60 cm refracting telescope in yellow light, are shown in Fig. 2.3 and with especially high detail in Fig. 2.4. Similar irregularities were found in photographic images of Venus, obtained in yellow light at the 38 cm Camichel refractor. In particular, a sharp brightening in the region of the horns—the ends

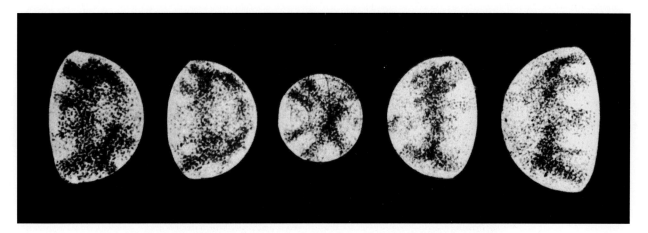

Fig. 2.3. Cloud features in the atmosphere of Venus in yellow light. Visual observations by A. Dollfus, Pic-du-Midi, refractor 60 cm, 1947–1953 (courtesy of A. Dollfus).

of the crescent shape—was detected. All these properties of the planet, in the visual spectral region, were interpreted as being caused by a cloud layer in the atmosphere, consisting of a more or less homogeneous aerosol. Detailed studies of the observable formations and their evolution gave rise, in particular, to notions concerning a two-level structure to the clouds, and the existence of irregular breaks in the upper layer over a more dense, unbroken layer (Dollfus, 1955).

In the 1920s, the first attempts were undertaken to observe Venus at other wavelengths. Infrared photography did not reveal any new details. However, photography in the near ultraviolet region, accomplished using the 60-inch reflector at the Mount Wilson Observatory, unexpectedly showed most interesting results (Ross, 1928). In the images in the $0.3-0.4$ μm spectral band, with a maximum at 0.365 μm, there are distinct dark bands directed approximately perpendicular to the horns (as opposed to the observed bands and arcs, in the visual range, which are parallel to the terminator or limb). These bands, which are higher in contrast than those observed in the visible range, suggested for Venus a certain similarity to the banded structure of Jupiter. The difference in the arrangement of the bands and other dark details from night to night reinforced the idea that they were formed in the atmosphere and clouds. These results found convincing corroboration in later observations of Venus (Kuiper, 1954; Richardson, 1955) and, especially, in a series of long-term, systematic investigations carried out by Boyer (1960; Boyer and Guerin, 1969) and Dollfus (1975). These studies led to the extremely important deduction of a four-day recurrence (periodicity) of a group of dark details in the disk of Venus, in the ultraviolet range (Figs. 2.5–2.7). The indicated periodicity, governed as we now know by atmospheric dynamics, was incorrectly attributed by Boyer to the intrinsic rotation of the planet about its axis.

The development of the technology of observational astronomy has allowed us to move from qualitative ideas about Venus to more specific, quantitative estimates of its physical characteristics. Obtaining such data became possible through the use of photometric, spectroscopic, and radiometric research techniques dating back to the end of the nineteenth century.

Some properties of a planet may be deduced from such basic measurements as the

Fig. 2.4. Cloud features of
Venus in yellow light. A
high-magnification visual
observation by A. Dollfus,
Pic-du-Midi, refractor
60 cm (courtesy of
A. Dollfus).

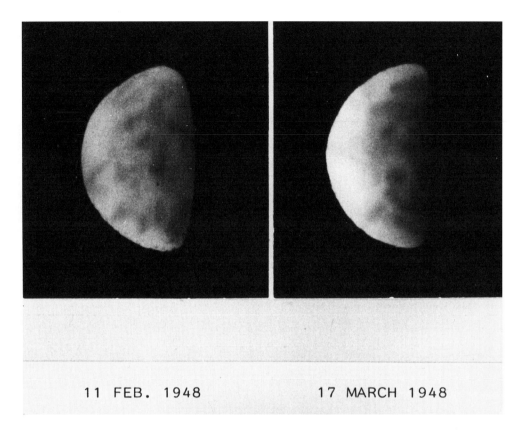

11 FEB. 1948   17 MARCH 1948

distribution of intensity over its disk, the overall brightness, the phase angle dependencies,
and the polarization of reflected light. Even the first measurements using visual stellar photometers, which provided the stellar magnitude of Venus (in comparison with bright stars),
led to the recognition of the lack of change in its photometric properties from year to year,
and of the sharp difference in its brightness curve from that obtained for bodies with no atmospheres, such as the Moon and Mercury (Zöllner, 1865; Müller, 1893; Müller, 1926).
On this basis, it was concluded that the observer most likely sees the cloud layer and not
the solid surface of Venus itself. More complete measurements, using the more sophisticated equipment and methods of integral photometry, by and large confirmed these initial
results (King, 1923, 1929; Danjon, 1949). Determinations of the color index, the difference between the photographic and visual stellar magnitudes of a body, enabled researchers to conclude (Russel, 1916; King, 1929) that the color of Venus, when compared
to the Sun as a star of known spectral class (G2), is somewhat yellow. This latter value is
called, in photometry, the color excess, or yellowness index, and is denoted D. Later, this
result was confirmed by a series of observations, and it was also shown that the color of
Venus does not vary noticeably with phase (Parshin, 1948; Barabaschev and Chikirda,
1950, 1952; Link and Neuzil, 1957; Sharonov, 1953b, 1963; Kozyrev, 1954a; Knuckles et
al., 1961). In addition, several differing quantitative estimates of the color index were presented, which find their explanation in the actual behavior of Venus' spectral curve: its
reflectance noticeably decreases at lower wavelengths, especially in the ultraviolet region of
the spectrum (see Fig. 1.8). Therefore, the resulting values of D depended significantly on

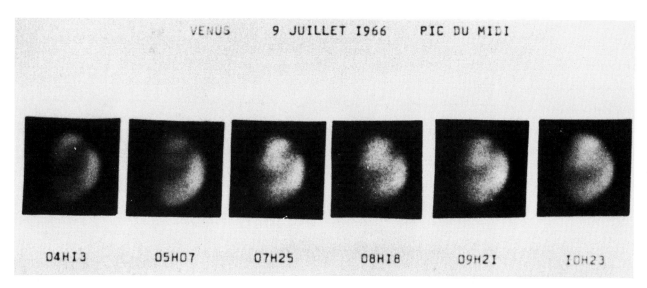

Fig. 2.5. Cloud features of Venus in the ultraviolet spectrum. A sequence of images by C. Boyer and P. Guerin, covering six hours, illustrating the rotation of the upper atmosphere in four days retrograde. Pic-du-Midi, reflector 107 cm (courtesy of A. Dollfus).

the spectral range covered, when determining the photographic stellar magnitude using data of different observers (Sharonov, 1963).

Along with the results of integral photometry, which provide a mean characteristic curve for the entire hemisphere of the planet, data were collected on the brightness distribution over Venus' disk. Such data, obtained by the methods of absolute and relative surface photometry, are of great value, because they enable us to draw conclusions about the surface properties of a celestial body. Here, the principles of photometric standardization—covered in detail, along with specific examples of their application, in a monograph by Sytinskaya (1948)—play an important role. In a series of such studies (Barabaschev, 1928; Barabaschev and Semeykin, 1935; Parshin; 1948; Sharonov, 1951; Ezersky, 1957), brightness distributions were measured along the equator, along with their changes with latitude, in several spectral regions. The results suggest that the brightness curve of portions of the disk, far from the terminator, can be approximated by a cosine dependence, in accordance with Lambert's law. In other words, the optically observable surface of Venus is neutral, although more complex relationships have been proposed (Koval, 1958). It was likewise confirmed that this surface has a very high luminosity, and that in the photographic region of the spectrum, including the ultraviolet, the brightness coefficient decreases.

Finally, measurements of the degree of polarization of light reflected by Venus are yet another source of information on the properties and nature of its surface of reflection. Here, it is especially important to measure the relationship between the degree of polarization (that is, the fraction of polarized light in the overall light flux) and phase angle, because for different kinds of surfaces (pure gas, aerosol, solid surface, dust, and their combinations) this dependence may be quite different (see, e.g., Coffeen and Hansen, 1974). The first measurements of this type did not provide any tangible results (Rosse, 1878; Landerer, 1890). Success arrived much later, when Lyot (1929), using a highly accurate visual polarimeter at the large refractor of Meudon Observatory, conducted a series of measurements and obtained convincing evidence that the polarization curve of Venus' disk is ex-

Fig. 2.6. Planisphere of the upper atmospheric features of Venus in the ultraviolet spectrum. Mercator projection with South up. *Top*: From telescopic observations by C. Boyer in 1966 (courtesy of A. Dollfus). *Bottom*: From Mariner-10 spacecraft in 1974.

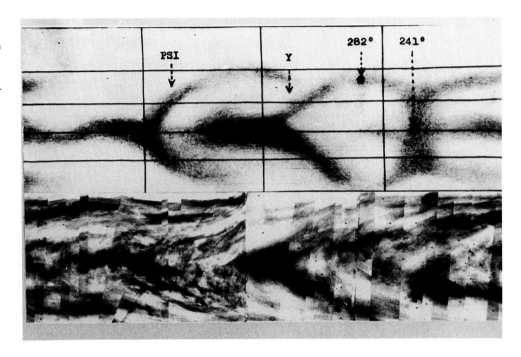

tremely complex, with several areas of inversion (that is, a transition from positive to negative polarization and back) at certain phase angles. Subsequent measurements (Gehrels and Samuelson, 1961) supported this result in the visual spectral range. At the same time, the corresponding curves at other wavelengths, primarily the ultraviolet, turned out to be considerably different. This allowed a conclusion to be made about the various sources governing the polarization process: molecular scattering in the shortwave region of the spectrum, and aerosol scattering (by particles of approximately micron size) at longer wavelengths.

The inhomogeneities in the polarization over the disk were studied by Dollfus (1956, 1957), who discovered individual "spots" with irregular contours, whose polarization values differ from the mean, positively or negatively, and which are sometimes accompanied by a rotation of the polarization plane by several degrees with respect to the mean. Such regions of anomalous polarization (not coinciding in position with barely discernible details on the disk) have not been specifically interpreted, although it seemed perfectly reasonable to link them with the dynamics of cloud formations. Meanwhile, attempts were made to link the results of polarimetric measurements not only with clouds but also with an abundance of fine dust over the whole atmosphere of the planet. Kuiper (1947), commenting on the shape of the polarization curves in the near-infrared region of the spectrum wrote, "One can assume that the stormy atmosphere of Venus is filled with dust, which in the absence of free O_2 may be slightly tinted, similar to white sand." Later, however, more complete polarization measurements, and their comparison with the results of detailed theoretical analysis, led to the unambiguous conclusion that the clouds are the source of the polarization. This served as the basis for evaluating the microphysical properties of the particles (such as size, and index of refraction) near the visible outer boundary of the cloud layer (Hansen and Arking, 1971; Hansen and Hovenier, 1974).

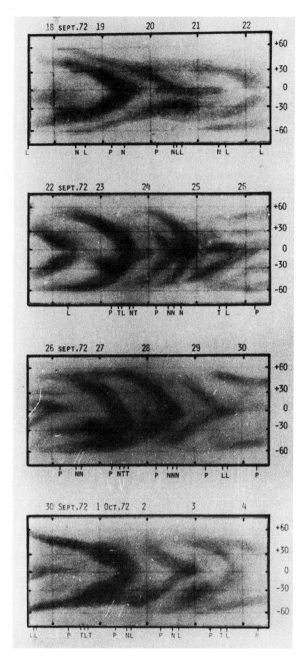

Fig. 2.7. Maps of the ultraviolet cloud configurations in Venus' upper atmosphere, covering the period from 18 September to 30 September 1972. Mercator projection with North up. Each map is separated from the previous one by a complete rotation of the upper atmosphere around the planet, in 4.0 days. From coordinated observations between Pic-du-Midi (P), Lowell Observatory (L), New Mexico University (N), and Table Mountain Observatory (T), as indicated below the maps (courtesy of A. Dollfus).

Determination of Rotational Elements

Rotational elements include the planet's period of rotation about its axis (and the corresponding length of the sidereal and solar days), the direction of rotation, and the orientation of the rotational axis in space, usually defined by the equatorial or ecliptic coordinates of the planet's North Pole on the celestial sphere. With prograde planetary rotation (coinciding with the planet's direction of motion along its orbit), the solar day is longer than the sidereal, and with retrograde rotation, shorter. Because of small periodic oscillations in the length of the solar day over a year, the concept of a mean solar day is introduced (which for the Earth is $24^h 03^m 56.55536^s$ sidereal time). The rotational characteristics of Venus have now been reliably established and are given in Chapter 1, but the path to determine them was long and thorny.

The traditional method for determining rotational elements—observing the visible movement of surface details across the disk—was used successfully for Mars, Jupiter, and Saturn but proved to be quite unsuitable for Venus. Unsuccessful attempts to establish the period and direction of its rotation with optical observations continued for three centuries, up to the 1960s. Attempts at visual observations, begun by Cassini in the seventeenth century, were especially numerous. A summary of these results, from 1666 to 1954, was presented in the well-known book on Venus by P. Moore (1961) and shows a huge spread in values of Venus' rotational period, from 23 hours to 225 days. This can be explained by the constancy of weakly delineated dark and light arcs parallel to the limb or terminator, essentially unchanging from day to day, whose positions were associated with the rotation of a solid planet surface. Some observers were inclined to view this immutability as proof that the synodic period of Venus' rotation (S, which is the interval between two successive passages of a point on a planet's surface through the central meridian relative to a terrestrial observer) was equal to an Earth day. Others attributed the persistence of details on the disk to the coincidence of Venus' rotational period with its orbital period about the Sun, approximately 225 days. The first to point out this possibility was Schiaparelli in 1887.

As far as the position of the defenders of a 24-hour diurnal period is concerned, a similar situation actually occurs when observing Mars; its rotational period is close to 24 hours, and therefore, at the same observational time (say, midnight) on consecutive nights, one can see an approximately identical pattern of details on its disk. The groundlessness of such a conclusion, however, is easily revealed by more long-term and careful observations, and by coordinated observations from different observatories. The fact is that, depending on the relative positions of Earth and another planet, the synodic period must exhibit oscillations, and because of the inexact correspondence between the synodic period and the duration of the terrestrial day, easily detectable changes (shifts) in the observed recurrence of details on the disk must occur. Through prolonged, multihour observations, it should be easy to discern a gradual migration of details from one edge of the disk to the other, as is actually observed for Mars and Jupiter, but not for Venus.

The position of defenders of the opposing viewpoint—the coincidence of the rotational period with the Venusian year—was also vulnerable. In principle, such a situation is quite possible; a typical example of a similar coincidence is the synchronous rotational motion of the Moon with respect to Earth. In this case, the position of details on Venus would remain virtually constant, not with respect to the center or the limbs (as in the case of 24-hour rotation) but relative to the terminator and subsolar point. However, when the planet was at its maximum angular distances from the Sun for a terrestrial observer (at the western and eastern elongations), the picture would be different, because different portions of the illuminated hemisphere would be turned toward the Earth. Evidence for such a difference was not obtained.

Finally, one must add the extreme skepticism of critics of both sides, who were dubious about making definite conclusions about Venus' rotational period from observations of weak, indistinct formations of unknown nature on its disk. P. Moore (1961) quite naturally concluded that "the direct observer has lost the battle" inasmuch as "hundreds of drawings, made to the present, have not shed any light on the subject under discussion." The use of more sophisticated observational methods and data analysis, by combining and obtaining composite images of a large number of drawings and photographs, also did not live up to expectations. Apparently, they suggested that the rotational and orbital periods of Venus coincided (Dollfus, 1953, 1955).

Contrary to expectations, the situation was not clarified by regularly photographing the planet in the ultraviolet, which was done especially intensely at the Pic-du-Midi Observatory in France, starting in 1918, and also in Brazzaville, the Congo. Combining a series of photographs encompassing almost 30 years of observations, and tracing the recurrence of the three most distinct bands, forming a horizontally oriented letter Y (see Fig. 2.6), led to the conclusion of a sidereal rotational period for this configuration equal to 97 hours, 38 minutes, in a retrograde direction (Boyer and Camichel, 1961). This period, however, did not reflect the true motions of the markings; only the direction of rotation was correct.

Spectroscopic methods were also used in an attempt to determine the length of the Venusian day. This practice was based on the possibility of detecting a shift in the position of spectral lines, which is different for various points on the disk of a rotating planet, owing to the Doppler effect. This additional component is superimposed on the overall radial velocity dictated by the motion of the planet, relative to Earth and the Sun. This method was first used at the beginning of the twentieth century at the Pulkovo Observatory and later at the Lowell Observatory (Belopolsky, 1900, 1903, 1911; Slipher, 1933); much later such observations were continued at Mount Wilson Observatory (Richardson, 1958). Analogous measurements of the well-known rotational period of Mars were used to verify the method. The results of these studies gave values for Venus' period from 1 to 15 days; later measurements indicated a likely retrograde rotation, with a period of 8 to 46 days.

An approximately 24-hour rotational period for Venus was inferred by the discovery of periodic "flashes" of radio radiation at a wavelength of 2 m, which were associated with lightning phenomena in the Venusian atmosphere, related to volcanic activity on the surface. In reality, however, these signals turned out to be unknown interference of terrestrial origin (Kraus, 1956, 1960).

Finally, to determine the direction of Venus' rotational axis, attempts were made to use certain indications of zonality in the atmosphere (such as the alternation of light and dark bands), which presumably must coincide with planetographic parallels. In principle, such bands can be detected in ultraviolet photographs, although far from unequivocally (sometimes they diverge from the terminator to the limb in a fan shape). The use of this method led to quite a large scatter in determining the coordinates of Venus' pole, primarily corresponding to an inclination of the equator from 14° to 39° (Kuiper, 1954; Zotkin and Chigorin, 1953; Tsesevich, 1955; Richardson, 1955).

Fundamentally new ways of studying the characteristics of planetary motion grew out of the use of radar techniques. Radar is used for determination of planetary distances,

Fig. 2.8. Doppler broadening of a spectrum upon reflection from a rotating planet. $\Delta\nu$ is the frequency shift due to the relative motion of Earth and the planet, $2\delta\nu_{max}$ is the change in frequency due to the rotation of the planet (the overall Doppler broadening of the spectral radiation reflected from the whole planet), and $\delta\nu = (2x\Omega/c)\,\nu_0$ is the Doppler frequency shift caused by the linear velocity, along the line of sight, of a line segment separated by x from the subradar point and parallel to the projection of the rotational axis onto the plane of the figure.

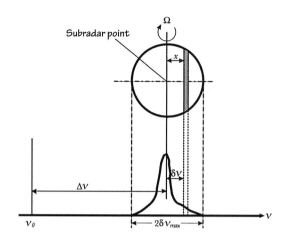

size and rotational elements, and surface properties. The method for computing the rotational period is based on the fact that a monochromatic signal reflected off the surface of a rotating planet is broadened owing to the Doppler frequency shift at the edges of the disk (Fig. 2.8). From the measured width of the band, that is, the difference in the radial velocity values at the points of intersection of the equator with the edge of the disk, one can determine the rotational period of the planet about its axis and the direction of the axis of rotation in space.

Radar investigations of Venus were begun in 1958 at the Millstone-Hill Observatory in the United States using a 25 m radiotelescope (Newbauer, 1959; Pettengill and Price, 1961). Since 1959, experiments have continued at the Jodrell Bank Observatory in England on the 76 m radiotelescope (Evans and Taylor, 1959; Thomson et al., 1961), and since 1961 at a series of other radio observatories.

The first successful measurements were carried out near the inferior conjunction of Venus in 1961, in the United States at the Millstone-Hill and Goldstone Observatories, and in the USSR using the large antenna of the Deep Space Network Center in the Crimea, in the decimeter range of radiowaves (Muhleman, 1961; Victor and Stevens, 1961; Pettengill et al., 1962; Kotelnikov et al., 1962, 1963). These measurements provided definite evidence of a slowly rotating Venus turning in a retrograde direction (in a clockwise direction as observed from above the North Pole). These measurements provided values for the rotational period within the range of approximately 100 to 300 days. Soon thereafter, a significant increase in receiver sensitivity permitted a substantial decrease in detection errors, which gave a mean period of 250 days (close to the actual value) and, at the same time, substantially refined the position of the rotational axis (Pettengill et al., 1962; Goldstein and Carpenter, 1963; Goldstein, 1964; Kotelnikov et al., 1965; Carpenter, 1966). Thus, by the mid-1960s the problem of Venus' rotational elements had already been conclusively solved. The results of these later measurements are presented in Table 2.1.

Table 2.1   Rotational elements of Venus

Period (Earth days)	Direction of rotation	Coordinates of the North Pole[a]	Year of measurement	Reference
243.0±0.1	retrograde	272.7±0.7 65.3±1.0	1964, 1967, 1969	Jurgens, 1970
242.98±0.04	retrograde	274.1±3 71.4 ± 1	1962, 1964, 1966, 1967	Carpenter, 1970
243.0185±0.04	retrograde	272.76±0.02 67.16±0.01	1990	Davies et al., 1992

Note: Data are according to radar measurements conducted near inferior conjunctions.

[a] Right ascension α and declination δ

In 1970, the International Astronomical Union accepted the rotational period of the planet as T = 243.0 days. With a retrograde rotational direction, this quantity corresponds to a duration for the Venusian day of 116.69 terrestrial days. In other words, in one Venusian year the Sun rises and sets on the planet twice. The measured coordinates of the north rotational pole enabled us to conclude that the axis is almost perpendicular to the orbital plane (Ward and De Campli, 1979). Note that the actual position of the axis turned out to be 15° less than the lower limit estimated from optical measurements (Kuiper, 1954). Most recently, Magellan data enabled the calculation of improved values for the rotational period and for the position of North Pole: T = 243.0185 ± 0.0001 days; right ascension α = 272.76 ± 0.02°; declination δ = 67.16 ± 0.01° (Davies et al., 1992).

The Chemical Composition of the Atmosphere

The first information on the chemical composition of Venus' atmosphere was obtained by spectroscopic observations. Information on the gases in a planet's atmosphere is obtained by studying its absorption spectrum, in which characteristic bands belonging to these gases are present. Absorption bands occur in the spectrum of solar radiation reflected by a planet, owing to selective absorption of the incident solar rays as they pass down to the level inside the planet's atmosphere where reflection occurs and then travel back to the observer. The task of identifying individual bands and lines within the band, which are usually quite weak, is complicated by contamination due to the superposition of telluric bands, caused by additional absorption in the Earth's atmosphere. To eliminate this effect, it is necessary to measure the Doppler shift of actual absorption bands, caused by the relative motions of Earth and the planet, or to use a comparison spectrum (which is often easier); it is convenient to take the reflection spectrum of the Moon for this purpose. By measuring the line intensities, one can estimate the content of specific gases in the planet's atmosphere, with allowance made for the propagation geometry of the rays. Such estimates are very difficult to obtain, owing to the strong effect of a priori unknown physical conditions (the thermodynamic state of the gas) on the formation of the bands, and to the existence of suspended particles (aerosols) or cloud condensates in the atmosphere. These particles or condensates make it necessary to take into account scattering of higher orders when calculating the line intensities; this substantially decreases the reliability of the estimates. As far as the physical conditions are concerned, one can estimate the pressure from the ratio of band intensities in the absorption spectrum, because the dependence of individual bands versus pressure is distinct. In turn, one can estimate the temperature of the absorbing layer by studying the fine structure of a band for a specific component in the absorption spectrum. This structure characterizes the distribution of given types of molecules over various vibrational and rotational states. The corresponding temperature is called the rotational temperature.

The first attempts to detect signs of absorption on Venus in reflected light by visual means were made using an ocular spectroscope during the second half of the nineteenth century (Huggins, 1863; Secchi, 1864; Vogel, 1874). The presence of oxygen and water vapor bands was reported, but later this was refuted by higher-quality observations made with photographic techniques (Slipher, 1909, 1933; see also Dunham, 1948). Attempts to find any sure signs of absorption in the violet and ultraviolet regions of the spectrum turned out to be unsuccessful (Wildt, 1940; Heyden et al., 1959; Spinrad, 1962a), but because expectations had been high, some results included the erroneous identification of organic matter (Kozyrev, 1954b). The question of oxygen on Venus again arose in the early 1960s, based on results of observations conducted at the Crimean Observatory. The discovery of weak Doppler companions in the wings of the telluric band lines of oxygen ($\lambda = 7660$ Å), which were interpreted as belonging to Venus' spectrum, pointed to an appreciable amount of oxygen in the atmosphere (Prokofiev and Petrova, 1963; Prokofiev, 1964, 1965). This result, however, was not confirmed by later measurements (Belton and Hunten, 1968; Spinrad and Richardson, 1965; see also Chapter 5).

Measurements in the infrared portion of the spectrum turned out to be considerably more effective. The first success came in 1932, when Adams and Dunham (1932; see also Dunham, 1948), observing at the 2.5 m telescope of Mount Wilson Observatory with a diffraction spectrograph, discovered distinct bands, within which lines were identified with the vibrational spectrum of carbon dioxide. Using the results of laboratory experiments, the CO_2 content was estimated to be quite high, equivalent to a column abundance of 3.2 km-atm (equivalent to a column of pure CO_2, at one atmosphere pressure, 3.2 km high). At the same time, an attempt was made to find oxygen bands and water vapor lines, but they were not detected in the resulting spectra. Later, with the appearance of higher-quality radiation detectors and photoelectric recording methods, many new CO_2 bands (up to 40) were identified in the near-infrared spectrum of Venus (1–2.5 μm), thus providing more accurate information on CO_2 content and, likewise, the content of C^{12} and C^{13} isotopes (Kuiper, 1947, 1962; Sinton, 1963a, 1963b; Moroz, 1963, 1964). The observations were conducted at the Kitt Peak, McDonald, and Lowell Observatories in the United States, and at the Crimean Observatory in the USSR. Theoretical and laboratory research, correlated with spectroscopic measurement data, led to an estimate of the overall amount of CO_2 of about 2 km-atm, which would correspond to a partial pressure of 0.25 atm. Taking into account the estimates of the overall pressure available at the time, this led to the conclusion that the CO_2 content did not exceed 4% of the entire mass of atmospheric gas (Herzberg, 1952). Although the planet apparently had an overall CO_2 content approximately three orders of magnitude greater than that in Earth's atmosphere, CO_2 seemed to be a relatively small component of Venus' atmosphere.

In 1968, however, Belton, Hunten and Goody, in interpreting new, higher-quality spectra of Venus, included the effect of multiple scattering of solar radiation by particles within the clouds (Belton et al., 1968; Belton, 1968). It was shown that, inasmuch as this effect significantly changes the width and intensity of CO_2 absorption lines, the previous estimates of CO_2 content were too low. A similar conclusion was made based on the analysis of high-resolution Fourier spectroscopy of Venus (Connes and Connes, 1966). In these studies it was often assumed that nitrogen, widespread in the solar system and chemically relatively inert, made up a significant portion, and possibly the bulk, of the planet's atmosphere.

Concerning the relatively minor atmospheric components, the question of the water vapor content has long remained controversial. Spinrad (1962b), observing at the 2.5 m telescope of Mount Wilson Observatory, established an upper limit for the H_2O content of 0.7 g/cm^2, which is about 1/200 the amount of water in Earth's atmosphere. Because the water vapor in Earth's atmosphere strongly affects the results of measurements of H_2O content on Venus and other planets, attempts have been undertaken to lift the measuring equipment as high as possible above Earth's surface. Measurements have indicated the presence of a characteristic H_2O band at a wavelength of 1.3 μm and, from the intensity of the band, have provided an estimate of the vapor content above the clouds, equivalent to a 19 μm layer of condensed water (2×10^{-3} g/cm^2), which is comparable to the amount of water above the tropopause on Earth but five orders of magnitude less than the amount over the whole thickness of Earth's atmosphere (Strong et al., 1960). Observations from the Swiss High-Mountain Observatory provided a value of $(1.-1.5) \times 10^{-2}$ g/cm^2 for the H_2O vapor content in the atmosphere above the clouds (Dollfus, 1964), which agreed well with Spinrad's estimate. Subsequent measurements provided a value of about 2×10^{-2} g/cm^2, which corresponds to a relative content on the order of 10^{-4}, assuming that the atmospheric pressure at the level of the cloud layer is approximately 0.1 atm (Bottema et al., 1964, 1965). A value close to this was obtained by Belton and Hunten (1966), Spinrad and Shawl (1966), and Owen (1967, 1968).

For the Venusian atmosphere above the clouds, the upper estimate of the H_2O content was obtained from an analysis of the microwave emission at $\lambda = 1.35$ μm, which corresponds to an intense water vapor absorption band. These measures became possible with the advent of radio astronomical studies of the planets (see, for example, Kuzmin, 1967). The lack of appreciable attenuation in the radiation intensity at this wavelength can be interpreted as an indication that the limiting value of the water vapor content must not exceed 20 g/cm^2 (which, as it subsequently turned out, on the whole agrees with the results of direct measurements).

As far as the other minor components of Venus' atmosphere are concerned, as with oxygen, initial attempts to detect carbon monoxide, oxides of nitrogen, ammonia, and certain hydrocarbons gave negative results (Kuiper, 1947). Later, however, the presence of CO was detected in small amounts, corresponding to 4 cm-atm (Sinton, 1963b). This result was confirmed by high-resolution near-infrared spectra of Venus according to which the relative concentration of CO does not exceed 10^{-5} (Connes et al., 1968). In these same measurements, hydrogen chloride (HCl) and hydrogen fluoride (HF) were found, in relative concentrations of 10^{-7} and 10^{-8}, respectively (Connes et al., 1967). Although the amount of these gases is negligible, their existence in an oxidizing medium such as Venus' atmosphere was unexpected. Corrections that take into account the special features of absorption line formation in an inhomogeneous cloudy atmosphere, according to the theory of growth curves, were incorporated into the estimates of abundances of minor components (Belton, 1968). In connection with this, the estimates for HCl and HF were lowered from their original values to those presented in Table 5.5.

An example of Venus' spectrum was obtained from high-flying aircraft (Kuiper and Forbes, 1967; Kuiper et al., 1969; Fig. 2.9). The absorption lines, responsible for the identified components of Venus' atmosphere, stand out, CO_2, H_2O, HCl, and HF being most pronounced. Among other possible components, carbon protoxide (C_2O_3), hydrocarbons—CH_4, C_2H_2, and C_2H_4—and sulfur compounds—SO_2, COS, and H_2S—and several others were identified. The limiting values of their content were based, however, on estimates of the upper limit corresponding to the resources of ground-based spectroscopic measurements, almost until the end of the 1960s. More recent contents and upper limits are presented in Table 5.5.

Temperature and Pressure

The first measurements of Venus' temperature were carried out in 1923 and 1924, using an infrared radiometer at a "window of transparency" in the Earth's atmosphere (8–13 μm), at the Mount Wilson and Lowell Observatories (Pettit and Nicholson, 1924; Coblentz and Lampland, 1924, 1925). The values obtained for the radiometric (brightness) temperature T_B, which were quoted in the literature for many years, were 250–270 K for the night side and 320–330 K for the daylight side. Because the measurable range of the spectrum was rather narrow, these values were calculated using Planck's formula, which conveys the monochromatic radiation intensity.[1] In physical terms the infrared brightness temperature T_B is the absolute temperature of a body (or a portion of its surface), which, given similar size, form, position in space, and similar exposure and observation conditions, radiates as an absolutely black body. To convert to the true temperature, one should use Kirchhoff's law. To do this, one must know the emissivity of the body's surface ε, which, for longwave radiation, is usually extremely difficult or even impossible to find. The situation is complicated by inadequate measurement techniques, especially by the difficulty in determining the true infrared flux at wavelengths indicative of deep and highly variable water vapor absorption bands. These difficulties were overcome about 30 years later, when Pettit and Nicholson published the reanalyzed results of their observations (Pettit and Nicholson,

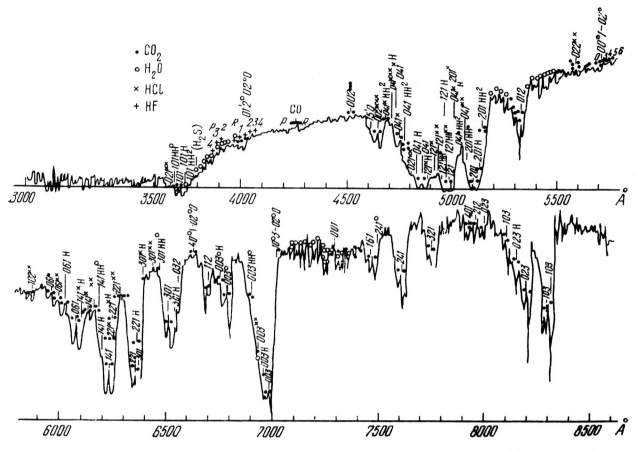

Fig. 2.9. Spectra of Venus in the infrared region, obtained from high-altitude aircraft (Kuiper, 1967, 1969), with the primary absorption lines identified.

1955). The mean value of the brightness temperature on the night side was found to be 240 K, and on the daylight side, 235 K. Taking into account measurement errors and the weak dependence of the obtained values on the phase angle, the authors concluded that no diurnal temperature variation exists.

The results of Pettit and Nicholson were confirmed by the measurements of Sinton and Strong (1960), who used the significantly more thorough method of infrared radiometry—the so-called double monochromator method. This method allowed them to distinguish clearly the intrinsic radiation of the planet above the background, in any spectral range, including those where intense absorption bands of water vapor are absent. The measurements, carried out in the 8–13 μm range, provided values of $T_B = 228$–238 K at the center of the disk, and an average of $\overline{T}_B = 234$ K. Scans along the terminator, as well as perpendicular to it, revealed a pronounced darkening toward the edge of the disk, satisfying a $\cos^{1/2}\gamma$ law, where γ is the angular planetocentric distance from the center of the disk. This result definitely indicated an increase in temperature with depth, which is simple to understand if one keeps in mind that as the radiation deviates from the normal to the surface (an increase in the angle γ), the optical depth of the layer along the normal, corresponding to the optical path of the ray reaching the observer, decreases. In other words, the linear thickness of the layer from which the radiation escapes decreases, and the amount of this decrease varies approximately according to a cosine law. So if the atmospheric temperature increases with depth, then the drop in brightness observable upon transit from the center of the disk to the edge is naturally explained by the fact that with an increase in γ, lower (hotter) layers are gradually "shut off," and the mean temperature of the radiating layer decreases. Accordingly, the brightness of the corresponding region of the disk also drops. Simultaneously, it was confirmed that a diurnal trend in the brightness temperature is absent on Venus. This was most convincingly shown by Sinton from measurements at the Lowell Observatory, conducted during the inferior conjunction of Venus in 1961 (Sinton, 1963a). The mean value of $\overline{T}_B$ was 236 K (see Table 1.4).

The most complete research of the temperature distribution over Venus' disk was carried out in 1962 by Murray et al., at the 200 in reflector at Mount Palomar Observatory (Murray et al., 1963). The achieved resolution of $1.5''$ corresponded to 1/30 of the diameter of the disk, or 400 km at the planet's surface. A special photodetector was used, based on germanium, which is sensitive in the $8-14$ μm range. Detailed scanning over arcs on the disk permitted the construction of isophots (isothermal contours) at $1°$ intervals (Fig. 2.10). They turned out to be quite distinct from circular, approaching ellipses with the major axis directed parallel to the equator. Along all radii of the disk, a monotonic decrease in the brightness temperature was noted, from 208 K at the center to 185 K at the edge (the actual values are systematically understated by 7–28 K, as the figure legend explains). No noticeable difference between the temperature distributions on the day and night sides is evident, except for a minor tendency toward higher nighttime values of T_B.

The measured brightness temperatures correspond to a region of thermal radiation and must relate to a level where the radiating gas layer becomes optically thick. Estimates of the radiative efficiency and systematic methodological discrepancies affect the absolute values of T_B. That the color temperatures in the infrared region of the spectrum do not differ essentially from the brightness temperature suggests that a temperature of $\sim$235 K is established at the cloud level.

Estimates of the rotational and vibrational temperatures, obtained by Adel and Slipher (1934) and Dunham (1948), who analyzed the fine structure of the CO_2 bands in Venus' infrared spectrum, resulted in values of 350 K and 300 K, respectively, which differ from the brightness temperatures. By more thoroughly analyzing the intensity distribution, with respect to the energy levels for the rotation-vibration bands of CO_2 in the 0.8 μm region, and by taking into account the strong scattering of reflected radiation by aerosols in addition to absorption by CO_2 molecules, Chamberlain and Kuiper (1956) obtained 285 ± 9 K. Spinrad (1962c), however, by studying the same photographic plates with spectra of Venus obtained at the Mount Wilson Observatory, later found that the rotational temperature has a spread between 215 and 440 K, varying from plate to plate. This result corresponded to different pressure values; in this case the highest temperature value was attributed to a level with a pressure of about 5 atm (which agrees well with the actual model of Venus' atmosphere). The observed scatter in the recorded temperatures could have been explained if one accepted the presence of breaks in the clouds, resulting in a change in the effective depth of band formation. It is important to keep in mind that the resulting estimates of the rotational temperature depend on the reflection model used and, primarily, on

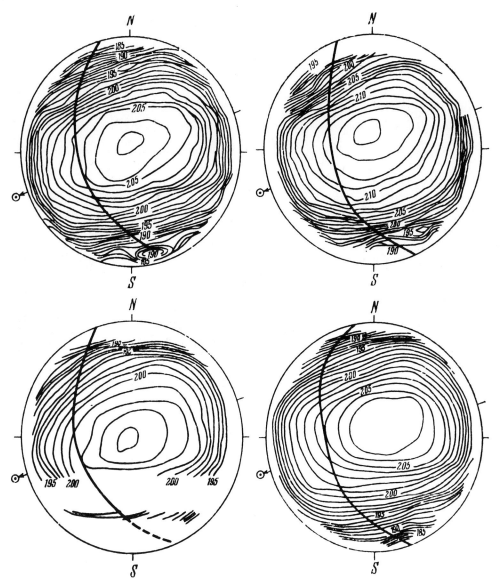

Fig. 2.10. Isophots of Venus' disk in the 8–14 μm wavelength range (according to Murray et al., 1963). The line of the terminator and the direction to the Sun are shown. The values of the brightness temperatures are systematically under-stated by 7°–28°, owing to unaccounted losses through calibration errors at the telescope and possible fluc-tuations in atmospheric ex-tinction. The decrease in the brightness temperature from the center of the disk to the edge, and the absence of noticeable differences in the day and night tempera-tures, are clearly visible.

the estimation of multiple scattering effects within the clouds. This problem was studied in detail by Spinrad (1962c) and a series of other investigators (Chamberlain, 1965; Moroz, 1968; Hunten, 1971). It was shown that for the simple reflection model (a transparent at-mosphere) the temperature value determined from the weak 8689 Å band in the reflection spectrum is $T_{rot} = 317 \pm 10$ K. At the same time, taking into account scattering (a turbid medium), $T_{rot} = 291 \pm 10$ K. Spinrad (1962c), using the $\lambda = 7820$ Å band, found T_{rot} of 338 K and 297 K, respectively, for these two limiting models.

Optical measurement data are limited to a certain depth in the atmosphere, located near the visible upper boundary of the clouds. Thus optical astronomy was incapable of addressing the question of the surface temperature. Estimates of Venus' surface tempera-ture were obtained from measurements, begun in the mid-1950s, of the intrinsic radiation of the planet at centimeter and millimeter wave intervals. In these intervals the intensity of thermal radiation, specified by the Rayleigh-Jeans law (a special instance of Planck's law where $hv \ll kT$), is directly proportional to the temperature of the radiating body. It is significantly weaker than in the infrared spectral region, within which the intensity maximum of the thermal flux falls at a temperature of several hundred degrees Kelvin. Therefore, recording such weak signals, and distinguishing them from the noise, have become possible only with the introduction of relatively large radio telescopes and the use of highly sensitive radiation detectors.

Radio emissions from Venus were first reliably recorded in May–June 1956, at a wavelength of 3.15 cm, using the 15 m radio telescope of the Naval Research Laboratory of the United States (Mayer et al., 1957, 1958). The result was truly sensational: unexpectedly, the radio brightness temperature proved to be exceptionally high, 620 ± 110 K in May and 560 ± 73 K in June. These values were confirmed shortly thereafter by measurements at 10.2 cm, which provided a mean value of 580 ± 160 K, and yet again at 3.15 cm (Mayer et al., 1960, 1962). The effort to ensure that the unusually high radio fluxes from the planet were real gave rise to a series of new experiments, involving many prominent researchers. Numerous measurements followed, in the broad wavelength interval from 0.4 to 70 cm. Especially detailed research, which included studies of the variation in the radio brightness temperature of Venus as a function of phase angle, was conducted by Drake (1962) at the 10 cm wavelength using the 27 m radio telescope of the Green Bank Observatory. Temperature values of 775 K and 540 K were obtained at the subpolar point and the antisolar point, respectively. On the whole, the results of all measurements, a selected summary of which is presented in Table 1.4, consistently gave mean values for the radio brightness temperature of Venus of about 600–700 K at centimeter wavelengths and approximately 300–400 K at millimeter wavelengths (see Fig. 1.10).

At the same time, attention was focused on determining the source of the planet's radio emission: was it associated with the surface or the atmosphere? The closeness of the results obtained at various frequencies in the centimeter wavelength range could be considered a serious argument for the observable radio emission having a thermal origin. Drake concluded that the energy distribution in the radio emission spectrum, in the 3–10 cm wavelength interval, was described by a frequency spectrum approximately corresponding to an inverse square relation ($\lambda^{-1.97}$); this satisfied the thermal radiation model of a dark, opaque surface more than the model of a hypothetical source, of nonthermal nature, in the planet's atmosphere.

This conclusion was also indicated by research at a resolution on the order of 1 angular minute (corresponding to the diameter of Venus at inferior conjunction) carried out at the 3.02 cm wavelength, using the large radio telescope of the Pulkovo Observatory; the telescope forms a parabolic boom 130 m in length (Korolkov et al., 1963). A study of the recorded radio emission curve, as the planet passed across the meridian of the radio telescope location, indicated the presence of a sharp maximum coinciding with the center of the disk and an abrupt drop in radio emission at a distance of 1.07 radii from the center of the disk (corresponding to an altitude of approximately 400 km above the cloud layer). At such a low spatial resolution, however, it was impossible to unequivocally reject the proposition that the atmosphere, through some non-thermal radiation mechanism, was the source of the high radio brightness temperature.

In research on Venus' radio emission at millimeter wavelengths, conducted using the 22 m radio telescope at the P. N. Lebedev Physical Institute of the Russian Academy of Sciences in Pushchino near Moscow (Kuzmin and Salomonovich, 1960, 1961, 1962; Kislyakov et al., 1962), primary attention was focused on the study of trends in the radio brightness temperature versus wavelength. At both 4 and 8 mm, similar values of about 400 K were obtained, with an uncertainty less than ± 100 K. It was tempting to assume that the temperature, which was lower than was found in the centimeter range, was due to atmospheric effects. Such an interpretation, however, depended on the assumption that Venus lacked other sources of microwave emission. If, for example, the atmosphere provided an additional contribution to the radio emission, then the surface temperature must be correspondingly lowered. If, however, the atmosphere weakened the emission from the surface (quantitative estimates of this are dependent upon its gas composition), then the true value for the surface temperature must be higher.

Generally, the microwave emission was thought to be of thermal origin, and thus the concept of a hot surface on Venus had the most support. There was even less definite information, however, regarding the surface pressure P_S and the thickness of the Venusian atmosphere, that is, how far below the clouds the surface was located. According to results of spectroscopic measurements using growth curves (the relationship between the effective absorption width W and wavelength λ), the effective pressure at the level of the upper cloud boundary was estimated at 0.1 ± 0.2 atm (Chamberlain and Kuiper, 1956). Independently, from the width of individual lines, a pressure of up to several atmospheres was derived (Spinrad, 1962c). These values correlate satisfactorily with the rotational temperatures.

Polarization measurements from observations in the optical region served as another independent source of information on pressure. Dollfus (1957), having discovered the difference in polarization between red and green spectral bands, interpreted the difference to be the result of molecular scattering in the atmosphere above the upper cloud boundary. The corresponding pressure at the boundary of the cloud layer turned out to be 0.09 atm, which is close to the spectroscopic estimates. A more detailed study of the pressure near the observable clouds was carried out later by Sagan and Pollack (1969), using the results of polarization measurements by Gehrels and Samuelson (1961), by Coffeen (1968), and by Coffeen and Gehrels (1969). At a phase angle of 90°, they obtained a pressure value of $P \simeq 130$ mb (for the 0.56–0.9 μm spectral band) and $P \simeq 35$ mb (for the 0.36–0.99 μm spectral band). If one takes into account that these estimates relate to different levels, the figures are in satisfactory agreement with almost identical optical thickness in the scattering of CO_2 in the appropriate spectral intervals. It nevertheless remained unclear how and to what extent one could extrapolate these values to depth, in order to obtain even approximate pressure values in the lower atmosphere.

At this stage, radio measurement data seemed to be more promising for judging the pressure at the surface. Most models were constructed with the goal of explaining the relationship between the radio brightness temperature and wavelength. From this point of view, the sharp drop in the radio brightness temperature T_B in the millimeter range was of great interest. Using laboratory data, however, the appropriate computations (see, for example, Plass and Stull, 1963; Ho et al., 1966) led to estimates of the surface pressure varying from several atmospheres to several hundred atmospheres. Such a wide range of uncertainty was primarily caused by the fact that the abundances of the components that absorb most intensely in the infrared and millimeter ranges (CO_2 and H_2O) were unknown. With a relatively small amount of these gases, a massive atmosphere is necessary (with a surface pressure greater than $200-300$ atm) in order to ensure the requisite opacity in the millimeter range due to induced dipole absorption. If, by contrast, CO_2 and H_2O were the principal components, then to explain the observed spectrum of microwave radiation, as well as the surface warming (caused by the greenhouse effect), it would be sufficient to have a surface pressure on the order of $2-5$ atmospheres.

The status of the problem of Venus' surface pressure, before space flights, as summarized by Deirmendjian (1968) is illustrated in Fig. 2.11, in which the ranges of expected P_S values are presented, according to the estimates of various authors from 1960 to 1968. The minimum value is approximately 3 atm, and the maximum reaches 1000 atm. Only a straightforward experiment inside the planet's atmosphere could provide an unambiguous answer to this disturbing question.

Climatic Conditions and Atmospheric Models

The unbroken cloud cover constantly enshrouding Venus made it an enigmatic planet and, for a long time, afforded a broad scope for the most exotic ideas about the conditions on its surface. For example, observations of refraction phenomena upon occultation of the Sun and stars by Venus suggested a small angle of horizontal refraction (approximately 20 angular seconds), which must correspond to a relatively thin and rarefied atmosphere. This conclusion, however, conflicts with the intense crepuscular phenomena observed. Attempts were undertaken to resolve this difficulty using a multilayer atmospheric model in which the above-cloud and below-cloud gas layers are responsible for the two phenomena, respectively (Sharonov, 1953a).

The study of refraction phenomena actually turned out to be extremely effective in investigating the structure of the above-cloud atmosphere. Observations of the occultation of Regulus (α Leo) by Venus made it possible to determine the optical radius of Venus and the height of the "optical" layer, above the cloud layer (Vaucouleurs and Menzel, 1960; Martynov, 1962; Martynov and Pospergelis, 1961). The altitude of this layer was determined to be within the $70-78$ km range. From an analysis of the brightness change in Regulus during occultation by Venus, values were obtained for the scale height in the atmosphere above the clouds, and the temperature was estimated; the certainty in determining the temperature depended substantially on the assumption of a predominant chemical composition for the atmosphere, that is, a mean molecular weight of the gas. As a result, the

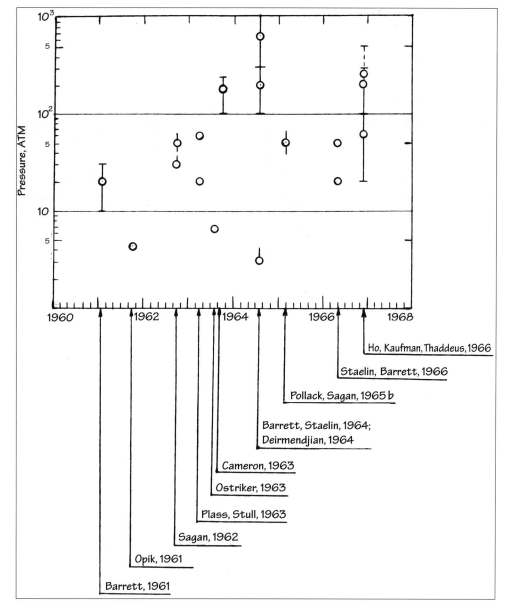

Fig. 2.11. Atmospheric pressure at the surface of Venus from estimates of various authors, during the period 1960–1968 (according to Deirmendjian, 1968).

question of how the above-cloud temperature fluctuates from the value $T = 234$ K determined by Sinton and Strong (1960) remained essentially open.

The presence of darkening toward the edge of the disk, convincingly demonstrated by Sinton and Strong (and later on also by Pollack and Sagan [1965]), suggested a positive temperature gradient deep in the atmosphere. This observation presented an opportunity for modelers. The theoretical principles of the mechanism of radiative energy transfer from the deep atmospheric layers of a celestial body to the outer layers were laid down in 1905 by K. Schwarzschild to explain the darkening toward the edge of the solar disk; subsequently the theory was developed in detail in the works of many researchers (see, for example, Chandrasekhar, 1950; Mustel, 1960). Relying on this theory, King (1964) considered the model of a semi-infinite, planar atmosphere with a measured infrared brightness curve, and calculated the various alternatives for temperature increase with depth, as a function of the optical thickness and the temperature gradient. Nevertheless, the high uncertainty in estimating these parameters, and the possible effect of non-radiative sources (convection) on the temperature trend, made it impossible to obtain any definite estimates of the temperature of the lower atmosphere.

The fact that spectroscopic measurements indicated an extremely limited water vapor content in Venus' atmosphere, which was attributed to the upper cloud boundary, did not hinder the hypothesis that a great abundance of water might be present in the underlying atmosphere and on the surface. To some extent, such ideas were built on the tradition of the first visual spectroscopic observations, which led to the conclusion of the presence of water vapor absorption bands in Venus' spectrum comparable in intensity to telluric lines. In particular, this conjecture gave rise to the assumption that, over the whole thickness of Venus' atmosphere, there might be six times more water vapor than in Earth's troposphere (Arrhenius, 1923). Even after the results of more reliable spectroscopic measurements in the 1930s revealed essentially no water vapor in Venus' spectrum and established an upper limit for its content at a level less than 2% that in Earth's atmosphere, the hypothesis of a humid atmosphere retained many adherents for a long time. It found indirect expression in the various models of water-ice clouds, which, at first glance, did not contradict data on the high aridity of the atmosphere above the clouds (see, for example, Menzel and Whipple, 1954, 1955; Deirmendjian, 1964). Moreover, results of spectrometry in the near-infrared spectral region (in areas free of CO_2 bands), as compared with the reflectivity of ice crystals 2.5 mm in size, seemed to allow the possibility of interpreting a whole series of specific features in Venus' spectrum as brought about by light scattering by such crystals; in particular, an attempt was made to identify these features with H_2O absorption bands at $\lambda = 1.9, 2.7,$ and 3.1 μm (Sinton, 1963a, 1963b). According to other data, however, no details characteristic of the spectral reflectivity of ice could be detected in the intensity curve of Venus' continuum spectrum (Moroz, 1963, 1964). These data generated skepticism regarding the water-ice cloud hypothesis. An attempt was made to circumvent these difficulties by employing models of clouds consisting of dust particles (Van de Hulst, 1948), such as ammonium chloride or ferrous chloride (Kuiper, 1969), hydrocarbons, or other organic compounds. Organic compounds were ruled out by Lewis (1968, 1969), however, because of their thermodynamic instability in a weakly oxidizing medium.

Especially great uncertainty, up to essentially the middle of the twentieth century, concerned the conditions on the surface of Venus. Among the three main hypotheses proposed, two especially popular ones were associated with ideas of a quite favorable climate on the planet. The first supposed that the current conditions on Venus corresponded approximately to those on Earth during the Carboniferous period, with an exceptionally humid climate and a rich organic landscape. The second hypothesis proposed that the surface of Venus was a continuous water ocean.

Underlying this second hypothesis were ideas about the predominant geochemical cycle of atmosphere-surface interaction, during which absorption of CO_2 gas and rock formation of carbonate composition occur. According to the authors of this hypothesis, the water ocean mass should have ensured the stability of the equilibrium composition of the atmosphere. Furthermore, belief in an abundance of hydrocarbons in the geochemical cycle gave support to the concept of a petroleum ocean on the surface. Finally, according to the third hypothesis, which turned out to be the closest to reality, Venus' surface was thought to be an absolutely waterless desert, devoid of any plant life.

The discovery of intense microwave radio emissions from Venus gave rise to a series of new atmospheric models in which attempts were made to ascertain the physical mechanism responsible for the observed spectrum of radio brightness temperatures. One can divide these models into those proposing a hot surface, whose emissions in the millimeter wavelength range are absorbed by the relatively colder atmosphere, and a model in which the radio emission of non-thermal origin is generated by the atmosphere, and the surface temperature only slightly exceeds that of the Earth. It was not a simple matter to admit the possibility that our neighboring planet, so similar to Earth, had a surface temperature of ~ 600 K, and to provide a convincing interpretation for such a possibility. As a result of this, perhaps, a cold surface model was first proposed, in which the observed microwave intensity was explained as caused by emission from Venus' hypothetical ionosphere, with an electron density on the order of 10^9 cm^{-3} (this exceeds the maximum electron concentration in the most dense F_2 layer of Earth's ionosphere by approximately three orders of magnitude), or as caused by gas discharges in the atmosphere. These mechanisms were subjected to persuasive criticism. It was shown that the observable microwave spectrum cannot be synchrotron or cyclotron radiation (Walker and Sagan, 1966). As for the hypothesis on gas discharges in the troposphere, it was difficult to reconcile it with the regular nature of the emission; moreover, owing to the release of heat by electrical discharges of the requisite energy, the atmosphere should have heated up to a temperature exceeding that of the radiating layer (Pollack and Sagan, 1967).

The model with a hot surface seemed to be much more reasonable (and better supported by the results of optical measurements). Greenhouse, eolospheric (frictional), and circulational mechanisms were proposed to explain the heating of the surface in this model. In the greenhouse model, it was proposed that the unreflected portion of visible solar radiation penetrates to the surface, where it is absorbed, and surface radiation in the infrared region is trapped by the atmosphere owing to the presence of such triatomic molecules as CO_2 and H_2O, which display strong absorption bands in this region. As a result, the surface and atmosphere are heated to a temperature T_S, approximately corresponding to the radio brightness temperature of the planet.

The first model of this type was that of Wildt (1940a), who considered heat transfer in the CO_2 atmosphere of Venus with parameters close to those of Earth and arrived at the conclusion that heating due to the greenhouse effect would not exceed about 50 K. Rather simplified ideas about the nature of absorption of thermal radiation by CO_2 were used in this case, however. The greenhouse model of Sagan (1960a, 1960b) was considerably more thorough. It was calculated using integral absorption functions contained in technical literature on boilers and furnaces, in an attempt to approximate band superposition within the infrared range under consideration. In approximation of a "gray" atmosphere (where the highly simplifying assumption that absorption does not depend on wavelength

Fig. 2.12. Atmospheric temperature at Venus' surface T_S (as the ratio T_S/T_E, where T_E is the effective temperature of the planet as a function of lg b, with $\mu = 0.50$ for various optical thicknesses τ, according to Wildt, 1966). See the text for further explanation.

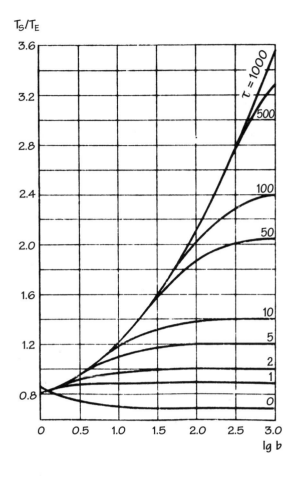

is used), estimates of the opacity required to decrease the escaping heat flux, for $T_S = 600$ K, to a value equal to the absorbed solar radiation were obtained. It was also shown that the necessary degree of heating may be attained in an atmosphere with a high CO_2 content, given 1–10 g/cm^2 water vapor, that is, less than 1%. The atmospheric opacity to infrared radiation turned out to be approximately 99% in this case. More detailed calculations of radiative transfer in a gray atmosphere (Jastrow and Rasool, 1963) showed, however, that the required opacity would have to be even higher. In this work there was no attempt to identify the appropriate opacity value with physical parameters of the actual atmosphere.

The supposition of the possible existence on Venus of atmospheric pressures in the tens of atmospheres forced a more careful evaluation of the likely changes in the absorption band characteristics under such conditions. Plass and Wyatt (1962) showed that, at high pressures, the opacity of CO_2 increases significantly in certain regions of the infrared spectrum. Estimates of the radiative transfer made by these authors, however, used approximately the same straightforward assumptions as in the early work of Sagan. In addition, cooling at wavelengths greater than 20 μm in the CO_2 atmosphere under consideration was not incorporated—an extreme oversimplification of the actual nature of radiative transfer in a CO_2 medium.

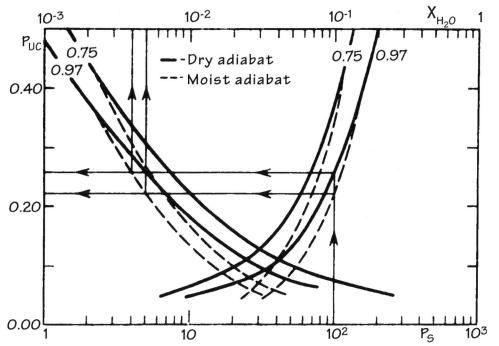

Fig. 2.13. The relation between the atmospheric pressure at the surface, P_S, and at the upper cloud boundary, P_{UC}, for various mixing ratios of water vapor X_{H_2O}. The solid lines correspond to a dry-adiabatic temperature gradient, the dashed lines to a moist adiabat (according to Pollack, 1969c).

Proceeding from the obvious notion that clouds play an important role in the heat exchange problem on Venus, Ohring and Mariano (1964) attempted to account for the clouds' effect on the radiative flux distribution. They arrived at the conclusion that in the presence of condensed clouds, considerably less opacity in the subcloud atmosphere is required. In this case, however, an adiabatic temperature gradient was assumed to exist over the whole atmosphere, which, in approximation of radiative equilibrium, was difficult to reconcile with the proposition of moderate opacity of the gas to thermal radiation and a simultaneously high surface temperature. In order to circumnavigate this difficulty, the existence of a "gray" absorber of unknown nature, located in the atmosphere below the clouds, was artificially assumed. This ensures a regime of radiative equilibrium.

The most rigorous treatment of the greenhouse effect in a gray atmosphere, assuming a local energy balance and strict radiative equilibrium, was carried out in a later model by Wildt (1966). He obtained an atmospheric temperature dependence as a function of the degree of the greenhouse effect intensity, defined by the ratio of gray absorption coefficients for solar and thermal radiation ($b = \varkappa_\odot / \varkappa_t \ll 1$) and as a function of the nature of the angular distribution of the radiation field, defined by the quantity $\mu = \cos\theta$, where θ is the angle between the direction of the ray along the path z and the normal to the surface, of an infinitely thick plane-parallel atmosphere. According to these computations, an example of which is shown for $\mu = 0.50$ in Fig. 2.12, a surface temperature corresponding approximately to the radio brightness temperature of Venus would be reached for a range of lg b from 2.5 to 3.0, and for an optical thickness, $\tau = \int_z^\infty \varkappa_t \, dz$, between 100 and 500.

Thus, the results of calculations showed that the greenhouse model requires considerable opacity to thermal radiation. The possibility of achieving such an opacity in Venus'

Fig. 2.14. The surface elevation profile in Venus' equatorial region, determined from radar ranging measurements. The numbers denote the number of revolutions of Venus relative to Earth, referenced to August 1967. An equatorial radius for Venus of 6050 ± 0.5 was taken as zero altitude (according to Campbell et al., 1972).

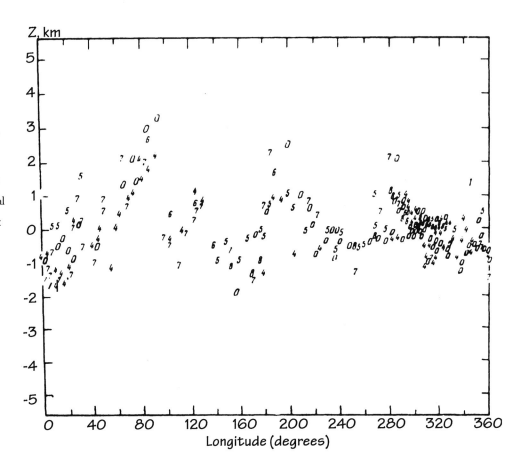

atmosphere, along with the necessary high transparency to shorter-wave solar radiation, was repeatedly questioned. The strongest argument by critics of the greenhouse model has focused on the presence of aerosols, which in sufficient quantity would prevent solar radiation from reaching the surface, although additional opacity to infrared radiation would also be provided.

The hypothesis of a large amount of dust on Venus brought about the eolospheric model, proposed by Öpik (1961). According to this model, solar radiation is absorbed at the top of the atmosphere, which is in a state of convective equilibrium, and is transferred to the planet's surface by means of the friction of wind-borne dust particles. Convection sustains the necessary amount of dust particles, consisting of calcium and magnesium carbonates, in a nitrogen–carbon dioxide atmosphere. In this case, the radio brightness temperature, if it is associated with the surface, must not exhibit any dependence on phase.

The obvious difficulty in the eolospheric model is that, in the local theory, the absorption of solar radiation at the upper boundary assumes the absence of radiative heat flux at lower altitudes and the existence of isothermal equilibrium in the atmosphere. The requirement for convective equilibrium must result in the transfer of absorbed energy to lower-lying layers by convective heat exchange, and in the establishment of a temperature profile close to adiabatic down to the very surface. From this point of view, the addition of a further source of heat release, owing to friction, may be unnecessary.

Note that the local theory considers energy exchange in a limited volume (for ex-

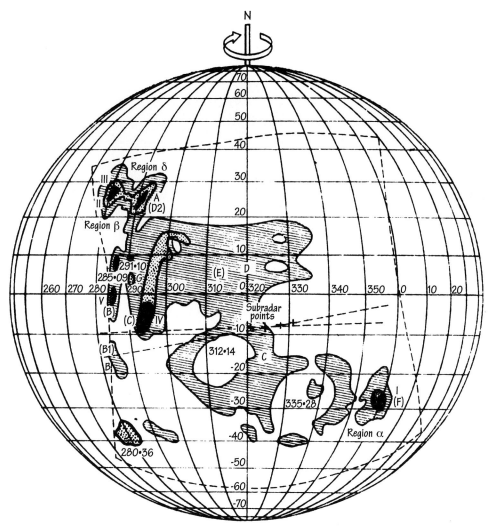

Fig. 2.15. A map of the surface reflective properties of Venus at a wavelength of 3.8 cm (according to Rogers and Ingalls, 1970). The darkened areas correspond to regions of maximum reflection. The dashed line is the boundary of the mapped region.

ample, inside a gas column of unit cross-section), in this case neglecting the effects of global circulation. Such an approach was fruitful for Earth's atmosphere, where global circulation mainly influences the temperature in the near-surface boundary layer (on the order of 50 m thick, in which, according to Goody [1964, 1969], the most noticeable diurnal temperature fluctuations occur). Global circulation may, however, play a very important role over the whole depth of atmospheric gas if solar radiation is absorbed over a fairly narrow layer of the atmosphere close to the subpolar point because of opacity in the shortwave region of the spectrum.

These considerations were taken into account by Goody and Robinson (1966), who, attempting to circumvent the indicated difficulties of the eolospheric model, proposed an original deep-circulation model for Venus, using the analogy of ocean circulation on Earth. In this model the solar radiation absorbed at the top of the atmosphere is transported to the night side through large-scale motions. The downward flow of gas in the region of the antisolar point, and upward flow at the subsolar point, cause adiabatic heating and cooling, respectively, and as a consequence establish an adiabatic temperature gradient in the atmo-

Fig. 2.16. A map of the surface reflective properties of Venus at a wavelength of 12.5 cm. The ratio of light and dark regions reaches 20 to 1. The most "bright" regions, α, β, and δ, have dimensions of approximately 1000 km. The circles denote areas of Venus mapped in 1977 using radio interferometry, from three ground-based receiving stations (according to Jurgens et al., 1980).

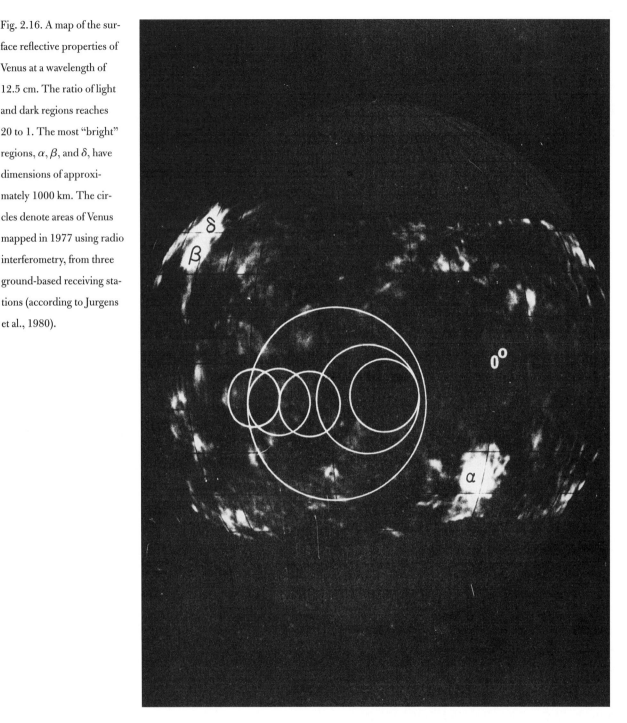

sphere. In turn, the temperature of the surface is determined by the depth of the atmosphere. The basic feasibility of such a model was confirmed by calculations based on treatment of a system of two-dimensional hydrodynamic equations for an incompressible viscous gas. Later, in a similar treatment, the problem of meridional circulation, which causes the temperature of the equatorial and polar regions to equalize, was considered (Goody and Robinson, 1970).

The models by Goody and Robinson and by Öpik proceeded from the powerful

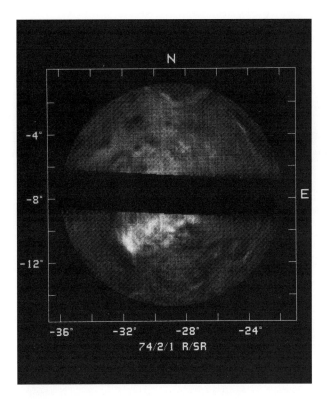 

assumption that solar radiation is absorbed in a narrow layer of the atmosphere at high altitudes (near the upper boundary of the clouds). This was contradicted, however, by estimates obtained in calculations by Samuelson (1967, 1968) and Pollack (1969a, 1969b). Samuelson examined radiative transfer in an atmosphere in a state of radiative equilibrium and "gray" in the two spectral intervals of visible and thermal radiation. The single scattering albedo values,[2] based on data from optical terrestrial observations of Venus, were taken to be $\omega_\nu = 0.989$ (minor absorption in the visible range, according to estimates by Sobolev [1944, 1964, 1968]) and $\omega_{IR} = 0.2$ (strong absorption in the infrared range). Solar radiation, in such a model, penetrates deep into the atmosphere despite the existence of a cloud layer. Yet the limiting surface temperature T_S, reached due to the greenhouse effect in this case, turned out to be no higher than 550 K. The reason for the difference from the higher radio brightness temperatures may be connected with the fact that the effective value of ω_ν, over the thickness of the atmosphere, is close to 1, as well as with the large extension of the asymmetry parameter g to the forward hemisphere.[3]

Pollack (1969b) solved the inverse problem of transfer theory, with the goal of determining the amount of CO_2, H_2O, and N_2 required to heat Venus to a temperature on the order of 700 K. From this solution the temperature gradient was determined in the region of the atmosphere from the surface to the base of the clouds. Corrections to the fraction of radiation transmitted by the clouds, by the subcloud atmosphere, and by the lower-lying molecular sublayer, were introduced into the solar radiant energy flux falling onto the top of a planar, stratified atmosphere; that is, scattering and absorption by cloud particles, and gas absorption in the ultraviolet, visible, and near-infrared regions of the spectrum were

Fig. 2.17. *Left*: A large-scale radar image of a circular region on the surface of Venus, from the results of radio interferometry in 1972. Latitude and longitude of the region are indicated along the axes. *Right*: An elevation map of the same area. The altitude difference between the contours is 200 m. The dark band in the center of both images is the interferometric fringe (according to Goldstein and Rumsey, 1972).

Fig. 2.18. The first space-craft launched to Venus, Venera-1. It stopped functioning before approaching Venus (courtesy of NPO Energiya).

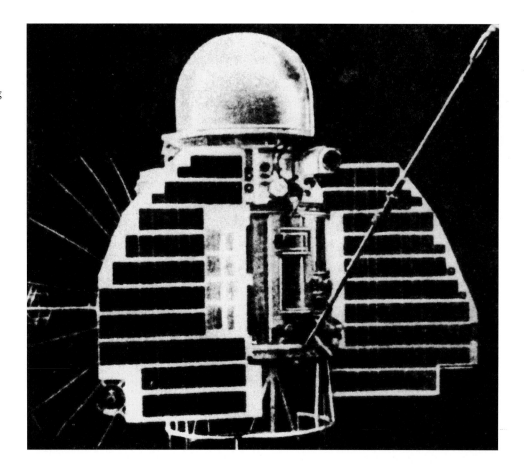

taken into consideration. The calculations were carried out to approximate the expected nitrogen atmosphere of Venus, with a surface pressure P_S of 15, 75, or 300 atm, and for a relative CO_2 and H_2O content of $X_{CO_2} = 0.1\%-10\%$ and $X_{H_2O} = 0.01\%-0.1\%$, respectively. But Pollack, taking into consideration reliable data that were already emerging at that time about the predominantly CO_2 composition of the atmosphere, asserted in the same paper that the results of the calculations would remain valid if the accepted values of P_S for a nitrogen atmosphere were decreased by 1.5 times.

This model showed that to ensure the opacity required to obtain high T_S and P_S values by means of the greenhouse effect, atmospheric parameters at the cloud level must exert a considerable influence. The connection between the surface pressure, P_S, and the pressure at the upper boundary of the clouds, P_{UC}, as a function of X_{H_2O} for an atmosphere with X_{CO_2} values of 97% and 75%, is shown in Fig. 2.13. The data presented in the diagram resulted in estimates of $X_{H_2O} = 0.005-0.01$ for the $P_S = 100$ atm case (for an atmosphere containing 97% CO_2 and for $P_{UC} = 0.2-0.3$ atm), although the calculations on which they were based were dependent on a whole series of assumptions, primarily concerning the structure of the atmosphere and its coefficients of opacity in the infrared spectral range, and generally allowed the broad range of conditions shown in the diagram. Therefore, the attractive possibility of realizing greenhouse model conditions remained hy-

Fig. 2.19. The Mariner-2 spacecraft, which achieved the first close fly-by of Venus in December 1962 (courtesy of NASA).

pothetical, along with the opposing idea, which was the basis of the Goody-Robinson model.

Doubts about the possibility of achieving the necessary atmospheric transparency in the visible range and opacity in the longwave spectral range gave rise to the idea of surface heating due to radioactive decay in the planet's interior (Kuzmin, 1965; Hansen and Matsushima, 1967). However, solutions of the thermal conductivity equation for soil, with parameters corresponding to those on Earth, led to the conclusion that unrealistically high temperature gradients would be necessary in the near-surface layer to achieve a high surface temperature. By making reasonable estimates of the heat flow from the interior, more or less comparable with the terrestrial value (which is approximately eight orders of magnitude less than the solar flux), the requisite surface heating may be achieved from the condition of thermal balance with the atmosphere only for very high opacity; in principle this may be ensured by a very high aerosol concentration. Maintaining the necessary amount of dust, however, requires the presence of mixing, which, under equilibrium conditions, cannot be realized within the framework of such a model. Therefore, the radioactive heating model turned out to be quite contrived, and substantially inferior to the greenhouse and circulation models.

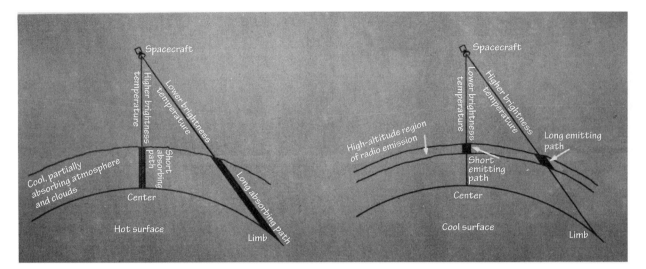

Fig. 2.20. The plan for conducting an experiment on Mariner-2, designed to determine the nature of the high radio brightness temperature on Venus. In the case of a hot, radiating surface (left), the absorption by the atmosphere at the center of the disk is low, whereas at the limb it is high; consequently, a darkening must be observed toward the limb of the disk. In the case of high ionospheric radio emission (right), the radiating region must be larger at the limb of the disk, and consequently the effect will be reversed; that is, brightening of the limb will occur, owing to the higher brightness temperature.

Size, Surface Properties, and Relief

Until the beginning of radio astronomical and radar studies, ideas about the relief and surface properties of Venus were as uncertain as the parameters of the lower atmosphere or the rotational elements. The radius of the planet's solid surface was unknown, and there was no data about its shape. Obtaining such information became possible only in the 1960s.

For a long time all information on the size of Venus came from results of optical measurements, which provided the so-called optical radius, corresponding to the upper boundary of the clouds. Martynov (1962) estimated it to be $R_{opt} = 6100 \pm 30$ km, and Vaucouleurs (1964) obtained $R_{opt} = 6120 \pm 7$ km. A more accurate value was later obtained by Dollfus (1968, 1972) from an analysis of his own measurements and the results of other authors. He took into account the dependence of the optical radius on the phase of Venus: with a decrease in the phase angle, the measured diameter decreases. This dependence was interpreted as the change in the visible size of the disk due to scattering by particles suspended in the atmosphere above the clouds; in this case the effect of scattering is at a maximum at inferior conjunction and at a minimum at superior conjunction. Dollfus determined the radius of the cloud layer to be 6115 ± 13 km and the radius of the scattering layer to be 6140 ± 13 km, which turned out to be close to the actual values. Note that the optical radius (taken to be equal to 6100 km) has been used, until quite recently, to calculate the ephemeris of the planet.

The first measurements of the surface radius of Venus were conducted by radio astronomy, using a radio interferometer with a high angular resolution. They provided a value of $R = 6057 \pm 55$ km (Kuzmin and Clark, 1965). This value was subsequently improved substantially by simultaneously processing an extensive series of systematic radar measurements of the distance to Venus and other planets. As a result, a mean value for the equatorial radius of Venus' surface was obtained, of 6052 ± 2 km (Ash et al., 1968), which correlates well with up-to-date data combining radar results with an analysis of spacecraft trajectories (see Table 1.3).

Fig. 2.21. An overall view of the Venera-2 or Venera-3 automatic station (courtesy of NPO Energiya).

Information about the shape of the planet, prior to spacecraft results, was essentially based entirely on the results of radar measurements at the radio telescope of the MIT Lincoln Laboratory at Haystack, with a dish diameter of about 36 m (Rogers and Ingalls, 1970; Rogers et al., 1972). The experiments were conducted at a wavelength of $\lambda = 3.8$ cm; the resolution at the surface was 0.5 km in altitude, 1000 km in latitude, and

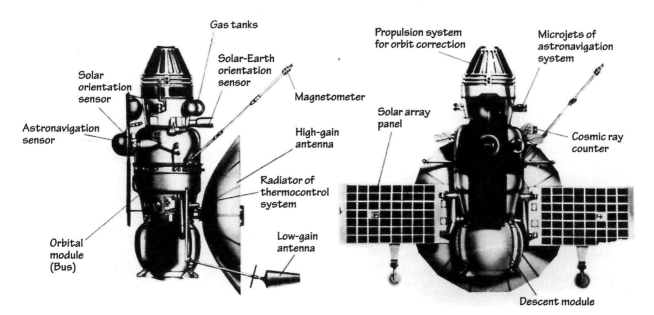

Gas tanks

Solar orientation sensor

Solar-Earth orientation sensor

Astronavigation sensor

Magnetometer

High-gain antenna

Radiator of thermocontrol system

Orbital module (Bus)

Low-gain antenna

Propulsion system for orbit correction

Microjets of astronavigation system

Solar array panel

Cosmic ray counter

Descent module

Fig. 2.22. The Venera-4 automatic station, whose descent module in October 1967 accomplished the first parachute descent into Venus' atmosphere and measured the planet's parameters (courtesy of NPO Lavochkin).

100 km in longitude. It was found that in the equatorial plane the planet's cross section is approximated by an ellipse; the difference in the semiaxes was 1.1 ± 0.35 km (for Earth this difference is about 200 m). The major axis of the ellipse makes an angle of $55°$ (clockwise) with the vector to Earth at inferior conjunction. Subsequently, data about the planet's shape were further refined, and the irregularity in the cross section in the equatorial plane was confirmed. This irregularity, however, cannot be approximated by an ellipse; the form of the cross section, briefly described in Chapter 1, is much more complex.

Studies of Venus' surface relief in the equatorial region were begun at the Haystack and Arecibo Observatories in 1967, at wavelengths of 3.8 cm and 70 cm, and continued until 1970 (Rogers and Ingalls, 1970; Smith et al., 1970; Campbell et al., 1972). The resolution in altitude was from 1 km to 150 m, and over the surface it varied from 200 to 400 km, in a band of latitudes approximately $\pm 10°$ from the equator. The most remarkable discovery proved to be an elevated region between about $60°$ and $120°$ longitude, with an altitude approximately 3 km above the mean surface level (referenced to a radius of 6050 km). It extended almost 6000 km in longitude and at least 500 km in latitude (Fig. 2.14). At the same time the reflection from this area turned out to be approximately 2.5 times higher than the surrounding terrain. The location of this high ground falls within the now well known highland region of Aphrodite Terra, shown on the map in Fig. 3.5. Independent analysis of the altitude profile in Venus' equatorial region was carried out by Rogers et al. (1972), from data on the change in the effective radar reflection cross section from the planet at a wavelength of 3.8 cm, as a function of the planetocentric longitude.[4] Radio emissions at this wavelength, although weak, are noticeably absorbed in Venus' atmosphere. Comparing the effective radar reflection cross sections at this and longer wavelengths (where the absorption is small), one could determine the amount of absorption in the atmosphere, its thickness, and, consequently, the surface elevation. The altitude profile

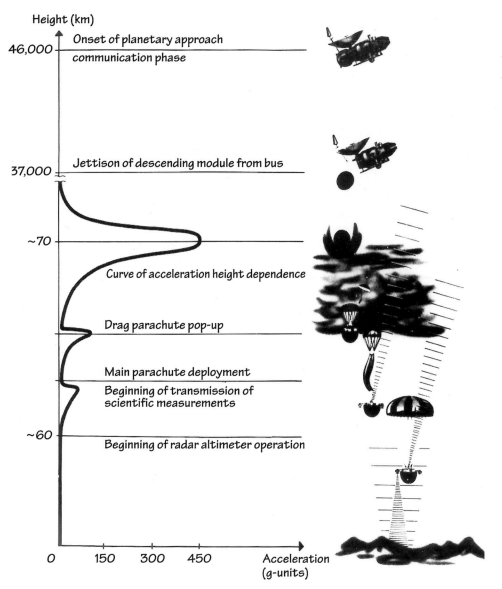

Height (km)

46,000 — Onset of planetary approach
communication phase

37,000 — Jettison of descending module from bus

~70 —

Curve of acceleration height dependence

Drag parachute pop-up

Main parachute deployment

Beginning of transmission of
scientific measurements

~60 —

Beginning of radar altimeter operation

0   150   300   450   **Acceleration**
(g-units)

Fig. 2.23. A diagram of the entry and descent into Venus' atmosphere of the Venera-4 descent module.

obtained by this technique was in good agreement with data for determining the relief using radar distance measurements.

The results of mapping the reflective properties of Venus' surface were of great importance for studying its topography and physical properties. The first experiments, conducted from 1962 to 1964, enabled the detection of local regions with higher than usual reflectivity, stretching for hundreds and thousands of kilometers (Goldstein and Carpenter, 1963). These areas received the designations α, β, and δ (Figs. 2.15 and 2.16). The presence of such peculiar regions was subsequently confirmed by a series of radar experiments conducted near the inferior conjunction of Venus in 1964 at the Jet Propulsion Laboratory in Pasadena and the Ionospheric Laboratory in Arecibo (United States), and at the Radio Electronics Institute (USSR). At Arecibo, using a wavelength of 70 cm, a high-reflectivity feature was discovered in the high-latitude region of the northern hemisphere ($64 \pm 2°$ latitude and $3.5 \pm 1°$ longitude). As we will see, this feature, named Maxwell, turned out to be one of the most remarkable features of Venus' relief.

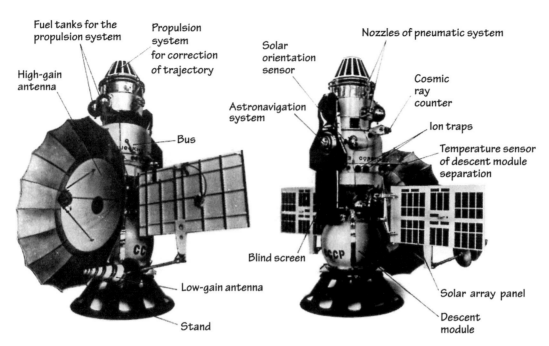

Fuel tanks for the propulsion system

Propulsion system for correction of trajectory

High-gain antenna

Solar orientation sensor

Nozzles of pneumatic system

Cosmic ray counter

Astronavigation system

Ion traps

Bus

Temperature sensor of descent module separation

Blind screen

Low-gain antenna

Solar array panel

Stand

Descent module

Fig. 2.24. An overall view of the Venera-5 or Venera-6 automatic station (courtesy of NPO Lavochkin).

Later, at the MIT Lincoln Laboratory, Rogers et al. (Rogers and Ingalls, 1970; Rogers et al., 1972) carried out interferometric measurements. In these experiments, the ambiguity between the northern and southern hemispheres (inherent in the frequency-time selection method) was solved. The width of the interference lobe at the 3.8 cm wavelength was 0.1 of Venus' disk. Based on five daytime observations, a map of the reflective properties of Venus' surface, encompassing 80° in longitude and 90° in latitude, was constructed (see Fig. 2.15). Two large circular formations, thought to be perhaps similar to the lunar mare, were discovered. The centers of these areas are located at $\lambda = 335°$, $\varphi = -28°$, and $\lambda = 312°$, $\varphi = -14°$, in a coordinate system in which the zero meridian crosses the α region, which has higher than usual radiowave reflection. The first formation was distinctly observed on each of the five days, but the second proved to be less well defined. Independent measurements of the distribution of reflective properties over Venus' surface were conducted by Goldstein et al., at the 12.5 cm wavelength, in 1969 and 1970 (Goldstein and Rumsey, 1972). The resolution over the planet's surface was about 50×50 km, although the ambiguity between the northern and southern hemispheres was not completely eliminated. In these same measurements it was shown that features with anomalous reflection (with dimensions several times greater than those on the Moon) had significantly greater contrast with the surrounding surface in depolarized reflection than in polarized. Thus, these data served to reinforce the idea of substantial variability in the physical properties of Venus' surface.

The most detailed measurement of the distribution of reflection and topographical features on Venus was produced near the inferior conjunction of 1972, which resulted in the sensational discovery of craters on Venus' surface (Goldstein et al., 1976). The reflected signals were received by two antennae, with diameters of 64 m and 2.6 m, spaced 21.6 km apart. The spatial resolution reached 10 km, and 200 m in altitude. This made it

Fig. 2.25. The descent
module of the Venera-4
spacecraft (courtesy of NPO
Lavochkin).

possible to produce a map of the reflective properties of the surface and stereoscopic images of several individual regions, 1500 km in diameter, close to the equator (in Fig. 2.16
they are designated by small circles). An example of such a large-scale image is shown in
Fig. 2.17 (left), and an altitude map of the same region indicating 200 m altitude contours
is shown in Fig. 2.17 (right). This and other similar images, produced later at Goldstone
(Goldstein et al., 1978) and also at Arecibo (Campbell et al., 1972, 1976, 1979; Campbell
and Burns, 1980), enabled the detection of tens of such circular structures ranging in diameter from 35 to 150 km, similar in appearance to lunar craters but considerably less
deep. It turned out that within each of the regions, 1500 km in size, the altitude differences
are less than 1 km. This suggested a high degree of obliteration of the relief on Venus' surface in the regions studied.

Studying the properties of the reflected signals also provided information on the
nature of Venus' surface rocks. The first estimate made of the soil's permittivity, equal to
$\varepsilon = 4.1$ (Pettengill et al., 1962), turned out to be in good agreement with the results of
polarized radio interferometric measurements (Kuzmin and Clark, 1965) and with later
analyses (see Kuzmin and Marov, 1974; Kuzmin, 1983). This value, also confirmed by
radar measurements ($\varepsilon = 4.7 \pm 0.8$), is considerably higher than for the Moon. For a solid

Fig. 2.26. Descent into Venus' atmosphere (drawing) (courtesy of NPO Lavochkin).

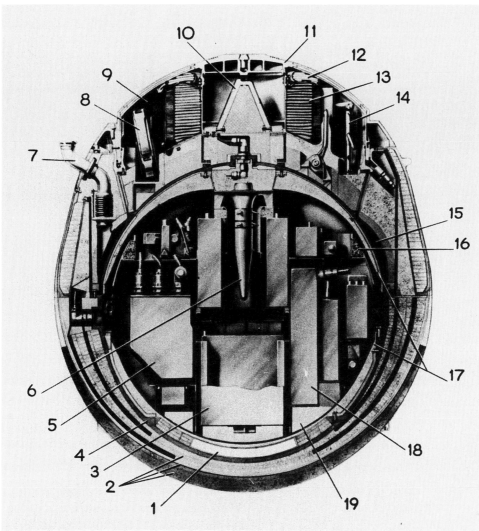

Fig. 2.27. The descent module from the Venera-8 automatic station, in cross section. The structural elements and onboard equipment are as follows: (1) the damper, (2) thermoinsulation (outer), (3) the radio-transmitter, (4) the payload case, (5) the commutation block, (6) the fan, (7) the tube of the thermocontrol system, (8) the ejected antenna of the radiotransmitter, (9) the parachute compartment, (10) the main antenna of the radiotransmitter, (11) the lid of the parachute compartment, (12) the drag parachute, (13) the main parachute, (14) the antenna of the radar altimeter, (15) the heat exchanger, (16) the thermoaccumulator, (17) thermoinsulation (inner), (18) the time-program device, and (19) the thermoaccumulator (courtesy of NPO Lavochkin).

surface, determining ε affords the possibility of estimating the density of the surface material ρ, by using the empirical dependence $\rho = (\sqrt{\varepsilon} - 1)/0.5$, obtained by Krotikov (1962) for various arid terrestrial rocks. A value of $\varepsilon = 4.7 \pm 0.8$ corresponds to $\rho = 2.3 \pm 0.4\,\mathrm{g/cm^3}$, characteristic of dry silicate rocks like basalt, granite, and other hard terrestrial rocks. Along with the measured high radio brightness temperature of Venus, this density clearly contradicted hypotheses of water or petroleum oceans on the planet's surface, and also did not require the presence of porous or finely granulated rock, similar to lunar rocks.

The First Space Missions to Venus

The era of space exploration of the planets in the solar system began with Venus. Without exaggeration, one can say that for more than a third of a century we have been witnesses to a virtual assault on our sister planet, which has gradually but unwillingly given up its secrets. Entire fleets of spacecraft have completed flights to Venus and have enriched us with information of fundamental importance, while posing many new problems. The chronol-

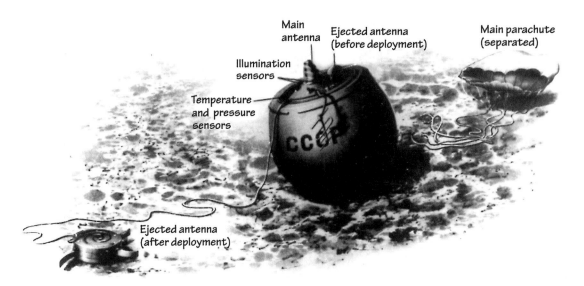

Main
antenna

Ejected antenna
(before deployment)

Main parachute
(separated)

Illumination
sensors

Temperature
and pressure
sensors

CCCP

Ejected antenna
(after deployment)

Fig. 2.28. The Venera-8 descent module on the surface of Venus (drawing) (courtesy of NPO Lavochkin).

ogy of the launches and the main characteristics of the missions are presented in Tables 2.2 and 2.3, whereas Table 2.4 summarizes the data on regions of descent and landing on Venus' surface.

The first automatic interplanetary craft, launched to Venus in February 1961, was the Soviet spacecraft Venera-1 (Fig. 2.18). Unfortunately, the craft failed to respond to commands shortly after insertion into the flight trajectory, and it passed by the planet at a distance greater than 100,000 km. In December 1962, the American spacecraft Mariner-2 (Fig. 2.19) accomplished a successful flight close to Venus and, using a miniature onboard radio telescope, conducted measurements of the radio emission from the planet at a wavelength of 1.9 cm. The goal of the experiment was to determine what mechanism causes the unusually high radio brightness temperature of Venus and to make a definite choice between the two extreme models: a hot surface and a dense atmosphere, or a cold surface and a strong radiating (hot) gas layer high in the atmosphere, associated with a very dense ionosphere. This goal dictated the choice of wavelength (on the order of a centimeter), which satisfies the requirements of both models. The decisive measurement for choosing between the models was the determination of how the brightness temperature varies as one scans the disk (at that time this could not be resolved using Earth-based radio telescopes); in other words, is it darkening or brightening that occurs toward the limb (Fig. 2.20)? The results favored the first model—darkening was found toward the edge of the disk—and consequently provided weighty, though indirect, evidence that the main portion of the microwave emission was generated by the surface, and that Venus' surface was indeed hot (Sagan, 1969).

After the 1961 and 1962 launches of the first spacecraft to Venus and Mars in the Soviet Union (which, unfortunately, were unsuccessful), new large-scale projects to develop rocket technology were embarked upon. These projects aimed to solve problems of ballistics and the energy associated with putting automatic stations into interplanetary trajectories and carrying out the necessary corrections to their trajectory; problems of ensuring

Table 2.2 Chronology of space missions to Venus

Spacecraft	Type of vehicle	Date of launch	Date of fly-by or landing
Venera-1	Fly-by	2/12/61	5/19/61
Mariner-2	Fly-by	8/27/62	12/14/62
Venera-2	Fly-by	11/12/65	2/27/66
Venera-3	Entry probe	11/16/65	3/1/66
Venera-4	Descent vehicle and fly-by	6/12/67	10/18/67
Mariner-5	Fly-by	6/14/67	10/19/67
Venera-5	Descent vehicle and fly-by	1/5/69	5/16/69
Venera-6	Descent vehicle and fly-by	1/10/69	5/17/69
Venera-7	Soft lander and fly-by	8/17/70	12/15/70
Venera-8	Soft lander and fly-by	3/27/72	7/22/72
Mariner-10	Fly-by	11/3/73	2/5/74
Venera-9	Soft lander and satellite	6/8/75	10/22/75
Venera-10	Soft lander and satellite	6/14/75	10/25/75
Pioneer-Venus Orbiter	Satellite	5/20/78	12/4/78
Pioneer-Venus bus and probes	One large and three small probes and an orbital bus	8/8/78	12/9/78
Venera-11	Lander and fly-by	9/9/78	12/25/78
Venera-12	Lander and fly-by	9/14/78	12/21/78
Venera-13	Lander and fly-by	10/30/81	3/1/82
Venera-14	Lander and fly-by	11/4/81	3/5/82
Venera-15	Satellite	6/2/83	10/10/83
Venera-16	Satellite	6/7/83	10/14/83
Vega-1	Lander and fly-by	12/15/84	6/11/85
Vega-2	Lander and fly-by	12/21/84	6/15/85
Magellan	Orbiter	5/4/89	8/10/90

Table 2.3   Main features of space missions to Venus

Spacecraft	Basic features of the mission	Fundamental results
Venera-1	Fly-by at a distance of 100,000 km from Venus without transmitting scientific data	Development of onboard systems and space navigation
Mariner-2	Fly-by at a closest distance of 34,833 km from Venus	Measurement of microwave emissions (1.9 cm); confirmation of the hot planetary surface model
Venera-2	Fly-by at a distance of 24,000 km from Venus without transmitting scientific data	Development of onboard systems and space navigation
Venera-3	Entrance into Venus' atmosphere without transmitting scientific data	Development of onboard systems and space navigation
Vencra-4	Parachute descent in the atmosphere with transmission of data to an altitude of 24 km, night side	First direct measurements of temperature, pressure, chemical composition of the atmosphere, and wind speed
Mariner-5	Fly-by at a closest distance of 4100 km from Venus	Measurement of atmospheric and ionospheric parameters using radio occultation method
Venera-5	Parachute descent in the atmosphere with transmission of data to an altitude of 16 km, night side	Direct measurements of temperature, pressure, and chemical composition of the atmosphere
Venera-6	Parachute descent in the atmosphere with transmission of data to an altitude of 16 km, night side	Direct measurements of temperature, pressure, and chemical composition of the atmosphere
Venera-7	Semisoft landing on the surface of Venus, night side	Direct measurements of temperature and wind velocity to the surface
Venera-8	Soft landing on Venus' surface, day side (morning terminator)	First direct measurements of the temperature and pressure of the atmosphere at the surface on the daylight side, of the intensity of illumination and its variation with altitude, and of wind velocity and the nature of the surface rocks

Table 2.3 (*continued*)

Spacecraft	Basic features of the mission	Fundamental results
Mariner-10	Fly-by at a minimum distance of 5785 km from Venus	Measurements of the atmospheric and ionospheric parameters using radio occultation method; television images of the cloud layer in the UV range; IR radiometric and UV spectrometric measurements
Venera-9	Soft landing on Venus' surface, day side; orbital parameters of the satellite: periapsis 1560 km, apoapsis 112,200 km, period 48 h 18 m, inclination 34° 10′	First black-and-white panoramas of Venus' surface; first measurements of the spectral illumination characteristics, the structure and microphysical properties of the clouds, and wind speed at the surface; measurements of atmospheric parameters, the nature of the surface rocks, emissions of the upper atmosphere, and characteristics of the ionosphere and its interaction with the solar plasma; measurements of the gravitational field
Venera-10	Soft landing on Venus' surface, day side; orbital parameters of the satellite: periapsis 1620 km, apoapsis 113,900 km, period 49 h 23 m, inclination 29° 30′	Same as for Venera-9
Pioneer-Venus Orbiter	Orbital parameters: periapsis <200 km, apoapsis 66,000 km, period 24 h, inclination 105°	Spectral, radiometric, and polarimetric studies of the upper cloud layer and above-cloud atmosphere; images of the cloud morphology in the UV range; measurements of the parameters and chemical composition of the upper atmosphere and ionosphere; surface mapping with a resolution of ~ 50 km; investigations of the near-planet plasma and electrical activity of the atmosphere; measurements of the gravitational field
Pioneer-Venus bus and probes	Hard landing on the day side (Large probe and one small probe), the night equatorial region (Night probe), and the night high-latitude (North probe) region of Venus' surface	Measurements of atmospheric parameters and chemical composition, solar flux and thermal radiation, cloud properties, and rate and direction of the wind; studies of the spatial variations in the parameters of the atmosphere and clouds

Table 2.3 (*continued*)

Spacecraft	Basic features of the mission	Fundamental results
Venera-11	Soft landing on Venus' surface, day side; fly-by at a minimum distance of 25,000 km	Measurements of the parameters and chemical composition of the atmosphere, the illumination and cloud properties, wind velocity, electrical activity of the atmosphere, and the nature of the surface rocks
Venera-12	Soft landing on Venus' surface, day side; fly-by at a minimum distance of 25,000 km	Same as for Venera-11
Venera-13	Soft landing on Venus' surface, day side; fly-by at a minimum distance of 36,000 km	First color panoramas of Venus' surface and measurements of the elemental composition of the surface rocks; measurements of the parameters and chemical composition of the atmosphere and solar radiation, properties of the clouds, electrical activity, and wind velocity
Venera-14	Soft landing on Venus' surface, day side; fly-by at a minimum distance of 36,000 km	Same as for Venera-13
Venera-15	Orbital parameters: periapsis 1000 km, apoapsis 66,000 km, period 24 h, inclination 92.5°	Radio mapping of Venus' northern hemisphere with a resolution of 1–2 km; research of the upper cloud layer and above-cloud atmosphere
Venera-16	Orbital parameters: periapsis 1000 km, apoapsis 66,000 km, period 24 h, inclination 92.5°	Same as for Venera-15
Vega-1	Soft landing on Venus' surface, night side; launch of a balloon that drifted at an altitude of 53–54 km for ∼20 hours; fly-by at a minimum distance of 24,500 km	Measurements of the chemical composition of the atmosphere, atmospheric dynamics, and illumination
Vega-2	Soft landing on Venus' surface, night side; launch of a balloon that drifted at an altitude of 53–54 km for ∼20 hours; fly-by at a minimum distance of 24,500 km	Same as for Vega-1, plus measurements of the atmospheric parameters and elemental composition of the surface rock
Magellan	Orbital parameters: periapsis 294 km, apoapsis 2000 km, period 3.26 h, inclination 85.5°	Radio mapping of essentially the entire surface at a resolution of better than 300 m

Table 2.4   Regions of descent and landing on Venus' surface

Spacecraft	Coordinates lattitude, degrees	Coordinates longitude, degrees	Local solar time, hours	Zenith distance of the Sun, degrees
Venera-4	19	38	4:40	110
Venera-5	-3	18	5:30	117
Venera-6	-5	23	5:30	115
Venera-7	-5	351	5:00	117
Venera-8	-10	335	6:20	85
Venera-9	31.7	290.8	13:15	36
Venera-10	16	291	13:45	28
Venera-11	-14	299	11:10	17
Venera-12	-7	294	11:16	20
Venera-13	-7.5	303.5	9:30	38
Venera-14	-13.4	310.2	9:55	32
Pioneer-Venus				
Large probe	4.4	304	7:40	66
Day probe	-31.7	317	6:40	80
Night probe	-28.7	56.7	0:05	151
North probe	59.3	4.8	3:35	108
Vega-1	8.1	176.7		169.3
Vega-2	-7.2	179.4		164.5

reliable functioning of the craft under conditions of a space vacuum, penetrating radiation, and drastic temperature differences; and problems of navigation in interplanetary space, flight control, and maintenance of radio communication at distances of hundreds of millions of kilometers, with the goal of transmitting control commands to the station and receiving telemetry, including television images (Keldysh and Marov, 1981). In addition, great importance was ascribed to the need to considerably refine some fundamental astronomical constants—the astronomical unit, the planet's mass, and its orbital parameters. With this goal in mind, a series of radar distance measurements to the planet were conducted, and data from highly precise measurements of spacecraft trajectories were also subsequently used. As a result, by the middle of the 1960s the error in determining the astronomical unit, which had previously amounted to tens of thousands of kilometers, was reduced by almost three orders of magnitude.

Given the limited thrust of existing rocket systems, which use chemical propellants, the success of interplanetary flights is dependent on the optimal choice of trajectory (the launch time versus the relative position of the planets) and maneuvers. The required rocket thrust is specified in this case not only by the magnitude of the velocity at the end of the acceleration phase but also by the angle of inclination of the velocity vector to the horizon.

Fig. 2.29a. The facility used to imitate the descent into Venus' atmosphere, showing (1) thermal insulation, (2) an electrical heater, (3) the descent module, (4) a high-pressure chamber (autoclave), and (5) injection of preheated CO_2 gas. Injection of nitrogen, from cylinders of N_2, provides counterpressure to the thermal insulation.

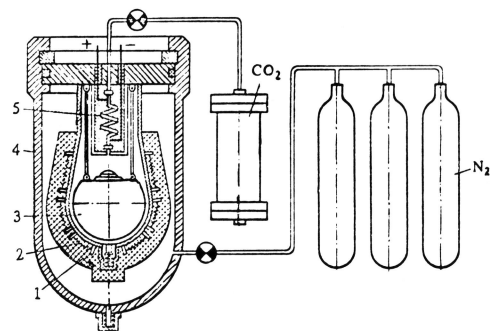

Fig. 2.29b. The temperature control system for the Venera spacecraft and descent module. The arrows indicate the induced circulation of the coolant. Shown are (1) the antenna-radiator, (2) the orbital unit, (3) the multilayer thermal insulation, (4) the supercoolant duct, and (5) the descent module.

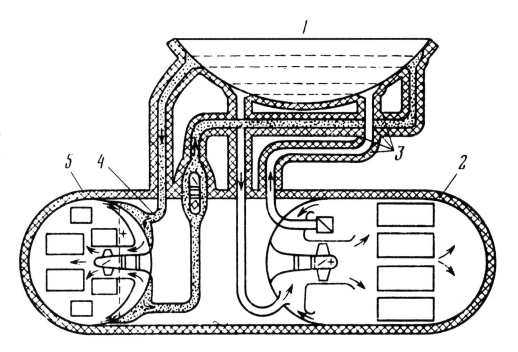

For a launch directly to the planet, this angle would usually be quite large but uneconomical in terms of energy. A launch profile with intermediate insertion of the satellite into Earth orbit, which was developed and realized in the USSR (see Keldysh and Marov, 1981), is considerably more efficient. The use of this type of ballistic maneuver essentially makes it possible to start the flight from the most optimal point on the globe.

Using this scheme, the unmanned Venera-2 and Venera-3 spacecraft were launched toward Venus in November 1965 (Fig. 2.21). The mass of each, increased by 300 kg com-

Fig. 2.30. Overall view of the photometric instrument installed on the Venera-8 descent module; during parachute descent, it measured, for the first time, the attenuation of solar scattered radiation as a function of altitude, from an altitude of 49 km, and after landing it measured the intensity of illumination at the surface. In the center is the instrument with some of the thermal insulation removed; to the left is the electronics assembly.

pared to Venera-1, was 960 kg. The range of instruments was considerably broadened, and a detachable vehicle, for landing on the planet's surface, was installed on Venera-3. The thermal regime required for the normal functioning of onboard systems, however, could not be maintained. Therefore, because of overheating during the approach to Venus, where the thermal flux from the Sun is approximately two times greater than at Earth orbit, the final communications with the spacecraft could not be carried out. From previously received trajectory data, it was reliably established that Venera-2 passed Venus at a distance of 24,000 km, and Venera-3 was the first to reach the planet's surface. The results of these flights provided the designers with additional important information on the performance of the spacecraft's subassemblies and onboard systems. These flights thus made possible new technical solutions and the necessary modifications that became the foundation of future successful missions to Venus.

On 18 October 1967, the unmanned Venera-4 station entered Venus' atmosphere and, after aerodynamic braking, began to descend by parachute (Figs. 2.22 and 2.23). For the first time, direct measurements of the temperature, pressure, and chemical composition of the planet's atmosphere were transmitted to Earth (Avduevsky et al., 1968a, 1968b; Vinogradov et al., 1968; Marov, 1972; Kuzmin and Marov, 1974; Keldysh and Marov, 1981). It was found that more than 90% of the Venusian atmosphere consists of CO_2 gas. The temperature and pressure increased smoothly until they reached 535 K and 18 atm, respectively. The casing of the lander was designed to withstand this pressure limit. Owing to the ambiguity in the readings from the radio altimeter on the lander (the presence of this ambiguity was recognized somewhat later), it was initially erroneously assumed that the device had reached the surface. More thorough analysis indicated, however, that the device

Fig. 2.31. The Mariner-10
spacecraft, with basic onboard systems indicated
(courtesy of NASA).

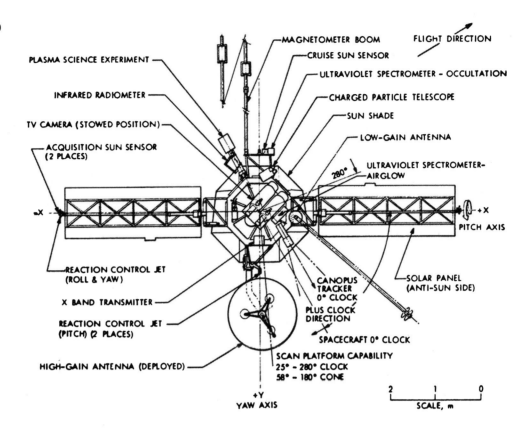

remained more than 22 km above the surface; the temperature and pressure at the surface
would be considerably higher. Higher temperature and pressure were also indicated by
data from the American Mariner-5 spacecraft, which flew by Venus literally the next day at
a minimum distance of 4100 km and conducted measurements of the atmospheric density
by radio occultation as it passed behind the planet's disk and emerged from behind it (from
the perspective of an observer or, more accurately, a receiving antenna on Earth). Such
measurements are possible to the point of critical refraction, that is, the level below which,
owing to the large distortion of the radio beam (the radius of curvature becomes less than
the radius of the planet) and defocusing of the signal, its signal can no longer be received on
Earth. By correlating the Mariner-5 data from the night and day sides of the planet via the
trajectory parameters (using the radar radius of its solid surface) and comparing them with
Venera-4 data (measured relative to the surface), it was apparent that the point of critical refraction (for a pressure near 7 atm and a temperature of 490 K) corresponds to an altitude
of about 35 km (Kliore et al., 1967, 1969; Fjeldbo and Eshleman, 1969). Data on the principal chemical composition of the atmosphere returned by Venera-4 were of great importance for this correlated analysis. Among the series of interesting hypotheses concerning
the minor chemical constituents of the atmosphere and clouds is a study of the nature of the
intrinsic radiowave absorption in Venus' atmosphere (Fjeldbo et al., 1971).

As a result of the flights of the Venera-4 and Mariner-5 spacecraft, the first information on the composition of the upper atmosphere and the density of the ionosphere were
also obtained. An extended hydrogen corona was discovered on Venus, and the upper limit

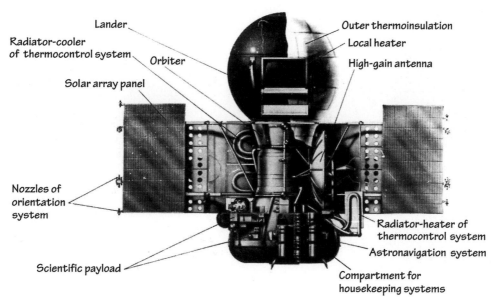

Lander

Radiator-cooler
of thermocontrol system

Orbiter

Solar array panel

Outer thermoinsulation

Local heater

High-gain antenna

Nozzles of
orientation
system

Scientific payload

Radiator-heater of
thermocontrol system

Astronavigation system

Compartment for
housekeeping systems

Fig. 2.32. The new-genera-
tion Venera spacecraft.
Shown is the Venera 9–10
station, which served as the
model for the Venera 11–16
and Vega 1–2 spacecraft.
The primary systems are indicated in the figure (courtesy of NPO Lavochkin).

for the intensity of the intrinsic magnetic field was established—about $1/(3 \times 10^3)$ of Earth's magnetic dipole moment (Dolginov et al., 1968, 1969; Kurt et al., 1972). This figure was significantly decreased later, indicating no intrinsic magnetic field of the planet (see Chapter 8). A special publication was devoted to the experimental results of Venera-4 and Mariner-5.[5] The flights of Venera-5 and Venera-6, which descended by parachute into the planet's atmosphere on 16 and 17 May 1969, considerably refined and supplemented the data from Venera-4 by expanding the area of direct measurements (Fig. 2.24). From the results of these flights, a model of Venus' atmosphere was created, which established the basis for designing the subsequent series of Venusian landers (Avduevsky et al., 1970a, 1972a; Marov, 1971). These flights proved that Venus possesses an unusually dense, hot atmosphere, almost completely made up of CO_2 gas. Measurements using onboard gas an-alyzers provided a value of 97% CO_2 (with a possible error of $\pm 3\%$) and established an upper limit for the nitrogen content of 3–5%. From measurements using resistance ther-mometers and aneroid manometers, it was concluded that the temperature at the surface reaches 735 K and the pressure reaches 90 atm. A thorough analysis of the results of these first space missions and a discussion of the peculiarities of Venus' environment were given by Marov (1972).

Landing under such difficult environmental conditions and transmitting scientific in-formation from the surface represents a complicated engineering problem, which was first resolved during the flights of Venera-7 and Venera-8. Venera-7 reached the surface on the night side on 15 December 1970, and Venera-8 descended to the illuminated side of the planet, close to the morning terminator, on 22 July 1972, and functioned on the surface for nearly one hour (Marov et al., 1971, 1973a, 1973b; Kuzmin and Marov, 1974). Because the instruments on board Venera-7 did not operate properly, the only measurements taken were the atmospheric temperature, as a function of time; the temperature at the surface; and the velocity of the Venera-7 lander during descent. Nevertheless, using well-known thermodynamic relations and the aerodynamics of descent, an altitude correlation of the

Fig. 2.33. The Venera 9–14 type of spacecraft, in the assembly building (courtesy of NPO Lavochkin).

Fig. 2.34. Overall view of the descent (landing) module for the Venera 9–10 spacecraft (courtesy of NPO Lavochkin).

measured temperature and an altitude profile of atmospheric pressure were produced. The program of Venera-8 measurements was completely carried out, and it provided fundamentally new information about the planet. For the first time, the temperature and pressure of the atmosphere at the surface, illumination at the surface, and a series of other parameters were measured on the daylight side of Venus.

The Venera 4–8 spacecraft are approximately the same in their construction. They

Fig. 2.35. The Venera 9–10
lander on the surface of Venus
(drawing) (courtesy of NPO
Lavochkin).

were created by gradual modification of the base spacecraft, Venera-3 (see Fig. 2.21). An
overall view of the Venera-4 lander, with its primary systems, is shown in Fig. 2.25. The
lander is located at the lower portion of the station shown in Fig. 2.24, and is the central
unit of the complex; it automatically carries out all the measurement and transmission of
the scientific information to Earth, after separation from the orbital section (bus). The orbital section transports the landing vehicle and ensures proper operation during all phases
of the interplanetary flight. The mass of the Venera 4–8 stations was within 1100–1180 kg,
and the mass of the landers was 380–475 kg.

Separation of the landers from the orbital section occurred at a distance of about
20,000–40,000 km from Venus, and the aerodynamic braking phase began at a velocity of
about 11.2 km/sec at an angle of approximately 50–70° to the local horizon. The para-

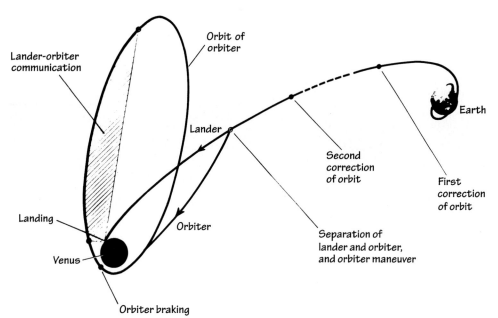

Fig. 2.36. The Earth-to-Venus flight path and orbital insertion of the Venera 9–10 artificial satellites; communication from the descent module to Earth was accomplished via a relay satellite.

chute was deployed when the velocity of the lander decreased to about 210 m/sec, with an atmospheric pressure of 0.5 atm. A single-stage parachute system, with a drogue parachute whose heat-resistant material was designed for a temperature up to 800 K, was used (Fig. 2.26).

The descent vehicle of the Venera-8 automatic station contains the radiotelemetry system, scientific instruments for measuring atmospheric and surface parameters, automatic control and heat regulation systems, and power sources (Fig. 2.27). The outside of the vehicle is enclosed by a heat shield, which protects it from aerodynamic heating during atmospheric entry, and thermal insulation, which prevents the intense inflow of heat from the hot atmospheric gas (Fig. 2.28).

Creating the descent vehicles for the Venera spacecraft required the solution of many complicated scientific-technical problems and the performance of numerous theoretical and experimental studies. These included creating a structure capable of withstanding excessive loads upon entering the atmosphere (greater than 300 g), designing a heavy-duty housing that could withstand pressures greater than 100 atm, rating and testing highly efficient thermal insulation and heat shielding, and finally, developing techniques and special facilities to test the devices in an environment simulating the Venusian atmosphere (Fig. 2.29a). Such simulation made it possible to choose the optimal structure and materials. Porous materials, formed into a honeycomb, were used as the most efficient thermal insulation; these materials also possessed the necessary strength to withstand the high external pressure. An important means of prolonging the lifetime of the instruments in the hot atmosphere and on the planet's surface was the use of heat accumulators. These were based on crystal hydrates of lithium nitrate ($LiNO_3 \cdot nH_2O$), which work like evaporative systems with a phase change at a temperature of about 30° C (their use made it possible to raise the overall heat capacity of the lander by 20%). A special device was likewise incorporated into the heat control system to cool the instruments to $-10°$ C before separation from

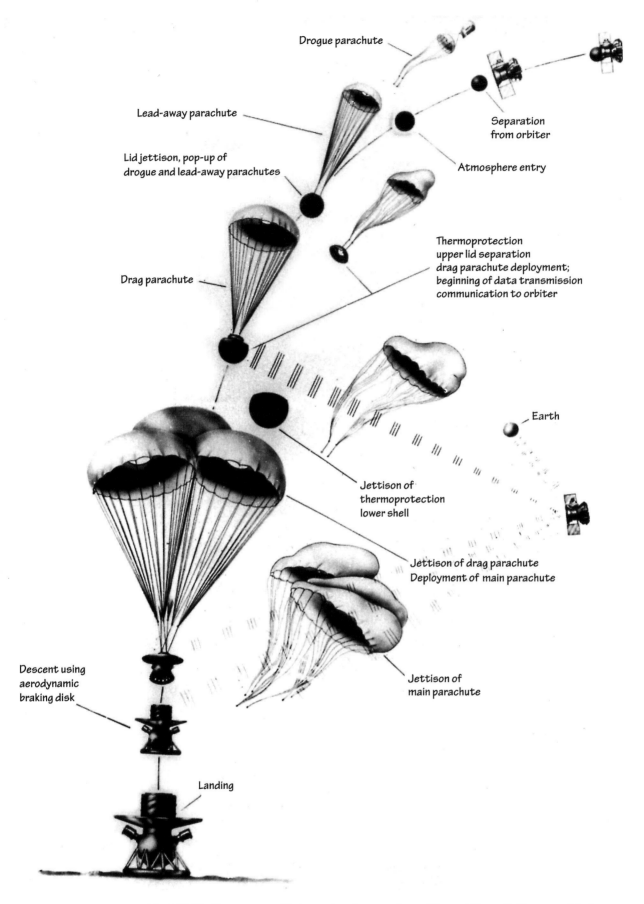

Drogue parachute

Lead-away parachute

Lid jettison, pop-up of
drogue and lead-away parachutes

Drag parachute

Descent using
aerodynamic
braking disk

Landing

Separation
from orbiter

Atmosphere entry

Thermoprotection
upper lid separation
drag parachute deployment;
beginning of data transmission
communication to orbiter

Earth

Jettison of
thermoprotection
lower shell

Jettison of drag parachute
Deployment of main parachute

Jettison of
main parachute

Fig. 2.37. The Venera spacecraft's descent into the atmosphere of Venus. A Venera 9–10 spacecraft is shown
as an example; a similar descent profile was realized for the Venera 11–12, Venera 13–14, and Vega 1–2 spacecraft. See the text for an explanation of the descent.

Fig. 2.38. Cup anemometers, which were used on the Venera 9–10 landers to measure the wind velocity at the planet's surface.

the bus (Fig. 2.29b). A large number of trials were conducted to test the devices' strength, the automatic descent system, and the reliability of the parachute system in Earth's atmosphere (Avduevsky and Marov, 1976; Keldysh and Marov, 1981).

The initial results proved the high effectiveness of these space missions. The temperature, pressure, and density of Venus' gas envelope, its thickness, and its chemical composition all became known. Research on direct measurements of the thermodynamic state of the gas in the atmosphere led to the important conclusion that, over the whole altitude range from the surface to ~50 km, the atmosphere is close to adiabatic; this means that convective-circulational mixing must occur (Marov, 1972). Estimates of the values for the vertical transport velocities were obtained from the characteristics of descent and atmospheric measurements. In turn, from the magnitude of the Doppler frequency shift of onboard radio transmitters (master oscillators) of all landers, starting with Venera-4, the horizontal component of atmospheric motions, that is, the wind speed, was determined; this information enabled us to reach several important conclusions associated with the problem of planetary circulation (Kerzhanovich et al., 1972; Kerzhanovich and Marov, 1974). With a gamma-ray spectrometer installed on Venera-8 (essentially the same technique employed on Luna-10), the first data on the nature of Venusian surface rock at the landing site were obtained (Vinogradov et al., 1973).

Venera-8, using a simple photometer (Fig. 2.30), measured for the first time the solar light flux penetrating the subcloud atmosphere, obtained an altitude profile of its attenuation, and estimated the illumination at the planet's surface (Avduevsky et al., 1973a, 1973b; Marov et al., 1973c). Such an experiment on the descent vehicle provided essentially the only opportunity to find out whether solar light penetrates to the surface or is completely absorbed at the level of the clouds. It turned out that the optical density of the atmosphere

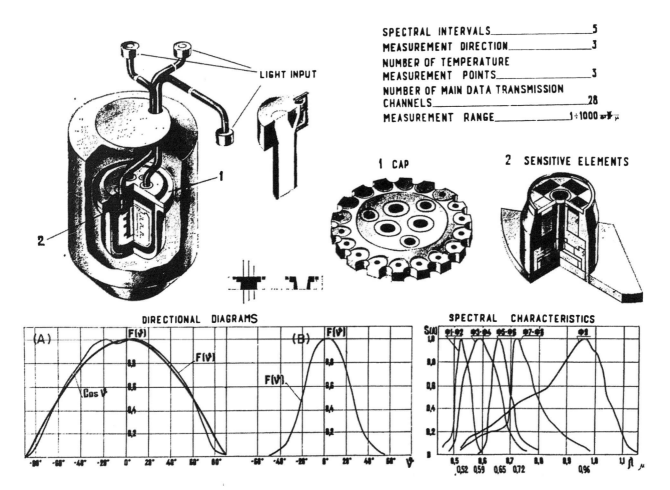

Fig. 2.39. Overall view and optical diagram of the multichannel photometers used on the Venera 9–10 landers to measure, for the first time, the solar flux in several spectral regions. Below are the upward (a) and downward (b) directional diagrams and spectral characteristics for five channels.

decreases with an increase in the molecular density, so that in the altitude range from the surface to 32 km the atmosphere is practically transparent, and the light in this region attenuates according to the molecular (Rayleigh) scattering law in CO_2 gas. Above 32 km, and especially above 40 km, considerably greater attenuation of the light was detected than in the lower-lying atmosphere; this was explained by the existence of cloud aerosol. These first measurements of the extent, structure, and density of the clouds were subsequently confirmed by new experimental data. Analysis of the overall light attenuation in the atmosphere and clouds led to the conclusion that not less than 2–3% of the solar constant at Venus' orbit finds its way to the surface and, consequently, it is quite bright at the surface — like an overcast day on Earth, when the Sun is completely obscured by clouds. This realization made it feasible to prepare an experiment to receive television panoramas from the surface of Venus.

In contrast to Soviet scientists, who concentrated their efforts on direct measurements in the atmosphere and on the surface of Venus using landing vehicles, American researchers steadily continued on a course begun with Mariner-2—that of conducting remote measurements from fly-bys. In February 1974, at a minimum distance of 5785 km from the surface, Mariner-10 flew by Venus, accomplishing a series of studies and, simulta-

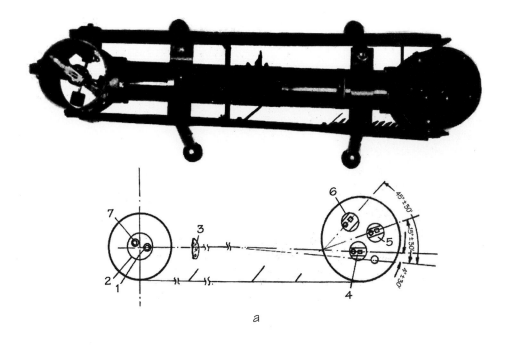

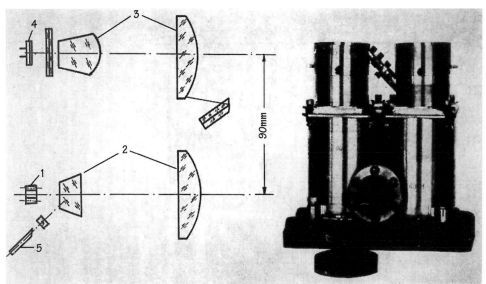

Fig. 2.40. Overall view of the nephelometers, which are used to measure the cloud structure and the microscopic properties of the aerosols. Shown are (a) the multichannel (small-angle) nephelometer and (b) the backscatter nephelometer. (a): A light beam emitted by an artificial source (1) placed in an obturator (2) and focused by a lens (3) is scattered by the medium and recorded by photoreceivers (4, 5, and 6) by means of the photoreceivers' mirrors arranged under angles of 15°, 30°, and 45°, respectively. (7) is a control light source used for calibration. (b): Light emitted by an artificial source (1) and focused by radiators' optics (2) is backscattered by a medium, collected by optics (3), and recorded by a photoreceiver (4). (5) is a control light source used for calibration.

neously, a perturbation maneuver in its gravitational field that assured a continued flight to Mercury (Dunne, 1974). This spacecraft, which had a mass somewhat greater than 500 kg and was equipped with triaxial stabilization (Fig. 2.31), produced television images of Venus' cloud cover in the ultraviolet, measured the characteristics of the thermal field using an infrared radiometer, obtained data on the chemical composition and temperature of the upper atmosphere using an ultraviolet spectrometer, studied certain properties of the near-planet plasma and its interaction with the solar wind, and measured ionospheric and atmospheric parameters by radio occultation. Television images, containing information on the nature and structure of zonal and meridional currents, revealed additional features of the four-day circulation at the cloud level (Murray et al., 1974).

Fig. 2.41. Flight plan of the Pioneer-Venus Multiprobe spacecraft near Venus, and separation of the Large probe and three small probes (courtesy of NASA).

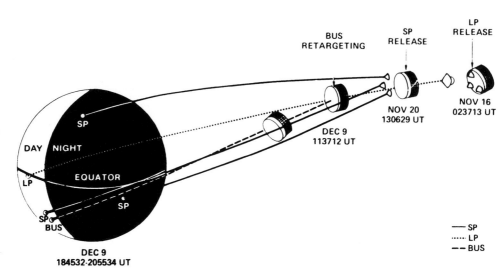

Broadening the Research Programs

A new stage in the study of Venus began in June 1975 with the launch of the Venera-9 and Venera-10 spacecraft. For the first time two satellites were put into orbit around Venus, and, simultaneously, landing vehicles carried out soft landings in different regions. In this way, a complex, integrated experiment to study the characteristics of the planet on its surface, in the atmosphere, in the cloud layer, and in the surrounding space was accomplished (Avduevsky et al., 1976a; Keldysh, 1977; Marov, 1978, 1979a, 1979b).

The two stations, similar in their construction and scientific equipment, belonged to a new generation of Soviet planetary robot spacecraft (Figs. 2.32 and 2.33). They had features in common with earlier automated spacecraft launched to Mars by the USSR in 1971 and 1973. By virtue of the completely different environmental conditions at Venus and Mars, however, the landing vehicles of these stations were fundamentally different. Also belonging to this new generation of spacecraft were Venera 11–12 and Venera 13–14, launched in September 1978 and October and November 1981, which were essentially identical, in their appearance and landing profile, to Venera-9 and Venera-10. Later came Venera 15–16 and Vega 1–2, which were further modified to accomplish radar mapping of Venus and a close fly-by of Halley's comet, respectively.

The landers of the Venera 9–10 spacecraft (Figs. 2.34 and 2.35) set down on the illuminated side of the planet (which was not visible from Earth at that time) on 22 and 25 October 1975. Transmission of the scientific information was carried out through a relay satellite (Fig. 2.36). Not only was the design of the landing vehicle changed, but so was the descent trajectory into the atmosphere (Fig. 2.37). Two contradictory requirements had to be satisfied: in order to study the cloud layer, the vehicle would have to descend slowly at high altitude where the density of the atmosphere is comparatively low, and then it had to pass rapidly through the greater mass of the hot atmosphere so there would be insufficient time for an intense buildup of heat. The first stage was achieved using a parachute system; after separation of the main parachute, a special aerodynamic shield was employed.

Fig. 2.42. Overall view of the Pioneer-Venus Orbiter and Multiprobe spacecraft during assembly (courtesy of NASA).

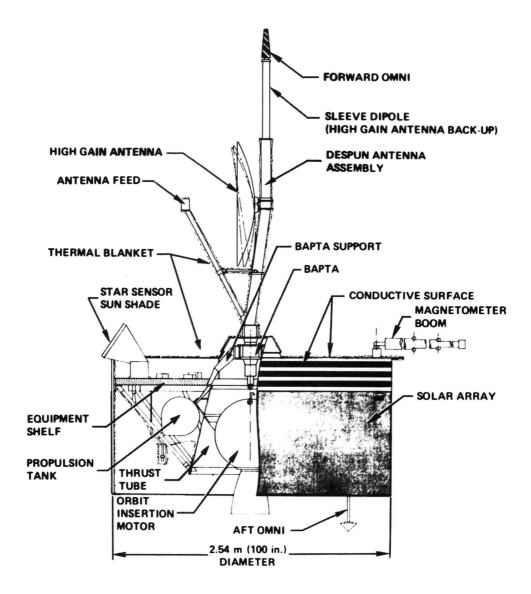

The descent is explained as follows (see Fig. 2.37). The descent vehicle enters the atmosphere at an angle of about 20°, and after aerodynamic braking at an altitude of almost 65 km, the cover of the parachute compartment is blown off, putting into operation the drogue parachute and the parachute designed to carry off the upper lid of the heat-shield housing. The rate of descent at this point decreases from 250 m/sec to 150 m/sec. Then the drag parachute is deployed, the transmitter is turned on, and the transmission of information begins. In 15 seconds, after the velocity has dropped to ~50 m/sec, at an altitude of about 62 km, the three-canopy main parachute, with an overall area of 180 m², deploys. Four seconds later the lower hemisphere of the heat-shield housing separates. The parachute descent continues for approximately 20 minutes, after which the parachutes are jettisoned at an altitude of about 50 km. At this point the rate of descent initially increases, but close to the surface, owing to the increase in atmospheric density, it falls to 7 m/sec. To lower the probability that the lander would tip over on a complex landscape, the landing was carried out on a supporting ring, which also served as a shock absorber. After insertion

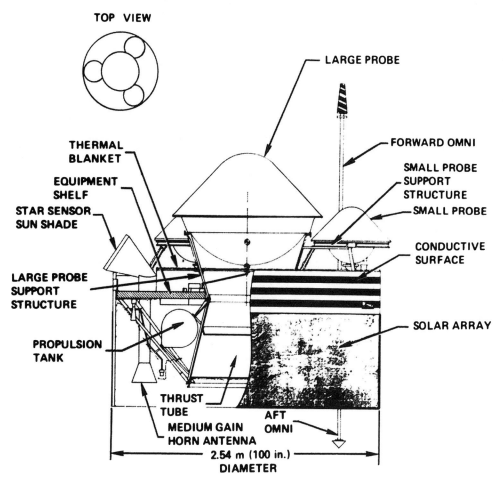

TOP VIEW

LARGE PROBE

THERMAL BLANKET

EQUIPMENT SHELF

STAR SENSOR SUN SHADE

LARGE PROBE SUPPORT STRUCTURE

PROPULSION TANK

FORWARD OMNI

SMALL PROBE SUPPORT STRUCTURE

SMALL PROBE

CONDUCTIVE SURFACE

SOLAR ARRAY

THRUST TUBE

MEDIUM GAIN HORN ANTENNA

AFT OMNI

2.54 m (100 in.) DIAMETER

Fig. 2.44. Layout of the Pioneer-Venus Multiprobe spacecraft, with main systems indicated (courtesy of NASA).

into a trajectory to Venus, the mass of each Venera 9–10 spacecraft was about 5 tons, and that of each descent vehicle with heat shield was 1560 kg. Information from the surface was transmitted for 53 minutes from the Venera-9 lander, and for 63 minutes from Venera-10.

As a result of these flights, a huge volume of data was obtained, which significantly expanded our knowledge of Venus. Panoramas of the surface from the two landing sites, separated by almost 2000 km, were transmitted to Earth for the first time, and new information was obtained on the nature and density of the constituent rock, shedding some light on the planet's geological past and present. Atmospheric parameters were measured, and the degree of diurnal temperature variation was estimated; these data were of great importance for understanding the characteristics of the thermal regime and atmospheric dynamics. Direct measurements using cup anemometers (Fig. 2.38) confirmed the finding, made earlier based on Doppler measurements, of low wind speeds near the surface (≤ 1 m/sec) and an increase of wind speed with altitude. The change in solar flux with altitude in several spectral ranges was studied in much more detail than on Venera-8. With onboard spectrophotometers and nephelometers, the first information was obtained on the structure of the clouds and their microphysical properties (Figs. 2.39 and 2.40a and b).

With spectrometers, radiometers, and photopolarimeters installed on the orbiters, the properties and spatial-temporal variations of the cloud layer and above-cloud

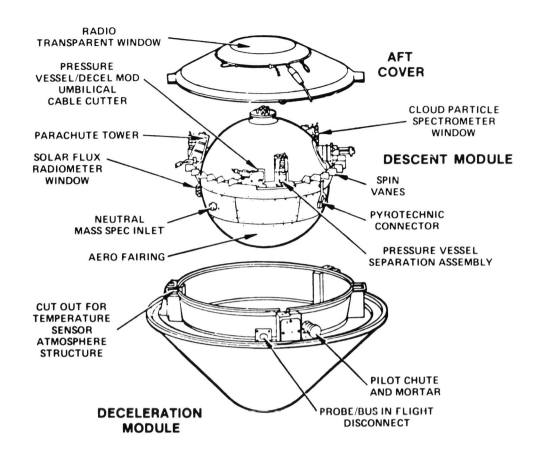

Fig. 2.45. Main components of the Large probe from the Pioneer-Venus Multiprobe spacecraft (courtesy of NASA).

RADIO TRANSPARENT WINDOW

PRESSURE VESSEL/DECEL MOD UMBILICAL CABLE CUTTER

AFT COVER

CLOUD PARTICLE SPECTROMETER WINDOW

PARACHUTE TOWER

SOLAR FLUX RADIOMETER WINDOW

DESCENT MODULE

SPIN VANES

PYROTECHNIC CONNECTOR

NEUTRAL MASS SPEC INLET

AERO FAIRING

PRESSURE VESSEL SEPARATION ASSEMBLY

CUT OUT FOR TEMPERATURE SENSOR ATMOSPHERE STRUCTURE

PILOT CHUTE AND MORTAR

PROBE/BUS IN FLIGHT DISCONNECT

DECELERATION MODULE

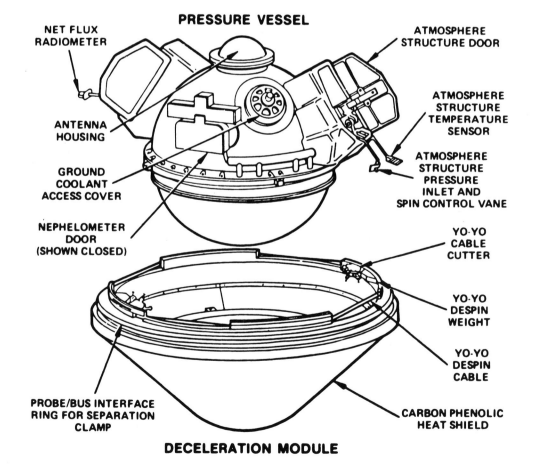

Fig. 2.46. Main components of the small probes from the Pioneer-Venus Multiprobe spacecraft (courtesy of NASA).

PRESSURE VESSEL

NET FLUX RADIOMETER

ATMOSPHERE STRUCTURE DOOR

ANTENNA HOUSING

ATMOSPHERE STRUCTURE TEMPERATURE SENSOR

GROUND COOLANT ACCESS COVER

ATMOSPHERE STRUCTURE PRESSURE INLET AND SPIN CONTROL VANE

NEPHELOMETER DOOR (SHOWN CLOSED)

YO-YO CABLE CUTTER

YO-YO DESPIN WEIGHT

YO-YO DESPIN CABLE

PROBE/BUS INTERFACE RING FOR SEPARATION CLAMP

CARBON PHENOLIC HEAT SHIELD

DECELERATION MODULE

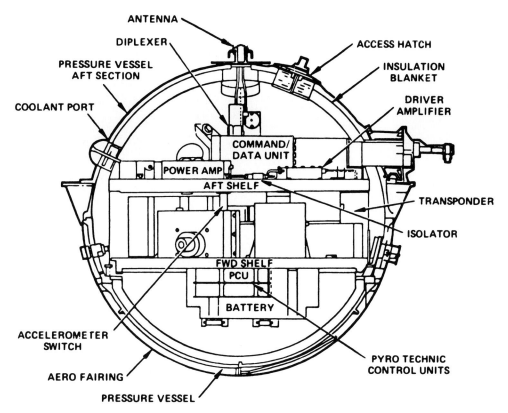

ANTENNA
DIPLEXER
PRESSURE VESSEL
AFT SECTION
COOLANT PORT
ACCESS HATCH
INSULATION
BLANKET
DRIVER
AMPLIFIER
COMMAND/
DATA UNIT
POWER AMP
AFT SHELF
TRANSPONDER
ISOLATOR
FWD SHELF
PCU
BATTERY
ACCELEROMETER
SWITCH
AERO FAIRING
PRESSURE VESSEL
PYRO TECHNIC
CONTROL UNITS

Fig. 2.47. Internal layout of the Large probe from the Pioneer-Venus Multiprobe spacecraft, in cross section (courtesy of NASA).

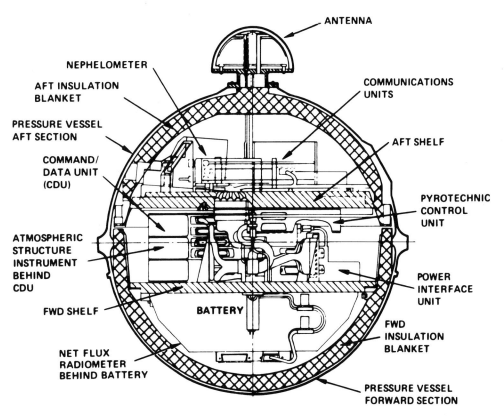

ANTENNA
NEPHELOMETER
AFT INSULATION
BLANKET
PRESSURE VESSEL
AFT SECTION
COMMAND/
DATA UNIT
(CDU)
ATMOSPHERIC
STRUCTURE
INSTRUMENT
BEHIND
CDU
FWD SHELF
NET FLUX
RADIOMETER
BEHIND BATTERY
COMMUNICATIONS
UNITS
AFT SHELF
PYROTECHNIC
CONTROL
UNIT
POWER
INTERFACE
UNIT
FWD
INSULATION
BLANKET
PRESSURE VESSEL
FORWARD SECTION
BATTERY

Fig. 2.48. Internal layout of the small probes from the Pioneer-Venus Multiprobe spacecraft, in cross section (courtesy of NASA).

Fig. 2.49. Overall view of the Venera 13–14 lander (courtesy of NPO Lavochkin).

atmosphere were studied. Spectra were also obtained of the intrinsic night-sky luminescence of the upper atmosphere; these spectra could be useful in determining the atmosphere's composition and in studying fundamental photochemical processes. Radio occultation experiments allowed measurement of the density of the atmosphere and ionosphere, and radio science was also used to estimate the parameters of Venus' gravitational field (Akim et al., 1978). A major goal was to study the magnetic fields, energy spectra, and electron and ion flux intensities, both near the planet where the solar wind streams past it, and in unperturbed regions. With this goal in mind, satellites were put into elliptical orbits

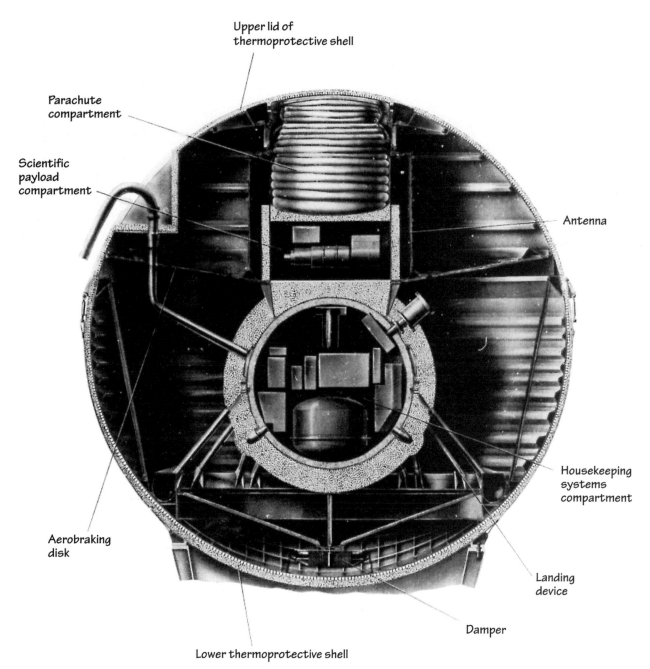

Upper lid of
thermoprotective shell

Parachute
compartment

Scientific
payload
compartment

Antenna

Housekeeping
systems
compartment

Aerobraking
disk

Landing
device

Damper

Lower thermoprotective shell

Fig. 2.50. Internal layout
of the Venera 13–14
lander (courtesy of NPO
Lavochkin).

oriented so that the pericenter of one corresponded to the apocenter of another. Accordingly, fundamental, characteristic regions were investigated: the quiescent solar plasma, the shock wave, the stagnant plasma near the planet, a zone of possible solar plasma interaction with the upper atmosphere of Venus, and zones of optical and corpuscular shadow. The data obtained have broadened our understanding of the structure and properties of the near-planet environment. The results of these experiments are contained in special editions of journals.[6]

The next major steps in the study of Venus were made in 1978, as a result of the American Pioneer-Venus spacecraft (Colin and Hunten, 1977) and the Soviet Venera-11 and Venera-12 landers. The rendezvous with Venus for Pioneer-Venus occurred on De-

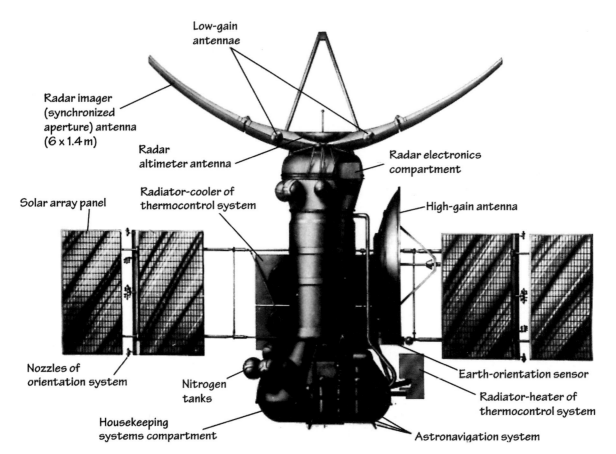

Low-gain
antennae

Radar imager
(synchronized
aperture) antenna
(6 x 1.4 m)

Radar
altimeter antenna

Radar electronics
compartment

Solar array panel

Radiator-cooler of
thermocontrol system

High-gain antenna

Nozzles of
orientation system

Nitrogen
tanks

Earth-orientation sensor

Radiator-heater of
thermocontrol system

Housekeeping
systems compartment

Astronavigation system

Fig. 2.51. The artificial satellites Venera 15–16, equipped with side-looking radar. At the upper portion of the spacecraft is the elliptically shaped antenna used to carry out radar mapping of Venus' surface (courtesy of NPO Lavochkin).

cember 4 and 9, while Venera-11 and Venera-12 landed on the planet on December 21 and 25. Pioneer-Venus consisted of two spacecraft: an Orbiter, which had an initial pericentric altitude of about 200 km, and a Multiprobe, which contained one large probe and three small probes installed on a delivery module (a bus); the probes were designed to descend through the atmosphere. The small probes were separated from the module 20 days before rendezvous with the planet, and were oriented to probe the atmosphere at various regions in longitude and latitude. Accordingly, they were named the Day, Night, and North probes. The module (bus) and Large probe, which separated from the module 4.5 days before separation of the small probes, carried out measurements of the day side, close to the equator (see Table 2.4). The flight plan of the Pioneer-Venus Multiprobe spacecraft during the near-planet phase and separation of the probes is shown in Fig. 2.41; an overall view of it and the Pioneer-Venus Orbiter are shown in Fig. 2.42. The configuration of the Pioneer-Venus spacecraft is presented in Figs. 2.43 and 2.44. The main components of the Large and small probes are presented in Figs. 2.45 and 2.46, and their internal layout is given in Figs. 2.47 and 2.48.

The experiments on the Pioneer-Venus and Venera 11–12 spacecraft were comprehensive and complementary. These flights enabled us to considerably refine our knowledge of the chemical composition of the atmosphere, through a series of measurements using onboard mass spectrometers and gas chromatographs (although their results were not in complete agreement, apparently owing to differences in calibration). Aside from the

Fig. 2.52. The Venera 15–16 spacecraft in the assembly area (courtesy of NPO Lavochkin).

main constituents—carbon dioxide gas and nitrogen—the content of minor atmospheric admixtures was measured, for example, carbon monoxide gas, water vapor, and also sulfur, chlorine, and compounds containing them. Of particular importance, from the point of view of planetary cosmogony and geological and atmospheric evolution, were data on the absolute content of noble gases—argon, neon, krypton—and their isotopic composition (see Table 5.5).

Our knowledge of the spectral characteristics of the scattered solar radiation in the Venusian atmosphere, and their change with altitude and cloud properties, was further refined; data on the water vapor content below the clouds, and altitude profiles of radiative heat flow, were also obtained. Intense electrical activity (low-frequency electromagnetic noise) associated with lightning was discovered in Venus' atmosphere, and new information was gathered on neutral particle and ion composition, temperature and density of the

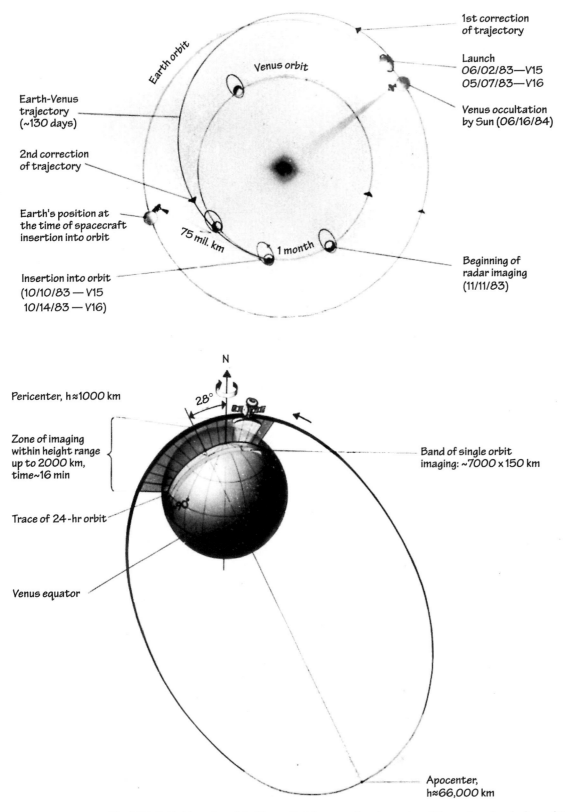

1st correction
of trajectory

Launch
06/02/83—V15
05/07/83—V16

Venus occultation
by Sun (06/16/84)

Earth orbit

Venus orbit

Earth-Venus
trajectory
(~130 days)

2nd correction
of trajectory

Earth's position at
the time of spacecraft
insertion into orbit

75 mil. km

1 month

Beginning of
radar imaging
(11/11/83)

Insertion into orbit
(10/10/83 — V15
10/14/83 — V16)

N

28°

Pericenter, h≈1000 km

Zone of imaging
within height range
up to 2000 km,
time~16 min

Band of single orbit
imaging: ~7000 x 150 km

Trace of 24-hr orbit

Venus equator

Apocenter,
h≈66,000 km

Fig. 2.53. Orbital insertion of the Venera 15–16 spacecraft and mapping of the northern hemisphere of Venus.

Fig. 2.54. Overall view of the Vega 1–2 spacecraft. In the upper portion is the descent (landing) module, designed to land on Venus' surface during the spacecraft's fly-by on the way to Halley's comet (courtesy of NPO Lavochkin).

upper atmosphere and ionosphere, their diurnal variations, and the thermal regime and atmospheric dynamics. A Venusian cryosphere, with the temperature continuously dropping with height above about 100 km, was detected on the night side, in contrast to the daylight hemisphere. Sharp differences in the values of neutral and electron or ion temperatures were found in both hemispheres. Important new mechanisms concerning the interaction of a planet without an intrinsic magnetic field with the solar plasma were identified. The results of all these experiments are contained in special editions of journals,[7] and appropriate references are presented in the following chapters.

Toward the end of 1981, two more spacecraft—Venera-13 and Venera-14—were launched to Venus, and they reached the planet and landed on its surface in March of the

Fig. 2.55. Flight plan of the
Vega 1–2 spacecraft.

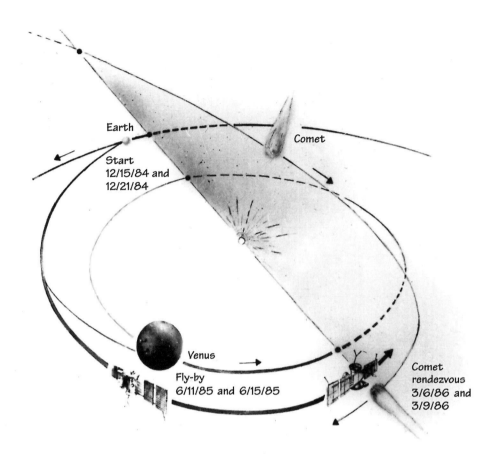

VENUS FLY-BY

Min. distance from Venus:
39 and 25 thousand km

Distance from Earth:
102 mil. km

Entry velocity of lander:
11 km/sec

Earth-Venus distance covered:
485 mil. km

COMET RENDEZVOUS

Min. distance from comet:
~10 thousand km

Distance from Earth:
~170 mil. km

Rendezvous velocity:
79 km/sec

Earth -comet distance covered:
1200 mil. km

following year. The scientific tasks of these missions were a logical continuation of previous experiments in the same spacecraft series. A lander, somewhat modified because of the expanded scientific research program, is shown in Fig. 2.49; the internal layout of its hermetically sealed compartment is presented in Fig. 2.50.

The traditional complex of experiments was performed during the descent stage: detailed studies of the chemical composition, atmospheric temperature and pressure; measurements of the solar radiation scattered in the atmosphere; and analysis of the structure and chemical composition of the aerosol component. As a result, we managed to further refine our knowledge of the abundances of minor components in the atmosphere, primarily water vapor and sulfur-bearing components, as well as noble gases. In particular, the isotopic ratio of neon was determined for the first time. A series of minute variations were

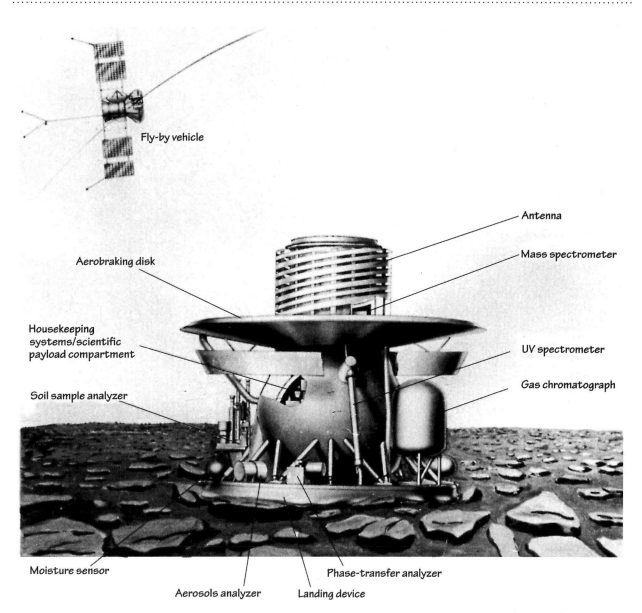

Fly-by vehicle

Antenna

Mass spectrometer

Aerobraking disk

Housekeeping systems/scientific payload compartment

Soil sample analyzer

UV spectrometer

Gas chromatograph

Moisture sensor

Aerosols analyzer   Landing device   Phase-transfer analyzer

Fig. 2.56. The Vega 1–2 lander on the surface of Venus (drawing) (courtesy of NPO Lavochkin).

revealed in the structure of the clouds, within which noticeable variations in the optical depth, governed by changes in the concentration and microphysical properties of aerosols, were detected. More reliable data on the relative content of sulfur and chlorine in the clouds were gathered. Once again, intense bursts of low-frequency electromagnetic radiation were recorded, and, by combined analysis with measurement data of whistling atmospheric disturbances from the Pioneer-Venus Orbiter, a more representative interpretation of their nature resulted.

The scientific program of the Venera 13–14 spacecraft was mainly devoted to experiments on the surface, which started immediately after touchdown and continued for 127 minutes for Venera-13 and 57 minutes for Venera-14. During this time each spacecraft transmitted two television panoramas, including color images. For the first time the elemental composition of the soil was studied using X-ray radiometric techniques. The type of constituent surface rock was then reconstructed. The physical-mechanical and optical properties of the surface were likewise measured, and the microseismicity in the

DESCENDING MODULE

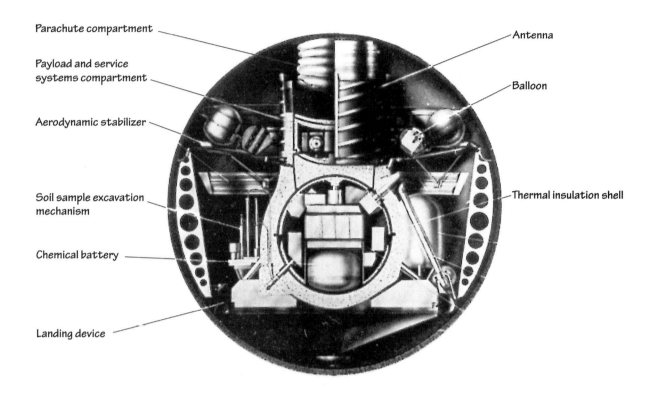

Parachute compartment

Payload and service
systems compartment

Aerodynamic stabilizer

Soil sample excavation
mechanism

Chemical battery

Landing device

Antenna

Balloon

Thermal insulation shell

THE MAIN CHARACTERISTICS:

Separation from the bus: two days before atmospheric entry

Atmospheric entry velocity: ~ 11 km/s

Extreme overload at atmospheric entry: ~210 g

Time in the atmosphere before the deployment of thermal insulation shell: ~1 min

Fig. 2.57. The Vega 1–2 lander in cross section (courtesy of NPO Lavochkin).

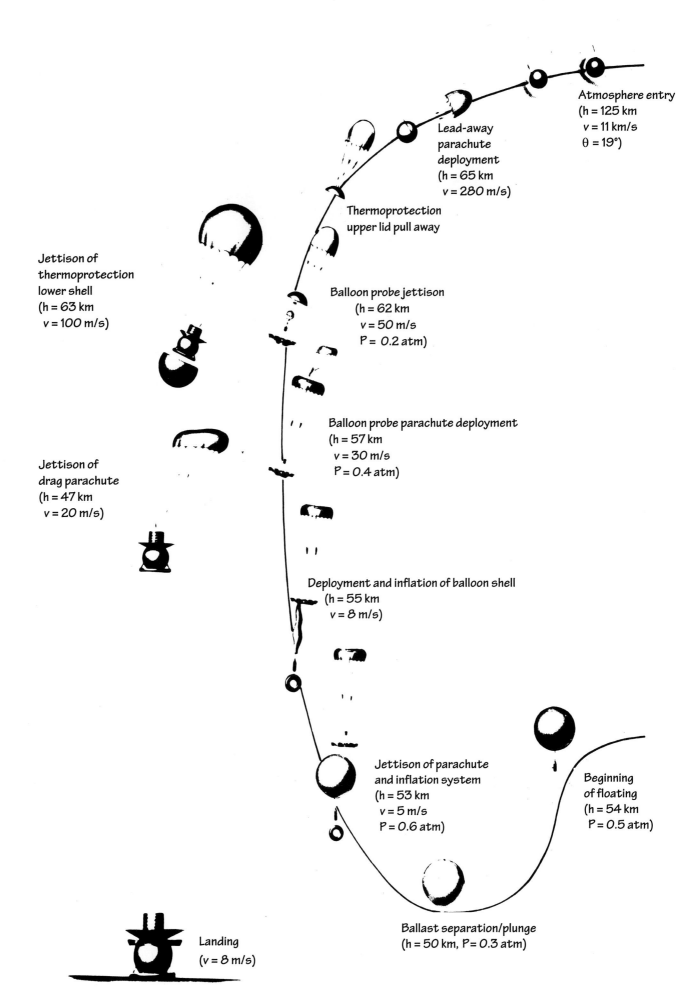

Atmosphere entry
(h = 125 km
v = 11 km/s
θ = 19°)

Lead-away
parachute
deployment
(h = 65 km
v = 280 m/s)

Thermoprotection
upper lid pull away

Jettison of
thermoprotection
lower shell
(h = 63 km
v = 100 m/s)

Balloon probe jettison
(h = 62 km
v = 50 m/s
P = 0.2 atm)

Balloon probe parachute deployment
(h = 57 km
v = 30 m/s
P = 0.4 atm)

Jettison of
drag parachute
(h = 47 km
v = 20 m/s)

Deployment and inflation of balloon shell
(h = 55 km
v = 8 m/s)

Jettison of parachute
and inflation system
(h = 53 km
v = 5 m/s
P = 0.6 atm)

Beginning
of floating
(h = 54 km
P = 0.5 atm)

Ballast separation/plunge
(h = 50 km, P = 0.3 atm)

Landing
(v = 8 m/s)

Fig. 2.58. Descent profile of the Vega 1–2 lander, and deployment of the balloon probe into the atmosphere (courtesy of NPO Lavochkin).

Fig. 2.59. A floating balloon probe in Venus' atmosphere (drawing) (courtesy of NPO Lavochkin).

Height of floating:	53–55 km (temperature ~ 30°C, pressure ~ 0.5 atm)
Lifetime:	~ 48 hr
Mileage:	~ 15 · 10³ km
Mass of balloon:	21 kg
Mass of gondola w/ payload:	7 kg
Diameter of shell:	3.4 m

landing areas was estimated. A special edition was devoted to the experimental results of Venera 13–14.[8]

Studies of Venus' surface relief, using spacecraft, were begun using bistatic radar from the Venera 9–10 satellites, which provided information on altitude differences in distinct, limited regions of the planet (see Keldysh, 1977). Much more complete measurements were accomplished using an onboard radio altimeter on the Pioneer-Venus Orbiter, which provided a spatial resolution of relief details of about 50 km and made it possible to construct a surface map of Venus in a latitude band from 65° S to 75° N (Pettengill et al., 1980).

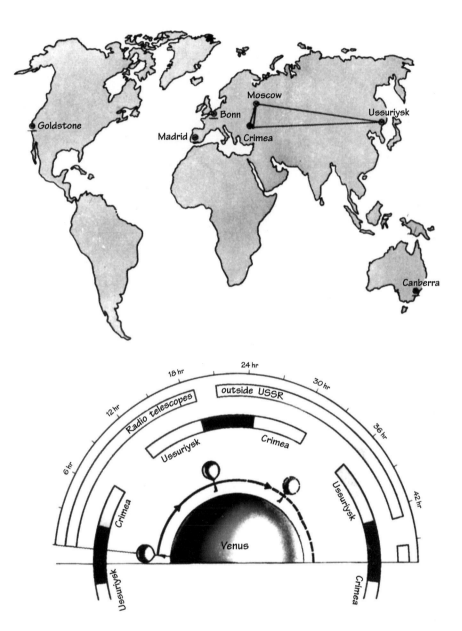

Fig. 2.60. *Top*: The international radio-interferometric network tracking the Vega balloon probes. Only the main ground stations are shown. The lines connect Soviet radio telescopes. *Bottom*: Visibility zones of the balloon probe from the ground-based radio telescopes.

A significant step in the study of the relief and physical properties of Venus' surface was the launch in 1983 of the Venera-15 and Venera-16 orbiters (Figs. 2.51 and 2.52). They were designed to obtain images of the surface using side-looking radar (see Chapter 3). With this goal in mind each spacecraft was equipped with an array of instruments, including radar devices and radio altimeters, and also special antennae of elliptical form with a maximum dimension of 6 m. The side-looking radar furnished images of the planet's surface, and the radio altimeter provided their corresponding altitude profiles and local reflection characteristics. High-quality images of Venus' surface were produced, in which the resolution of the relief details reached 1–2 km (see Rzhiga, 1988). The spacecraft's flight path to Venus and its operation in orbit about Venus is presented in Fig. 2.53.

Located in near-polar orbits with pericenters close to the North Pole, the Venera 15–16 spacecraft, over the course of a year's active operation, performed radar mapping of the

Fig. 2.61. Primary systems
of the Magellan spacecraft.

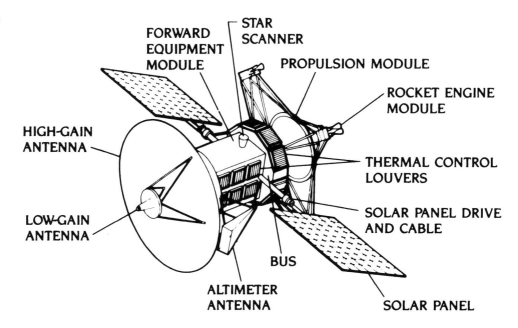

northern hemisphere, including polar and middle latitudes (to ~30°) over an area of about 100 million km², that is, approximately one-fourth of the planet's surface. This mapping enabled scientists to study these regions in great detail and to learn much about their morphological features, in correlation with geological processes and surface properties.

The next stage in the Venus research program was the series of experiments conducted in June 1985 by the Vega-1 and Vega-2 spacecraft, which detached descent vehicles during Venus fly-by en route to rendezvous with Halley's comet (Figs. 2.54, 2.55, 2.56, and 2.57). The landings were performed on the night side near the equator, in the region of Aphrodite Terra (Fig. 2.58). The instrumentation to measure atmospheric parameters and surface properties was approximately the same as on previous Venera spacecraft, with the exception of a television camera to transmit panoramas. The elemental soil analysis in the Vega-2 landing area (a similar experiment planned for the Vega-1 station could not be carried out owing to equipment breakdown), and gamma-ray spectra of the natural radioactivity of the rock, confirmed conclusions made earlier based on Venera 13–14 results, and the rocks analyzed exhibited more definite similarities with certain characteristic basalts of the Moon. More elaborate analysis will be given in the next chapter.

During this mission a fundamentally new tactic was used: the placement of balloon probes, each of which drifted in Venus' atmosphere at an altitude of about 55 km for approximately 15 hours (Fig. 2.59). Tracking their drift from Earth, with a network of radio telescopes and long-baseline interferometry, considerably broadened our information about the nature of motion at the level of the clouds on Venus (Fig. 2.60). Along with data from direct measurements of atmospheric parameters, this information made it possible to clarify the nature of dynamic processes on the planet (Sagdeev et al., 1986). The experience gained by using international cooperation to develop the balloon probes opened new horizons for their further use to investigate other planets, primarily Mars.

Fig. 2.62. Assembly of the Magellan spacecraft and testing in the assembly building.

Fig. 2.63. Insertion of the Magellan spacecraft on a trajectory toward Venus, after separation from the space shuttle.

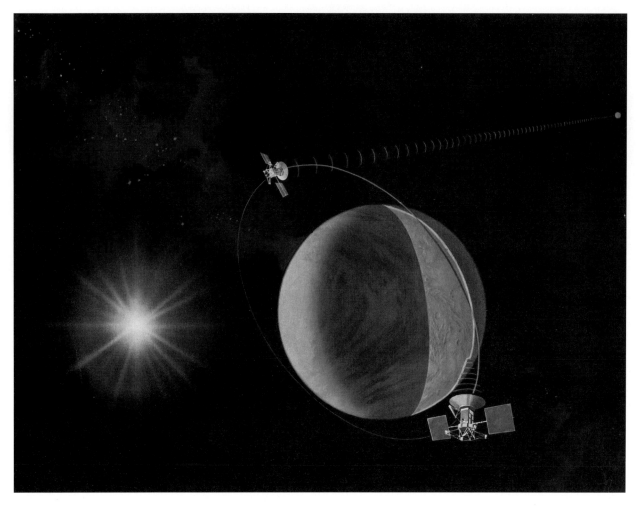

Fig. 2.64. Operation of the Magellan spacecraft in orbit around Venus, with a period of 3.5 hours (artist's rendering).

On 4 May 1989 the Magellan spacecraft was launched from the Space Shuttle Atlantis (Figs. 2.61 and 2.62). It entered into a highly elliptical, nearly polar orbit around Venus on 10 August 1990 (Figs. 2.63 and 2.64). Soon after entering orbit, the spacecraft experienced a disturbing loss of signal to Earth, and for a brief time it was uncertain whether the mission could be successful. The technicians at the Jet Propulsion Laboratory, however, were impressively resourceful under pressure and managed to fix the spacecraft. The mapping mission thus began in September 1990. The Magellan synthetic-aperture radar operated at a wavelength of 12.6 cm (S-band). The primary mission of radar mapping of 70% of the planet's surface with a resolution on the order of 0.3 km was completed by April 1991. The second and third mapping cycles were subsequently successfully completed, and more than 97% of the surface was covered with a resolution close to 100 m, including more than 35% of the surface in stereo, using both right- and left-looking radar imaging (Saunders, 1992; Saunders et al., 1992; see Chapter 3).

Beginning in May 1993 the periapsis of the Magellan probe was lowered from 290 km to place the spacecraft in a near-circular orbit with an altitude of 140 km. This adjustment allowed more sensitive gravity measurements of the whole planet.

On 10 February 1990, while Magellan was en route to Venus, the Galileo spacecraft made a close fly-by of Venus, in order to achieve a gravity boost that would eventually send it to Jupiter after two fly-bys of Earth. The broad range of sophisticated instruments on board, designed for an extended study of Jupiter and its satellites, was used to make many observations during the Venus fly-by (Johnson et al., 1991). These included studies of the clouds and atmosphere with the Near Infrared Mapping Spectrometer (NIMS); studies of the clouds in the visible, infrared, and ultraviolet with the CCD solid-state imaging system (SSI); and studies of the upper atmosphere with the ultraviolet spectrometer (UVS) (see Chapters 6 and 8).

Now that we have reviewed the history and major findings of the study of Venus, we shall turn in the following chapter to more detailed considerations of the nature of our neighboring world.

The Surface: Relief, Composition, and Geology

3 The facial expression of a planet is found in its surface features. The structural formations of the surface, distinct relief features, and physical-chemical properties of individual, characteristic regions (labeled "provinces" by geologists) contain important information about a planet's past and present, revealing the mechanisms and chronology of events that have led to its present appearance. Often, processes that occurred in remote epochs are manifested in particular features of the relief, even when the original structures prove to be strongly camouflaged by subsequent endogenic and exogenic processes. Primary among these processes are tectonic-volcanic activity, impact cratering, sedimentation, and various types of erosion. Aside from the surface relief, the study and comparison of the elemental and mineralogical composition of the constituent rocks provide the most important information for deciphering and reading the stone chronicles—for determining what the surface is like.

The sequence of events forming the surfaces of the terrestrial planets dates back to the final phase of accretion, when the shower of planetesimals falling on the surface was close to exhausted. The large, partially modified craters of the lunar highlands, craters of similar morphology on Mercury, and the most ancient, highly eroded craters on Mars are attributed to this period. Traces of this stage may have been completely destroyed on Venus by subsequent geological activity. On Earth, both ancient and more recent structures have generally been obliterated or buried below a thick sedimentary mantle owing to the existence of our hydrosphere and biosphere.

Until now, of all the planets only Earth has been accessible to investigation by the diverse arsenal of tools available to modern geology, geophysics, and geochemistry. Nevertheless, as recently as the 1960s we were unfamiliar with the appearance of approximately two-thirds of the surface of our own planet, hidden as it was beneath the watery layer of oceans and seas. Hydroacoustic methods have allowed us to see the bottom of the oceans. Numerous ships, equipped with special devices—echo sounders, which determine depth by measuring the propagation time of an acoustical signal reflected from the bottom—have studied the underwater relief in detail, revealing numerous phenomenal structures that have no analogs on Earth's continents (Fig. 3.1). Among these are a system of mid-ocean ridges, bands of deep-sea trenches and island arcs, and numerous underwater mountains and volcanoes.

The system of mid-ocean ridges, with broken relief, is global in nature. This system is a well-defined band, which stretches along the bottom of all oceans for almost 60,000 km and partially branches out onto the continents. The ridges of the Indian and Atlantic oceans, whose uplifts rise 3.5–4 km above the ocean floor, are the most typical.

Large-scale, tectonically active fault structures, known as rift zones, extend along the centers of the mid-ocean ridges. These are narrow canyons with steep walls from one to

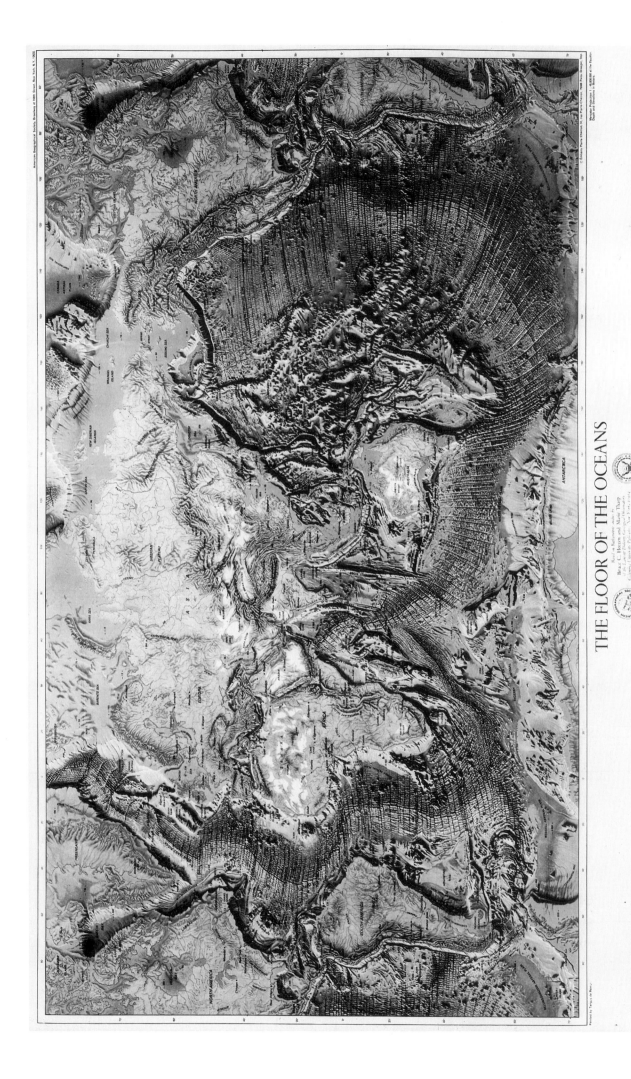

THE FLOOR OF THE OCEANS

Fig. 3.1. Surface relief of the ocean floor and continents of Earth (according to B. Hisen and M. Tarp, 1974).

tens of kilometers in width. They are framed by mountain chains (horsts) divided by inter-montane troughs (grabens), with relative relief differences of 2–3 km. A high degree of seismic activity is characteristic of the rift zones in the central ridges.

The island arcs and their associated deep-sea trenches are likewise characterized by high seismicity. Volcanic eruptions occur in these arcs and trenches more frequently than anywhere else on Earth. A system of volcanic island arcs borders the western Pacific, northeastern Indian, and west and south Atlantic oceans. A large area of the ocean floor is also filled with rugged and flat plains, separated by rises, ridges, and volcanic island chains. Numerous volcanic seamounts are scattered on the basins of the oceans. The summits of some volcanoes protrude above the water, forming isolated islands.

The geological structure and development of Earth is well described by the theory of global tectonics, or lithospheric plate tectonics, which became widely accepted toward the end of the 1960s. Within the framework of this theory, persuasive explanations have been developed involving the drift of terrestrial continents, which was hypothesized at the beginning of the twentieth century by Wegener (1924).

According to the theory of lithospheric plate tectonics, seismic belts, within which not less than 95% of all earthquakes are concentrated, divide the hard stony crust of Earth (the lithosphere) into separate large blocks or plates ranging in size from 2000–3000 km up to 10,000 km in diameter. Among the largest plates are the Eurasian, African, Indian (including Australia), Pacific Ocean, North American, and Antarctic. The lithospheric plates are 20–50 km thick below the oceans and up to 100 km or thicker below the continents. They lie on the asthenosphere, a softened layer of lower rigidity between the crust and mantle (see Chapter 4).

The seismic bands and, consequently, the plate boundaries are diverse. They can be classified into three fundamental groups. The first of these are the mid-ocean ridges. Here only shallow-focus earthquakes develop, at the foci of these earthquakes extensional stress conditions are detected, and numerous gaping fissures on the floor also provide direct evidence of extension. The lithospheric plates, sliding along the asthenosphere, move apart in both directions from the axes of the mid-ocean ridges. The rift zones are formed in areas of plate separation (spreading) where the open space is filled with basaltic lava. During cooling and crystallization, new oceanic crust is formed. The process of oceanic lithosphere regeneration occurs constantly.

The global system of oceanic trenches, which is most pronounced along the periphery of the Pacific Ocean, forms a second category of plate boundaries. Here the hypocenters of earthquakes run off obliquely under the continents to a depth of about 700 km, indicating that the rigid lithospheric plates are plunging into Earth's mantle. At the foci of these earthquakes compression dominates, and many geological structures have resulted from this compression. Thus, in underwater trenches the lithospheric plates come together, and the oceanic plate dives under the continental plate. Along the boundaries of plate convergence and subduction—the so-called Benioff zones—volcanism is widely developed. The famous Ring of Fire of the Pacific Ocean is governed by zones of plate subduction. It is in these subduction zones that remelting of sinking material and accretion of the continental crust take place.

In the third group of plate boundaries are the large-scale cracks that divide the mid-ocean ridges from the ocean floors (and in some places even continents). These are the so-called transform faults, along which the lithospheric plates slip relative to one another, and along which the interaction of plates is transformed from one boundary to the other.

So the movement and interaction of lithospheric plates on Earth is strongly correlated on a global scale. As the plates move apart at the axes of mid-ocean ridges, allowing the oceanic lithosphere to expand there, so they converge in the deep-sea trenches, where subduction of the ancient oceanic lithosphere occurs. A constant regeneration of the ocean floor takes place. At the edges lighter materials, incapable of sinking into the mantle, accumulate, building up the continents. The boundaries of the lithospheric plates are regarded as the dominant elements of Earth's surface structure. The most dramatic forms of surface relief are generally confined to these areas. They contain the dominant portion of magma production and seismicity.

The migration of the lithospheric plates over Earth's surface occurs at rates not exceeding 15 cm per year. This is the maximum observed velocity, characteristic of spreading centers at the middle ridge of the Pacific Ocean. The spreading velocity of the mid-Atlantic ridge is less, within the range of several centimeters per year. Over geological timescales of hundreds of millions of years, such movements amount to thousands of kilometers. In this way, the motion and interaction of lithospheric plates have determined Earth's appearance in the geological past and continue to do so in the present.

Space exploration of the Moon, Mercury, Mars, Venus, and the satellites of the giant planets has revealed a variety of phenomenal surface structures on these bodies and has begun a new stage in the study of their geology and evolution. This work laid the foundation of comparative planetology and enabled us to arrive at a fundamentally important conclusion: in contrast to Earth, such relatively small celestial bodies as the Moon, Mercury, and Mars are characterized by a single unsegmented lithospheric plate, which stabilized during the early history of the solar system (Sclater et al., 1980; Head and Crumpler, 1987). The situation on Venus is much more complex. Venus is similar to Earth in size and density but differs strongly in its topography, surface temperature, the absence of oceans, and other features (Figs. 3.2 and 3.3). Questions about what causes tectonic processes on our neighboring planet, to what extent the lithosphere is mobile, and how heat is transferred, have not been answered unequivocally.

Radiophysical Methods

For Venus, which is enveloped by a dense gas covering and clouds opaque to optical wavelengths, radio techniques are the most effective means of study. However, the width of the window of transparency for radiation emanating from the surface or sounding radiation is limited. Radio waves in the range from approximately 3 cm to 30 cm pass almost unimpeded through the Venusian atmosphere. At shorter and longer wavelengths consider-

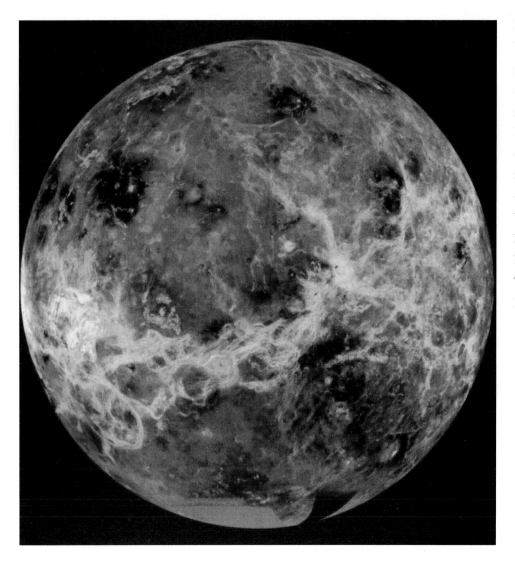

Fig. 3.2. A global view of Venus developed from Magellan radar mosaics (first mapping cycle only). Data gaps were filled by using Pioneer-Venus Orbiter data. A color version of the image was composed based on information from the Venera-13 and Venera-14 landers and is used to enhance small-scale structure. The view is centered at 180° E longitude.

able absorption is observed. It was in the middle of this range that the radio brightness temperature of Venus' surface T_B was determined most accurately.

Since the beginning of the 1960s radar has been used successfully to investigate the rotational periods and directions of planets and, subsequently, the relief and physical properties of their surfaces. If, to determine the rotational periods, we analyze the shift and broadening of a spectral line of reflected radiation (the echo signal), caused by the Doppler effect (see Fig. 2.8), then to study the profiles and properties of the surface we need data on the intensity of the reflected radiation and on the intensity distribution over the spectrum, taking into account the time lag of the signals arriving at the receiver and the Doppler frequency shift. Data on the degree of polarization of radio waves reflected by the planet also provide important information on the microstructure of the surface.

Unfortunately, ground-based radar studies are informative mainly for low-latitude regions; upon transition to higher latitudes (further from the region closest to the sub-Earth region), the measurement errors and the ambiguity in interpretation rise sharply. The working frequency range for ground-based radar stations, specified by minimum ab-

Fig. 3.3. A global view of Venus produced in the same manner as Fig. 3.2 but centered on the North Pole. 0°, 90°, 180°, and 270° E longitude are, respectively, at 6 o'clock, 3 o'clock, 12 o'clock, and 9 o'clock.

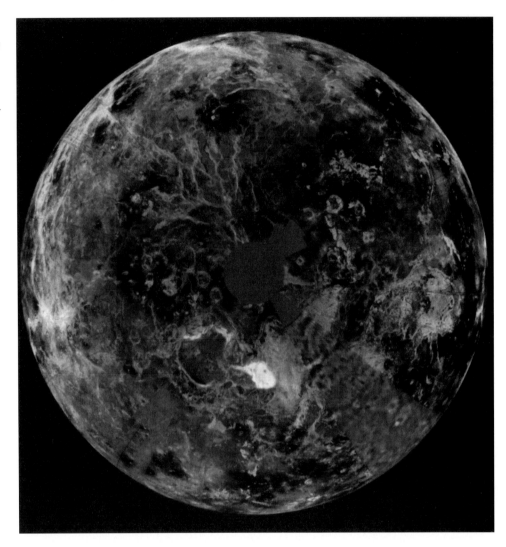

sorption in Earth's atmosphere, encompasses a broad range of wavelengths, from several millimeters to several meters. In radar astronomy, centimeter wavelengths are predominantly used.

The intensity of the reflected planetary signal depends on the coefficient of reflection, k (expressed as a percentage), which is directly related to the physical properties of the surface (primarily the density of the surface layer at a depth on the order of several wavelengths of the sounding radiation) and to the nature of the constituent surface rocks. These properties determine the permittivity, ε, of the material from which the electromagnetic wave is reflected. At the same time, one can estimate the scattering properties of the surface and the angles of inclination of areas comparable in extent to the wavelength. The greater these angles, the greater the surface roughness or, in other words, the more irregular the microrelief. Usually, the mean square values of the inclination angles $\bar{\theta}$ are determined; for Mars, for example, they lie within the range of 0.5°–4°, which is considerably less than for either the Moon or Mercury. For Venus, intermediate values of 2.5°–5° were obtained from Earth-based observations.

For various dry terrestrial rocks a simple empirical dependence was found between

Table 3.1 Surface characteristics of the Moon and terrestrial planets

	k, %	ε	ρ, g/cm^3	$\bar{\theta}$, deg.
Moon	5.7–6.3	2.6–2.8	1.2–1.3	6–7
Mercury	5.8–8.3	2.7–3.3	1.3–1.6	5–8
Venus	11–18	4–6	2–3	2.5–5
Mars	3–14	1.4–4.8	1–2.5	0.5–4

Source: Radar data.

the permittivity and density: $\rho \simeq (\sqrt{\varepsilon} - 1)/a$, where a is usually $\sim$0.5 (Krotikov, 1962). Thus by measuring ε, one can estimate the planet's soil density ρ. This method was first used successfully in lunar research. Radar investigations of Mars revealed variations in the permittivity of its surface over a broad range, from approximately 1.5 to 5, which corresponds to densities of 1–2.5 g/cm^3. These estimates were later confirmed by measurements using radiometers in the centimeter range on board the Mars-3 and Mars-5 spacecraft (Basharinov et al., 1973, 1976). The wide range of values obtained suggests variations in the properties of the Martian surface from hard rock to highly granulated soils, as analysis of television images and other data returned by the Viking landers and orbiters confirmed.

From the coefficient of reflection of Mercury in the radio range, which proved to be almost the same as for the Moon, a mean value for the permittivity, $\bar{\varepsilon} = 3$, which corresponds to a surface layer density of about 1.4 g/cm^3, was obtained. This value is intermediate between the densities of surface rock on the Moon and on Earth (and, note, much less than the mean density of Mercury; this has important implications for the planet's internal structure). For Venus the scatter in the ε values is less than for Mars but larger than for either the Moon or Mercury. In absolute value ($\varepsilon = 4$–6), however, it is significantly higher than for these celestial bodies (Pettengill et al., 1982, 1988). A summary of radar results for the Moon, Mercury, Venus, and Mars is presented in Table 3.1 (Marov, 1986, 1994).

In the radio range, one can not only study the reflective properties of the surface but can also actually "see" what it looks like. To do this, a high spatial selectivity of the ground-based receiving antenna is required. This is usually achieved by reception of the reflected signal at two antennae, separated by a known distance, working on the principle of a radio interferometer. From the phase difference of the received signal, the position of each section of the planet, relative to some mean surface figure, is measured in succession. The higher the spatial selectivity of the antenna, the higher the resolution at the surface of the celestial body under consideration and, consequently, the higher the image quality. In due course, images of the Moon that are difficult to distinguish from normal photographs using ground-based optical telescopes were obtained using this method, which is called the time and frequency selection method. This method was used for Venus starting in the early 1970s. The most detailed investigations were conducted at the Goldstone and Arecibo observatories.

A surface map for the $\pm 70°$ latitude band, provided from measurements at Gold-

Fig. 3.4a. A mosaic of Venus surface images, from radar results at the 12.6 cm wavelength, at the Arecibo Observatory, Puerto Rico, in 1975 and 1977 (according to Campbell and Burns, 1980). The resolution at the surface is 10–20 km. The discontinuity in the images within 30° N is a consequence of the unfavorable angle of incidence of the radio waves and the lack of the requisite resolution. The black band near the equator is an artifact of the data processing.

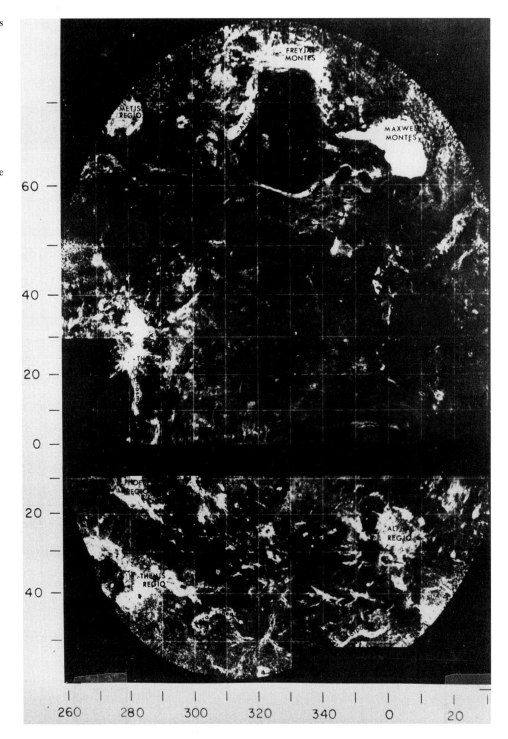

stone, is shown in Fig. 2.16; a fragment of a map for high latitudes of Venus' northern hemisphere, from measurements at Arecibo, is given in Fig. 3.4a (Jurgens et al., 1980; Campbell and Burns, 1980). The ratio of reflection intensities between light and dark regions reaches 20 to 1. The most extensive bright regions at approximately 30° S and N latitude (0° and 280° longitude, respectively), and a region close to the zero meridian at 65° N latitude (Maxwell) are most clearly visible; they were discovered at the beginning of radar studies of Venus (α, β, and δ). The zero meridian runs through the radio-bright region of

α; it coincides with the direction to Earth about one month after inferior conjunction. A region of approximately similar brightness is located somewhat higher in latitude, at longitude $310°-340°$.

What could cause such significant differences in the reflective properties of the Venusian surface? There are three possible factors. The first is the relief: the higher the region, the less atmosphere above it, and hence the lower the degree of radio absorption. Even relatively small variations in the relief may introduce a substantial contribution to the distribution of reflective properties over the surface simply because the atmosphere is most dense close to the surface. In fact, altitude differences of $2-4$ km vary the thickness of the atmosphere's absorbing layer, at wavelengths shorter than 10 cm, by approximately 15%–25%, and for elevations up to 10 km, the effective thickness is decreased by almost a factor of two.

The second cause is the microstructure of the surface, that is, its roughness on the wavelength scale of radar measurements. Usually two components of the reflected radiation are differentiated: specular and diffuse. In reflection functions there is a sharp maximum near the angle of emergence of radiation, $\theta* = 0°$, corresponding to quasi-specular reflection, and a large portion of the reflected signal energy is contained in the quasi-specular component. For a rough surface the contribution of the diffuse component to the reflected signal energy is higher. However, it is usually no more than 10%–15% of the overall reflected energy for angles of $\theta* = 0°-50°$ (Kuzmin and Marov, 1974).

Finally, the third cause is the physical-chemical (electrical) properties of the surface, on which the permittivity of the material, ε, depends: with an increase in ε, the degree of reflection grows. The high ε value for Venus (see Table 3.1) can be interpreted as indicating that its surface is covered by an electrically conducting layer that, in several regions, is shut off by less reflecting material with lower ε.

The microstructure of the surface strongly affects the reflective properties, as well as the degree of radiation polarization. It was found that polarization of the major portion of radio energy reflected by Venus corresponds to specular reflection, and that for Venus the ratio of depolarized reflection to polarized reflection is significantly less than for the Moon (Kuzmin and Marov, 1974). This indicates that, compared to the Moon, Venus' surface microstructure is on average smoother, whereas large irregularities are linked with regions of higher-than-usual radiowave reflection. This scenario is suggested by the estimated mean-square angles of inclination of the surface irregularities (see Table 3.1).

In spite of the powerful spacecraft techniques that have become available in recent years, ground-based radar measurements, which enabled us to detect for the first time a whole series of representative relief and morphological features on Venus' surface, have not lost their importance. Owing to improvements in equipment and research methods, studies of individual regions of Venus' surface, with a surface resolution increased from several tens of kilometers to 1.5 km, are being carried out at the Arecibo Observatory (Campbell et al., 1983, 1984, 1991).[1] An example of a radar image obtained for Alpha Regio and the surrounding terrain in 1988 is shown in Fig. 3.4b. Several prominent features in the image,

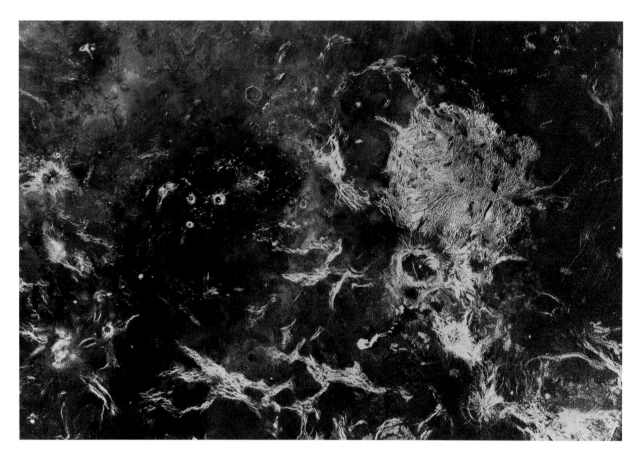

Fig. 3.4b. A radar image of Alpha Regio and surrounding terrain on Venus made in 1988 with the 12.6 cm wavelength radar system on the 305 m antenna of the Arecibo Observatory in Puerto Rico. The image, which has a resolution of 1.5 km (0.9 miles), covers longitudes 320° to 20° and latitudes 45° S to 10° S—an area approximately 5500 km (3400 miles) by 3700 km (2300 miles). Prominent features in the image are three volcanoes (Hathor, Innini, and Ushas Montes) down the left (west) side of the image, and three large impact craters (Danilova, Aglaonice, and Saskia) in the dark area left of center, dubbed the "crater farm" from Magellan observations. Alpha Regio, a 1000 km area of disrupted, probably ancient, terrain, is to the right of center.

such as volcanoes, can be distinguished, and specific features of the morphology of the whole terrain can be revealed.

The most detailed structures, however, were discovered using radio telescopes placed on satellites orbiting Venus. The benefits of this technique are exceptionally great, and with the successes in radar mapping of Venus from the Pioneer-Venus, Venera-15, and Venera-16, and especially Magellan spacecraft, we have obtained radar images of practically the planet's whole surface. These images are in no way inferior to television pictures of Mercury, Mars, and the systems of Jupiter, Saturn, Uranus, and Neptune transmitted from the Mariner, Viking, and Voyager spacecraft.

The importance of producing television and radio images to study the morphology, physical properties of the surfaces, and geological history of planets is difficult to over-estimate. The research possibilities have been immeasurably expanded, not only owing to the high resolution, which exceeds the best attainable resolution from Earth-based photography by factors of several hundred, but also because of global coverage, including access to polar regions and other areas unfavorable for Earth-based observations. In particular, for optical or radar observations of Venus from Earth, periods close to inferior conjunction are the most convenient times to observe, when the same side of Venus is always turned to Earth. Therefore, its other hemisphere and polar regions had remained essentially unstudied until recently. The use of satellites has solved this problem.

The first attempts to study the relief of Venus on the Venera-9 and Venera-10 spacecraft, using the bistatic radar technique, turned out to be only partially effective. This technique involves exposing the planet to radio waves from an orbiting satellite and receiving the reflected signals on Earth. Several regions with non-uniform relief were discovered in Venus' southern hemisphere, stretching in the latitudinal direction for several hundreds of kilometers. The altitude differences in these regions reached about 3 km, and their surfaces turned out to be almost smooth (Keldysh, 1977).

The radio mapping technique used on the Pioneer-Venus spacecraft was considerably more effective. In this case the spacecraft emits the radio waves and receives the reflected signals without the assistance of ground-based telescopes. The results are then transmitted to Earth. The radio waves from the satellite, at a wavelength of 17 cm, are emitted along the vertical, as well as at a certain angle with respect to the surface area under study. Sounding along the vertical enables us to measure the altitude and the physical properties of the surface along the flight path of the satellite from the time lag and intensity of the reflected signals. In turn, the rotation of the satellite, with its fixed antenna, makes it possible to perform oblique (side-looking) sounding, where the radio beam illuminates the surface from the side of the orbital plane. In other words, owing to rotation of the satellite, the radio beam skims along the surface on both sides of the orbital plane, passing through the vertical. Separation of the reflected signals by oblique sounding, according to lag time and Doppler frequency shift, ensures spatial selectivity even with a relatively small antenna and thus provides the possibility of constructing a radio image of the surface. The image quality and resolution, however, proved to be quite low.

The spacecraft was stabilized in space with a rotational period of 12 seconds, and the antenna of the radio altimeter was positioned perpendicular to the rotational axis. Altitude measurements were carried out each time the antenna was directed toward the planet's surface. The soundings thus took place once every 12 seconds, during which time the spacecraft flew approximately 100 km. The orbital period of the craft was 24 hours, and therefore for each successive approach to Venus the planet rotated by approximately 1/243 of its circumference, that is, by about 1.5°, which is 150 km at the equator. Consequently the track of the orbit on the surface shifted by the same distance. The measurement frequency and the chosen orbit determined the resolution of the surface map constructed for Venus. The resolution was 30–50 km, and the map covered the surface between 65° S latitude and 75° N latitude (Pettengill et al., 1979, 1980, 1988) (Fig. 3.5). At this resolution characteristic morphological features on a continental scale are clearly detected; however, this resolution is insufficient for identifying smaller-scale structures, which provide fundamentally important information on the geological structure of the planet and its evolution-

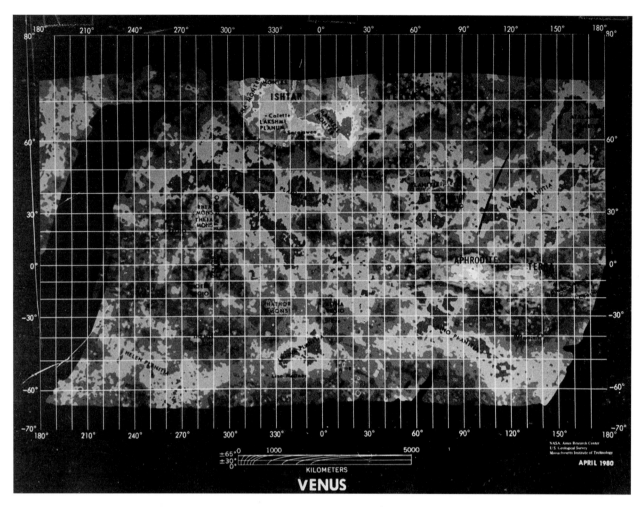

Fig. 3.5. Relief map of Venus' surface, compiled from radar altimetry from the Pioneer-Venus spacecraft, in Mercator projection. The dark sections are lowlands; the light are highlands (according to Mazursky et al., 1980).

ary path. Nevertheless, these results considerably clarified the nature of the reflective properties and relief features of most of Venus' surface (Masursky et al., 1980; McGill et al., 1983; Head et al., 1985; Bindshadler and Head, 1988, 1989; Garvin et al., 1985; Senske, 1990).

The launch of Venera-15 and Venera-16 provided new opportunities for studying the relief and structure of Venus' surface. The complex of onboard equipment (radar and radio altimeter) developed for these experiments ensured the reception of high-quality images by the side-looking technique using synthetic-aperture radar and ensured altitude measurements of imaged surface regions. Although the latest Arecibo images are comparable in quality to those obtained by the Venera-15 and Venera-16 spacecraft, they are constrained by Venus' position relative to Earth and are mainly of the hemisphere facing Earth at inferior conjunction. Only equatorial and mid-latitudes can be studied from Earth, whereas satellites in orbit around Venus have no limits on spatial coverage. The method used by the spacecraft will now be examined in greater detail.

A scheme for radar photography is shown in Fig. 3.6 (Rzhiga, 1988). The spacecraft S, at the chosen radar sounding wavelength $\lambda = 8$ cm, periodically illuminates a section of the planet's surface located within the range of the antenna's beam pattern (1) at an angle of

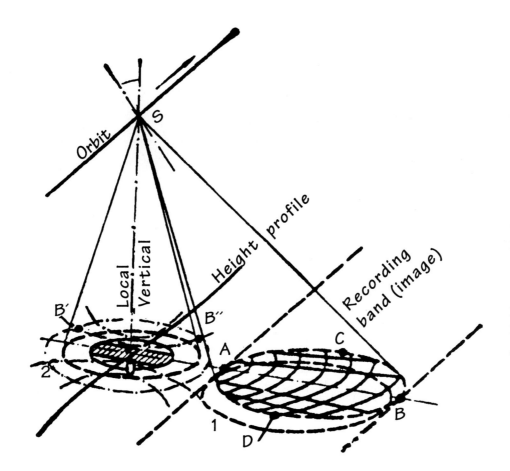

Fig. 3.6. The side-looking radar method, which was used on board the Venera 15–16 spacecraft. S is the satellite, region 2 is a spot on Venus' surface irradiated by the radio altimeter (altitude profile), and region 1 is a spot on the surface at an angle of 10° relative to the local vertical, irradiated by the side-looking radar (photographed band). For an explanation of points A, B, B′, B″, C, and D, see the text and Rzhiga, 1988.

10° relative to the local vertical. Simultaneously, using a radio altimeter, the altitude SO is measured at the spot illuminated below the spacecraft (2). The altitude profiles are determined using highly accurate measurements of the spacecraft's orbital parameters and are constructed from the difference between its distance from the planet's center, at a given time, and altitude SO.

The surface elements inside the illuminated spot (1) are located at various distances from the radiation source and have different radial velocities relative to it. Therefore the reflected signals received at the satellite have different lag times, and different frequencies because of the Doppler effect. So, for example, signals from point A arrive earlier than those from point B, and signals from point C, which is approaching the source, have a higher frequency than those from point D, which is receding. Much as in Earth-based radar methods, the set of points of equal lag forms concentric circles with the center at 0, and points of equal Doppler shift form hyperbolae on the planet's surface. The actual radar image is constructed based on the difference in the lag and Doppler shift of the reflected signals. Sequences of these images are combined into a continuous image band as the spacecraft moves across the planet. All that has been said about this method would remain valid while the spacecraft's antennae were directed along the local vertical (2). In this case, however, it would prove impossible to distinguish signals arriving from surface elements distributed symmetrically relative to the flight path, having identical lags and equal Doppler shifts (see the dotted-dashed line in Fig. 3.6, points B′ and B″). Deflection of the

Fig. 3.7. The synthetic-aperture radar method, which enables us to increase the effective diameter of the onboard antenna by *n* times and also to increase the resolution.

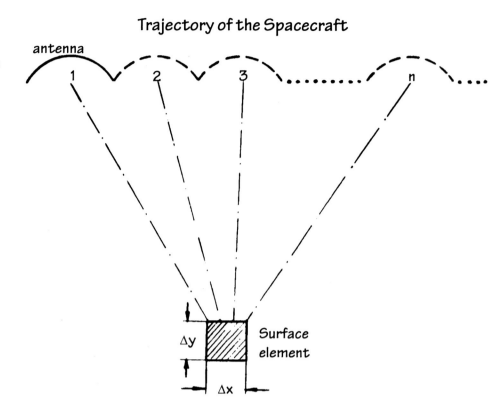

antenna to the side of the flight path enables us to eliminate this uncertainty; thus the concept of "side-looking" radar.

Each day, as the satellites passed close to the planet, a surface strip about 150 km wide and 8000 km long was imaged. The pericenters of the satellites were located close to the North Pole, and the imaging was conducted from an altitude of 1000–2000 km. Owing to the intrinsic rotation of Venus and the 24-hour orbital period for the satellites, a new band adjacent to the previous one was imaged each time. More than half the area of the northern hemisphere was imaged with a local resolution of 1–2 km. This is approximately the same spatial resolution one would expect with the unaided eye from an altitude of 1000–2000 km if the planet were not obscured by a dense atmosphere and clouds. But to achieve this resolution with a radiometer it was necessary to employ the synthetic-aperture method, which is based on the principle of synthesizing an antenna *n* times longer than that installed on the spacecraft (Fig. 3.7; see Rzhiga, 1988). Moving along the satellite's orbit, the antenna subsequently occupies positions 1, 2, . . . *n*. Simultaneous processing of the reflected signals furnishes an "equivalent" antenna with a diameter *n* times larger than the actual antenna, with a corresponding increase in the resolving power Δx. So, with an onboard antenna 6 m in length and 1.4 m in width on Venera-15 and Venera-16, the equivalent of an antenna 70 m in diameter was achieved by this method. In the direction perpendicular to the flight path (Δy), resolution was achieved, as usual, by modulating the sounding signal, making it possible to distinguish the reflected signals by their arrival time at the antenna.

The signals were stored in a high-capacity memory and were subsequently trans-

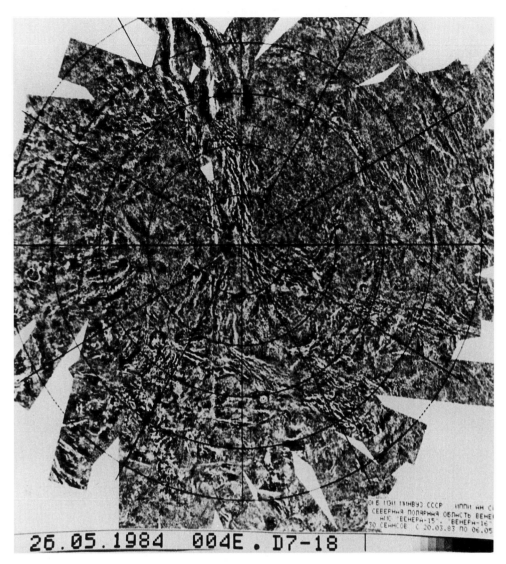

Fig. 3.8. An image mosaic of Venus' northern hemisphere from radar imaging results of the Venera 15–16 spacecraft.

mitted by radio link to Earth. Using a special computer system at the Institute of Radio Technology and Electronics of the Russian Academy of Sciences, radio images were synthesized. Exactly as with ground-based radar methods, the brightness of the photo tones in the images is determined by three fundamental factors on which the intensity of the reflected signals depends: the local relief (the various orientations of the surface elements with respect to the radar antenna), the microstructure of the surface, and the electrical properties of its constituent rock. Here, the concept of light and shadow are treated not in the customary sense of solar illumination but in terms of radiowave illumination with respect to the satellite. By correlating the measured intensity of radiowave reflection with the degree of darkening of the photographic paper, a photoimage is constructed. Good-quality maps of Venus' northern hemisphere were compiled based on these images. The initial material, obtained from the Venera-15 and Venera-16 spacecraft after processing, is shown in Fig. 3.8 in the form of an image mosiac.

Essentially the same technique was used for the radio mapping of Venus with the Magellan spacecraft. The Magellan spacecraft carried four antennae: a parabolic high-gain

Fig. 3.9. Magellan in the cargo bay of the Space Shuttle Atlantis. The parabolic, dish-shaped high-gain antenna (HGA) is seen on top. The medium-gain antenna is the cone-shaped structure mounted to the top side of the equipment bus. The low-gain antenna is mounted on a platform held by struts above the HGA. Both these antennae augment the HGA and are useful when the HGA cannot be pointed directly at Earth (courtesy of NASA).

antenna used both as the main antenna for radar operations and for communication with Earth, a medium-gain antenna, a low-gain antenna, and an altimeter antenna (Fig. 3.9). The initial highly elliptical mapping orbit had an apoapsis altitude of 8458 km and a periapsis at 289 km. During each orbit, four spacecraft attitude maneuvers were needed. The antenna was pointed toward space to orient the spacecraft precisely with respect to the stars. Then it was pointed first toward Earth for data transmission and next toward the surface of Venus for radar mapping. It was then again pointed toward Earth for further data transmission. The large numbers of maneuvers were accomplished with an electrically driven reaction wheel. Each mapping orbit imaged a "ribbon" of the surface approximately 25 km wide and 16,000 km long. With each orbit the spacecraft, which remained nearly fixed in an inertial frame, imaged a new swath of the slowly rotating planet (Saunders et al., 1992).

Magellan data is superior in both coverage and resolution to that obtained earlier with Pioneer-Venus, Venera-15 and Venera-16, and ground-based radar measurements. Nonetheless, these data by no means diminished the importance of the results of these precursors. Pre-Magellan results gave us the first opportunity to look at the planet through the shroud of its heavy gas-cloud envelope, revealing many characteristic surface features and generating important ideas about its morphology and geology. That is why it seems reasonable (and instructive) to offer in the following paragraphs a historical account. Such an account enables us to show how current ideas of the geological evolution of Venus were gradually developed. In essence, the pre-Magellan epoch created the basis for the contemporary understanding of these processes; Magellan tremendously extended this knowledge, while also affecting general concepts of comparative planetology.[2]

Relief of the Surface and Geology of Venus

A surface relief map of Venus is shown in Fig. 3.5, with a mean resolution of about 100 km, compiled from radar mapping data from the Pioneer-Venus spacecraft. Nearly 90% of Venus' surface lies within ±1 km of the mean level, corresponding to a radius of 6051.5 km, with the large highlands occupying less than 8% of the surface area. (Magellan data have refined the mean radius to 6051.84 km.) Thus, on the whole Venus has proven to be the planet closest to spherical in shape. But Venus also has prominent plateaus, on the scale of terrestrial continents, with mountainous massifs. These are Ishtar Terra (named after the goddess associated with Venus in Assyrian-Babylonian mythology) and Aphrodite Terra (named after the ancient Greek goddess of love and beauty). Among the other major highlands is Beta Regio.

An overall idea of the locations of the principal relief features is given in an altimetric map of Venus (Fig. 3.10). Thematic maps of Venus incorporating Venera data are presented in the *Atlas of the Terrestrial Planets and Their Satellites* (Marov, 1992). The results of radio altimetry from Pioneer-Venus, although of low resolution, provided a first representation of features on a global scale; the portion of the northern hemisphere imaged from Venera-15 and Venera-16 was the most detailed before Magellan mapped essentially the whole globe. A section of low and middle latitudes with good resolution, limited to a region in view close to Venus' inferior conjunction, became accessible owing to progress in ground-based radio interferometry by the end of 1989.

The fragment of the surface shown in the upper right of Fig. 3.4a contains Ishtar Terra. Essentially all this region is an immense uplift, rising above the basaltic lowlands.

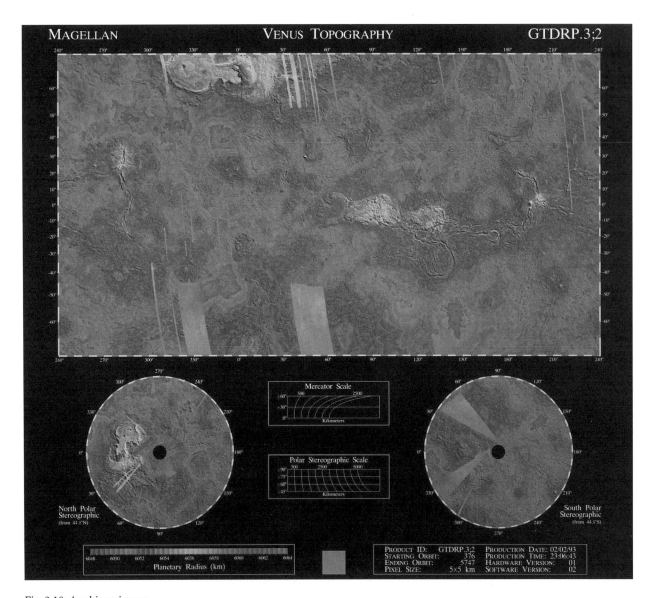

MAGELLAN   VENUS TOPOGRAPHY   GTDRP.3;2

Fig. 3.10. An altimetric map of Venus, from Magellan data (courtesy of NASA).

Images from Venera-15 and Venera-16 (Figs. 3.11 and 3.12) enabled us to examine Ishtar Terra in detail, detect numerous features of the rugged relief, and formulate opinions as to the geological processes that formed it.[3]

Maxwell Montes and two mountainous massifs (Frejya Montes and Akna Montes) to the west and north stand out especially on this extensive, dark, pear-shaped plateau stretching more than 5000 km (see Fig. 3.5). This whole mountainous region is isolated from the adjacent lowlands by steep escarpments. One of the summits, in the center of the Maxwell massif, reaches an altitude of 11 km above the mean surface level, exceeding the highest Earth summit, Mount Everest, by almost one and a half times. On the slope of this mountain is a huge double-ringed impact basin, 95 km in diameter, named Cleopatra patera. It is obvious that the entire Ishtar Terra region was formed as a result of powerful tectonic processes, with volcanic activity also playing an important role.

Venera mapping revealed an extensive zone of linear folds in a system of nearly parallel ridges and valleys with widths ranging from 5–10 km to 30–50 km. These were thought

Fig. 3.11. A photographic map of Venus in the Ishtar Terra region. The projection is a Lambert conformal projection. The entire region is an immense uplift, rising in the middle of basaltic plains. At the left is Maxwell Montes, with a highest altitude of 11.5 km, and on its slope is the Cleopatra crater, 95 km in diameter. This map was compiled from the results of Venera 15–16 radar imagery.

perhaps to be analogous to terrestrial orogenic belts, and they form the mountain systems of Maxwell, Frejya, and Akna Montes on Ishtar Terra. Framed by these systems is the immense plateau Lakshmi Planum, approximately equal in size to Tibet, in whose central region are two flat-bottomed craters about 150 km in diameter, later identified by Magellan as being of volcanic rather than impact origin (Fig. 3.13). The plateau itself, whose altitude is about 4 km above mean planetary radius and which is apparently covered by basalts, drops off abruptly to the lowlands adjoining Ishtar Terra from the south. Here also linear dislocations were observed, passing along the Vesta escarpment, possibly formed by extrusion of material from under the plateau at its boundary, and also by strike-slip faults. It was proposed (Barsukov et al., 1986a; Basilevsky et al., 1986), that the framing ridges are partially pushed up onto the plateau, or, conversely, that the edges of the plateau thrust under the ridges. There are, however, no structures here that resemble trenches. This region of Venus could therefore hardly be considered similar to zones of transition from oceanic to continental crust, or subduction zones, by analogy with the theory of lithospheric plate tectonics on Earth.

Analysis of Magellan mapping has not yet provided an unequivocal explanation for the origin of Ishtar Terra. Proposed models include mantle downwelling beneath a thickened crust (Bindschadler et al., 1992a) and the surface expression of an upwelling mantle

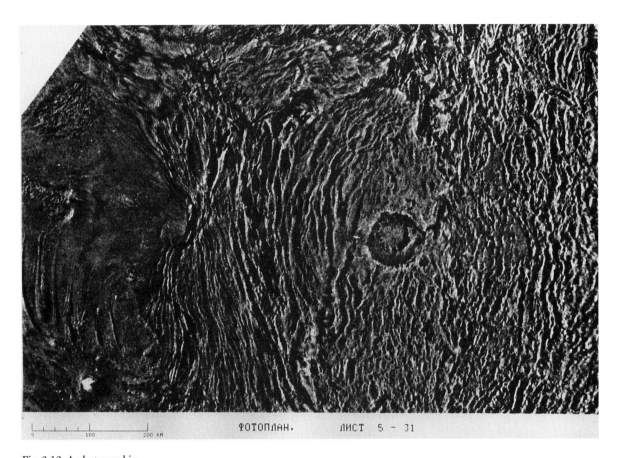

ФОТОПЛАН.   ЛИСТ   5 - 31

Fig. 3.12. A photographic map of Venus' surface, in the region of Maxwell Montes. The crater Cleopatra is clearly visible. The mountain massif is separated from the neighboring plain, to the left, by steep escarpments (image from Venera 15–16 radar data).

plume (Grimm and Phillips, 1991). In the second hypothesis, the mountain belts are seen as resulting from incipient mantle return flow. These models are discussed by Phillips and Hansen (1994).

The characteristic signatures of tectonic and volcanic activity, in the form of extended folded structures, unusual ring-shaped forms, extensive outflows of basalts, and other lava flows, are clearly visible on the multiple images returned by Venera-15 and Venera-16. Judging from the variety of relief forms, it became evident that the crust of Venus has undergone intense deformation processes, and that extensive lava eruptions have also occurred, obliterating traces of previous intense meteoritic bombardment (Figs. 3.14 and 3.15; see Basilevsky et al., 1981, 1985, 1987).

Over vast areas of the northern hemisphere having roughly homogeneous radio reflectivity, extensive flat plains with low mountain ranges, hills, depressions, and ridges are widely represented in Venera images. Analysis of the Venera maps led to the conclusion that these vast plains were probably formed by basaltic lava outflow. Magellan mapping confirmed this conclusion and showed that plains units cover roughly 85% of the planet (Fig. 3.16). These plains units contain tens of thousands of small shield volcanoes (Head et al., 1992). Preliminary attempts at detailed stratigraphic analysis of the plains have revealed a complex history of many episodes of volcanism and deformation (Phillips and Hansen, 1994). Coherent patterns of structural features—such as ridge and fracture belts, wrinkle-ridges, and extensional fractures—over thousands of square kilometers suggest

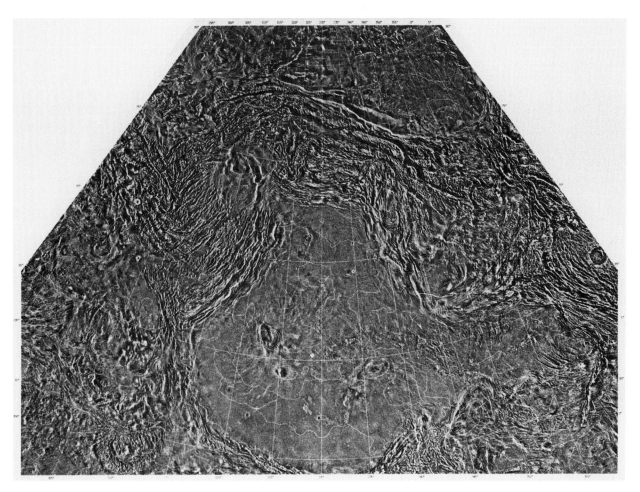

Fig. 3.13. Lakshmi Planum, framed by the Maxwell mountain systems, Frejya (from the north) and Akna (from the west), on Ishtar Terra. These systems are an example of linearly folded fault forms in Venus relief (image from Venera 15–16 radar data).

that they record a direct crustal response to mantle dynamics (Pronin, 1986; Billoti and Suppe, 1992). Venus thus may have a very different global tectonic style from Earth, where deformation is strongly concentrated at plate boundaries.

Venera images provide examples of powerful crust deformation. These examples reflect various types of disturbances of the original crustal strata (referred to as dislocations) as a result of which folds, normal faults, overthrust and underthrust faults, and other tectonic patterns form. Based on these data, geologists (see, for example, Barsukov et al., 1985, 1986a; Basilevsky and Head, 1988; Basilevsky, 1986, 1989) distinguished three types of dislocation: linear folded-fault patterns (see Figs. 3.13 and 3.14), systems of chaotic deformations in the form of extended ridges and valleys intersecting diagonally or orthogonally, reminiscent of tile or parquet flooring (tessera) (Figs. 3.17, 3.18, and 3.19), and unique ring structures with diameters from 150 km to 600 km, formed by concentric mountain chains and grooves, among which the largest were identified as coronae or ovoids (Fig. 3.20) and the smallest as spiders or arachnoids (Figs. 3.21 and 3.22).

Vast areas with systems of intersecting ridges and valleys (called "parquet" and, later, tessera) are widely represented in Venera images in a series of regions on Ishtar Terra, Tethus Regio, Sedna Planitia, and elsewhere (see Figs. 3.17, 3.18, and 3.19). Sometimes they are split by fractures (faults) upon crustal stretching, as a result of internal stresses.

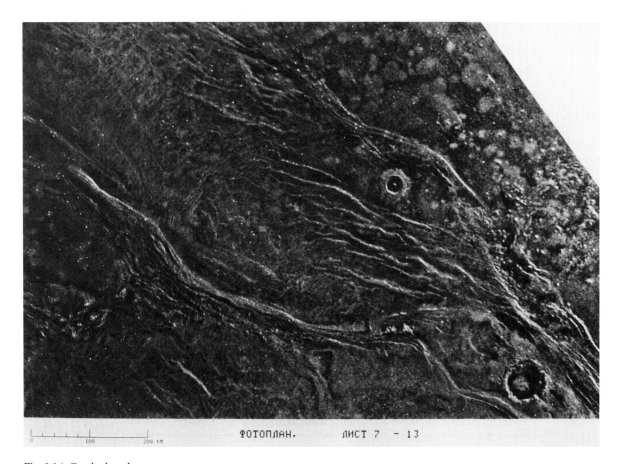

ФОТОПЛАН.   ЛИСТ 7 - 13

Fig. 3.14. Gentle sloped ridges Lukelong Dorsa on the Louhi Planitia. Impact craters to lower right and right of center are Dickinson and Rudneva (about 60 km and 20 km across, respectively), with ejecta blankets clearly visible. The extended central region of smooth plain is replaced by rolling plains to the northeast and southwest, where many small volcanoes are concentrated (image from Venera 15–16 radar data).

Venera 15–16 mapping showed that tessera comprise around 15% of the area north of 30° N (Sukhanov, 1986; Kreslavsky et al., 1988). These structures were interpreted as being formed by horizontal displacements of the surface strata, on the order of 1–2 km thick, probably in the direction of slight slopes around local uplifts (Sukhanov et al., 1986, 1988). The tessera are more ancient than the young smooth lavas and apparently formed over a geologically long time period. Magellan imaging confirmed that tessera are a widespread unit over the whole globe. Tessera tend to be concentrated in highland regions, and the larger units tend to be at higher altitudes. Most tessera regions are characterized by compressional ridges and troughs upon which younger extensional features are superimposed. They appear to be regions of thickened crust that have deformed as a result of regional stress patterns (Vorder Bruegge and Head, 1989).

In addition to globally widespread volcanic plains, Magellan imagery revealed a large number of volcanic constructs with a wide variety of morphologies, as seen in Fig. 3.23 (Head et al., 1992). Many of the large mountains are Hawaiian-style shield volcanoes, ranging in size up to several hundred kilometers across and 4 km high (Fig. 3.24). These frequently have concentric or circular central features and occasional radial fracture patterns. They show evidence of multiple episodes of radial lava flows, the most radar-bright of which appear to be quite recent and possibly currently active. Concentrations of small volcanoes less than 20 km in diameter have been termed shield fields. More than 550 of these regions have been mapped. About 25 intermediate-sized volcanoes with radiating

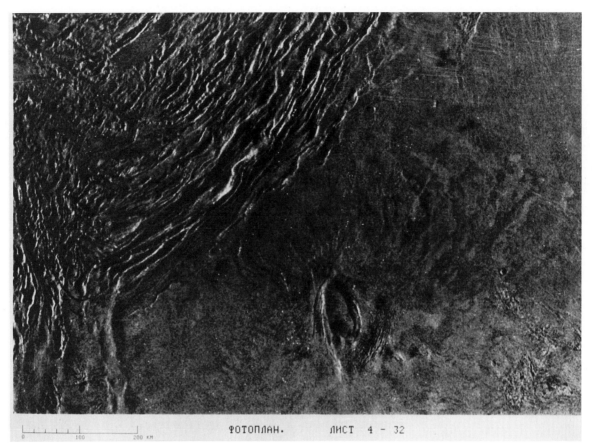

ФОТОПЛАН.   ЛИСТ 4 - 32

Fig. 3.15. A section of the Lakshmi Planum, formed by extensive outpourings of basalts. At the bottom center is a flat-bottomed caldera, 150 km in diameter, formed by ancient volcanism (photographic map constructed from Venera 15–16 radar data).

petal-like lava flows have been mapped, and termed anemone (Fig. 3.25). Also in the intermediate size range are the steep-sided domes: flat-topped, circular constructs often characterized by summit pits and sometimes by radial and concentric fracture patterns. These features have been widely referred to by the informal name "pancake domes" (Fig. 3.26). Their shape is similar to that of rhyolite domes on Earth, but they are typically more than an order of magnitude larger than these terrestrial features, which are usually less than 1 km in diameter. Steep-sided domes often occur in clusters and are commonly associated with tessera terrain. More than 150 of these have been mapped. Another type of intermediate volcano has been termed "ticks." These have radial ridges or spurs extending away from a rim surrounding a circular interior. The similarity of these features to a class of terrestrial seamounts has been noted (Head et al., 1992). Also seen in Magellan images are numerous volcanic calderas not associated with well-defined volcanic rises, which are often surrounded by concentric patterns of fractures. These are typically 60 km–80 km in diameter. A remarkable class of features discovered by Magellan is the sinuous lava channels (Fig. 3.27; see Head et al., 1992; Baker et al., 1992). These features are distributed globally and range up to several thousand kilometers in length. Hypotheses for the identity of the channel-eroding fluids include ultramafic silicate melts, sulfur, and carbonates (Kargel et al., 1994).

Very little evidence was found for pyroclastic features on Venus, which is consistent with the prediction that the high-pressure atmosphere would inhibit volatile exsolution

Fig. 3.16. A Magellan image of a portion of Leda Plani-tia, centered at 41° N, 52° E. The image is 220 km × 275 km in area and shows a variety of geologic terrains. Rising above the volcanic plains from the upper left is a region of old, highly frac-tured highlands referred to as tessera. The more recent volcanic flood plains and subsequent lava flows sur-round and embay the an-cient tessera, and also embay the remains of an impact crater.

(Head et al., 1992). It has been suggested that the large size of many volcanic features is in-dicative of very large magma reservoirs. Maps of the global distribution of volcanic features show that they are much more broadly distributed than on Earth, where they tend to be narrowly concentrated along lithospheric plate boundaries. A zone of increased concentra-tion of volcanic features covering approximately 40% of Venus' surface has been found in the Beta, Atla, and Themis regions (Crumpler et al., 1993). A concentration of fracture belts is also found in this area, and these two facts have been interpreted as evidence of mantle upwelling. An annulus of interconnected low plains with features indicative of crustal shortening—such as ridge belts, mountain belts, and ridged plains—surrounds this area and is seen as a possible site of mantle return flow and downwelling. Most of the recent volcanism on Venus seems to be concentrated in the Beta-Atla-Themis area, and most of the recent tectonism appears to be associated with the equatorial highlands. This impression is strengthened by the finding that a region comprising approximately 33% of the planet's surface within 30 degrees of the equator, and bounded by longitudes 60° E and 300° E, contains twice as many heavily fractured craters and 1.4 times as many lava-embayed craters as the planetary average (Strom et al., 1994).

Among the unusual volcano-tectonic structures found on Venus are coronae, repre-sented for example by the ring structure Nightingale, located on the Atalanta Planitia (see Fig. 3.20). This is a massive ring wall; a non-continuous trench stretches along its outside flank. Its ridge is intersected by a series of short transverse fissures. In places it is covered by gently sloping volcanoes and lava flows, evidently flowing down from the ridge wall onto the adjacent lowlands. On the inner slope in the northern region of the corona, traces

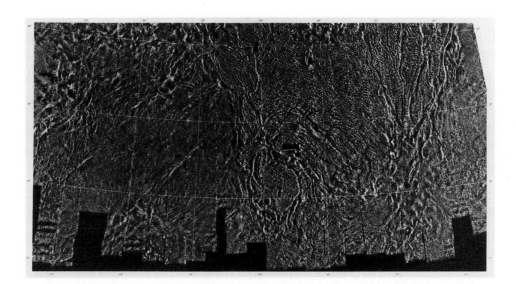

Fig. 3.17. A large-scale image of powerful crustal deformations, in the form of diagonally, and orthogonally intersecting structures (parquet). Traces of crustal extension, as a consequence of internal stresses, are visible (photographic map constructed from Venera 15–16 radar data).

of material creep and the formation of chaotic, tessera-like relief are visible. The inner platform of the whole formation has only partially subsided below the surrounding terrain; the remainder of the platform is located at the level of the surrounding terrain. It is separated from the inner slopes by steep escarpments and thus could have subsided from an original, higher position.

It was originally proposed (Barsukov et al., 1985a, 1985b; Nikolaeva et al., 1986) that coronae were formed during the final stage of intense meteoritic bombardment soon after the completion of planetary accretion. Obvious traces of impacts, however, have not been preserved. Moreover, the structure of the coronae differs strongly from similar ring, basin-type formations on the Moon, Mercury, and Mars (Kropotkin, 1989); Venus' coronae have more similarities with Precambrian ring structures on ancient platforms of Earth. It was concluded, on the basis of Venera data, that these structures are more likely to be older tectono-magmatic structures, resulting from the formation of regional tectonic-volcanic complexes.

Coronae are unique to Venus, and they are quite common; more than 360 have been recorded (Stofan et al., 1992). These average 250 km in diameter (Head et al., 1992). Coronae are defined by a concentric structure consisting of an annulus of ridges or fractures, an interior that may be topographically positive or negative, and a peripheral moat or trough. Volcanic and tectonic landforms in the interior are quite common (Fig. 3.28). Detailed geological analyses of these features using Magellan images have revealed that radial fracturing is the first stage in corona development. This evidence is consistent with domical uplift caused by a rising diapir. Later stages of corona development may be due to radial spreading of the diapir followed by cooling and gravitational relaxation. Volcanism may accompany all these stages (Squyres et al., 1992a; Janes et al., 1992).

About 150 impact craters were identified on Venus from Venera 15–16 radar imagery (Basilevsky et al., 1987; Ivanov et al., 1986). Many features that were previously identified as candidate impact structures on the basis of Earth-based images from Arecibo were not

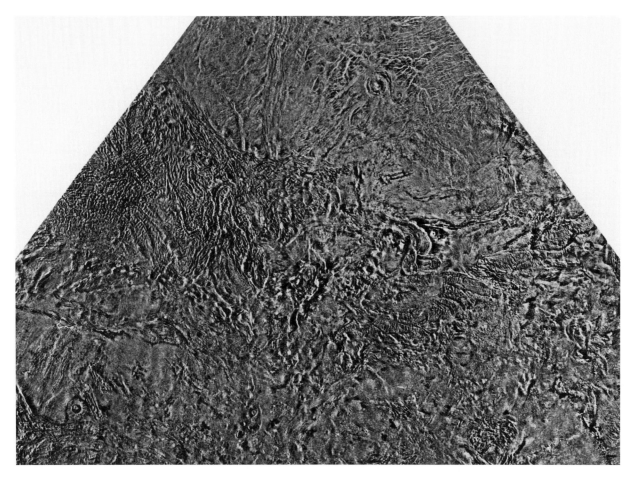

Fig. 3.18. The Tethus Regio. This morphologically complex surface is representative of different types of relief forms. To the left are diagonally intersecting ridges, or tessera (image from Venera 15–16 radar data).

recognized as such in the more detailed Venera images. From this Venera data, Ivanov and Basilevsky (1987) estimated an average surface age of 300 to 900 million years.

To first order, impact craters appear to be uniformly distributed across the surface of the planet. On the basis of Magellan mapping, roughly 900 impact craters have been identified on the planet's surface (Schaber et al., 1992; Strom et al., 1994). This number indicates an average surface age of about 500 million years. The size-frequency distribution of craters larger than 35 km in diameter resembles that found on other young surfaces of the inner solar system (such as lunar maria or the Martian plains). At smaller diameters the numbers fall off rapidly owing to atmospheric filtering, and few craters are smaller than 3 km (Zahnle, 1992; Herrick and Phillips, 1994). Although large craters have the familiar progression of morphologies with diameter, from bowl-shaped to complex, small craters tend to be irregular or multiple, indicating atmospheric disruption of impacting bodies. Approximately 400 dark splotches, interpreted to be the signatures of small bodies that disrupted before impact, have been identified (Fig. 3.29; see Zahnle, 1992; Strom et al., 1994). Some apparently young Venusian craters are surrounded by radar-bright or -dark parabolic arcs opening toward the west, which have apparently formed from the fallout of ejecta entrained in the globally superrotating winds (Campbell et al., 1992). Another feature that distinguishes many craters on Venus is extensive, apparently low-viscosity fluid outflows originating in or under the continuous ejecta (Fig. 3.30). These have been

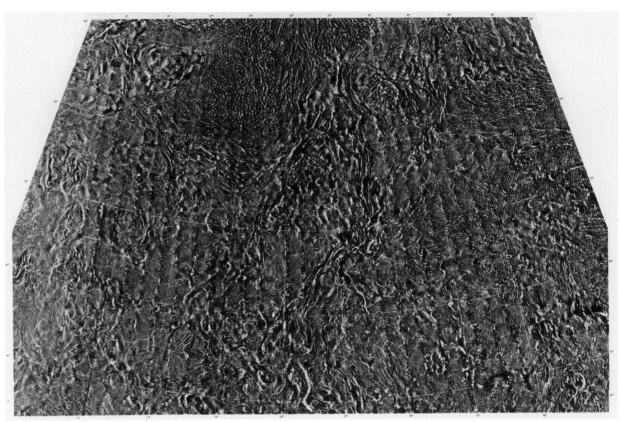

Fig. 3.19. The Sedna Planitia, located between Ishtar Terra and Laima Regio. In the lowland are numerous traces of geological activity in the form of folds, extended belts (left and right of center), and tessera (in the upper-left and lower-right sections of the image) (photographic map constructed from Venera 15–16 radar data).

interpreted as being indicative of large volumes of impact melt created by impact into the hot surface rocks, and also as being affected by interactions of the thick atmosphere with fluidized debris, vaporized material, or melt (Asimow and Wood, 1992; Ivanov et al., 1992).

The crater population on Venus is unusual in that a large number (84%) of craters appear to be pristine, that is, unmodified by volcanic or tectonic processes (Schaber et al., 1992; Phillips et al., 1992; Strom et al., 1994). Estimates of the number of craters that have been embayed by volcanism (see Fig. 3.16) range from 2.5% to 7% depending, at least in part, on slightly different definitions of embayment. Approximately 12% of the craters appear to be fractured (Fig. 3.31). The highly preserved state of most of the craters is surprising on a planetary surface dominated by volcanism and rife with tectonic features, if one starts with a uniformitarian assumption. Clearly the formation of most craters postdates the bulk of volcanic and tectonic activity. Although exotic hypotheses, such as tidal disruption of a Venusian moon, have been invoked to explain a relatively recent surge of impacts on Venus (Bills, 1992), a more likely explanation is that the rates of resurfacing have been highly variable over the lifetime of the visible surface. The extreme version of this hypothesis, in which all previous craters were removed by a massive and short-lived resurfacing episode, has been termed catastrophic resurfacing. An alternative hypothesis is that secular cooling of the interior, combined with the exponential dependence of strain rate on temperature, led to a rapid cessation of tectonism roughly 500 million years ago, resulting in the pristine state of many craters (Solomon, 1993). It is clear that resurfacing rates have

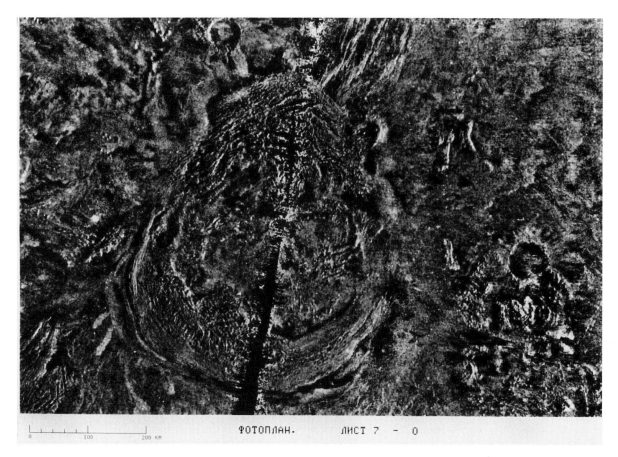

Fig. 3.20. The Nightingale Ovoid on the Atalanta Planitia, with a maximum dimension of 600 km. Numerous short transverse grooves are clearly visible, which intersect the massive ring wall along the contour of the ovoid. On the inner slope in the northern section, traces of material creep and the formation of chaotic relief are evident (image from Venera 15–16 radar data).

declined over the age of the visible surface, yet there is abundant evidence for volcanic and tectonic activity that is much more recent than the average surface age. Monte Carlo simulations have been used to constrain the resurfacing rate, since the proposed global resurfacing, that would result in the observed number of modified craters (Bullock et al., 1993; Strom et al., 1994). Results range from 0.01 to 0.4 km^3 per year. Although the spatial distribution of craters, based on crater data alone, is indistinguishable from a random distribution (Phillips et al., 1992), when combined with geological data areas of low crater density can be seen to correlate with concentrations of volcanoes, coronae, rifts, and lava flow fields (Fig. 3.32; see Namiki and Solomon, 1994; Price and Suppe, 1994). Thus, activity in these areas apparently postdates the plains formation epoch, which created most of the current surface and is most likely continuing at present.

The Global Tectonics of Venus

The discovery, made with Venera and Pioneer-Venus data, of the extensive smooth and rolling lowlands, probably formed by outflows of liquid basalts, led to the hypothesis that a mechanism of heat transfer from the interior by "scattered volcanism," that is, by several thousand small volcanoes (Morgan and Phillips, 1983), could be dominant on Venus. Competing theories relied on limited evidence of the presence of a mechanism similar to plate tectonics. Indeed, a study of the topography and morphology of Aphrodite Terra, es-

Fig. 3.21. The Metida
Regio, with complex
patterns of geological structure including arachnoids
and coronae (image from
Venera 15–16 radar data).

pecially its western section (Crumpler and Head, 1987; Head and Crumpler, 1987),
identified numerous features similar to those characteristic of diverging lithospheric plate
boundaries in the mid-oceanic ridges on Earth (Fig. 3.33a). This study was based on
analysis of Pioneer-Venus Orbiter data and radar profiling of equatorial latitudes from the
Arecibo Observatory. The presence of mutually symmetric topographic elements (oriented from northwest to southeast) along both sides of the central uplift within individual
blocks, extending 500–1000 km, pointed in favor of these ideas (Fig. 3.33b). It was noted
that in the orthogonal direction these blocks are separated from one another by zones

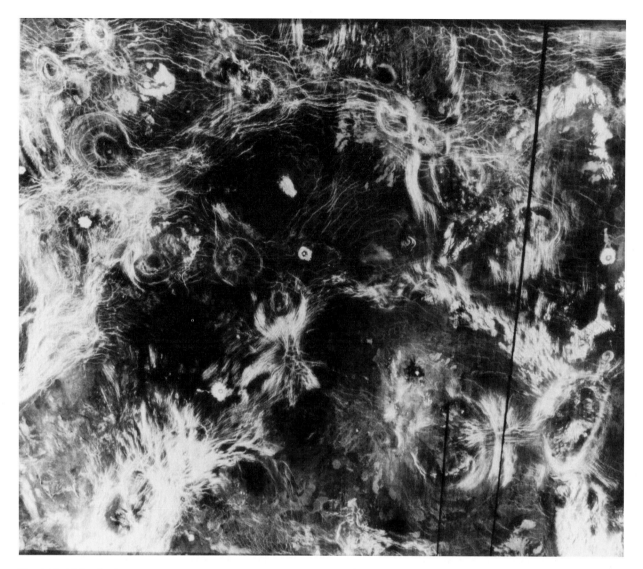

Fig. 3.22. A Magellan image of an area in Bereghinya Planitia, 1843 km × 1613 km, centered at 45° N, 11° E. The area displays a wide variety of volcanic features, the most prominent of which are arachnoids: structures formed by concentric ridges and fractures, cross-cut by radial fractures that may connect a string of these features. Arachnoids range in size from 60 km to 250 km.

whose morphology resembles zones of transform faults and normal faults, and it was inferred that these topographical elements (linear cross-strike discontinuities) emerged at the crests of the ridge, in whose central area a rift-type ravine is located, and gradually moved apart to either side owing to plate separation. At the same time it was assumed that magma from the interior entered this central section and the temperature of the thermal boundary layer dropped as it receded from the ridge axis (see Figs. 3.33a and 3.33b).

Similar morphological structures were identified through analysis of images of Venus' northern hemisphere from Venera-15 and Venera-16 data (Sukhanov and Pronin, 1988; Markov et al., 1989). On Ishtar Terra, a series of large kilometer-deep grabens whose floors are filled with lava was clearly seen, along with flexures hundreds of kilometers wide. On the axis of one such flexure a central ridge with large lengthwise fissures was observed. In addition, there are systems of grooved-layered bands. These are confined to the primary "continental" uplift and run from it to the southeast, onto the vast lava lowlands of "oceanic" size. On the anticlines of such bands, fractures, grabens, depressions, and nar-

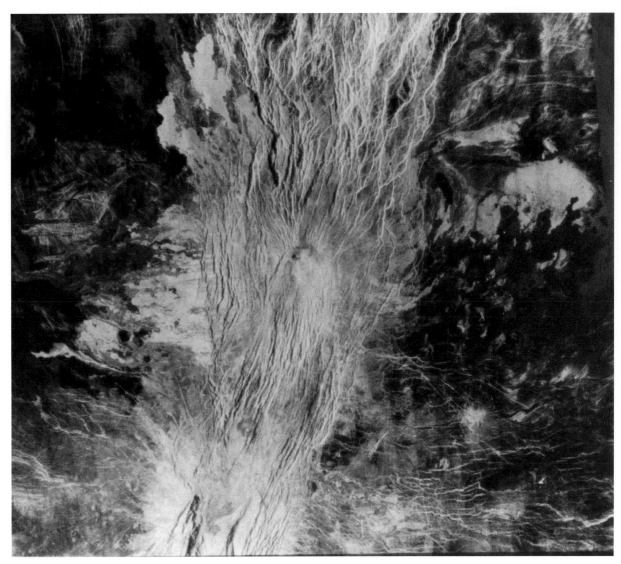

Fig. 3.23. A Magellan image centered at 288° E on the equator, showing the complex deformation associated with faulting in Devana Chasma. The zone of deformation ranges from 100 km to 180 km in width, and the faults seen here have a relatively uniform spacing of about 5 km. A group of bright lava flows, 70 km × 280 km, can be seen on the raised west flanks of the Devana Chasma rift valley.

row, long mountain chains are visible; shield volcanoes are found in the expansions of the bands (see Fig. 3.24). From the bands to the lowlands, the mountain chains form horse-tail-type structures, cross and polygonal systems, and they were interpreted as surface manifestations of parallel dikes and linear extrusions. The morphological sequence of (1) grabens, (2) large grabens with central ridges, and (3) large flexures with systems of subparallel grooved-layered bands was suggested as manifesting a possible linkage with spreading.

In the central part of Beta Regio, Venera 15–16 found topographical elements close in morphology to those found in the Aphrodite Terra, which were interpreted as rift zones and their associated volcanoes (Head and Crumpler, 1987). This young and most active region is, as was recognized on the basis of Venera mapping, a promising candidate for a hot spot.

An attempt was made to trace the fan of perceived spreading and rifting structures and to determine an overall symmetry, in order to distinguish the primary spreading axis

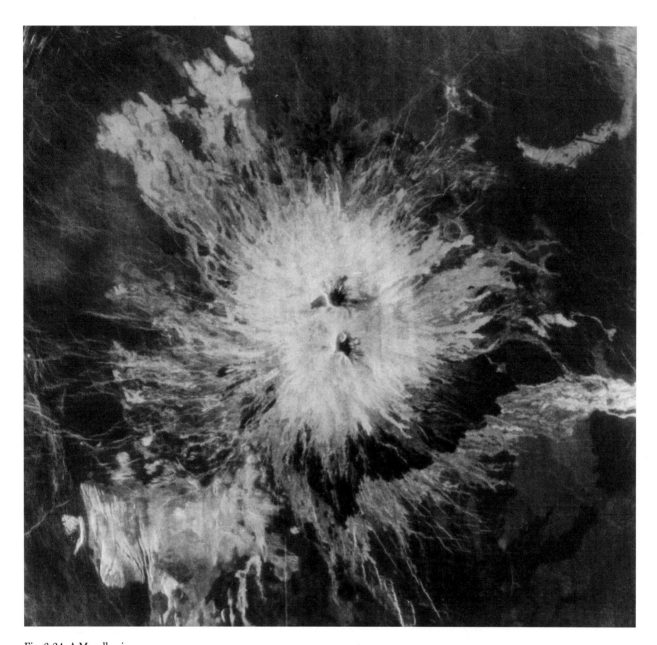

Fig. 3.24. A Magellan image of Sapas Mons, a shield volcano located on Atla Regio. The volcano is about 400 km across and 1.5 km high, and it is composed of numerous overlapping lava flows. Many of the flows appear to have erupted along the flanks of the volcano, rather than from the summit, as is common on the Hawaiian volcanoes on Earth.

by associating it with other, less distinct formations in the northern hemisphere. The proposed position of this spreading axis is shown on the schematic map (Fig. 3.34).

If one accepts that spreading zones are located near the equator, it would be natural to expect convergence zones at high latitudes. In other words, it would be important to detect evidence of the presence of characteristic zones of plate convergence and their associated regions of compressive deformation. It is obvious that for the scheme in Fig. 3.33, the situation would have a more complex nature but would be approximately symmetric relative to the main axis. However, no structures that could be associated with trenches and subduction zones were found. A proposed explanation of this peculiar phenomenon was that, unlike on Earth, the high-temperature lithosphere of Venus is lighter than the asthenosphere and by virtue of its buoyancy cannot be deeply submerged. In other words,

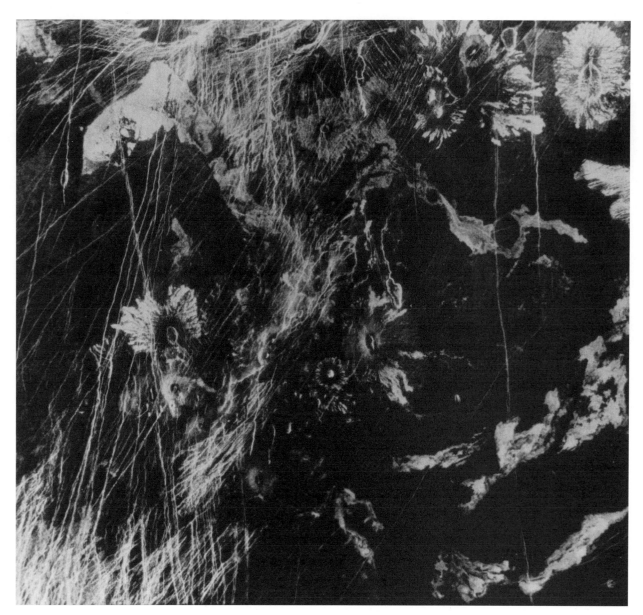

Fig. 3.25. A Magellan image from Atla Regio showing several types of volcanic features transected by surface fractures. The region shown is about 350 km across and is centered at 9° S, 199° E. The lava flows emanating from the circular pits or linear fissures form flower-shaped patterns and are known as anemone.

we are considering the tectonics of thin and light basaltic plates that move like ice floes and that, upon collision, do not plunge one below the other but crumble and form folded patterns.

Arguments concerning the presence of spreading as a manifestation of plate tectonics on Venus supported the idea that heat loss from the interior was not due to scattered volcanism but was more likely to occur through the creation and separation of plates as the primary mechanism of heat transfer (Solomon and Head, 1982). It was also noted that the positive correlation found between the topography on Aphrodite Terra and the gravitational field (Sjogren et al., 1980) could point to a more direct connection on Venus than on Earth between large-scale convection in the mantle and the divergence of plates. At the same time, the data available were insufficient to provide definite conclusions about the similarity of the lithosphere of Venus and Earth, that is, about the presence of segmentation

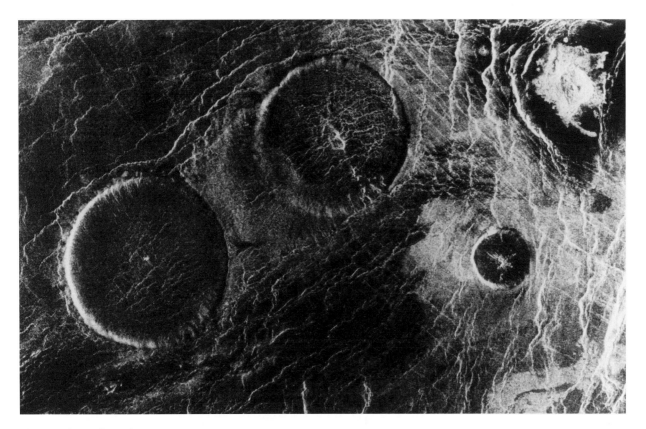

Fig. 3.26. A Magellan radar mosaic centered at 12.3° N, 8.3° E, showing a section of Eistla Regio 160 km × 250 km in area. The prominent circular features are known as pancake domes; they are about 65 km in diameter and less than 1 km high.

and the recycling processes of plate movements. To accomplish this, detailed imaging of the planet's surface by the Magellan project was required.

Let us now summarize some main findings and ideas about global tectonic style from the eve of Magellan mapping of the planet. We had learned that the surface of Venus is divided into several types of landscape: quiescent regions not subjected to much observable deformation; regions of uplift with what appeared to be mutually symmetric topographical elements, possibly associated with spreading centers; and highly deformed stress bands, also possibly linked with spreading. The data available at that time gave some support to a hypothesis about the existence on Venus of a primary spreading axis on a planetary scale. Hence a definite analogy with Earth was perceived, while the absence of trenches, coupled with the presence of compression systems, testified that the movement of lithospheric plates occurs without subduction.

Yet, it was also argued that global tectonics on Venus were more fundamentally different from terrestrial plate tectonics than was indicated just by the lack of trenches. Attempts were made to link the possibly large-scale horizontal tectonic motions, in the absence of recycling processes, with heat loss due to thermal anomalies and hot spots. This description of Venus resembles that of a one-plate planet such as Mars or the Moon, although Venus showed considerably more recent activity than is found on these worlds (Solomon and Head, 1982). It was unknown whether the intriguing geological activity observed had ceased on Venus in its relatively recent past or was continuing in the present. One could only maintain that Venus is undoubtedly more similar to Earth in its variety of complex

Fig. 3.27. A Magellan image at 49° S, 273° E, of an area 130 km × 190 km. Running through the image is a 200 km segment of a sinuous channel on Venus. The channel is approximately 2 km wide.

geological processes than is Mars (to say nothing of the Moon and Mercury), whose major tectonic activity was completed during the formation stage of the giant shield volcanoes and canyons, no less than 1.5 billion years ago, with some minor, highly localized volcanism continuing to more recent times.

Some of these perceptions of the Venusian relief and their association with the underlying geological hypotheses remained essentially unchanged by the enormous bulk of new

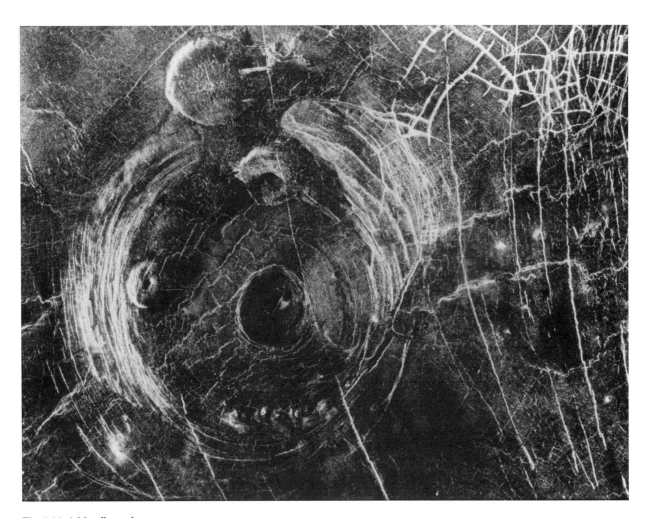

Fig. 3.28. A Magellan radar image of a region 300 km across, located in the vast plain to the south of Aphrodite Terra. The image is centered on 59° S, 164° E, and prominently displays the circular fractures, 200 km in diameter, that form Aine Corona. Pancake domes are found just to the north and interior to the corona annulus.

data provided by the Magellan mission. However, the much higher resolution Magellan images made it possible to clarify substantially these geological concepts and to reject some ideas. Mapping of nearly the whole globe has promoted a more coherent understanding of the overall tectonic and volcanic processes occurring on the planet. Although many first-order questions still remain as to the overall tectonic style and history of Venus, we are now much more confident of our answers than was possible before the Magellan mission.

Detailed mapping, based on Magellan images, of the equatorial highlands of Venus shows that previously inferred similarities with terrestrial spreading centers were misleading. Aphrodite Terra, the largest of these highland areas, is actually a complex structure with wide variations along its length. Aphrodite is built largely from a series of quasi-circular volcanic rises and large massifs composed of tessera. There is no compelling evidence for a history of axial spreading. Gravity data indicate that this area is compensated at great depth and is perhaps dynamically supported. This evidence has led to the idea that Aphrodite may have been created by a series of hot spots, although the possibility that it is a site of mantle convergence and downwelling has not been ruled out. Magellan data have generally supported the idea that other large volcanically dominated rises, such as Beta Regio, are likely to be created by hot spots. An analysis by Smrekar (1994) of Magellan

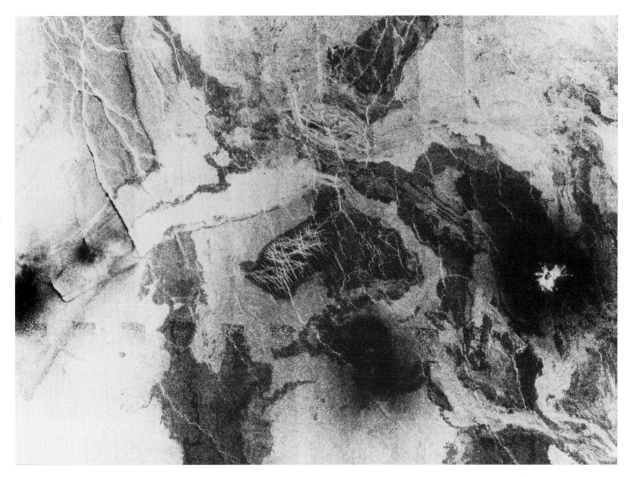

Fig. 3.29. A Magellan image of an area in the eastern Lakshmi region of Venus, centered at 55° N, 348.5° E.
The image covers an area 300 km × 230 km and shows a flat plains region composed of many lava flows.
Superimposed on these flows are three dark splotches, which are thought to be the signature of impacting
bodies too small to penetrate the thick atmosphere of Venus intact. The splotch to the far right (east) has
a central crater, indicating that the impactor was not completely destroyed before reaching the surface. The
dark splotch on the far left has been modified by wind.

line-of-sight gravity and topographic data for four large volcanic swells (Beta, Atla, Bell,
and Western Eistla Regiones) found that active mantle plumes were the most likely source
of support for these features. An analysis of spherical harmonic representations of the gravity field (Phillips, 1994) led to the same conclusion.

Another observation has further dispelled the notion of a simple analogy between
global tectonics on Venus and terrestrial-style plate tectonics: the global distribution of
both volcanic constructs and features indicative of regional tectonic deformation reveal a
pattern of volcanic and tectonic activity distributed over broad zones (Crumpler et al.,
1993) rather than the terrestrial pattern of concentrated volcanism and tectonic activity
along the margins of rigid lithospheric plates. This observation has led to a tentative picture of a global tectonic system on Venus containing a large number of less rigid lithospheric plates than on Earth. Whether this is considered to be a different style of plate

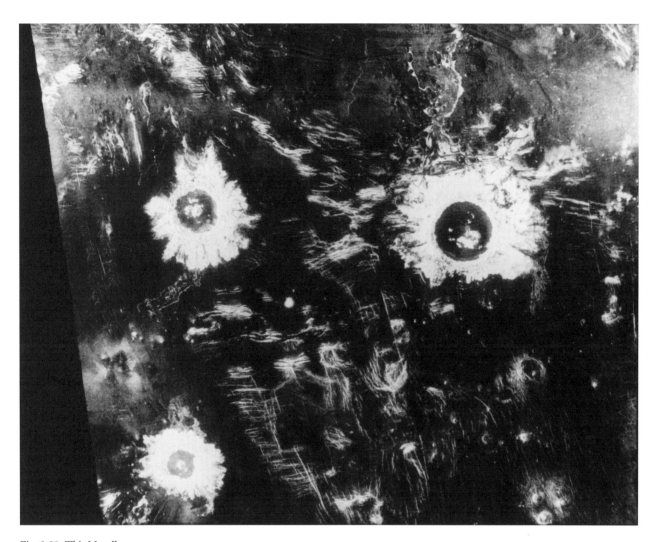

Fig. 3.30. This Magellan mosaic shows a fractured plains region on Venus, with three large impact craters whose diameters range from 37 km to 50 km. The rough ejecta blankets appear radar-bright, whereas the crater interiors are dark and smooth, having been flooded with lava. Each crater has a bright central peak.

tectonics or simply the absence of plate tectonics may be a task of definition that was not required of planetary scientists prior to the Magellan mission.

Another major complication hinders definition of the global tectonic style of Venus. Compelling evidence suggests that the style and rates of tectonism and volcanism have changed radically over the age of the surface. Thus the uniformitarian assumption, implicit in many early papers on Venusian tectonics, that processes operating during the present epoch have persisted for most of the history of the planet, is clearly, to some extent, false. The distribution, numbers, and modification spectrum of impact craters imply that resurfacing rates have declined radically since the episode, or epoch, of flooding that created most of the globally distributed plains. Volcanism has clearly continued since that time but at a much lower rate, in styles more favorable to volcanic constructs than plains formation and with a much more limited geographical distribution, concentrated in the Beta-Atla-Themis region (Crumpler et al., 1993; Head et al., 1992; Bullock et al., 1993).

Several hypotheses have been advanced to explain this radical change (from a terrestrial viewpoint) in resurfacing style. One idea is that because of the high surface temperature, the thermal evolution of Venus has been more rapid than that of Earth and that an

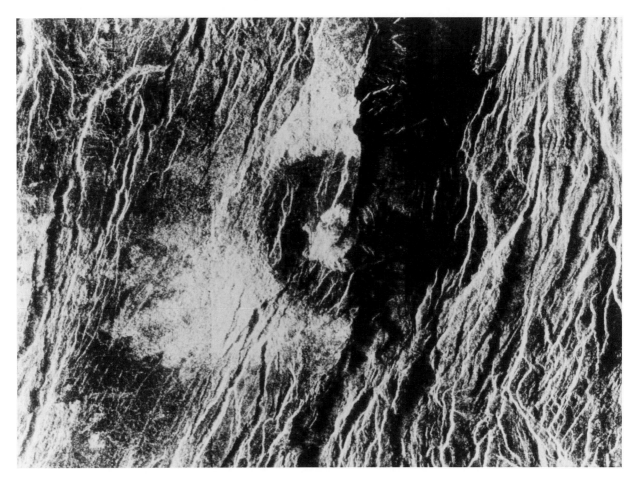

Fig. 3.31. This image shows the split crater called Somerville, located in the rift between Rhea and Theia Montes in Beta Regio. Somerville, about 37 km in diameter, is cut by many fractures, the most notable of which is the fault valley 20 km wide that partially destroyed the eastern portion of the crater.

earlier epoch of higher heat flow (and associated resurfacing rates) ended within the past billion years, as the interior cooled to a point where this heat flow could no longer be supported (Arkani-Hamed and Toksoz, 1984; Arkani-Hamed et al., 1993). Other hypotheses propose that the change in tectonic-volcanic style could be cyclic or episodic in nature. Parmentier and Hess (1992) have proposed that cyclic crustal overturn could be an inevitable consequence of chemical differentiation in a one-plate planet. In this scenario, new crust created by partial melting is at first supported by compositional buoyancy. This layer thickens and cools, creating negative thermal buoyancy. On Earth, subduction and mixing into the mantle prevents a large negative thermal buoyancy from developing. On Venus, however, in the absence of subduction, thermal buoyancy may grow until it exceeds the compositional buoyancy, leading to a dramatic period of crustal overturn with global resurfacing occurring. Model results show that this process may recur with a timescale of 300–500 million years.

Turcotte (1993) has proposed that episodic plate tectonics may be operating on Venus. In this model, episodes of rapid plate tectonics are separated by periods of relative surface quiescence, such as the present epoch. Turcotte suggests that for the past 500 ± 200 million years (the age of the visible surface) the lithosphere of Venus has been cooling and thickening as a one-plate planet. This theory is supported by abundant evidence for a thick lithosphere, provided by gravity data and by analyses of flexure at coronae. Such a

Fig. 3.32. Maps showing the association of impact crater density with geologic features on Venus. (a) A smoothed map of crater density (10^6 km^2) and crater locations. (b) A map of tectonic and volcanic units including 38×10^6 km^2 of tessera, 150,000 linear km of rifts, 364 coronae, 130 arachnoids, 46 novae, 123 calderas, 128 large volcanoes, and 48 flood-type lava flow fields. The largest light-gray area contains mostly radar-dark plains consisting of overlapping volcanic flows inferred to be products of a global resurfacing. (c) A combined map of crater density and geologic features. Low-density areas correlate with concentrations of rifts, coronae, volcanoes, and lava flow fields. This correlation indicates that the craters are not spatially random with respect to geology and suggests that volcanism and tectonism since the global resurfacing have reduced the crater density (according to Price and Suppe, 1994; courtesy of Maribeth Price).

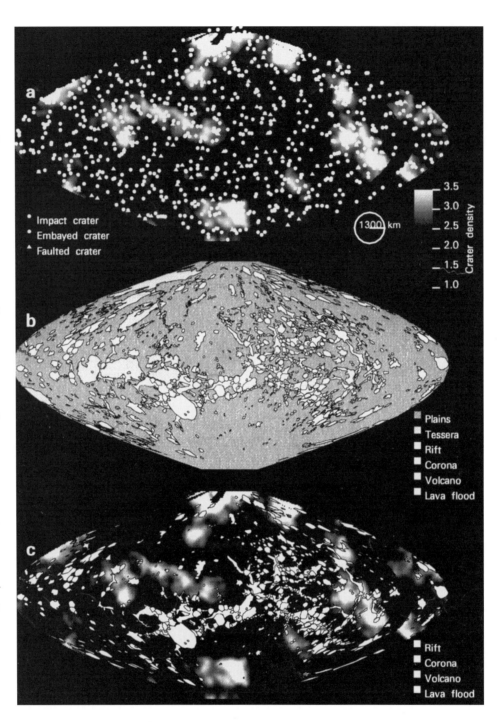

lithosphere, however, would have a thickness of approximately 300 km. This is an order of magnitude greater than the thickness that one derives by assuming steady-state conductive heat loss. Thus, the heat flow through the current thick lithosphere may be insufficient to balance interior heat generation, causing the interior to heat up. In Turcotte's model, periods like this are punctuated by brief periods of crustal overturn with very high heat flow, during which rapid global tectonics, perhaps not unlike terrestrial plate tectonics, occurs. The apparent largely uniform age of most of the surface may reflect the time since the most recent of these events. These models present us with the intriguing possibility that 500 mil-

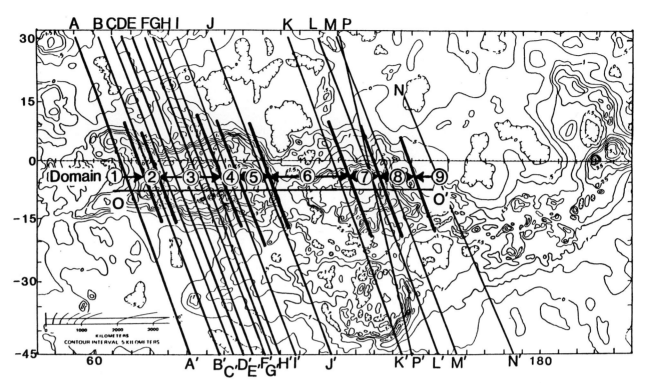

Fig. 3.33a. A map of Venus' near-equatorial region, with altitude contours at 500 m intervals, constructed from Pioneer-Venus and Arecibo Observatory radar measurements. The solid straight lines, oriented approximately from northwest to southeast and denoted A–A', B–B', ... N–N', mark the positions of transverse valleys in the relief of Aphrodite Terra, along which a break in the continuity of topographic elements (linear cross-strike discontinuities) is observed; the corresponding regions between these lines (domains) are denoted by numbers in circles and are separated by boldface lines (according to Crumpler and Head, 1988).

lion years ago, while Earth was experiencing its Cambrian explosion, the surface of Venus was dramatically different from its present state.

A Close-Up View of the Surface from Venera Landers

Hypotheses about volcanic processes on Venus and possible mobility of the Venusian crust were put forward as early as 1975, after the first television panoramas of the surface were transmitted by the Venera-9 and Venera-10 landers. They enabled us to see directly, for the first time, the landscape of our neighboring planet (Keldysh, 1979).

To transmit the television images, two cameras with a 37° × 180° field of view were installed on each station (Fig. 3.35). With the cameras 0.9 m above the surface, nearly full circular coverage of the landing area was assured. Because the panning axis is tilted by 50° from the vertical, the transmission of the landscape and the resolution of details on the terrain are not uniform: the middle portion of the panorama contains a surface image directly in front of the camera, whereas the outer portions are images of the most distant areas up to the horizon, seen as an inclined line with a fragment of the sky above it. Unfortunately, only one panorama from one side of each of the Venera-9 and Venera-10 landers was received (Fig. 3.36). The photographic program was accomplished in full by the Venera-13 and Venera-14 spacecraft, that is, a circular survey was made and the quality of the transmitted images was considerably improved. By registering images obtained through blue, green, and red filters, color panoramas were synthesized—one complete (180°) and one within a 60° angle on opposite sides of each lander (Selivanov et al., 1976, 1983a, 1983b). The lander's circular support, with a notched framework in the field of view of the Venera-13

Fig. 3.33b. Altitude profiles along the transverse faults C–C′ . . . H–H′, shown on the map in Fig. 3.33a. The vertical (dotted-dashed) line denotes the position of a probable axis of symmetry (O–O′ in Fig. 3.33a), relative to which the similarity in the relief elements is evident. Such an axis is clearly evident at distances of about 3000 km along both sides of O–O′ (denoted by the vertical dotted lines). The altitude scale is indicated to the left. The discovery of such symmetry infers the presence of spreading in the direction orthogonal to the O–O′ axis (according to Crumpler and Head, 1988).

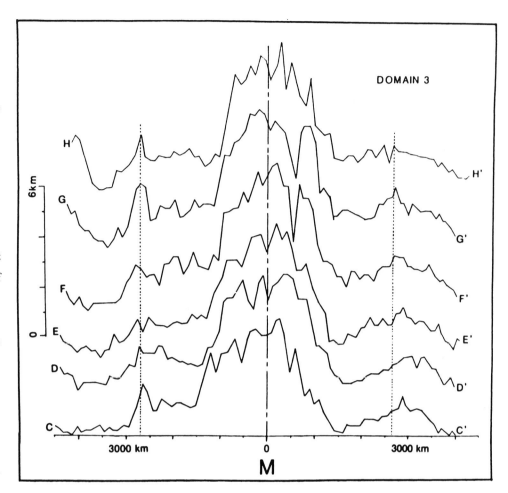

and Venera-14 cameras, was intended to increase its stability without a parachute during the descent phase (the distance between the notches is 5 cm). The same configuration but without notches is seen in the frame of panoramas gained by Venera-9 and Venera-10. The latticework of an instrument for determining the physical-chemical properties of the soil (60 cm in length), and the discarded semi-cylindrical covers of the camera illuminators, 20 cm in diameter and 12 cm high, are also visible. These numbers give an idea of the size scales on the surface. The exact values and relative position of details were determined by photogrammetric analysis. On the Venera-13 and Venera-14 landers a color test band was also used that can be seen to the right on the panoramas.

The Venera-9 landing site is located at the northeast end of Beta Regio, known for its high radiowave reflection (Fig. 3.37). In the panorama transmitted from this region, large, sharp-edged stones fill about half the surface. The width of the largest rocks reaches 50–70 cm, whereas the height is quite small, 15–20 cm. A plate-like form and the presence of tapered broken-off fragments are characteristic here; they are also clearly visible in the Venera-10 panorama, for example, in a segment in the lower portion of the photograph on three flat rocks to the left of the spacecraft's landing ring (see Fig. 3.36). The space between the stones is filled with comparatively light material, apparently fine-grained soil, which formed during disaggregation and deformation of the surface rocks and was probably

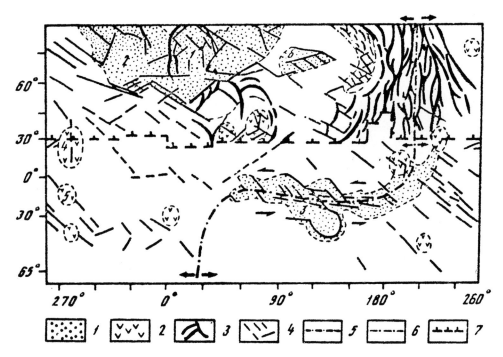

Fig. 3.34. A schematic map of Venus' surface, indicating the location of regions with various geological structures. Key to symbols: (1) highlands with broken relief, (2) volcanic-tectonic uplifts, (3) belts of mountain chains and grooves, (4) structural lines, (5) a spreading axis, (6) a regional boundary of distributed spreading, and (7) the photographic boundary of the Venera 15–16 spacecraft. Numerals on the map: (1) Ishtar Terra, (2) Lakshmi Planum, (3) Aphrodite Terra, (4) Beta Regio, (5) Phoebe Regio, and (6) Tethus Regio. The assumed spreading axis is indicated by the dotted-dashed line (according to Sukhanov and Pronin, 1988).

transported by winds. This fragmented material covering the hard bedrock can perhaps be likened to the lunar regolith. On individual rocks, dark spots are noticeable; these are possibly the traces of erosion. Magellan images show that, like the Venera-9 landing area, this part of Beta Regio is composed of tessera and plains that embay the tessera. Wrinkle-ridges are not observed here, however, and a fine-scale fracture system is seen to cover much of the plains.

Venera-10 landed in an area located to the southeast of Beta Regio and separated from the Venera-9 site by about 2000 km—a flat, rocky desert without any appreciable altitude changes (Fig. 3.38). The large stone block on which the spacecraft landed, no less than 3 m in diameter, is speckled with dark spots, similar to those seen in the Venera-9 panoramas, which are small cavities filled with soil. The block itself, like similar blocks far from the lander, is embedded in darker soil. The whole landscape is apparently magmatic bedrock that has undergone significant modification. In the central portion of the block, obvious traces of a cleft and a series of additional fissures at the edges are visible. Their formation may be associated with internal processes that occurred on Venus, as well as with the impact of the lander on the surface; it is impossible to establish the real cause. The differences in the microrelief of the surface (the presence of honeycombs, small ridges, and fissures on the blocks) and the degree of filling of individual irregularities by the eroded material reflect compositional differences in the rocks, such as their differential resistance to such erosive processes as chemical weathering, which probably plays a prominent role. Magellan mapping of the Venera-10 landing site shows that the geology of the area includes tessera and mottled plains that embay the tessera (Weitz and Basilevsky, 1993). The plains in this area contain prominent wrinkle-ridges. The panorama suggests that the spacecraft landed on the plains and not the tessera.

Fig. 3.35. The television camera setup on the Venera 9–10 (and Venera 13–14) lander, showing its main components: (1) the television camera, (2) the lander's thermal insulation chamber, (3) the cylindrical illuminator, (4) the scanning mirror, (5) the objective, (6) a rotating mirror, (7) the stop aperture, (8) the light detector, (9) the lander support ring with a color reference pattern, and (10) the artificial light sources (absent on Venera 13–14).

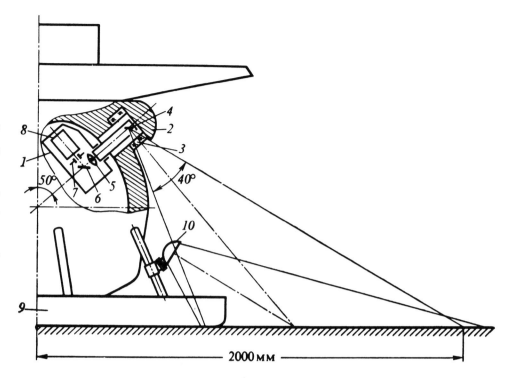

The origin of the landscape in the Venera-9 panorama (see Fig. 3.36) was most likely the destruction of rocks in cliffs due to internal movements and faults in the planet's crust. As a result, a rocky talus formed on the slope of the hill where the spacecraft landed. The grade of this slope is about 30°; the horizon, in contrast to the Venera-10 panorama, is located only several tens of meters away. The rocks are apparently quite solid; they have not undergone noticeable deterioration. One would think that this landscape is quite typical of many highland and lowland regions of Venus (Garvin et al., 1984) and was formed quite recently (on a geological timescale). It has also been suggested that there could be periodic movement of the rocks down the slope as a result of possible seismicity (Marov, 1978; Florensky et al., 1979; Ksanfomality, 1983a). An analysis of Magellan mapping of this area has led to the suggestion that the spacecraft landed on one of the slopes of the numerous fractures seen in the plains in this area.

The Venera-13 and Venera-14 landing sites are also located in the immediate vicinity of the mountainous plateau of Beta Regio, in Phoebe Regio where, based on the data of radar mapping, rolling hills and smooth lowlands dominate. Even a cursory survey of these panoramas (Figs. 3.39 and 3.40) will show that the two regions are similar and do not differ strongly in their morphology from the Venera-9 and Venera-10 panoramas. Clearly defined projections of hard bedrock in the Venera-13 panoramas, raised slightly above the surface, resemble the stone slabs in the Venera-10 panorama. The depressions between the projections are filled by a layer of unconsolidated soil, which causes these surface areas to appear darker. Horizontal bedding of the rock at the Venera-13 landing site, similar to that seen on chips of individual rocks of the stone talus in the Venera-9 panorama, is quite noticeable.

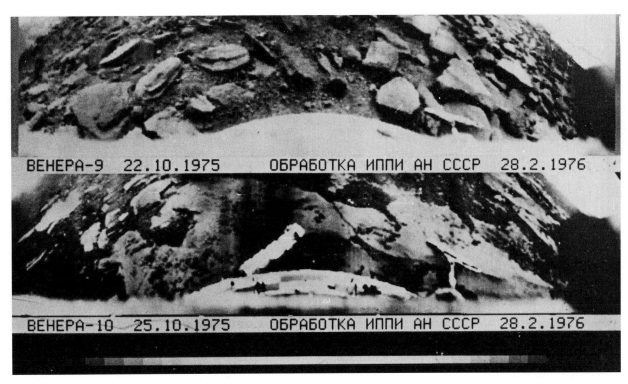

Even more distinctly visible is the multi-tiered, horizontal bedding in the Venera-14 panorama, where light and dark layers alternate in places and there is almost no unconsolidated soil. The thickness of individual layers does not exceed several centimeters, and their number reaches ten and more. The uniformity of the microrelief to a distance of tens of meters from the lander suggests cyclical deposition of material, differing in its chemical composition or granularity, causing the dissimilar reflective properties of the layers. In particular locations traces of deterioration (tapering) are perceptible, sometimes covered by older sediments. It has been hypothesized (Florensky et al., 1983a, 1983b) that such a surface texture is characteristic of sedimentary bedding. The sources of such sediments could be volcanic eruptions with subsequent deposition of ejecta (ash) onto the surface in a quiescent atmosphere, or stratification of magma flows. The second hypothesis seems less likely because of the thinness of individual layers. It is possible that the sedimentation products are tuffs, well known on Earth, which correspond to basaltic-type volcanic rocks in their composition. Although more specific thoughts about the origin and age of these rocks are not currently feasible, we have one more apparent confirmation that active geological processes have occurred on Venus in the not-too-distant past, at least within the past billion years.

Magellan images of the Venera-13 landing area show that it is dominated by radar-dark plains containing wrinkle-ridges (Fig. 3.41). These plains are darker than those at the other landing sites, which may be caused by smoothing by fine debris (Ivanov et al., 1992; Weitz and Basilevsky, 1993). These plains are cut by a northwest-southeast-trending lineament belt. The presence of wind streaks around craters in the area suggests a local origin

Fig. 3.36. The first panoramas of Venus' surface, transmitted from the Venera 9–10 landers, from areas located to the north and southeast of Beta Regio. The light segment at the bottom is the lander support ring; to the right of it is the probe for measuring the soil density. The illuminator cover, which shielded the television camera and was removed automatically after landing, falls in the field of view of the Venera-10 panorama (in the center). See the text for an explanation of landscape features.

Fig. 3.37. A Magellan image of the Venera-9 landing site. The image is centered at 31.01° N, 291.64° E, and is 538 km × 614 km.

for the debris found in the panoramas at this site. Both a steep-sided dome and a corona-like feature located to the southeast could be possible sources for the lavas composing the platy bedrock observed here.

The Venera-14 landing circle is centered on the eastern flank of a volcano 75 km in diameter with gentle slopes, and the landing site is dominated by lava flows from this volcano (Fig. 3.42).

Synthesized color images have led to the conclusion that the surface of Venus has a reddish-orange hue. Such coloring is caused by the incoming diffuse solar radiation whose spectral makeup is shifted toward the red because of the high absorption and scattering of blue rays by the thick Venusian atmosphere (Selivanov et al., 1983b; Pieters et al., 1986). Simulated color of some Magellan images approximates hues that might be seen by the human eye, based on color images recorded by the Venera-13 and Venera-14 spacecraft.

Fig. 3.38. A Magellan image of the Venera-10 landing site. The image is centered at 15.42° N, 291.51° E, and is 538 km × 614 km.

Composition and Characteristics of the Surface

For analysis of Venus' geological present and past, data on the chemical composition of the constituent surface rocks and their physical-mechanical properties are of great importance. The first such information became available with the beginning of ground-based radio measurements (see Kuzmin and Marov, 1974), which provided some limited information on the surface properties of the planet on a global scale. More complete information on several local regions was provided by the geochemical and physical experiments conducted on the Venera landers. These results enabled us to draw important conclusions about the degree of planetary differentiation and the efficiency of erosion processes, and to estimate the mineralogical composition of the surface. This information is important for understanding the conditions for the formation of geological structures, the nature of atmospheric interactions with the surface, and so on. Ultimately, along with the study of the surface morphology (including age estimates from the density of impact structures), data

Fig. 3.39. Panoramas transmitted from the Venera-13 landing site in Phoebe Regio (Navka Planitia). In the lower portion of each panorama, a notched segment of the support ring, with inscribed photometric marks, is visible. To the left is a deployed probe with a device at the end to measure the physical-mechanical properties of the soil; the white semicircle in the center is a cover, which shielded the television camera. The panoramas were taken through blue, green, and red filters and were synthesized color images. The band to the right of the support ring was a color test strip, with several standard colors; it was deployed after landing. See the text for an explanation of landscape features.

on chemical composition and physical properties provide an approach for answering fundamental questions about the role of exogenic and endogenic (magmatic) processes, and about the formation of the planet's surface.

Using the mean value of the permittivity of the Venusian surface $\varepsilon = 4.5$ (see Table 3.1), an estimate of the density, 2.5 g/cm^3, was obtained. This figure is considerably higher than for the Moon and individual regions of Mars and, evidently, suggests the absence of highly porous or finely granulated rock in the surface layer of Venus. A value of $\varepsilon = 4.5$ corresponds to dry silicate rock, like basalt or granite, whose density lies in the range of 2–3 g/cm^3 (Kuzmin and Marov, 1974). Direct measurements of the constituent surface rocks by the Venera landers confirmed these notions. Generally, the mean reflectivity of Venus' surface, at a level of $\sim$0.1, agrees well with the modeled reflection from a stony, electrically conducting surface with a small covering of fine-grained, lower-conducting material. In a number of the highland regions (such as Maxwell Montes in the western Ishtar Regio and Sif Mons in the western Eistla Regio), however, the reflectivity was considerably higher ($>$25), suggesting that the surface layer in these regions has an unusually high level of electrical conductivity.

Magellan radar experiments confirmed previous findings of extremely low values of radio-thermal emissivity. Some areas are as low as 0.3, and the global mean is 0.845 (Pettengill et al., 1992). This figure is consistent with dry basalts. Several highland regions show very low values of emissivity, corresponding to the high values of permittivity mentioned previously (Campbell, 1994). The large range in pressure and temperature over elevation differences encountered on Venus has led to several hypotheses that explain this behavior by a change in surface materials related to local ambient conditions. This change may be due to minerals composed of loaded dielectrics (Pettengill et al., 1988; Klose et al., 1992). Iron pyrite is an excellent candidate, owing to the correspondence of the equilib-

Fig. 3.40. Panoramas transmitted from the Venera-14 landing site in Phoebe Regio. See the caption to Fig. 3.39 for details.

rium point of its transition to magnetite with the temperatures and pressures found in the Venusian highlands (Fegley and Treiman, 1992). Other proposed explanations include kinetically inhibited destruction of perovskite at higher elevations (Fegley et al., 1993), vapor transport of metal halide or chalcogenide volatile phases from low to high altitudes (Brackett et al., 1995), the presence of small amounts of ferroelectric mineral phases in the highlands (Shepard et al., 1994; Arvidson et al., 1994), and a thin, high-altitude frosting of the element tellurium (Pettengill, 1995).

Experiments to constrain the composition of the surface rock were initiated using gamma-ray spectrometers, which measured the natural radioactivity of the rock at the landing site by recording the intensity of hard gamma radiation for several characteristic lines of uranium, thorium, and potassium. The radiation is caused by the radioactive decay of these elements (radionuclides), which are always present in the crust of a planet. The ratio of their intensities serves as a good indicator of the predominant type of rock making up the surface layer under study. This method was applied successfully for the first time on the Luna-10 artificial satellite, enabling us to deduce the basaltic nature of the lunar rock (Vinogradov et al., 1966). Using satellites is only possible, however, when the celestial body has essentially no atmosphere (as on the Moon) and the gamma radiation can reach the orbit of the satellite unimpeded. This technique can be also applied for a planet with relatively thin atmosphere, as on Mars. Such experiments were carried out from the Mars-3 and Mars-5 satellites (Vinogradov et al., 1975). For Venus the only way to conduct such measurements is to land on the surface. But the method has its advantages: although gamma radiation from natural radionuclides penetrates the shell of the lander and can be recorded by the internal scintillation gamma-ray spectrometer, the dense atmosphere completely shields primary and solar cosmic rays, which thus do not generate induced background noise on the recorded spectra. This makes interpretation easier.

The first such measurements were accomplished by the Venera-8 lander in the south-

Fig. 3.41. A Magellan image of the Venera-13 landing site. The image is centered at 7.55° S, 303.69° E, and is 538 km × 614 km.

ern hemisphere. Subsequently, they were carried out in the northern hemisphere, near Beta Regio, by the Venera-9 and Venera-10 spacecraft, and in the Rusalka Planitia plains in the foothills of Aphrodite Terra by the Vega-1 and Vega-2 spacecraft. The first experiments provided definite evidence that the surface rocks of Venus resemble various terrestrial basalts (Vinogradov et al., 1973; Surkov et al., 1976).

Sample spectra of the natural radioactivity of the rocks are shown in Fig. 3.43. The elemental content was determined using ground calibrations conducted under field conditions at rock outcrops in a natural bedding with known amounts of uranium, thorium, and potassium. The gamma-ray spectrometer was mounted on a mock-up of the lander so that the working conditions most closely approximated those on the surface of Venus. The results of all the measurements conducted by Venera spacecraft are summarized in Table 3.2.

The classification of Venusian rocks according to their petrochemical composition, based on data about the amount of natural radioactive elements, can be accomplished using the supplementary data in Table 3.3 (aside from terrestrial rocks, data for the Moon and

Mars are presented for comparison). For this purpose Surkov et al. (1987a) developed a special classification: various families of magmatic rock are distinguished on the main alkali and silica classification diagram (Fig. 3.44). Note that, taking into account errors in measuring the uranium, thorium, and potassium content along with a certain scatter in the amounts of these elements in the rocks corresponding to the selected petrochemical families in the diagram, the estimates obtained for the rock under study have a random character in any particular family. Nevertheless, such techniques have made it possible to narrow significantly the range of possibilities, even within classes of Venusian basaltic rock. It was concluded that the rocks at the Venera 9–10 and Vega 1–2 landing sites, which are fairly similar in their amount of natural radioactive elements, most probably correspond to terrestrial gabbros and tholeiite basalts. The rock at the Venera-8 landing site, which is enriched in potassium, is closest not to granites (as was assumed in the initial interpretation) but to alkali basalts (Surkov et al., 1987a). The relief patterns of the Venera-8 landing site as

Fig. 3.43. Spectra of natural gamma radiation from Venusian rocks, measured by the Vega 1–2 landers. The arrows, with the corresponding energy values (in megaelectronvolts, Mev), indicate the positions of characteristic spectral features of radioactive elements exhibiting the features, from which the uranium (U), thorium (Th), and potassium (K) contents are determined.

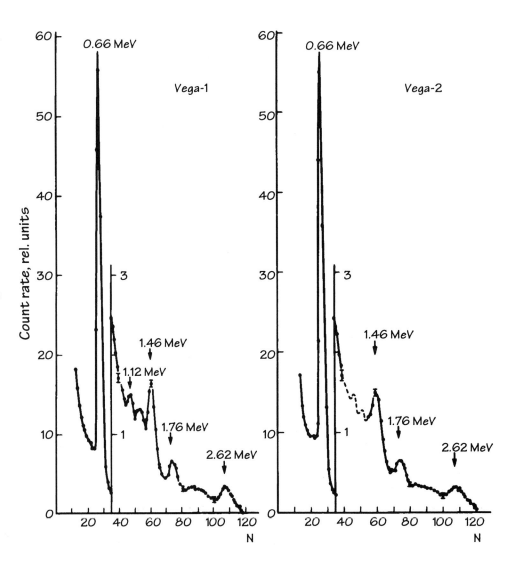

revealed by Magellan, and their possible correlation with this peculiar rock composition, were analyzed by Basilevsky et al. (1992).

   The most complete information concerning the kinds of rock that make up the surface of Venus has been provided by measurements of the elemental soil composition by the X-ray fluorescence technique on the Venera-13, Venera-14, and Vega-2 landers (Barsukov et al., 1984, 1986b; Surkov et al., 1984). The method is based on the relation between the intensity of the characteristic radiation excited by the radioisotope source and the amount of the analyzed element in the sample rock. To imagine the technical complexity behind such an experiment, it is sufficient to recall the surface conditions on Venus and to keep in mind that by the given method, soil analysis can be conducted only at room temperature and with pressure not exceeding 0.1 atm. Therefore, a special soil-intake system was created, which includes a miniature drilling unit and sluice compartment to convey the selected sample into the lander (Fig. 3.45; see Barmin and Shevchenko, 1983). The automated system then achieved a hermetic seal (separating the sample from the external environment), evacuated the chamber, and provided thermostatic control. The sample ma-

Table 3.2 Uranium, thorium, and potassium content in Venusian rocks, % by mass

Spacecraft	Year	K	$U \times 10^{-4}$	$Th \times 10^{-4}$
Venera-8	1972	4.0±1.2	2.2±0.7	6.5±0.2
Venera-9	1975	0.47±0.08	0.60±0.16	3.65±0.42
Venera-10	1975	0.30±0.16	0.46±0.26	0.70±0.34
Vega-1	1984	0.45±0.22	0.64±0.47	1.5±1.2
Vega-2	1984	0.40±0.20	0.68±0.38	2.0±1.0

Source: Data from Surkov et al., 1983, 1986.

Table 3.3 Thorium and uranium content in rocks on Earth, the Moon, and Mars

	Type of rock	Th, g/t[a]	U, g/t[a]	Th/U
Earth	Ultrabasic rocks	0.08	0.03	2.7
	Oceanic tholeiitic basalts	0.18	0.1	1.8
	Alkali olivine basalts	3.9	1.0	3.9
	Platform raft basalts	2.5	0.8	3.1
	Geosynclinal basalts	2.4	0.7	3.9
	Granites, granodiorites, granite-gneisses	15.6	3.9	4.0
Moon	Highland rocks of the ANT group[b]	0.8	0.21	3.8
	Creep material	9.3	2.8	3.4
	Mare basalts with high Ti and K content	3.98	0.68	5.8
	Mare basalts with moderate Ti and K content	1.18	0.64	1.9
	Mare basalts with high Ti and moderate K content	0.61	0.16	3.9
Mars	Volcanogenic	5.0±2.5	1.1±0.8	4.5
	Highland	0.7±0.35	0.2±0.14	3.5

[a]1 g/t = $10^{-6} \equiv 10^{-4}$ %
[b]Anorthosite-Norite-Tractholite regolith-type rocks
Source: Data from Surkov and Moskaleva, 1990.

terial delivered to the receiving tray was irradiated by radioactive sources using isotopes of plutonium (^{238}Pu) and iron (^{55}Fe). The X-ray fluorescence spectra excited by these isotopes were recorded by a set of detectors. An overall view of the X-ray fluorescence spectrometer is shown in Fig. 3.46.

During operation on the surface, each station took several tens of spectra and transmitted them to Earth (Fig. 3.47). These spectra made it possible to determine the content of basic rock-forming elements, from magnesium to iron, and thereby more reliably iden-

Fig. 3.44. Compositional trends for terrestrial rocks, as a function of the percentage ratio of silica (SiO$_2$) and alkalis (K$_2$O), and their comparison with measurement data of the elemental composition and content of radioactive elements at the Venera and Vega landing sites (measurement errors are indicated by boldface lines and arrows): (I) the low-alkali trend, (II) the calcium-alkali trend, and (III) the trend of alkali basalts (according to Barsukov et al., 1986).

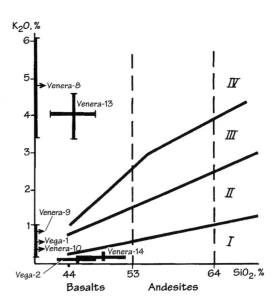

tify the studied sample with known minerals and terrestrial rocks, whose spectra are well studied.

The chemical composition of Venusian rock, according to measurement data from Venera-13, Venera-14, and Vega-2, is presented in Table 3.4 (Barsukov et al., 1984, 1986b). Over the whole series of constituents the results are quite similar, but there are distinct differences. At the Venera-13 landing site the rock is rich in potassium and apparently does not differ greatly from the rock at the Venera-8 landing site. Magellan imagery found both these sites to have unusual volcanic features. Judging by the composition, by analogy with Earth, this rock actually resembles a potassium-rich, high-magnesium alkali basalt, in accordance with estimates from data analysis of the natural radioactivity. This type of igneous rock is quite rare on Earth, found mainly on oceanic islands and in continental rifts of the Mediterranean. The Venera-14 rock (which is enriched in titanium like the rock near Venera-13, and in contrast to the Vega-2 rock) is apparently similar in its composition to rocks at the Venera-9 and Venera-10 landing sites. It thus is consistent with our notions about its correspondence to tholeiitic basalts, which are widely distributed in Earth's oceanic crust and are also representative of ancient lava flows on the Moon.

We can conclude that the rocks in the Venera-13 and Venera-14 landing sites, located to the east of the relatively young volcanic formations Beta and Phoebe, belong to the most typical geomorphological provinces of Venus' surface, with rather smooth relief characteristic of the rolling plains. Because similar provinces occupy more than 80% of Venus, one would expect these types of rock to be representative of most of the planet's surface. In contrast, at the Vega-2 landing site, which is in a region of gradual transition from hilly plains to high mountains, the rock proved to be comparable to terrestrial gabbros of normal alkalinity. The elevated silicic acid content and other petrochemical features may indicate that this rock formed upon differentiation of some initial melt containing no more than 1% water (by mass). Moreover, it is difficult to detect an alkaline trend from these data (see Fig. 3.44).

It is interesting to note that a great deal of sulfur is present in all the Venusian rock

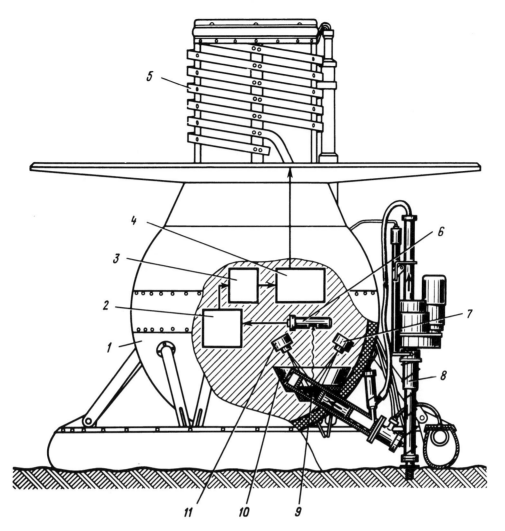

Fig. 3.45. The position of functional components for the intake and elemental analysis of soil on the Venera 13–14 landers (the Vega 1–2 spacecraft were similarly equipped): (1) the station housing, (2) the amplification and commutation unit, (3) the pulse amplitude analyzer, (4) the telemetry system, (5) the antenna, (6) the detector for X-ray fluorescence analysis, (7) the ^{55}Fe radioisotope source, (8) the soil-intake system, (9) the soil sample, (10) the soil receptacle, and (11) the ^{238}Pu radioisotope source.

samples studied with the X-ray fluorescence technique, especially those from Vega-2. An attempt to explain these high concentrations was made using thermodynamic calculations of the mineral composition of the Venusian soil (Barsukov et al., 1982, 1986c). The possible chemical weathering reactions that would be caused by the interaction of the near-surface gases with the original rock were considered, and the most likely reactions were determined using thermodynamic considerations. These calculations led to the conclusion that the high abundance of sulfur is probably caused by long-term contact of the surface rock with Venus' atmosphere. Anhydrite ($CaSO_4$) was identified as the most likely compound of sulfur with calcium (see Chapter 5).

Unfortunately, the available data have not yet provided any possibility of confidently determining the mineralogical composition of Venus' surface. The existing models, which rely on thermodynamic calculations, have nevertheless enabled us to set specific limits. In particular, it was shown (Barsukov et al., 1982, 1986c) that in weathered rocks of basic and acidic composition, preferential accumulation of sulfur is expected in the form of sulfates, whose content may account for ~10% of the mass. Solid phase-saturated compounds are presented in Table 3.5. The regions of the planet (lowland, mountainous) in which the cor-

Table 3.4   Chemical composition of Venusian rocks, % by mass

Element	Venera-13	Venera-14	Vega-2
MgO	11.4±6.2	8.1±3.3	11.5±3.7
Al_2O_3	15.8±3.0	17.9±2.6	16±1.8
SiO_2	45.1±3.0	48.7±3.6	45.6±3.2
K_2O	4.0±0.63	0.2±0.07	0.1±0.08
CaO	7.1±0.96	10.3±1.2	7.5±0.7
TiO_2	1.59±0.45	1.25±0.41	0.2±0.1
MnO	0.2±0.1	0.16±0.08	0.14±0.12
FeO	9.3±2.2	8.8±1.8	7.74±1.08
SO_3	1.62±1.0	0.88±0.77	4.7±1.5
Cl	<0.3	<0.4	<0.3
Na_2O	2.8±0.5	2.5±0.4	0.2
Total	98.1	98.7	95.8

Source: Barsukov et al., 1986b.

Table 3.5   Mineral composition of Venusian rocks, % by mass

Mineral	Venera-13	Venera-14	Vega-2
Clinoenstatite, $MgSiO_3$	28.1	18.0	40.1
Diopside, $CaMgSiO_2O_6$	1.1	5.0	6.8
Aportite, $CaAl_2Si_3O_8$	22.7	37.9	23.8
Albite, $NaAlSi_3O_8$	—	19.8	—
Microcline, $KAlSi_3O_8$	18.7	1.2	0.7
Quartz, SiO_2	—	3.4	4.2
Nepheline, $(Na, K)AlSiO_4$	12.2	—	—
Magnetite, Fe_3O_3	10.1	9.5	3.5
Sphene, $CaTi(SiO_4)O$	4.0	3.1	—
Anhydrite, $CaSO_4$	2.8	1.3	6.9
Marialite, $3NaAlSi_3O_8NaCl$	—	0.5	1.7
Andalusite, $Al_2(SiO_4)O$	—	—	12.0

Source: Data from thermodynamic calculations, Barsukov et al., 1986c.

responding minerals may be mainly found is significant, because their stability, along with the rates of geochemical reactions that affect their formation, are functions of temperature and pressure and therefore significantly depend on altitude (Surkov and Moskaleva, 1990; Nozette and Lewis, 1982).

From everything that has been said about the surface composition of our neighboring planet, we can conclude that basaltic rocks, which make up a substantial portion of Earth's crust, are also characteristic of Venus. One would imagine, however, that the surface of Venus is more homogeneous than that of Earth, not only in a structural-morphological

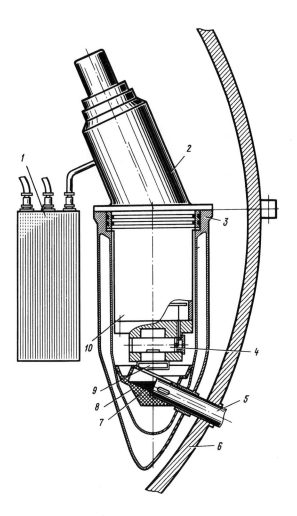

Fig. 3.46. The X-ray fluorescence spectrometer on the Venera 13–14 landers: (1) the pulse amplitude analyzer, (2) the shield for the detector unit, (3) the detector unit housing, (4) the counter, (5) the transport conduit, (6) the lander framework, (7) the sample under study, (8) the soil receptacle, (9) the radioisotope source, and (10) the electronics unit.

sense but also in its chemical composition. Perhaps because of the scarcity of water on Venus, the basaltic melts that formed the planet's crust apparently did not undergo any significant metamorphic modifications. Therefore, compared to Earth, the surface rock of Venus may be distinguished by much lower density (see Barsukov et al., 1983, for further discussion).

We shall now touch briefly on the physical-mechanical properties of Venus' surface. Included in these are the density, the bearing capacity, strength, and electrical conductivity of the soil. The first successful attempt to determine the density of Venusian rock was undertaken on the Venera-10 lander, using a radiation densiometer. The density turned out to be 2.7 ± 0.1 g/cm^3 (Surkov et al., 1976), which is in good agreement with the estimate of the mean density obtained from the permittivity, ε, derived from radar measurement analysis. Evidence was subsequently obtained, however, that this value (which corresponds to underdense basalts) is not representative of the whole surface of Venus and that the density values in various regions lie in the range of 1.2–2.7 g/cm^3. Upon surface impact by the densiometer sensor, the probe was deployed and an estimate was made of the strength of the rocks, which, as is obvious in the Venera-9 and Venera-10 panoramas (see Fig. 3.36), reside in rocky and semirocky landscapes. The strength turned out to be equal to several hundred kg/cm^2 (Kemurdzhian et al., 1978).

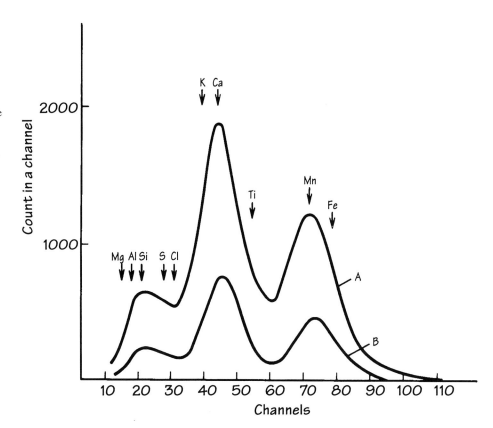

Fig. 3.47. Fluorescence
spectra of Venusian rocks
measured by the Vega-2
lander (according to
Barsukov et al., 1986). The
arrows indicate the posi-
tions of corresponding elements, with respect to the
recording channel number
on the horizontal axis. A
and B are two types of
counters.

Rock strength and electrical resistance were studied more completely by the Venera-13, Venera-14, and Vega-2 spacecraft using a dynamic penetrometer. The soil-bearing capacity measured by these spacecraft turned out to be 2.6 and 10 kg/cm^2, 62–250 kg/cm^2, and 115 kg/cm^2, respectively (Kemurdzhian et al., 1983). Based on control experiments on terrestrial rocks of various types, these values correspond in the first case to products of chemical reprocessing and weathering of hard rocks (heavy clays, loams, compacted dusty sand) and, in the second and third cases, to volcanic tuffs, pumices, and eroded hard rocks. Such analogies tie in well with the landscapes observed in the panoramas (see Figs. 3.36, 3.39, and 3.40) and, moreover, do not contradict the results of independent analysis of the impact dynamics (the g-forces) during the Venera-13 and Venera-14 landings. Another independent estimate is provided by data on the depth of rock drilling, and by the current required by the electric motor of the soil-intake system, while gathering the samples for X-ray fluorescence analysis (Barsukov et al., 1984). These data agree well with the estimates presented previously. The electrical conductivity measurements, which provide information on the composition and density of the surface rock, gave substantially differing results: from an unusually high value (10^2 ohm · m) in the areas of Venera-13 and Venera-14 to very low (10^6 ohm · m) in the area of Vega-2. The second value is close to the electrical conductivity of certain types of basaltic rocks at a temperature of 735 K (Kemurdzhian et al., 1983), while the first value is close to electrical conductivity of some highland regions with anomalous high radio reflectivity, mentioned earlier. The fundamental data on the physical

Table 3.6   Physical characteristics of Venusian rocks and their terrestrial soil analogs

Experiment	Venera-13	Venera-14
Measurement of physical properties	Bearing capacity was 2.6–10 kg/cm², heavy clays, compacted dusty sand analog	Bearing capacity was 62–250 kg/cm², volcanic tuffs, crumbling hard rocks analog
Studies of the impact dynamics of the lander	Bearing capacity was 4.0–5.0 kg/cm², foam concrete analog	Bearing capacity of the layer to 5 cm was 2 kg/cm²
Analysis of the operation of the soil-intake system	Compacted ash, volcanic tuff-type material	Compacted ash, volcanic tuff-type material
Analysis of the TV panoramas	Volcanogenic stratified rock, covered in places by unconsolidated soil	Volcanogenic stratified sedimentary-type rock

Source: Data from Kemurdzhian et al., 1983.

properties of the soil, obtained by experiments on Venera-13 and Venera-14, and an attempt to identify them by drawing upon terrestrial analogs, are summarized in Table 3.6.

Thus, on the whole, we arrive at the conclusion that in the landing areas of the Venera and Vega spacecraft, the surface material can be identified (according to its physical-mechanical properties) with rocks of medium density and strength, like volcanic tuffs or highly eroded basalts. Confirmation of these ideas is obtained by analyzing the Venera-13 and Venera-14 panoramas, and Magellan images of these same areas. Volcanogenic rocks, along with weakly associated products of chemical reprocessing and weathering of bedrock, are surely present in these locations.

The Internal Structure and Thermal History

4 The development of the present appearance of planets and satellites is directly connected with processes occurring in their interiors and ultimately is governed by certain common mechanisms and stages in the chemical evolution of planetary material. Moreover, the experimental data already accumulated, and the theoretical models that fit these data, enable us to put forward a series of well-founded hypotheses concerning the geological present and past of the planets, and to better visualize the features of their internal structure.

Unfortunately, we are still restricted in our ability to develop a representative model of the interior of Venus. We can, however, present some general ideas on planetary interiors (especially on the planet most similar to Venus—our Earth) and basic procedures used for their modeling, and we can examine which constraints on interior structure and evolution emerge. We shall first direct our attention to several general cosmogenic and cosmochemical concepts, which furnish an approach for understanding the nature of celestial bodies and their fragments (that is, meteoritic material). The need to draw on such information for the study of planetary interiors was shown in the classic monograph by Urey (1952).

The initial stage of solar system formation is attributed to a period about 4.6 billion years ago. One usually proceeds from the idea that the initial composition of material in the protoplanetary nebula was identical with the entire region surrounding it and corresponded to solar composition, or to cosmic abundances of elements. It is known that, with a minor exception, a governing principle is observed in the abundance of elements: with an increase in the atomic number, the abundance decreases exponentially (up to $Z = 40$), and for heavier elements the abundance remains almost constant.

Among the materials that make up the Sun, planets, and meteorites, there are fundamental groups of elements that formed, according to current concepts, during galactic nuclear synthesis (nucleosynthesis) not less than 10 billion years ago. Nucleosynthesis is now also linked with processes that occur in explosions during supernova outbursts, and during the ejection of material from the non-equilibrium layer of neutron stars in close binary systems. In general, the formation of elements heavier than iron evidently occurs as a result of slow and rapid neutron capture arising in the star's interior, the so-called s- and r-processes (see Burbidge et al., 1982; Cameron, 1982; Ulrich, 1982; Schramm, 1982).

It is possible that during the formation of the solar system there continued a natural synthesis of certain radioactive and stable chemical elements, whose evolutionary paths are reflected in their abundances and isotopic ratios. A graph representing the chemical composition of the Sun, according to Ross and Aller (1976), contains information on the relative abundance of elements (Fig. 4.1). It is easy to convince oneself that, in a description of

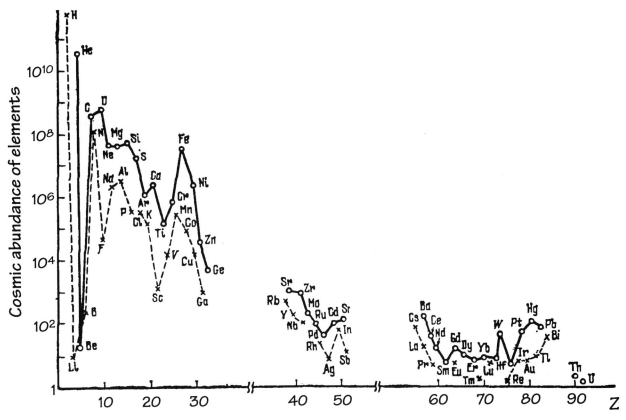

Fig. 4.1. Cosmic abundance of elements (according to Ross and Aller, 1976). The solid curve distinguishes elements with an even atomic number Z, the dashed line distinguishes elements with odd Z.

the chemical composition of the universe, about ten elements are of primary importance, among which hydrogen and helium play a dominant role. Along with neon, they form the most volatile group of substances (the gas component). Less volatile are water, ammonia, and methane, which form the ice component of condensing materials, whereas non-volatile substances (metals, silicon, and their oxides) form the heavy (or, as it is still known, rocky) component, which enters into the composition of rocks.

Hydrogen and helium serve as the fundamental building materials of our solar system, which was originally a rotating gas-dust disk from which a central fragment—the Sun—originated. The heavy and ice constituents formed the dust component of this disk. Because the temperature decreased throughout the disk away from the Sun, the most refractory compounds (forming the rocky component) condensed out closer to the Sun, whereas low temperature condensates (ices) emerged at much greater distances. Eventually, the heavy fraction of elements of solar composition were largely retained as the terrestrial planetary group, or as the relatively small cores of the giant planets, either surrounded by extended icy mantles (as on Uranus and Neptune) or shrouded by massive hydrogen-helium envelopes (as on Jupiter and Saturn).

Initial Premises and Model Constraints

The mechanical properties of planets—mass, size, figure, and rotational motion—are fundamentally affected by the nature of their internal structure. The mass distribution in a planet's interior determines its gravitational potential and moments of inertia. The actual gravitational potential differs from the simple Newtonian potential, which describes a spherically symmetric density distribution as a function of radius, and may be represented as an expansion of spherical functions. The leading term in the series corresponds to the gravitational potential of a sphere with a mass equal to that of the planet, whereas terms of second and higher order (so-called zonal and tesseral harmonics) reflect the details of its internal structure (see Zharkov and Trubitsyn, 1980).

Even and odd harmonics represent deviations in the density distribution, from a spherically symmetric state and a state of hydrostatic equilibrium, respectively. They are characterized by the multipole moments J_n, which are determined from the orbital perturbations of natural and artificial satellites, and the trajectories of passing spacecraft. The first second-order correction term, which is proportional to the quadrupole moment J_2 and accounts for compression of the planet, provides the main contribution. Consequently J_2, defined as $J_2 = [C - \frac{1}{2}(B + A)]/M\bar{R}^2 \approx (C - B/M\bar{R}^2$, is of the same order as the compression α (all notations are given in Chapter 1).[1] The hydrostatic portion of the quadrupole moment J_2^0 is extracted, and from the difference $\Delta J_2 = J_2 - J_2^0$, one can judge the deviation of the planet from hydrostatic equilibrium. Subsequent zonal moments in a body in a state of hydrostatic equilibrium decrease in proportion to the second and successive degrees of compression. Moments higher than J_4 are primarily affected by the outer layers of a planet. They thus say less about its internal structure.

Another important parameter, which reflects the trend in density variations with depth, is directly associated with the J_2 moment. This is the dimensionless moment of inertia, $I = \bar{I}/M\bar{R}^2$, which can be expressed in terms of the mean moment of inertia, $\bar{I} = (C + 2A)/3$, and the mean radius, $\bar{R}$, or in simplified form, in terms of the moment of inertia with respect to the rotational (polar) axis C and equatorial radius R_{eq}, that is, $\bar{I} \approx C/MR_e^2$. If the density over the whole thickness remains constant (the model of a homogeneous sphere), then $I = 0.4$. If the density increases with depth, then $I < 0.4$; in the opposite case $I > 0.4$. So, for Earth, the experimental value of $I = 0.3308$ corresponds to a significant increase in density toward the center. Nearly the same value, $I = 0.334$, is accepted for Venus, while for Mercury the value is somewhat less, $I = 0.324$, and for Mars it is considerably greater, $I = 0.375$ (see Zharkov, 1983; Schubert et al., 1992).

The multipole moments impose limiting conditions on the position of equipotential surfaces, which can be calculated from the theory of the figure of gravitating bodies. For planets in a state of hydrostatic equilibrium, the equipotential surfaces are characterized by identical values for the pressure (P), density (ρ), temperature (T), and other thermodynamic properties of the material. The degree of deviation from hydrostatic equilibrium can likewise be judged by the value of the first odd moment, J_3. For Earth, it unexpectedly proved to be quite large, on the order of the square of the compression. This suggests that the figure of Earth deviates from equilibrium (on the order of $\bar{R}\alpha^2$ at the surface) and that along with radial stresses, caused by pressure, there are shear stresses in its interior. Their intensity is several orders of magnitude less than the radial stresses but nevertheless represents several tens of kilograms per square centimeter (Zharkov, 1983).

Some deviation from hydrostatic equilibrium is characteristic of the other planets of the terrestrial group. One consequence of this is that the difference between their principal moments of inertia located in the equatorial plane (A and B) is not equal to zero. It is usually denoted by the parameter f, which is expressed in the form $f = (B - A)/M\bar{R}^2 = 4(C_{22}^2 + S_{22}^2)^{1/2}$. The smaller f is, the less the deviation from hydrostatic equilibrium and the closer the density distribution in the planet's interior to axisymmetric. For Earth the value is $f = 7.2 \cdot 10^{-6}$.

The Jovian planets, in contrast to Earth, Venus, and the other terrestrial planets, are closer to hydrostatic equilibrium (consequently, there is no appreciable contribution to the gravitational potentials from the J_3 moment). This is indicated by excess radiation of energy, as a consequence of heat flow from the interior, in which convective heat transfer plays a decisive role. The dimensionless moment of inertia, I, for these planets is considerably less than for the terrestrial planets: for Jupiter $I = 0.262$; for Saturn $I = 0.227$; and I is even lower for Uranus (0.212) and Neptune (0.200).

Successes in the field of high-pressure physics, and the development of the theory of the figure of gravitating bodies, have significantly contributed to progress in modeling the structure of planetary interiors. The advances in high-pressure physics are primarily associated with refinements in the equation of state, which determines the relationship between pressure, density, and temperature. This is necessary because at high pressures and temperatures the standard equation of state, $P = P(\rho, T)$, for materials under normal conditions, proves to be unsuitable. It is also necessary to take into account the pressure dependence in the interior versus the concentration of components, or the ratios of the compounds formed by them, which additionally complicates the description of the thermodynamic state of such a medium. Finally, a theoretical description of the conditions necessary for the formation and stability of phases at high pressures and temperatures represents a complex, independent problem. Matter may acquire unusual properties—an example is the transition of hydrogen into a metallic state, as occurs on Jupiter and Saturn.[2]

The theory of the behavior of matter under extreme conditions is based on concepts from statistical physics and quantum mechanics. In the most general case, the thermodynamic state of matter at great depths is described by the equation of state in the so-called Debye approximation, for a known or subsequently calculated dependence of the characteristic (Debye) temperature, Θ, versus density. This temperature is expressed in the form $\Theta = h\nu/k$, where ν is the maximum oscillation frequency of the crystal atoms, propagating along the crystal as waves, each of which can be represented as a quasi-particle—a quantum of thermal oscillations in a solid, or a phonon. Here the product $k\Theta$ (k is the Boltzmann constant) characterizes the energy of the shortest-wave phonons in the substance. In other words, the Debye temperature is the energy (in degrees) of the limiting Debye phonon (see Landau and Lifschitz, 1964). With seismological data, one can determine the change in Debye temperature, Θ, as a function of depth, ℓ.

The concept of a phonon enables us to study the thermal and other properties of solids and matter that occur in regions of ultra-high pressures, using techniques of kinetic gas theory. In calculating the deepest layers, where temperatures reach tens of thousands of degrees Kelvin, sometimes high-temperature corrections are introduced, which incorporate permitted deviations in the atomic oscillations in the crystals from the quasi-harmonic approximation, and also the effect of thermally excited conduction electrons. In the quasi-harmonic approximation, another important variable for calculating internal structure models is introduced: the so-called Grüneisen constant γ, which characterizes the change in oscillation frequency as a function of density and is usually written as the derivative of the logarithm of the Debye temperature with respect to density: $\gamma = (d \ln \Theta)/(d \ln \rho) = [\rho(\ell)/\Theta(\ell)]/(d\Theta/d\rho)$. Along with the Debye temperature, it completely determines the thermodynamic state of matter for the corresponding model of a solid body.

These basic concepts are the basis of model calculations for the solid planets of the terrestrial group, and for the gas-liquid giant planets. Differences in the bulk densities among the terrestrial planets (see Table 1.3), along with the results of model computations, gave rise to the hypotheses about the inhomogeneous chemical composition of the parent material. One explanation for such inhomogeneity is provided by the equilibrium condensation model (Lewis, 1972, 1974), which is based on the assumption of a systematic drop in temperature as one moves farther from the protosun and, consequently, fractionation of elements. We are primarily talking about metallosilicate fractionation due to differences in condensation temperatures of hot gas of solar composition, which resulted in dissimilar amounts of iron and silicon at various distances from the Sun. Iron and sulfur fractionation (more precisely, fractionation of siderophile and chalcophile elements) is also pertinent here. Consequently, one could imagine that with an increase in heliocentric distance, the difference in the chemical composition of the parent material from that of iron meteorites would grow, owing to greater oxidation of iron (with the formation of silicates) and, simultaneously, an increased sulfide content.

The relatively small amount of protoplanetary nebula material from which the planets and asteroids close to the Sun formed was most likely similar to meteorites in composition—this comparison is often applied when speaking about fractionation of the parent material. The overwhelming majority of meteorites that have fallen on Earth are approximately 4.6 billion years old, as determined from the isotopic ratios of lead (^{207}Pb/^{206}Pb), and strontium to rubidium (^{87}Sr/^{87}Rb), and more recently, using the more accurate method, from the ratio of samarium to neodymium (^{147}Sm/^{144}Nd) (Wasserburg and De-Paolo, 1979). Meteorites, which are collisional fragments of asteroids (Simonenko, 1985), have thus preserved information about the raw material of the solar system. It is not an accident, therefore, that some planetary models involve compositions based on the most common meteorites—chondrites (Wood, 1963; Morgan and Anders, 1980). In these models the material of planetary interiors was assumed to consist of combinations of the five fundamental groups of elements, which have been subjected to various degrees of fractionation.

Meteorites can be divided into three broad classes: iron, stony-iron, and stony, as a function of the proportion of two dominant phases, metal (iron-nickel) and stony (silicate).[3] In the iron group, the metal phase dominates (up to 94%, including iron and magnesium, in a free as well as an associated state), and in the stony group, the stony phase dominates (up to 90%), whereas in the stony-iron group almost equal amounts of iron and silicates are found. Aside from these two fundamental phases, in each class of meteorites there is also a sulfide phase, entering into the composition of ferrous sulfide (troilite) and a series of other rock-forming minerals (see, for example, Kaula, 1968; Wood, 1963; Marov, 1986).

The concentration of the most abundant and chemically active elements—hydrogen, carbon, oxygen, magnesium, silicon, sulfur, and iron (see Fig. 4.1)—evidently depended on variations in the temperature and pressure in the initially chemically homogeneous protoplanetary disk. With an increase in temperature, the relative iron content should rise, as a result of the loss of silicates and the dominance of reduction processes over oxidation from the loss of volatiles. Ordinary chondrites, rich in iron, form the high-iron (H) group, and those depleted in iron, the low iron (L) and very low iron (LL) groups. Enstatitic chondrites, in general consisting of a mineral from the magnesium silicate group, enstatite ($MgSiO_3$), and nickel iron, possess the highest degree of reduction, and the group of carbonaceous chondrites, or the C group, in which almost all the iron is bonded in magnetite (Fe_3O_4), have the highest degree of oxidation. The C group is apparently closest to the original chemical mixture from which the terrestrial planets and asteroids subsequently formed. These chondrites are also distinguished by the highest concentration of volatile (atmophile) elements and therefore, from a cosmogonic point of view, are of particular interest.

The chemical classification and structure of meteorites are thus directly linked with their origin. The common genealogy of meteorites and asteroids enables us to postulate that the parent bodies themselves were found in various phases of evolution, depending on their size. If chondrites were derived from relatively small, chemically undifferentiated bodies, which formed by condensation of the source material at various distances from the Sun, then iron meteorites and achondrites are, in all probability, debris of larger asteroids, whose material underwent differentiation in their interiors. In turn, from the degree of iron oxidation in the meteorites, one can speculate about the condensation conditions—mainly about the temperature in various portions of the protoplanetary nebula. One can imagine that iron meteorites and enstatitic chondrites, which are characterized by the most complete degree of reduction, were formed primarily at high temperature close to the Sun, approximately within Mercury's orbit, whereas the most oxidized carbonaceous chondrites formed at considerably lower temperatures, primarily beyond the orbit of Mars. Intermediate conditions were characteristic of regions of space corresponding to the orbits of Venus and Earth. Likewise, most bodies in the asteroid belt and most likely the cores of the giant planets (which presumably consist of a mixture of metal oxides and hydrated silicates) were formed from chondrites. Accordingly, the condensation conditions predetermined the differences in the composition and mean density of the terrestrial planets, and hence, the course of their subsequent thermal evolution.

No single theory explains the paths of chondrite material formation, or the reasons for the well-known isotopic anomalies in their composition. In particular, the condensation model does not explain the presence of grains of refractory metals and minerals in chondrites, whose parent bodies are located at distances of 2–4 AU from the Sun, where the temperature probably did not attain high values. Attempts to circumvent these difficulties have been made based on a model of systematic condensation, which would have occurred as material of the protoplanetary nebula moved from the hot region of the protosun to the periphery of the solar system. In essence, this is a model of initial compression of the solar nebula and subsequent "spreading" of material from the protosun inside the accretion disk. Such a model was examined in detail by Cassen and Summers (1983) and has been developed by Izakov (1986) as part of a mechanism of successive condensation.

In addition, an alternative to the equilibrium condensation model has been proposed (Ringwood and Anderson, 1977). It points to evidence that the terrestrial planets, primarily Earth and Venus, were formed from material of identical chemical composition (including an abundance of sulfur) and suggests that the differences in their mean density can be completely explained by different degrees of iron oxidation and, consequently, different relative FeO concentrations. Similar ideas were developed by Sorokhtin (1974). This approach, however, seems less convincing to us.

More recently, numerical models of planetary accretion have pointed toward a phase during the later stages of accretion that was characterized by giant impacts and a large amount of mixing of materials from different heliocentric condensation zones (Wetherill, 1985). If these results truly reflect the process of planetary accretion, then any theory that attempts to simply relate planetary composition to distance from the Sun is invalid.

We will pursue the concept that the basic chemical transformations of the parent material apparently occurred at relatively high temperatures, in the nearest environs of the Sun. Here, oxidation-reduction processes strongly dependent on temperature played an important role, whereas far from the Sun at low temperatures, reactions were retarded and the composition of the parent material evidently remained almost unchanged. Radioactive elements (short-lived and long-lived radionuclides) are found in the composition of the high-temperature fraction of condensed solid particles, from which the terrestrial planets mainly accreted; they served as a source of subsequent heating. Therefore, it is on these planets that the most immense changes took place during evolution, which resulted in differentiation of the constituent material and the formation of secondary gas envelopes— atmospheres.

Unfortunately, the establishment of a rigorous quantitative basis for all these complex processes, grounded in theoretical modeling techniques, is complicated by our lack of knowledge of many initial conditions in the evolutionary process, such as the initial mass and moment of the protoplanetary nebula, the distribution of temperature and component concentrations, the pressure at which initial condensation of solid materials occurred, the rates of chemical reactions, the intensity of material mixing during the late stages of accretion, the rate of de-gassing from the interiors, and the effectiveness of bonding gas components with solids. These uncertainties poorly constrain the selection of cosmogenic models that contain the initial cosmochemical prerequisites for calculating models of the interior structure of planets.

Composition and Internal Structure of Earth and the Moon

Research on the internal structure of the terrestrial planets is made easier to a certain extent by the presence of a reference (Earth) for which a powerful technique to study the interior is available: the velocity distribution of seismic wave propagation. This technique reveals the change in elasticity characteristics with depth and in this way enables us to detect the most characteristic features of the internal structure of a body, which are specified by the material's chemical composition, its phase state, and its thermodynamic parameters. This method was also used effectively to explore the interior of the Moon.

The similarity in the geometric and mechanical characteristics of Venus and Earth naturally gives rise to the assumption of extreme similarity in the fundamental features of their internal structure. Such an approach was reflected in the first model of Venus' interior, created by Jeffreys (1937) and based on Bullen's classical model of Earth (1936). With the appearance of the aforementioned monograph by Urey (1952), the need to draw on data of planetary cosmogony and cosmochemistry, involving primarily meteoritic material, to provide a more well-founded theory of Venus' composition became obvious. With further accurate observational data and the development of the physics and hydrodynamics of planetary interiors (see Toksoz et al., 1978; Schubert, 1979; Zharkov, 1983), definite differences between the physically more realistic models of Venus' internal structure and Earth-like models were recognized (Ringwood and Anderson, 1977; Kalinin and Sergeeva, 1979; Kozlovskaya, 1982; Zharkov and Zasursky, 1982). But it is easier to understand the interior structure of Venus, and the principles governing the distribution and extent of separate layers, by first becoming acquainted, at least in general terms, with the structure of our Earth and the rocks that make up its depths.

Modern concepts dealing with the internal structure of Earth are shown in Fig. 4.2a, where fundamental zones are distinguished, characterized by the propagation velocities of longitudinal (p) and transverse (s) waves. Such waves are called body waves. They originate at the foci of earthquakes and are generated by seismic energy resulting from tectonic processes. Penetrating deep in the Earth, body waves change in velocity, undergoing refraction and reflection at interfaces (of shells) with different physical properties. For p waves, these properties are characterized by the compression modulus and shear modulus of the material, but for s waves, they are characterized by only the shear modulus, because those waves produce only oscillations perpendicular to the direction of wave propagation (just like electromagnetic waves). For liquid media the shear modulus is zero, and therefore the velocity of transverse waves also goes to zero. From this it is clear that if a medium does not let these waves pass, it should be considered liquid. This is exactly the situation observed in Earth at the core boundary.

Within each of the three basic shells of Earth—the crust, mantle, and core—a series of additional features (zones) can be identified, which are associated with velocity changes in the seismic waves passing through them. The uppermost solid shell—the crust—is separated from the underlying layers of the mantle by a discontinuity, where the velocities of p waves jump from 6.7–7.6 km/sec to 8–8.3 km/sec, and s waves from 3.6–4.2 km/sec

Fig. 4.2a. A model of Earth's internal structure and the propagation velocities of longitudinal (v_p) and transverse (v_s) seismic waves.

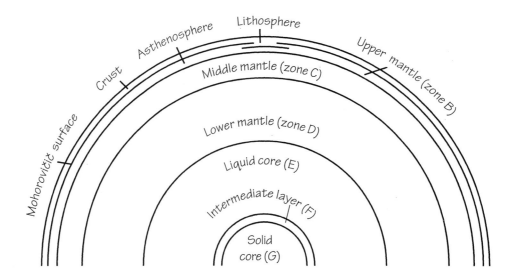

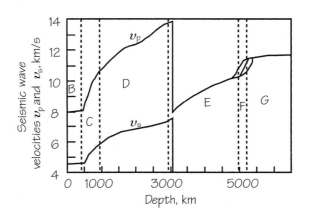

to 4.4–4.7 km/sec. This boundary is called the Mohorivičič discontinuity, or simply the Moho. It runs at different depths owing to the varied thickness of Earth's crust, from 30–60 km under the continents to 5–10 km under the oceans (see, for example, Monin, 1977). It comes closest to the surface in the axial zones of mid-oceanic ridges, where rift valleys are found. Keep in mind that mid-oceanic ridges and their rift zones are associated with regions of new oceanic crust formation. Volcanic activity and the epicenters of numerous earthquakes are concentrated in rift zones; anomalously high heat flow from the interior and a decrease in the density of the upper mantle are also detected there.

The upper mantle (**B** zone) is located under the crust. Its upper layer, immediately underlying the crust, is sometimes called the substratum. Along with the crust, it forms the lithosphere, the most rigid shell of Earth, below which is a somewhat liquefied layer of reduced strength, the asthenosphere. The asthenosphere is often identified with the Gutenberg layer, in which a noticeable decrease in the velocity of transverse seismic waves is observed. The reasons for their slowed transmission are, apparently, a large geothermal gradient in the asthenosphere and a significant decrease (by two or three orders of magnitude) in the viscosity of the material, compared with the lithosphere. The lower boundary of the asthenosphere lies at a depth of 250–350 km, and its upper boundary comes closest to the surface under the axes of mid-oceanic ridges. Upon transition to zone **C**, called the middle mantle or Golitsyn layer, seismic wave velocities increase up to a depth of nearly 1000 km, where the boundary with the lower mantle (zone **D**) is located; the increase in velocities falls sharply in the lower mantle. Between the lower mantle and core a small transitional layer **D′** (about 200 km thick), in which there is an additional small decrease in p wave velocities, stands out.

The increase in seismic velocities in zone **C** is caused by phase transitions, due to the rebuilding of minerals into structures with a denser packing of atoms. In contrast to the acidic and basic rocks of the crust, the mantle consists of ultrabasic rocks, containing the least amount of silica (SiO_2) in the form of quartz and, at the same time, the largest amount of magnesium oxide in the composition of several types of minerals. The primary rock-forming minerals of basic and ultrabasic rocks (basalts, dunites, gabbros, peridotites, diabases, and so on) are iron and magnesium-bearing silicates, that is, olivines and pyroxenes, which have formulas $(Mg, Fe)_2SiO_4$ and $(Mg, Fe)SiO_3$, respectively. The original mantle of Earth presumably consisted of olivine-pyroxene rock (so-called pyrolite) before differentiation of the constituent planetary material.

According to current notions, supported by laboratory petrochemical research on minerals under conditions in Earth's interior (see Gutenberg, 1963; Zharkov and Trubitsyn, 1980), olivines in the upper mantle at the **B** and **C** zone interface (400–420 km) are reorganized, as a result of polymorphic phase transition, into a modified spinel structure found at the magnesium end of the olivine series—a structure with the densest cubic packing of oxygen ions. This particular transition explains the increase in p wave velocities at this level. In turn, at a depth of almost 70 km, pyroxenes are already crystallizing into orthopyroxenes and, in the presence of aluminum oxide Al_2O_3 (alumina, corundum), transform into garnets, whereas quartz subsequently transforms into structures of coesite and stishovite—a mineral 62% more dense than normal quartz. In the middle mantle, beginning at a depth of approximately 700 km, yet another phase transition is inferred, from a spinel zone to a perovskite zone—this being a mineral with a well-defined cubic cleavage, that is, an especially dense cubic packing. Here, the structure of corundum can be reorganized into ilmenite by exchanging aluminum atoms with iron and titanium atoms. In the homogeneous **D** layer, which is apparently an entirely perovskite zone, seismic wave velocities increase (although with a lesser gradient) merely owing to material compression under the pressure of overlying layers, and to increased density.

In the **E** zone the longitudinal wave velocities decrease by approximately a factor of two, while transverse waves do not pass through this layer at all. Longitudinal wave velocities increase once again in the central region (the **G** zone), which is separated from the **E** layer by the minor **F** layer (about 150 km thick), where a small increase in p wave velocities is observed. These facts provide a basis for identifying the **E** layer with an external liquid core approximately 3460 km in radius, whereas the **G** layer is identified with an internal solid (or, more likely, partially melted) core 1250 km in radius. The mass of the entire core represents about 30% of the mass of Earth, including the mass of the inner core (which alone represents approximately 1.2%). On the whole, such a mass distribution in Earth agrees with model calculations of its interior, which are constructed to take into account the dimensionless moment of inertia value, I.

In its composition, Earth's core is evidently close to that of iron meteorites and was formed from a nickel-iron melt, abbreviated Ni-Fe (approximately 89% Fe, 7% Ni, and 4% FeS). Models assuming a metal-sulfide internal core also exist. Until recently, a competing hypothesis to the nickel-iron core idea was that Earth's core may consist of metallized silicates formed as a result of phase transitions in silicates into a metallic state, at pressures on the order of 1 Mbar. This hypothesis, however, has not been confirmed by shock compression experiments under conditions corresponding to physical conditions in the core. The most convincing hypothesis concerning core formation conforms to an iron core composition. This formation would occur by precipitation of the compositionally nearly homogeneous early Earth from a melt during gravitational differentiation. This situation is reminiscent of the well-known process of melting iron in blast furnaces: the iron, brought to a metallic state, settles to the bottom, forming a dense liquid phase. The remaining lighter silicates float to the surface to form a slag.

The basic tendencies in the distribution of chemical elements in Earth's shells during its formation, with the participation of lithophile, siderophile, chalcophile, and atmophile elements, are shown graphically in Fig. 4.2b. A model with a metal-sulfide internal core is hypothesized here. For other planets of the terrestrial group, a somewhat different distribution among these element groups is evident; Mercury has a high concentration of siderophiles, and Mars a high chalcophile content. This difference apparently complies with theories predicting a change in the planets' mean density with distance from the Sun. In this scheme, presumably, Venus occupies a place not very different from Earth. In turn, the giant planets are highly enriched in atmophile elements.

A knowledge of the distribution of seismic wave propagation velocities with depth makes it possible to determine pressure and density trends and, therefore, the equation of state of material in the interior. The two most important parameters that define the thermodynamics of a planet's shells depend on density; these parameters are the Debye temperature and the Grüneisen constant, which are easy to compute when seismic data are available. Thus, it proves to be relatively simple to calculate a physical model for the internal structure of Earth, and also to calculate the fundamental thermodynamic coefficients that characterize its structure (such as heat capacity, heat transfer, and compressibility coefficients). Computing the actual temperature trend, however, is much more difficult. From experiment, a geometric gradient directly linked with heat flow from Earth's interior exists only in the uppermost layer, amounting to an average value of 20 K/km at Earth's surface, with pronounced variations in various regions. But the temperature increase slows with depth. The actual melting temperatures of the materials—those mineral associations that exist at extreme depths—impose boundary conditions. Therefore, the temperature distribution along the melting curves serves as its own type of criterion, which determines the position of transition boundaries from one modification to another.

The view that in a liquid core temperatures must follow an adiabatic law serves as

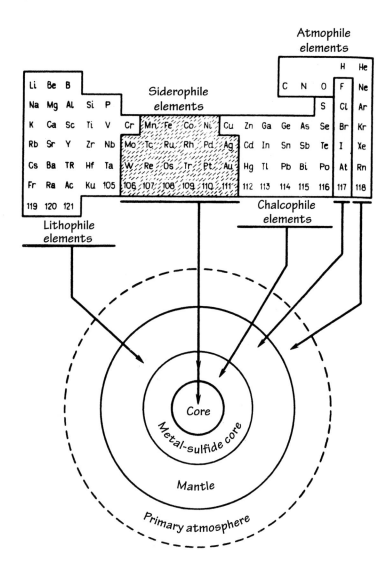

Fig. 4.2b. Distribution of the fundamental groups of elements in shells of Earth, in accordance with current concepts about the structure and chemical composition of planets of the terrestrial group (according to Voytkevich, 1988).

another criterion. With such a temperature gradient, the main heat transfer mechanism is convection. The convective state of Earth's interior, not only in the core but in the mantle as well, is beyond doubt. The presence of widespread convection in the core is usually used to explain Earth's large magnetic moment, as a result of electromagnetic induction in a moving medium. According to the hypothesis of a hydromagnetic dynamo, the magnetic moment of Earth is produced by the motions of a conducting liquid, causing self-excitation of the magnetic field similar to the way a current and a magnetic field are generated in a dynamo. According to current ideas, the temperature in the central portion of Earth is approximately 6000 K at a pressure of 3.65 Mbar and, at the interface between the core and lower mantle, 4300 K and about 1.4 Mbar (Zharkov, 1983).

To complete the picture, we will compare the Moon's internal structure with that of Earth, in order to understand the limitations imposed on models of internal structure by data on the figure of a celestial body, and especially by the nature of p and s wave propagation in that body.

The actual figure of the Moon, determined by analyzing the orbits of artificial satellites, turned out to be close to a state of spherical equilibrium. However, it strongly di-

Fig. 4.3. A model of the
Moon's internal structure
and propagation velocities
of longitudinal (v_p) and
transverse (v_s) seismic
waves.

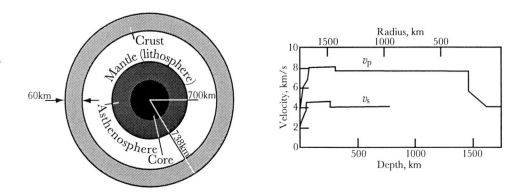

verges from dynamic equilibrium, defined by the level surface of the lunar gravitational potential (selenoid). The deviation is almost an order of magnitude greater than for Earth (for which it is on the order of the square of the compression, α^2). With the lower lunar gravitational acceleration, this deviation leads to shear stresses comparable with those on Earth, which must be withstood by the lunar lithosphere. The Moon's dimensionless moment of inertia, $I = 0.391$, was obtained from studies of the Moon's gravitational field. This figure is close to the value corresponding to the moment of inertia of a homogeneous sphere and thus implies that the density of the Moon is almost constant—that is, in contrast to Earth, the Moon has no large concentration of mass in the center.

The internal structure of the present Moon, determined from seismic data (Fig. 4.3), has been likened to a snapshot of an early stage of Earth's evolution (Toksoz and Johnston, 1975). The uppermost layer is represented by the crust, whose thickness, which was determined only in basin regions, is 60 km. It is extremely likely that the crust is approximately one and a half times thicker over the extensive highland regions of the lunar far side.

The lunar crust is composed of familiar igneous crystalline rocks—basalts. The basalts of the highland and mare regions, however, have distinct differences in their mineralogical composition. The most ancient highland regions of the Moon are formed principally from bright rock—anorthosites, almost entirely consisting of neutral and basic plagioclase, with minor amounts of pyroxene, olivine, magnetite, titanomagnetite, and so on. The crystalline rocks of the lunar mare, similar to terrestrial basalts, are composed, in general, of plagioclases and monoclinic pyroxenes (augites). They formed as a result of cooling on and near the lunar surface, that is, they are volcanic rocks. Compared with Earth's basalts, however, lunar basalts are less oxidized, which means that they crystallized with a lower ratio of oxygen to metal. Furthermore, compared to Earth rocks, a lower content of certain volatile elements is observed in lunar basalts, concurrent with a high concentration of refractory elements. Owing to admixtures of olivines and ilmenite in particular, the mare regions appear darker, and the density of their constituent rocks higher, than in the highlands (Vinogradov, 1975; Wood, 1975).

Underlying the crust is the mantle, where, as with Earth, one can distinguish upper, middle, and lower sections. The thickness of the upper mantle is about 250 km, that of the middle mantle is approximately 500 km, and its boundary with the lower mantle (asthenosphere) is located at a depth of nearly 1000 km. Down to this level the velocities of transverse waves are almost constant, and the interior material is found in a solid state, constituting a thick and relatively cold lithosphere. Because of the great extent of the lithosphere, whose material is highly depleted in volatiles, the Moon has a high mechanical quality factor, Q.[4] Therefore, seismic oscillations on the Moon (excited, for example, by a meteorite impact on the lunar surface) do not die down for a long time; it is said that the Moon rings like a bell.

The composition of the upper mantle is primarily olivine-pyroxene, whereas spinel and the mineral melilite, occurring in ultrabasic alkali rocks, are present at depth. At the interface with the lower mantle, temperatures approach melting temperatures and, as a result, vigorous absorption of seismic waves begins. This region is the lunar asthenosphere. In the very center there is apparently a small liquid core, less than 350 km in radius, through which transverse waves cannot pass. The core may be ferrous sulfide or iron; if it is iron it must be smaller, which is more in line with estimates of the density distribution with depth. Its mass, probably, does not exceed 2% of the entire mass of the Moon (Levin and Maeva, 1975; Zharkov, 1983).

A Few Remarks on the Problem of Thermal Evolution

The processes that give rise to separation of planetary material into a silicate mantle and a heavy, dense core, as well as determining the nature of the core, depend significantly on the accretion process and the following processes of thermal evolution, which were influenced by the relative original concentration of elements and the initial chemical-mineralogical complexes. Many uncertainties remain, especially when one attempts to generate models tracing the thermal history of Earth and its nearest neighbors.

Differentiation of constituent material begins during the accretion stage of a terrestrial planet, under the influence of the gravitational energy of accretion, and is followed by the release of energy of radioactive decay. Among other sources of heating during the initial and subsequent phases, the following may play an important role: tidal dissipation; release of heat associated with adiabatic compression of internal layers and, more important, with continuing impacts from planet-forming bodies; radiant and corpuscular energy from the Sun; and Joule heat. Accretional heating, however, appears to be the key mechanism responsible for the early heating and differentiation of the terrestrial planets, including the global process of separation into shells, whereas their post-accretional evolutionary paths (considered as secular cooling) were varied and depended on the size of the body. The post-accretional phases strongly influenced the existing structures and thermal state of their interiors and atmospheres.

Our views of planetary evolution were profoundly changed beginning in the mid-1980s as the result of the acceptance of Mars as the parent body of SNC (shergotite-nakhlite-chassognite) meteorites (McSween, 1984; Bogard et al., 1984; Becker and Pepin, 1984). An analysis of U/Pb isotopic composition of these meteorites showed that core formation of Mars took place about 4.6 billion years (Gyr) ago, or within a few hundred million years (Myr) thereof, in other words, within the accretion phase (Chen and Wasser-burg, 1986). Therefore, an earlier widely accepted idea that the core of Mars—and of the larger terrestrial planets, Earth and Venus—emerged subsequent to the accretion phase, after radioactive heating of the initially cold interiors (see, for example, Toksoz and John-ston, 1975; Toksoz et al., 1978; Solomon, 1979), was abandoned. Instead, the new con-cept argues that these planets began their post-accretional evolution fully differentiated and hot (Kaula, 1979; Wetherill, 1985; Shubert et al., 1988, 1992). Models that fully account for the heat of accretion, including the buried heat of large impactors and the development and subsequent thermal effects of an impact-induced atmosphere (Zahnle et al., 1988) show that the terrestrial planets became quite hot during accretion, and at least Earth and Venus developed magma oceans. This high temperature would have facilitated early core formation, which in its release of gravitational potential energy provides yet another significant heat source to a young terrestrial planet.

The post-accretional evolutionary path is determined by the balance between the in-tensity of thermal energy release (taking into account the heat of fusion) and cooling due to convection and thermal conductivity. One of the main sources of energy is radiogenic heat, generated by long-lived isotopes of uranium, thorium, and potassium, which belong to the group of lithophile elements: ^{238}U, ^{235}U, ^{232}Th, and ^{40}K. In modeling a planet's thermal evolution, one usually starts with an initial concentration of these elements, which satisfy a particular condensation model for a protoplanetary nebula. The most accurate quantitative criteria have been established for Earth, the Moon, and meteorites, for which the ratio of refractory lithophile elements, uranium and thorium, are approximately the same (~ 3.5), whereas significant discrepancies are observed in the relative potassium content. The high-est ratio, $K/U \sim (8 \cdot 10^4)$, is found for carbonaceous chondrites (similar in their composi-tion to the mean elemental content in the Sun). This ratio is nearly an order of magnitude less in terrestrial rocks and even half a magnitude less in lunar rocks, which in their chemi-cal composition turned out to be much closer to achondrites. As far as the other planets are concerned, the initial abundance of long-lived radioactive isotopes for Venus is assumed to be similar to Earth, and for Mars the rocks are intermediate between terrestrial and chondritic.

In the earliest evolutionary stage of the parent material of the terrestrial planets, short-lived radioactive isotopes may have played an important role—primarily isotopes of aluminum ^{26}Al, and also of beryllium ^{10}Be, iodine ^{129}I, chlorine ^{36}Cl, and certain trans-uranic elements, plutonium ^{244}Pu and curium ^{247}Cm. Because ^{26}Al and most of these other isotopes have a half-life of no more than 1–10 million years, all these isotopes are consid-ered extinct. Apparently, however, they promoted rapid heating of condensed, large mete-oritic bodies and protoplanets, having rapidly accelerated the start of their chemical differentiation. This helps to explain the similarity in the ages of meteorites of different composition. In contrast to isotopes of heavy elements, whose formation is governed by processes of nuclear synthesis, light isotopes (^{26}Al, ^{10}Be) are apparently products of super-nova explosions or the exposure of the protoplanetary nebula to corpuscular radiation from the young Sun (Wasserburg and Popanastassioo, 1982; Wetherill, 1981).

Release of radiogenic heat by short-lived isotopes was probably most active during the formation of planetesimals, and with small-sized bodies the heat would be rapidly radiated to the surrounding space. In a forming planet, the most intense heat release caused by accretion and the liberation of potential energy during gravitational differentiation of the constituent material would be later accompanied by heat release of long-lived isotopes (Wasserburg et al., 1964). The overall heat release inside Earth from this source over the past 4.6 billion years is approximately $2.5 \cdot 10^{38}$ ergs, according to estimates by Monin (1977), and the overall losses due to heat flow (given a current value of $61.5-80$ ergs $\cdot$ $cm^{-2}sec^{-1}$, or in average, about $1.8 \cdot 10^{-6}$ cal $\cdot$ $cm^{-2}sec^{-1}$) would not exceed $0.54 \cdot 10^{38}$ ergs. Consequently, approximately $1.8 \cdot 10^{38}$ ergs of stored heat would cause the heating and melting of Earth's interior. In such a case melting would have been achieved only in the vicinity of the core, because for complete melting to occur at all levels, nearly two times more energy (about $3.2 \cdot 10^{38}$ ergs) would be required. One could thus assume that Earth did not pass through this stage and that the mass average initial temperature of Earth's interior did not exceed 1700 K. However, a full accounting of accretional heating in the work of Zahnle et al. (1988) suggests that a significant portion of Earth's interior may have exceeded 2500 K during this epoch. This points to a sharp contrast between the first 1 Gyr and subsequent 3.5 Gyr of planetary thermal history.

The release of heat must be accompanied by heat exchange and cooling. The coefficient of thermal conductivity of the outer shells of the planet is small, and therefore equalization of temperatures by this path occurs extremely slowly and inefficiently. The primary mechanism of heat transfer, which determines the intensity of internal cooling, is convection. Core cooling by thermal or chemical compositional convection and generation of a magnetic field probably dominated the first stage of planetary evolution. A vigorously convective high-temperature mantle is assumed to be characteristic of this primordial stage, with convective transfer rapidly slowing down on the timescale of only a few hundred Myr. The subsequent heat exchange was balanced between radioactive heat production decay and convection, until the radiogenic isotopes were concentrated into shallow crust.

The presence of convective mixing and the removal of heat from depth are associated with melting. Because the melting temperature of silicates, but not of iron, is especially sensitive to increases in pressure, a certain optimal depth for their melting and solidification must exist, most likely in the middle and upper mantle. At this depth, light and heavy fractions of the melt material of chondritic composition drift to the surface and center, respectively. In this multistage process, essentially encompassing all interior regions, the formation of a core is completed, along with formation of the planet's lithosphere and asthenosphere. Because long-lived isotopes of the lithophilic elements uranium, thorium, and potassium have an affinity for silicates (that is, they have the ability to replace atoms in crystalline lattices constructed of silica SiO_2) they, along with silicates, drift upward by gravitational differentiation. They thus accumulate mainly in the basic rocks of the crust, and upon transition to the ultrabasic, less acidic rocks of the mantle their concentration drops sharply. The heat generated by these radioactive isotopes is evidently removed for

Fig. 4.4. Release of energy per unit of mass, as a function of time, during the thermal evolution of the Moon and planets of the terrestrial group (according to Toksoz and Johnston, 1975).

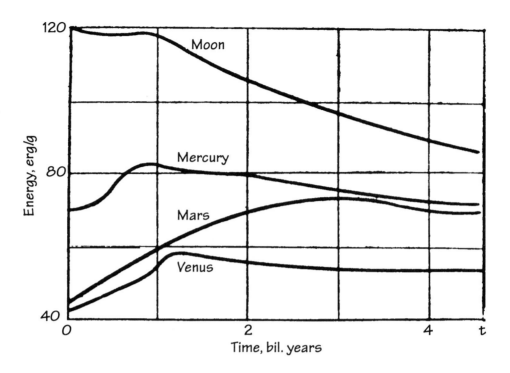

the most part by radiation from the surface, forming an observable heat flow, and is not expended to heat material located at great depth.

The described scenario is generally applicable to the Moon and other terrestrial planets whose thermal evolution is exhibited in the model shown in Fig. 4.4 (Toksoz and Johnston, 1975). The early thermal history of the Moon was no doubt dominated by the conditions of its peculiar origin. Although the origin of the Moon remained a fundamental mystery until long after the Apollo missions, with no one theory satisfying the basic geochemical and physical constraints, a new theory has rapidly gained favor. This is the so-called giant impact hypothesis (see Hartmann and Vail, 1986). It proposes that a Mars-size body struck Earth during its accumulation and that fragments from this collision formed an orbiting ring of material around Earth, from which the Moon accreted. The heat from lunar accretion, along with heating from the intense bombardment occurring at that time, resulted in a stage of widespread melting and volcanism, with the export of the light fraction of the melt out onto the surface and formation of a feldspar crust. After this, a period of continuous magmatic activity followed, which could be dated between 4.4 and 3.9 billion years ago, with the formation of rocks rich in aluminum and calcium, that is, anorthosites. At this stage, detrital rocks—breccias—formed, while partial or complete remelting of magmatic rocks occurred, owing to meteorite impacts and metamorphism of the ancient crust.

The most intense bombardment of the lunar surface by large meteoritic bodies continued until about 3.8 billion years ago (Wetherill, 1975). The meteorites ruptured the thin crust, which, upon the much later development of chambers of basaltic melt, led to filling of the cavities and, later still, their subsequent limited subsidence, with the formation of local concentrations of mass—mascons. Mascons are linked with anomalies in the lunar gravitational field and with places where internal shear stresses in the lithosphere are especially great. The asymmetrical distribution of lunar basins, which are the flattest regions of the surface, can be explained by the large concentration of basaltic melt on the visible side of the Moon and the relatively lighter and thicker crust on the far side, where meteorite bombardment merely led to the formation of deep craters in the hard rock.

This period was apparently the most intense in the Moon's evolution, corresponding to the maximum energy release and magmatic activity on the largest scale. At this time, the distribution of heat sources varied with depth, owing to the movement of silicate magma, enriched in isotopes of lithospheric elements, to the surface, and at the same time the settling of heavier elements to the center. A consequence of radiogenic heat release and gravitational differentiation of material was a rise in the temperature of the mantle, resulting in its melting. The filling of the lunar mare, in all probability, was largely completed about 3 billion years ago. This corresponds to the age of the youngest crystalline rocks, based on soil samples brought back to Earth, which were dated as being 3.16 billion years old. The period of melting was replaced by rapid cooling and the formation of an extended solid lithosphere, whose thickness grew, according to existing estimates, at a rate of 200–300 km/billion years. Consequently, the region where the melt occurred in the mantle moved deeper so that the zone of partially melted asthenosphere could now be preserved only near the center (see Fig. 4.3).

The fundamental differences in the theoretical models of the Moon's internal structure are associated with the existence of the necessary prerequisites for forming a metal core in the Moon. These models differ in their assumptions regarding the initial concentration of radioactive isotopes and, consequently, the activity of the heat sources, their uniform or non-uniform distribution with depth, the role of convection in heat transfer during the early stage of evolution, and the degree of solidification of the core as a consequence of cooling of the lunar interior. The temperature in the core depends on its composition and is apparently within the limits of 1300–1900 K. The hypothesis that the heavy fraction of the protomaterial is enriched in sulfur, primarily in the form of sulfides, and that the core formed from a eutectic of Fe-FeS with a melting temperature (which depends only slightly on pressure) of about 1300 K, corresponds to the lower limit. The proposition that the lunar protomaterial is enriched in light metals (magnesium, calcium, sodium, and aluminum), which, along with silicon and oxygen, enter into the composition of the most important rock-forming basic and ultrabasic minerals, pyroxenes and olivines, is more in line with the upper limit. The fact that the Moon has a lower-than-expected concentration of iron and nickel, which is definitely indicated by its low mean density, is favorable for this second hypothesis and for the most plausible model of the Moon's origin, the impact model.

The existence of a thick, cold lithosphere, capable of withstanding the stresses created by mascons, explains the Moon's present tectonic inactivity. The limiting value for the liberated seismic energy (with zones of activity at depths of 700–1200 km), according to estimates, does not exceed 10^{13} ergs/year, while in Earth it reaches 10^{25} ergs/year, which, on average, is $3 \cdot 10^{10}$ Watts.

Mercury shares many common topographical features with the Moon. In Mercury and the Moon, unique examples of the most ancient structures are preserved, not altered by subsequent processes. The condensation nature of Mercury's parent material was evidently different, however, represented principally by a relatively high-temperature fraction of iron meteorites. In fact, its mean density is considerably higher than the Moon's and only somewhat less than that of Earth. Material in Earth's interior is under significantly greater pressure owing to differences in the masses of the two planets; consequently, to achieve nearly the same mean density, Mercury must contain a relatively greater amount of heavy elements. Taking cosmic abundances of elements into consideration, the most important of these must be iron. According to earlier estimates (Urey, 1951, 1952; Lewis, 1972, 1973), the Fe/Si ratio of Mercury must be approximately two times greater than that of Earth and five times greater than that of Mars. Later models argue for a sulfur concentration of about 2%–3% to accommodate both the geologic and magnetic constraints.

Traces of early volcanism on Mercury's surface suggest that the process of material differentiation occurred early, soon after the completion of the primary accretion phase. Mercury probably had a hot start, with core formation followed by gradual cooling via mantle convection. The high conductivity of iron, which makes up about 70% of Mercury's mass, may have strongly affected the further thermal evolution of this planet. According to contemporary models reviewed by Schubert et al. (1988), the iron-nickel inner core of Mercury began to grow and solidify soon after accretion, and this process continues to date. The inner core is surrounded by a relatively thin outer core containing a light element such as sulfur, and it is this outer core that is responsible for the intrinsic magnetic field of the planet.

The thermal history of Mercury is intimately related to the origin of its magnetic field. Indeed, the observed magnetic field assumes particular importance as an indirect indicator and possible test of evolutionary models. These models assume an early magnetic field dynamo driven by thermal convection in a largely molten core, and preservation of a subsolidus Fe-S eutectic in a liquid outer core allowing for convection and dynamo generation of the observed magnetic field. The convection could be driven and maintained by release of gravitational energy and latent heat due to continuing growth of the inner core, which is estimated to make up about 3/4 of the diameter of Mercury (that is, approximately equal to the size of the Moon).

It is as yet uncertain whether the criteria for dynamo generation are currently satisfied. Nonetheless, a fluid outer core overlying the inner one is a likely feature of the present structure. The outer core is surrounded by a thin mantle, probably consisting of magnesium silicate rocks such as olivines. Along with the upper crustal layer, this forms Mercury's solid lithosphere. Lithospheric thickness, presumably, has increased from ~200 km about 3 billion years ago to ~ 500 km presently. The great thickness of the lithosphere provides a basis for inferring a low level of current tectonic activity on Mercury. One can assume, however, that as a consequence of more prolonged cooling, global tectonic processes and ancient volcanism may have encompassed a greater period of Mercury's history than that of the Moon. Hence, it follows that the age of the youngest rocks in the lowland regions of the planet, inside craters and basins that underwent filling by outflowing lava, must be considerably less than the minimum age of 3.16 billion years found for lunar rocks.

In size, Mars occupies an intermediate position between the Moon and Mercury, on the one hand, and the Earth and Venus, on the other. The characteristic features of geological structures on its surface serve as a good criterion for evaluating the calculated evolutionary models and suggest that very early differentiation of the interior material occurred on Mars. This is indicated by the aforementioned SNC meteorites, as well as by traces of original magmatic activity preserved on individual, most ancient areas of the surface, and by the chemical composition of surface rocks. Because accretional heating and core formation were essentially contemporaneous with planetary formation, the early history of Mars was characterized by high internal temperatures, rapid differentiation into a core and mantle, and high surface fluxes of heat and magma (Schubert et al., 1992).

Rapid convective cooling of Mars' interior and relatively fast exhaustion of radionuclides led rather soon to a globally thick lithosphere, and may have been accompanied by global contraction, documented in the pervasive formation of wrinkle-ridges in the ancient terrain. The limited tectonic and volcanic activity of Mars in the past 3.5 Gyr is characteristic of its continuing cooling since the post-accretion phase.

During the period of about its first 1–1.5 Gyr, Mars probably reached the peak of its surface activity—characterized by intense volcanism and tectonic activity, and by the formation of basaltic lowlands and volcanic shields, after which the planet gradually started to cool. An almost constant level of thermal energy is maintained: heat flow, in the current epoch, is estimated to be 40 ergs/cm^2 · sec, nearly the same as in Precambrian shields of Earth. The thickness of the lithosphere evidently reaches several hundreds of kilometers, of which about 100 km makes up its upper layer—the Martian crust. The comparatively large thickness of the lithosphere provides a basis for inferring moderate seismic activity on Mars at present.

Understanding of thermal evolution of Mars and its core would be strongly constrained by detection of its intrinsic magnetic field, yet the current evidence is still ambiguous. The problem is closely related to the uncertainties in the temperature in the central interior of Mars. An almost completely melted though nonconvecting core, or a merely weakly thermally or chemically convecting core, could explain the lack of a present magnetic field or the existence of only a very weak one. The situation hinges on the assumption of a critical sulfur concentration in the core: about 15 wt % S probably would not crystallize a solid inner core, this estimate being close to the sulfur content from elemental abundance of SNC meteorites (Shubert et al., 1992). Let us note that it is difficult to satisfy the requirement of a high temperature in the central interior of Mars if we consider only metalsilicate fractionation of the original material—which is what enables us to explain the low mean density of the planet, owing to the overall depletion of iron. We can overcome this difficulty if we consider the probable fractionation of iron and sulfur and the retention of a higher than usual percentage of chalcophile elements at the relatively low condensation temperatures found at the orbit of Mars. This possibility adds additional support for the assumption that the core is a mixture of iron and ferrous sulfide, taking into account also that the eutectic Fe-FeS remains liquid at relatively low temperatures (about 1300 K). Assuming also that potassium entered into a sulfide phase, it is possible that heat sources survived in the core, owing to the decay of ^{40}K.

Because a significant portion of the iron may have combined with sulfur, one can imagine that the Martian mantle is likewise rich in ferrous sulfide, and that in the composition of its silicates there are more minerals with a higher concentration of iron than magnesium. The high abundance of iron, which is greater than found in rocks on Earth and Venus, was discovered in the constituent material of surface rocks. This leads to the hypothesis that gravitational differentiation of Martian material was not as profound and complete as on other terrestrial planets. It is precisely this situation—the insufficiently complete segregation of metallic iron—that is linked with iron's higher than usual content in Martian rocks, while the overall relative iron content in Martian material apparently does not exceed $\sim 25\%$, which is considerably less than on Earth, Venus, and, of course, Mercury. The value of the generally accepted dimensionless moment of inertia ($I = 0.375$) places a severe limitation on the degree of differentiation in Mars: it indicates a relatively small deviation from a homogeneous density distribution, which is consistent with ideas about the presence of a relatively small iron sulfide core. If the core radius is estimated to be approximately 800–1500 km, then the mass of the core would be less than 9% of the total mass of the planet (Zharkov et al., 1981; Schubert et al., 1992).

This comparative study of Venus' neighbors in the solar system provides additional support for the accretional scenario and cosmochemical scheme, which treat Venus as a planet that also differentiated very early. Accretional considerations favor early core formation and initial melting of much of the outer portions of the planet. Thermal history models for Venus admit present core and mantle states essentially similar to Earth because Venus presumably contains interior radiogenic heat sources (uranium, thorium, and potassium) quite similar to Earth's. At the same time it is reasonable, in calculating models of Venus' interior, to start with a protomaterial composition at its orbit similar to that of Earth (although the use of the equilibrium condensation model [Lewis, 1974] or oxidation state model [Ringwood and Anderson, 1977] assumes definite differences). In particular, Venus could accrete less volatile-rich material than Earth. The recently popular accretion scenarios involving late stages of accretion dominated by giant impacts suggest the possibility of a large stochastic component to initial composition of terrestrial planets.

Remaining uncertainties concerning cosmochemical characteristics may still exert a greater influence on the differences in the internal structure of Earth and Venus than is predicted by the insignificant difference in the mean density of the two planets, which (taking into account the correction for compression) is less than 3% (Phillips and Malin, 1983). In other words, this difference lies within the tolerances and limitations imposed on the cosmochemical (and petrological) models of both planets, which specify the chemical composition of their constituent material. Nevertheless, one could imagine that the postaccretional thermal evolution of Venus, whose interior structure we will now analyze in greater detail, was roughly similar to Earth's.

A Model of the Interior of Venus

In contrast to Earth and the Moon, seismic data is not yet available for Venus. Therefore, models of its internal structure are limited by the need to make additional assumptions. The most important limitations on the choice of computational internal models are imposed by the currently known observational characteristics of the planet. The essential characteristics (with the exception of the equatorial radius, mass, and mean density, contained in Table 1.3) are presented in Table 4.1. Most of these were mentioned earlier: they are the dimensionless moment of inertia, I, the first coefficients of the gravitational potential expansion over the spherical functions, J_2, C_{22}, and S_{22}, the dynamic (α) and geometric (e) compression of the planet, and the quantity, f. Aside from these, we include the value of the so-called small parameter of the theory of figures, q, which is the ratio of the centrifugal acceleration at the equator, $\omega = 2\pi/T_{rot}$, to the gravitational acceleration $q = GM/R_e^2$ ($q = \omega^2 R_e^3/GM$), and the ratio J_2/q. Obviously, the greater q, the more vigorously the centrifugal forces expand the body of the planet in the equatorial plane and the more it is compressed along the polar axis. The quantity q is inversely proportional to the square of the rotational period, $q \sim T_{rot}^{-2}$.

From a comparison of the parameters of Venus with those of Earth (see Table 4.1), one important conclusion about the planet's equilibrium state can be made. That the figure of Venus deviates from hydrostatic equilibrium is known from Chapter 1. Now we will return to a more rigorous investigation of this fact. From theory it follows that, for a planet in equilibrium, J_2 and q are approximately of the same order, which is actually observed for Earth. Yet, for Venus, J_2 exceeds q by factors of (65 ± 25), or (96 ± 5.7), depending on the accepted initial value of J_2 (see Table 4.1). This means that the internal structure of Venus is in a state of extreme disequilibrium—in fact, among the planetary bodies in the solar system, it is one of the farthest from equilibrium (Zharkov, 1985). In other words, its figure conforms to considerably more rapid rotation (by several tens of times) than its current value. It has been hypothesized that this is not a random fact but that in the distant past Venus' rotation could have been slowed by tidal friction. Consequently, the value of q may be a relic of this era in Venus' history when, up to the moment of rotational slowing, the viscosity (or rigidity) of its mantle grew, as a consequence of appreciable cooling, so that the figure of the planet could not achieve an equilibrium form corresponding to the modern value of T_{rot}.

The value of f, determined from the gravitational potential coefficients C_{22} and S_{22} measured using spacecraft, proves to be even less than for Earth: for Venus, $f = 3.8 \cdot 10^{-6}$. This suggests that the density distribution in its interior is close to axisymmetric and provides additional support for the proposition that the large disequilibrium value, J_2/q, for the planet is a relic. From the values of J_2 and f, it is easy to estimate the difference $C - B$ and $B - A$, which turn out to be essentially indistinguishable from one another $(3.98 \cdot 10^{-6}\, MR_\venus^2$ and $3.78 \cdot 10^{-6}\, MR_\venus^2$, respectively). Hence, it follows that the figure of Venus (its outer equipotential surface) is closer to a triaxial ellipsoid than to an ellipsoid of revolution.

Table 4.1   Figure parameters for Venus and Earth

Parameter	Venus	Earth
I	0.334	0.3308
q	$6.5 \cdot 10^{-8}$	$3.47 \cdot 10^{-3}$
$J_2, 10^{-6}$	4.0 ± 1.5^a	1082.64
	5.87 ± 0.35^b	
J_2/q	65 ± 25^a	0.31
	96 ± 5.7^b	
$J_2^{\circ} 10^{-6}$	—	1072
$\Delta J_2 = (J_2 - J_2^{\circ})\, 10^{-6}$	4.0 ± 1.5^a	10
	5.87 ± 0.35^b	
α^{-1}		298.26
e^{-1}		298.18
$C_{22}\, 10^{-6}$	0.9147 ± 0.13	1.565
$-S_{22}\, 10^{-6}$	0.2392 ± 0.11	0.894
$f\, 10^{-6}$	3.8	7.2
$g\, \mathrm{cm} \cdot \mathrm{c}^{-2}$	887	982

Source: Data from Zharkov, 1985.

Notes: See the text for further explanation.

$I = \overline{I}/MR^2 \approx C/MR_e^2$ dimensionless moment of inertia (M = mass of a planet; R = radius; R_e = equatorial radius; C = polar moment of inertia.)

$q = \dfrac{\omega^2 R_e^3}{GM}$ ratio of centrifugal acceleration at the equator to acceleration due to gravity, GM/R_e^2.

$\alpha = 3/2\delta_2 + 1/2\, q$ dynamical oblateness

$e = \dfrac{R_e - R_p}{R_e}$ geometrical oblateness

[a] According to Akim et al., 1978.
[b] According to Williams et al., 1983.

In deriving models of Venus' interior, one usually starts from a condition of hydrostatic equilibrium, despite the planet's distinct disequilibrium state, assuming that at the yield strength of rock, $\geq 10^3$ bars (taking into account high temperatures and geological time intervals), the disequilibrium stresses over the greater part of the planet's interior are much less than the hydrostatic pressure $P(r)$, where r is the distance from the center. Such an approach has found broad application, mainly in constructing Earth-like models of Venus' internal structure. In the more representative physical models, large-scale shear stresses, caused by the non-hydrostatic state of the planet and maintained by rigid zones of the crust and mantle, are taken into account. In this case, a two-layer model, capable of bearing non-hydrostatic loads somewhere at the mantle-core interface, is closer to reality than a homogeneous elastic model over the entire interior. We shall consider both these modeling approaches, based mainly on the results obtained by Zharkov (1985).

In hydrostatic, Earth-like models of Venus' interior, the equation of state, $P(\rho)$, used for interior models of the Earth is initially used. This is associated with the reasonable assumption that the temperature distribution inside both planets, starting at depths greater than ~ 200 km, is nearly identical, and therefore the effect of this distribution on the equation of state is identical (Dziewonski et al., 1975; Zharkov and Trubitsyn, 1980; Dziewonski and Anderson, 1980). The largest correction to the equation of state, corresponding to the well-known Parametric Earth Model (PEM), takes into account a certain reduction in the density of the mantle, owing to an assumed systematic decrease in the iron content of mantle silicates as one moves from Mars to Mercury. An a priori estimate of the amount of material extracted from the upper mantle (whose boundary corresponds to the second phase transition level) and, consequently, the thickness of Venus' crust, is essential for this model. This estimate depends on the degree of crustal material exchange between the crust and mantle or, in other words, on the question of the existence and extent of global tectonics on the planet (see Chapter 3).

In calculating the temperature and pressure trend in the core, we likewise infer that its material is similar to terrestrial material and, thus, the same equation of state is used as for Earth. Possible errors in such an assumption are related to uncertainties in the estimates of the ferrous sulfide (FeS) content in an iron core, if one holds to the condensation model of the protoplanetary cloud. If the admixture of FeS actually increases as one moves from Mercury to Mars, then definite differences may exist between the core material of Earth and Venus. Obviously, the limiting models will be a completely Earth-like core (containing an admixture of FeS, on the order of 10%) and a purely iron core, which are described by slightly different equations of state. Phase diagrams for systems of magnesium- and iron-bearing silicates (olivines, pyroxenes), in the form of Mg_2SiO_4-Fe_2SiO_4, $MgSiO_3$-$FeSiO_3$, and $MgSiO_3$-Al_2O_3, are also used in calculations of Venus' mantle separated into mineralogical zones, following the works of Akimoto et al. (1976) and Liu (1977).

An example of an Earth-like model calculation for Venus' interior using this approach is shown in Fig. 4.5, and Table 4.2 points out the primary mineralogical zones, corresponding to the derived T and P distributions. Mineralogical composition of the outer silicate shell of Venus is assumed to be generally similar to that of Earth. In Venus' mantle, olivines and pyroxenes dominate, while at greater depth their high-pressure modifications are more prevalent. A temperature dependence versus depth was obtained, based on a priori temperature estimates: at the lithosphere-mantle interface, $T \approx 1200$ K (for a lithosphere 70 km thick), and at the mantle-core boundary, $T \approx 3500$ K. Under conditions of adiabatic equilibrium in the core, the temperature at the planet's center turns out to be 4670 K. This is considerably higher (by almost 1200 K) than the results of earlier calculations by Toksoz et al. (1978); although the temperature distributions in the mantle are similar according to the two models, the $T(r)$ trend in the core is significantly different. At present it is difficult to resolve this uncertainty. Nevertheless, taking into account Earth's internal structure, an adiabatic temperature trend in the core and the presence of convective heat transfer in the planet's interior seem to be the most correct, from a physical point of view.

In contrast to the Earth-like models we have been describing, physical models of Venus' interior are constructed based on a fundamental analysis of the thermodynamics of

Fig. 4.5. An Earth-like model of Venus. To the left along the y-axis, the temperature T (K) and gravitational acceleration (cm · sec^{-2}) are displayed; to the right are the pressure (Mbar) and density (g · cm^{-3}); along the x-axis, the distance from the center r is shown (in relative units with respect to the radius $R_{\venus}$). The sharp bends in the curves correspond to the core-mantle boundary. Temperatures in the core follow along an adiabat.

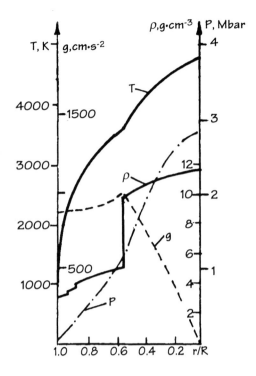

the mantle and core and the limits imposed by observational data on the gravitational field and figure of the planet. Initial characteristics here are the calculated Debye temperature, Θ, and the Grüneisen constant, γ. With the framework of the Debye model, one can compute (separately for the mantle and core) the γ versus depth, ℓ, dependence, and all fundamental thermodynamic coefficients, such as the adiabatic compression modulus, thermal pressure, the coefficient of thermal expansion, and the melting temperature curve of the planet's interior and, consequently, one can obtain the slope of the phase equilibrium curve. Moreover, like the simple Parametric Earth Model (PEM), one can also construct a simple Parametric Venus Model (PVM), in which the density, $\rho(x)$, and velocity distributions of longitudinal and transverse body waves, $v_P(x)$ and $v_S(x)$, are defined as piecewise continuous analytical functions of the dimensionless radius $x = r/R$. The Grüneisen constant, γ, is calculated to fit this distribution.

The $\Theta(\ell)$ relationships, the adiabatic temperatures, $T_a(\ell)$, the "experimental" smoothed temperature profiles, $T_f(\ell)$, and the melting temperature of the mantle, $T_M^m(\ell)$, are shown in Fig. 4.6a. The temperature variation does not differ greatly from $T(\ell)$, calculated in an Earth-like model. The boundary between the upper and lower mantle is located at the level of the second phase transition, at a depth of $\ell = 760$ km. In Fig. 4.6b, possible adiabatic temperature distributions in the core whose thermodynamic parameters are also computed using the Debye model (Zharkov and Zasursky, 1980, 1982), are presented. The upper boundary of the liquid convective core is assumed to be at the same level as in Earth-like models, and, moreover, it is assumed that Venus, like Earth, has a solid inner core, whose boundary is found at a depth of ~ 5000 km. At approximately this level in Earth, the adiabat of the convective outer core intersects the melting curve. In Fig. 4.6b, the "experimental" adiabats T_{af}^c and melting curves T_M^c, corresponding to this condition,

Table 4.2   Mineralogical zones of Venus

Zone	Depth, km	Basic mineral phases	
Upper mantle (zone **B**)	70 olivine zone 480	olivine (alpha-phase)	pyroxene + Al_2O_3 garnet → garnet
Transition zone (zone **C**)	spinel zone 760 ilmenite and perovskite zone 1000	beta-phase → spinel ilmenite → perovskite + ferropericlase (Mg, Fe)O	(gamma-phase) → garnet ilmenite — ilmenite perovskite — perovskite
Lower mantle (zone **D**)	perovskite zone 2840	perovskite + (Mg, Fe)O	perovskite — perovskite
Core	iron zone 6050	Fe(+FeS?)	

Source: Data from Zharkov, 1985

are compared with the experimental band for the melting curve of iron T_M^{Fe}. For a core with a composition of almost pure iron, only the upper curve satisfies the specified condition; this means that the temperature at the liquid core-mantle interface must be extremely high ($T_a^{cm} \simeq 4500$ K). The melting temperature at this boundary will be 3900 K in this instance, whereas the temperature at the boundary of the inner core is 5850 K. Other melting curves lie below the experimental band for iron, and, consequently, the validity of their corresponding adiabats with a lower temperature at the core-mantle boundary (4000 and 3500 K) turns out to be problematic and is not in line with the hypothesis of a solid inner core.

Based on an examination of the amount of heat flow from Venus' interior owing to convective heat transfer, one could conclude that at the core-mantle boundary (just as at the lower and upper mantle boundary) a zone of heating, relatively small in extent, with a superadiabatic temperature gradient is created. The temperature differential in these zones, 300–600 km thick, may reach 600–800 K. Because adiabatic temperatures at the mantle-core interface, according to the results of thermodynamic calculations for the mantle, are $T_a^m = 2900$ K (see Fig. 4.6a), the actual temperatures at this interface (taking into account the heating zone) may, apparently, lie in the 3000–4000 K range. The problem of how to reconcile these temperatures (and even higher ones) with a planetary mantle that is convective throughout remains unsolved, along with the question of the size of Venus' liquid core and the existence of a solid inner core.

Note that this question is directly related to the explanation of why Venus does not now have an intrinsic magnetic field. Studies of the conditions under which a magnetic

Fig. 4.6. (a) Calculated temperatures in the mantle of Venus depending on the depth l: Θ is the Debye temperature; T_a are adiabatic temperatures; T_f are "experimental" smoothed temperatures; T_M^m is the melting temperature of the mantle; and γ is the Grüneisen constant, according to the PVM model. (b) "Experimental" adiabatic temperatures T_{af}^c (solid curves), and melting temperatures of the core T_M^c (dashed curves) depending on depth l. The shaded region is an experimental band for the iron melting curve (according to Zharkov, 1985).

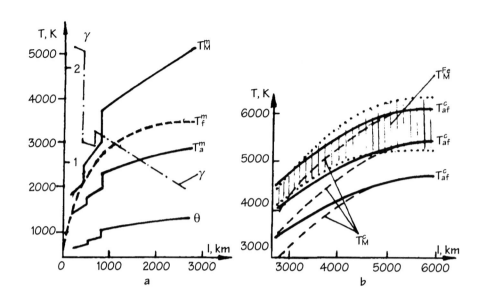

field is generated in a planet's core by a hydromagnetic dynamo mechanism, given thermal or chemical convection (Stevenson et al., 1983), have led to the conclusion that several core states may possibly be realized, as a function of minor changes in the parameters for the two-layer model under consideration. The case of a convective outer and growing inner core does not satisfy the present conditions, but it may have been realized in Venus' distant past by vigorous cooling of the originally hot planet. The lack of magnetic field generation is guaranteed by the following alternatives: a crystallized core with a thin liquid outer shell, or a liquid but non-convective core. The second of these variations, graphically shown in Fig. 4.7, seems more likely, although it does not completely conform to the model for formation of an adiabatic temperature profile in the planet's interior.

The structure of the Venusian interior is directly linked with estimates of heat flow liberated by the body of the planet and thus, in turn, is associated with the nature and intensity of tectonic processes on its surface. Under conditions of gravitational differentiation in the planet's interior long since completed, and in the absence of material exchange between the upper and lower mantles (as is thought to be the case for Earth, according to O'Nions et al., 1979, and Wasserburg and De Paolo, 1979), on Venus the primary internal heat sources will be radioactive mixtures in the crust and mantle, primordial heat, and heat flow from the core to the mantle. Radiogenic heat sources, which migrated to the basaltic crust as it melted out from the mantle, play a fundamental role in heat generation; this hypothesis is supported by results determining the intrinsic radioactivity of Venus' surface rocks at spacecraft landing sites. In turn, heat flow from the interior, q_o^V, is determined by the temperature at the lower boundary of the lithosphere, T_M^m, and the crustal thickness, d_{cr}.

For $T_M \approx 1700$ K (see Fig. 4.6a), a lithospheric thickness $d_L = 200$ km, and $d_{cr} = 70$ km, the mean value is $q_o^V \approx 30$ ergs $\cdot$ cm^{-2}sec^{-1}, which is 2–3 times less than the heat flow from Earth's interior (recall that $q_o^E = 61.5$–80 ergs $\cdot$ cm^{-2}sec^{-1}), and approximately the same factor less than the corresponding estimate for Venus by Toksoz et al. (1978;

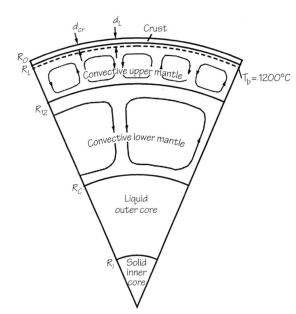

Fig. 4.7. A model of Venus' interior, which assumes the presence of a liquid non-convective outer core and a solid inner core. Convection occurs separately in the lower and upper mantles; the zone where melting begins, at a temperature of T_b, is denoted by a dashed line. The boundaries between the upper and lower mantles (R_{12}), the mantle and core (R_c), and the inner and outer core (R_i) correspond to the calculated relations in Fig. 4.6. d_{cr} is the crustal thickness, and d_L is the thickness of the lithosphere (according to Zharkov, 1989).

77 ergs $\cdot$ cm^{-2}sec^{-1}). The reason for this is the assumed greater thickness of the lithosphere and crust in Zharkov's (1985) model.

Both these models are difficult to reconcile with a lithosphere on Venus thick enough to ensure passive isostatic compensation of topographic irregularities on Venus' surface. The required depth of the compensating layer can be estimated, from the relationship between the coefficients of viscosity and rigidity over a typical timescale, on the order of 10^7–10^8 years—the so-called Maxwell time (Phillips and Malin, 1981, 1983). Another possibility is dynamic compensation with a thinner lithosphere, in which uplifted areas of relief are supported by less dense roots, retreating deep under the lithosphere. Here, dynamic support is furnished by buoyancy forces in regions of rising convective motions in the planet's interior, either directly or by heating and expansion of the overlying lithosphere.

Turcotte (1993) finds that a lithosphere thick enough to support the observed topography would result in a surface heat flux too low to be in a steady state with internal radiogenic sources. Alternatively, a heat flux that is simply scaled to Earth's mean heat flux, 62 mW m^{-2}, would lead to a thin lithosphere that could not support the topography and would allow the relaxation of large impact craters, which is not observed. This led Turcotte to propose episodic plate tectonics on Venus where periods of high heat flux and active plate recycling with a thin lithosphere are interspersed on a 500 million-year timescale, with tectonically and volcanically quiescent periods (such as the present epoch) of lower heat flow and thicker lithospheres.

Finally, accounting for Venus' non-hydrostatic state enables us to estimate the shear stresses in its interior, specified by the quantity ΔJ_2. Calculations presented for the two-layer model of Venus (Zharkov and Zasursky, 1981) have indicated that these stresses in the lower mantle do not exceed 3–10 bars. In other words, because of high temperatures, the lower mantle is incapable of withstanding non-hydrostatic loads over large time intervals, and it behaves like a liquid. In Venus' upper mantle and asthenosphere, stress levels are apparently several times lower. If we keep in mind that the thickness of the seismically active surface layer on Earth is ~ 15 km, then for a geothermal gradient at Earth's surface of $(20-30 \text{ K} \cdot \text{km}^{-1})$, the temperature at the base of the seismically active layer turns out to be $570-720$ K, that is, lower than the surface temperature of Venus $(T_S = 735$ K$)$. This could mean that the planet does not possess an external seismically active layer. If, moreover, we allow for low shear stress levels in its interior, then one can infer that Venus may be largely aseismic.

The Lower and Middle Atmosphere

5 Although the atmospheres of the terrestrial planets are thin compared with the solid bodies, and so can be likened to the skin of an apple, their importance can hardly be overestimated. The atmospheres are largely responsible for planetary environmental conditions and their dramatic differences. The principal new development in atmospheric science during the last decades of the twentieth century has been the rapid and exciting progress in our knowledge of the atmospheres of planets other than Earth, especially the neighboring planets Venus and Mars. Many advances in understanding of the physical processes occurring in planetary atmospheres have come from using the approach of comparative planetology. A wealth of information about specific atmospheric features has provided insight into the thermal regime and dynamical processes in the atmosphere on the local and global scales responsible for weather and climate. Moreover, planetary atmospheres are intimately linked with surface and interior processes, and thus these accessible envelopes make it possible to look into the past and learn about the origin and evolution of the planets.

The gases making up the atmosphere can be divided into major and trace gases. Major gases compose the bulk of the atmosphere and remain generally in the same proportion through time and from place to place. In Earth's atmosphere they are nitrogen (78%), oxygen (21%), and argon (about 1%). Trace gases are usually variable in both time and space. The main variable gases in Earth's atmosphere are water vapor, carbon dioxide, and ozone. These gases play a tremendous role in the heat balance, dynamics, and weather, and help to provide favorable conditions for life on the home planet. In turn, the carbon dioxide composing the principal part of the atmospheres of Venus and Mars is responsible for the completely different conditions existing on our neighboring planets.

Three quantities — temperature, pressure, and density — together with chemical composition, define the state, structure, and properties of a planetary atmosphere. Based on a variety of changes in the structure and properties that occur with height, the atmosphere is commonly divided vertically into several zones, beginning with the low and dense troposphere (where weather occurs) and then up to the stratosphere and mesosphere (together referred to as the middle atmosphere) and at the top the very rarefied thermosphere and essentially unconfined exosphere (together forming the upper atmosphere). The decrease of temperature with height (temperature lapse rate) in the lower troposphere of Earth normally averages 3–5 degrees C/km and reaches 6–7 degrees C/km in the middle troposphere. Comparable lapse rates are pertinent to the tropospheres of Venus and Mars. Temperature inversions can occur under certain conditions, which affects the associated vertical density distribution and thus the atmospheric stability. Such inversions develop periodically at all altitudes of Earth's atmosphere, within several hundred meters of the ground, under clear skies at night when the surface cools more rapidly than the air above it. These inversions are common in polar regions throughout the year and can also develop in the case of advection of a warm air mass over a cold surface. A similar situation regularly occurs at the near-surface layer of the thin Martian atmosphere (except when the atmosphere is very dusty) but is not pertinent to the extremely thick and hot atmosphere of Venus.

In simple mathematical form the relationship between the main atmospheric parameters is clearly defined by the hydrostatic equation, which essentially states that upward acceleration of gas molecules due to the pressure decrease must exactly balance downward acceleration due to the force of gravity. When coupled with the gas equation of state, this results in the barometric formula, which defines the exponentially decreasing character of pressure and density in the gravitational field of a planet. The exponent in this formula, and thus the rate of atmospheric thinning with height, is determined by the so-called scale height, which depends on temperature, gas molecular weight, and acceleration due to gravity. We shall explain these equations in more detail when discussing the inverse problem of height determination over the Venusian surface from direct temperature and pressure measurements made by descending Venera probes.

Venus is distinguished among the terrestrial planets primarily by its massive gaseous envelope consisting mainly of carbon dioxide. Because of its unusual atmosphere, Venus' surface is heated to a temperature of several hundreds of degrees Celsius, the pressure at the surface is almost two orders of magnitude greater than on Earth, and the atmospheric density is only approximately an order of magnitude less dense than that of water. These unique properties give rise to fundamental differences between the meteorology of Venus and that of the other terrestrial planets that possess atmospheres, Earth and Mars.

By the lower atmosphere of Venus we mean the entire troposphere, that is, the altitude range from the surface to the upper boundary of the clouds (65–70 km); the middle atmosphere, sometimes called the stratomesosphere, extends from the troposphere to approximately 130 km. The overlying regions will be classified as the upper atmosphere (see Chapter 8).

Much data has been accumulated about the chemical composition, structure, and properties of Venus' lower atmosphere. Nevertheless, much concerning key problems of the meteorology and genesis of the Venusian atmosphere remains unclear. The gaps in knowledge are largely associated with the limited spatial-temporal coverage of the experimental data, and, in a number of cases, with the relatively low measurement accuracy. The second problem concerns both measurements of structural parameters and compositional measurements, especially regarding the concentration of minor components and their distribution with altitude. First we shall review available measurements, briefly touching on the methodological side of the question, and then we shall turn our attention to the structure and properties of Venus' atmosphere.

Measurements of the Temperature and Pressure

Direct measurements, conducted by Venera, Vega, and Pioneer-Venus spacecraft (in all, sixteen cutaway views of the atmosphere), played a decisive role in determining the parameters of Venus' troposphere and the temperature and pressure at the surface; the structure of the upper troposphere and stratosphere was measured using the radio occultation technique from the Mariner-10 fly-by mission and the Venera, Pioneer-Venus, and Magellan spacecraft. These regions of the atmosphere were also measured by several Venera landers and Pioneer-Venus probes using accelerometers. (A chronology of these flights, indicating landing coordinates and local solar time, is presented in Tables 2.2 and 2.3; Table 5.1 provides the measurement methods used and the corresponding altitude range for each spacecraft. Methodological questions concerning the measurement techniques, along with a detailed analysis of the results obtained, can be found in the following works: Avduevsky et al., 1968a, 1968b, 1969, 1979; Kuzmin and Marov, 1974; Marov, 1972, 1978, 1979a, 1979b; Avduevsky, Marov et al., 1983a; Kliore et al., 1967, 1969; Howard et al., 1974; Yakovlev et al., 1978, 1982; Seiff et al., 1980; and Seiff and Kirk, 1982a). Vertical cross sections of the atmosphere have been studied in the regions indicated in Table 2.3, covering altitudes from approximately 120 km to the surface on the night and daylight sides and at various latitudes. In addition, the horizontal structure of the atmosphere in the cloud layer has been studied at about 55 km in the near-equatorial zone from the drift of balloon probes jettisoned by the Vega-1 and Vega-2 spacecraft (Sagdeev et al., 1986; Linkin et al., 1986a, 1986b). Data on the above-cloud atmosphere and its variations were obtained from remote sensing results in the infrared spectral range, by the Pioneer-Venus (Schofield and Taylor, 1983) and Venera-15 and Venera-16 satellites (Ertel et al., 1984), providing almost global coverage.

The first direct temperature and pressure measurements by Venera-4, Venera-5, and Venera-6 terminated at altitudes of 24–16 km; surface temperature and pressure, T_S and P_S, were extrapolated assuming an adiabatic atmosphere. These values were confirmed soon thereafter by direct temperature measurements from Venera-7 and temperature and pressure measurements from Venera-8. Thus, Venera-7 and Venera-8 reliably established that the surface of Venus is heated to 740 K and the pressure reaches 92 atm. This temperature proved to be even higher than one would expect from interpreting radio astronomical measurements (conducted assuming emission on an absolutely black body). Subsequently, measured temperatures and pressures at the surface and their altitude relations to an altitude of ~60 km were confirmed by similar experiments on the Venera 9–14 and Vega-2 spacecraft.

Direct T and P measurements by the Pioneer-Venus probes, like the Venera, were begun at approximately 60–65 km but, for as yet unknown reasons, ceased at an altitude of 12 km above the surface.[1] Therefore the values on the surface, as with the first Venera, were extrapolated for an adiabatic temperature profile (which was established conclusively before these experiments were conducted). Moreover, during these experiments atmospheric parameters were measured to a high accuracy, up to 0.2% for the temperature (1 K

Table 5.1   Measurement methods of atmospheric parameters and altitude ranges of space missions to Venus

Spacecraft	Date of measurement (day/month/year)	Local solar time on Venus	Coordinates of landing (descent)	Measured parameters and method[a]	Altitude range (km)
Venera-4	18/10/67	4:40	19°; 38°	T, P	51–24
Venera-5	16/05/69	5:30	−3°; 18°	T, P	55–16
Venera-6	17/05/69	5:30	−5°; 23°	T, P	49–16
Venera-7	15/12/70	5:00	−5°; 351°	T	55–0
Venera-8	22/07/72	6:20	−10°; 335°	T, P	55–0
				accelerometry	110–65
Venera-9	22/10/75	13:15	31.7°; 290.8°	T, P	62–0
				accelerometry; radio occultation	110–76
Venera-10	25/10/75	13:45	16°; 291°	T, P	62–0
				accelerometry; radio occultation	110–63
Pioneer-Venus					
Large probe	9/12/78	7:40	4.4°; 304°	T, P	65–12.5
				accelerometry	118–68
Day probe	9/12/78	6:40	−31.7°; 317°	T, P	67–12.5
				accelerometry	128–70
Night probe	9/12/78	0:05	−28.7°; 56.7°	T, P	65–12.5
				accelerometry	
North probe	9/12/78	3:35	59.3°; 4.8°	T, P	61–12.5
				accelerometry	120–64
Venera-11	25/12/78	11:10	−14°; 299°	T, P	61–0
				accelerometry	100–65
Venera-12	21/12/78	11:16	−7°; 294°	T, P	61–0
				accelerometry	100–65
Venera-13	1/03/82	9:30	−7.5°; 303.5°	T, P	65–0
				accelerometry	98–65
Venera-14	5/03/82	9:55	−13°; 310°	T, P	65–0
				accelerometry	98–65
Vega-1	11/06/85		8.1°; 176.7°		63–0
Vega-2	15/06/85		−7.4°; 179.4°	T, P	63.6–0

[a] The temperature T and pressure P designations correspond to direct in situ measurements.

at 500 K) and up to 0.4% for the pressure (in the middle of the measurement range); differences in the absolute calibration of the temperature sensors between various probes was in the range of 1–2 K (Seiff et al., 1980; Seiff, 1983). This accuracy enabled us to recognize a series of additional effects in the thermal structure of the atmosphere. On the Vega-1 lander, temperature measurements were conducted using thermistors, which have less drift and somewhat higher accuracy (~ 1 K) compared to resistance thermometers with an accuracy of 2–3 K, which had been installed in the Venera 4–14. The pressure on the Venera was measured using diaphragm-type (aneroid) manometers, with an error of $\pm 1.5\%$ of the measurement range.

The initial results of T and P measurements on the Venera as a function of time t were approximated by nth degree polynomials of the form $\Sigma_{i=0}^{n} a_i t^i$, while the least squares method was used to find the closest fit of the polynomial curves (usually for n $\leq$ 3) to the experimental points. Statistical weights of separate measurements, which depend on the accuracy of individual sensors, were incorporated in deriving the polynomial coefficients a_i. The individual points correspond to an average over approximately 50 measured values (see Avduevsky et al., 1983a).

Correlating the T and P values measured on all Venera with altitude was accomplished using the same measurement data for the atmospheric parameters as a function of time. To do this, a method was proposed (see Avduevsky et al., 1968b; Kuzmin and Marov, 1974), which applied well-known relationships between the atmospheric gas and the aerodynamics of spacecraft descent by parachute. This technique provided a convenient means for verifying the internal consistency of the measurements. Because, in our opinion, this question is of particular interest, we shall provide a brief account of its essentials, supplementing the discussion with elementary mathematical expressions.

We will assume that Venus' atmosphere is in a state of hydrostatic equilibrium, and that the coefficient of aerodynamic drag for the spacecraft remains constant during descent. Then, because the gas composition of the atmosphere is known, the rate of descent and the path traversed by the lander, under parachute, as a function of descent time, $V(t)$ and $z(t)$, may be accurately calculated from the measured $T(t)$ and $P(t)$, using hydrostatics and the gas state equations, and independently using the equation of quasi-uniform parachute descent. The assumption that in a quiescent atmosphere the lander moves at a rate essentially coinciding with a quasi-uniform descent rate is completely justified. A non-stationary condition may show up, insignificantly, only at high velocities before the deployment of the parachute, and its effect decreases as the cube of the velocity.

The equations have the following form:

$$dP = \rho g_{\venus}\, dz = \rho g_{\venus} v\, dt \tag{5.1}$$

$$P = \rho R / \bar{\mu} T \tag{5.2}$$

$$g_{\venus} M(1 - \rho/\rho_L) = C_x F \frac{\rho V^2}{2} \tag{5.3}$$

Here: $g_{\venus} = g_{\venus}^0 [R_{\venus}/(R_{\venus} + Z)]$ is the gravitational acceleration of Venus at altitude Z, $g_{\venus}^0$ is the gravitational acceleration at the surface, M is the mass of the lander, ρ_L is the mean density of the lander, $C_x(\Psi)$ is the coefficient of aerodynamic drag of the parachute-lander system for an angle of attack, equal to the trim angle Ψ (according to ground-based tests for parachutes used on the Venera, the quantity Ψ was negligible, that is, the longitudinal axis of the parachute-lander system essentially coincided with the direction of the gas flow), F is the cross sectional area, v is the descent rate relative to the surface, V is the rate of descent relative to the gas, ρ is the atmospheric density, $\bar{\mu}$ is the mean molecular weight, R is the gas constant, z is the path traversed along the vertical, and t is time. The factor in parentheses in the left-hand side of Eq. (5.3) accounts for the effect of buoyancy, which turns out to be substantial in the lowest layers of the atmosphere.

From Eqs. (5.1) and (5.2) it immediately follows that

$$v = \frac{dz}{dt} = \frac{d}{dt}\left(\frac{P}{\rho g_\female}\right) = \frac{d}{dt}\left(\frac{RT}{\bar{\mu} g_\female}\ln P\right) = \frac{d}{dt}(H \ln P) \tag{5.4}$$

where H is the scale height

$$z = \int_{P(t_0)}^{P(t_e)}\frac{dP}{\rho g_\female} = \frac{R}{\bar{\mu}}\int_{P(t_0)}^{P(t_e)}\frac{T}{g_\female}\frac{dP}{P} \tag{5.5}$$

Integration is carried out from the start (t_0) to the end (t_e) of the measurement interval, or any intermediate value (t_e').

In turn, from Eq. (5.3) we have:

$$V = \left[\frac{2g_\female M(1 - \rho/\rho_L)}{\rho C_x F}\right]^{1/2}$$

In this case the vertical and horizontal velocity components are $V_v = V\cos\gamma$ and $V_h = V\sin\gamma$, where γ is the glide angle. But during the parachute descent phase, one can assume $\gamma = 0$ and $V_v = V$, that is, one can assume the natural motion of the spacecraft, in the absence of wind, to be vertical; this is not contradicted by the ground-based test results.

Consequently we obtain:

$$z = \int_{t_0}^{t_k} V\,dt = \left(\frac{2M}{C_x S}\right)^{1/2}\int_{t_0}^{t_k}[\rho g_\female^2 (1 - \rho/\rho_L)]^{1/2}\,dt$$

or, taking $\rho/\rho_L \approx 0$ and $g_\female \approx const$, a simpler approximation:

$$z = \frac{2AR}{\bar{\mu} g_\female}\int_{t_0}^{t_k}\left(\frac{T}{P}\right)^{1/2}dt \tag{5.6}$$

where A denotes all constants:

$$A = \left(\frac{Mg_\female^3 \bar{\mu}}{2C_x FR}\right)^{1/2}$$

The error in determining the velocity and the path traversed, by each of these methods, usually represents fractions of a meter per second and fractions of a kilometer, respectively.

Combining Eqs. (5.1), (5.2), and (5.3) provides a convenient means for verifying the internal consistency of the atmospheric parameter measurements through the aerodynamics of parachute descent, because the construction parameters of the spacecraft (C_x, F, M) are known to a high accuracy (Kuzmin and Marov, 1974; Avduevsky et al., 1983a).

Altitude correlation of the measurement data from the Pioneer-Venus probes was also achieved using Eqs. (5.1) and (5.2). At the same time corrections were introduced to the readings owing to dynamic pressure (Seiff et al., 1980), specified by the right-hand side of Eq. (5.3). Dynamic pressure exerts a more noticeable effect during the aerodynamic descent phase (as with the Pioneer-Venus probes), as opposed to parachute descent.

Because the descent phase for Venera-7 could not be calculated, the altitude z as a function of time was obtained by another original method (aside from direct integration of the descent velocity $v(t)$ measured using the Doppler method, with respect to the master oscillator signal emitted by the onboard transmitter). To do this, it was sufficient to use the first law of thermodynamics $T\,dS = di - dP/\rho$, the hydrostatic equation $dP = \rho g_\varphi\,dz$, and the isoentropic condition $dS = 0$, which corresponds to an adiabatic atmosphere. Here S is entropy, i is the enthalpy of the gas, and the remaining notations are well known. Then, in finite increments one can write $\Delta z = (1/g_\varphi)\,\Delta i$, and because $\Delta i = C_V \Delta T$ (C_V is the specific heat for a constant volume), one can obtain a T versus z dependence (Marov et al., 1971; Kuzmin and Marov, 1974).

The results of temperature and pressure measurements as a function of altitude from the first Venera spacecraft are shown in Fig. 5.1. In the altitude range of 20–50 km, where all measurements overlap, considerable scatter is observed, as much as 50 K at 35 km. The data from Venera-8 and Venera-10 differ especially strongly, and although they refer to different latitudes and times of day, such large discrepancies could hardly be attributed to changes in insolation or other temporal variations. For the most part, they are more likely associated with errors in calibration and in the measurements themselves, including drift effects in the detectors and their correlation with altitude above the surface. Evidently, the later measurements from Venera-10 are closer to reality, as they coincide better with data from the subsequent Venera 11–14 spacecraft, and with the measurements from the Pioneer-Venus Large probe. This is clearly seen from a comparison of the set of measurements in Fig. 5.2. Measurements from the Pioneer-Venus Large probe provide values approximately 10–15 K lower, compared to Venera 9–14 at almost all altitudes from 20 to 60 km. Somewhat smaller, but noticeable, differences are observed when comparing the temperature measurements from the small probes and Large probe on Pioneer-Venus, conducted at various latitudes and at different times of day (Fig. 5.3). Temperature discrepancies (lower values by 20–40 K) are especially great for the high-latitude Night probe, starting at an altitude of about 50 km (Seiff, 1983; Seiff et al., 1985). These differences are, apparently, real; this is confirmed by measurements of the overall thermal radiation flux, carried out at the same time by the same small probes (Sromovsky et al., 1985; Seiff et al., 1985). It is interesting to note yet another peculiarity in comparing data from the Pioneer-Venus Large probe with that of the Day probe and Venera-10. In both cases a tendency toward lower temperatures is observed in the equatorial atmosphere, compared

Fig. 5.1. Temperature and pressure dependences in Venus' atmosphere versus altitude from the first direct measurements by the Venera spacecraft: (1) Venera-4, (2) Venera-5, (3) Venera-6, (4) Venera-8, (5) Venera-9, and (6) Venera-10.

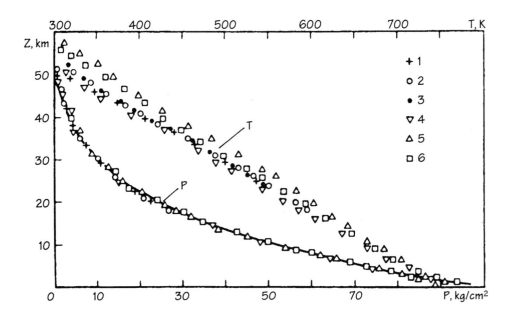

to the atmosphere at middle latitudes below 20 km. If one were to exclude measurement error, then it would be tempting to link this tendency to large-scale horizontal vortices in the lower atmosphere, which could cause heat transfer to high-latitude regions and create temperature irregularities (see Chapter 7).

Measurement results from all Venera in the lowest portion of the atmosphere (below ~ 10–15 km) and especially at the surface exhibit good mutual agreement, with errors in the range of $\Delta T_S = \pm 6$ K from the mean temperature $T_S = 735$ K, referenced to a surface level with a pressure of $P_S = 92$ bar (corresponding to a mean radius for Venus of $R_{\female} = 6052$ km). Adiabatic extension of the Pioneer-Venus measurements, from an altitude of 13 km, lies within the limits of the indicated error ΔT_S at the planet's surface. From a physical point of view, such a value is reinforced by independent theoretical estimates suggesting that diurnal variations in the surface temperature must not exceed several degrees (see Kuzmin and Marov, 1974) or, according to other estimates, must be even less than 0.1 K (Stone, 1975), owing to the high thermal inertia of the atmosphere (see Chapter 7). Values of T_S, P_S, and density ρ_S, calculated from them (for $\bar{\mu} = 43.4$ corresponding to an atmosphere consisting of 97% CO_2 and 3% N_2), are given in Table 5.2 from all available measurements.

Measurements of atmospheric parameters using accelerometers were conducted by all Venera spacecraft, beginning with Venera-8, as they entered the atmosphere, and by all Pioneer-Venus probes (Avduevsky et al., 1983a). The technique is quite simple, involving recording the acceleration of the spacecraft (naturally, with an inverse sign) as a function of time $n_x(t)$. Obviously, $n_x = Q/G$, where Q is the drag force, specified by the right-hand side of Eq. (5.3), and $G = Mg_{\female}$ is the mass of the craft on Venus. To determine the direction of this force, it is necessary to take into account the inclination of the trajectory, specified by the spacecraft's angle of entry into the atmosphere $\zeta(t)$ (the inclination of the trajectory to the local horizontal at time t). It is easy to convince oneself that the system of motion

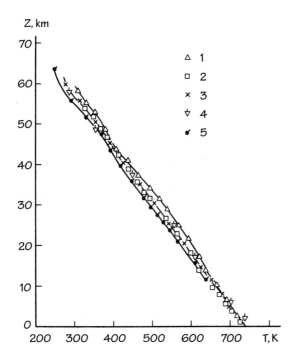

Fig. 5.2. A comparison of altitude temperature profiles in Venus' atmosphere from direct measurements by the Venera spacecraft and the Pioneer-Venus Large probe: (1) Venera-9, (2) Venera-10, (3) Venera 11–12, (4) Venera 13–14, and (5) the Pioneer-Venus Large probe.

equations (taking into account accelerometer calibration on Earth with an acceleration $g_\oplus$) can be written as follows:

$$\frac{dV}{dt} = n_x g_\oplus - g_\varphi \sin \zeta \tag{5.7}$$

$$\frac{d\zeta}{dt} = \left(\frac{V}{R_\varphi + Z} - \frac{g_\varphi}{V} \right) \cos \zeta \tag{5.8}$$

After integration (using boundary conditions from the moment of atmospheric entry $t_0(V_0, \zeta_0, z_0)$ according to ballistic data), $n_x(t)$ and the desired dependence $\rho = 2n_x M/C_x F V^2$ as a function of time are specified (Avduevsky et al., 1983a; Seiff, 1983).

Because, upon entering the atmosphere, the decelerations change from fractions to several hundred g, with a distinct maximum on the $n_x(t)$ curve (see Fig. 2.23, as an illustration), accelerometers with 2–3 measurement ranges were used (on the Venera and Vega), as were multirange accelerometers (on the Pioneer-Venus probes). Consequently, in the first case measurements were started from an altitude of less than 100 km, but in the second case, from approximately 125 km (Fig. 5.4).

In the ~80 km to ~35 km altitude range, atmospheric parameters were also determined using radio occultation. The essence of this method is as follows. As a spacecraft passes behind a planet from the observer's (receiving antenna's) point of view (Fig. 5.5), the increasing density of the atmosphere causes an increase in the effective path of the radio beam (electromagnetic wave) as a result of refraction. Along with this, defocusing of the signal occurs due to changes in the gradient of the index of refraction, and, as a consequence, there is "blurring" of the beam over a large solid angle. The combination of these effects is registered at the ground-based receiving antenna, in the form of phase-frequency variations in the radio signal and attenuation of its power.[2] The results of these measurements, along

Fig. 5.3. Venus' atmospheric temperature dependences versus altitude from direct measurements by the Pioneer-Venus probes: (1) the Large probe, (2) the Day probe, (3) the Night probe, and (4) the North probe (according to Seiff, 1983, 1985).

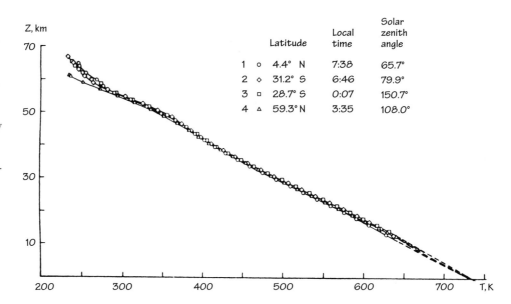

		Latitude	Local time	Solar zenith angle
1	○	4.4° N	7:38	65.7°
2	◇	31.2° S	6:46	79.9°
3	□	28.7° S	0:07	150.7°
4	△	59.3° N	3:35	108.0°

with trajectory data and predictions of the spacecraft's motion, as it passes behind the planet and comes out from behind it, provide the necessary initial data for computing the characteristics of the atmosphere.

A representation of the geometry of radiowave propagation and the relationship between the angles, which establish an altitude correlation, is provided by Fig. 5.6. In simplified form, without taking into account the actual geometry of the relative motion between the spacecraft and the receiving station on Earth, the change in frequency of an electromagnetic wave Δf_D is related to the velocity of the source by the simple relation

$$\Delta f_D = f_s \frac{v \cos \Phi}{c} \tag{5.9}$$

where f_s is the emitted frequency of the radio signal; v is the velocity of the spacecraft; Φ is the angle between the velocity vector and the direction to the receiving station, taking into account distortion of the beam due to refraction α in Fig. 5.6; and c is the speed of light. In turn, the relationship between the frequency change and the index of refraction n is expressed by the relation:

$$\Delta f_D = \frac{f_s}{c} \frac{d}{dt} \int_{-\infty}^{+\infty} (n - 1) \, d\ell \tag{5.9'}$$

where integration is conducted along the line of sight ℓ.

The measurable value Δf_D is, in general, the sum of two refraction effects: positive in a neutral atmosphere and negative in an ionized medium. Using the well-known electromagnetic field equations (Maxwell's equations), we can obtain an expression for the complex index of refraction, whose real portion, contained in Eq. (5.9'), can be written as

$$n^2 = n_n^2 - n_e^2$$

Table 5.2   Temperature and pressure on the surface of Venus, from spacecraft measurements

Spacecraft	T_SK	P_S, kg $\times$ cm^{-2a}	ρ_S, kg $\times$ cm^{-3}	Remarks
Venera-4	735	86	59	Extrapolation[b] (Z_0 for $R_\female$ = 6054 km)
Venera-5 and Venera-6	772	99	64	Extrapolation[b] (Z_0 for $R_\female$ = 6050 km)
Venera-7	747 ± 20	92 ± 15	63	Extrapolation[b] (Z_0 for $R_\female$ = 6050 km); P_S and ρ_S are calculated values based on T and rate of descent V
Venera-8	743 ± 8	93 ± 1.5	64	Taking into account altitude variations $\Delta Z_0 = \pm1$ km from $Z_0 = 0$ for $R_\female = 6052$ km
Venera-9	728 ± 5	85 ± 3	60	Same as for Venera-8
Venera-10	737 ± 5	91 ± 3	63	Same as for Venera-8
Pioneer-Venus				
Large probe	731	92	65	Extrapolation; $Z_0 = 0$ for $R_\female = 6052$ km
Day probe	732	92	65	Extrapolation; $Z_0 = -1$ km
Night probe	734	92	64	Extrapolation; $Z_0 = -0.6$ km
North probe	729	87	61	Extrapolation; $Z_0 = +1$ km
Venera-11	734 ± 5	91 ± 3	63	Same as for Venera-8
Venera-12	735 ± 5	92 ± 3	64	Same as for Venera-8
Venera-13	735	88.7	61	$Z_0 = -0.1$ km from Z_0 for $R_\female = 6052$ km
Venera-14	738	94.7	64	$Z_0 = -0.7$ km from Z_0 for $R_\female = 6052$ km
Vega-2	733 ± 0.5	91 ± 0.5	62	$Z_0 = \pm0.5$ km from Z_0 for $R_\female = 6052$ km

[a] Here we use kg $\cdot$ cm^{-2} as a measure of pressure; other widely used units are atm and bar, which are interrelated as follows: 1 kg $\cdot$ cm^{-2} = 0.968 atm or 0.981 bar.

[b] For a temperature gradient corresponding to the termination of atmospheric temperature measurements.

Fig. 5.4. The temperature of Venus' stratomesosphere as a function of altitude from the results of accelerometer measurements by the Pioneer-Venus probes and the Venera-11 (1) and Venera-12 (2) spacecraft, as they entered the atmosphere. Below, the squares, diamonds, and triangles with tails (') are direct measurement data from the respective probes, starting from an altitude of 65 km (according to Seiff, 1983).

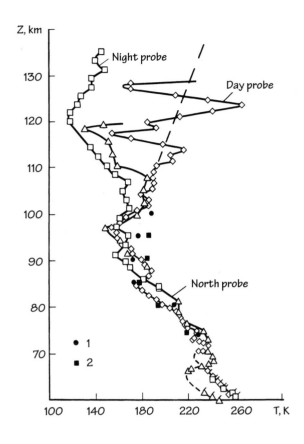

whereupon

$$n_e^2 = 1 - \frac{e^2 \mathcal{N}_e}{\pi m_e f_s}$$

Here n_n and n_e are the coefficients of radiowave refraction in a neutral and ionized medium, respectively; m_e and e are the mass and charge of the electron, and $\mathcal{N}_e$ is the electron density along the line of sight.

If the spatial coordinates of the transmitter, planet, and Earth-based receiver are known, then from frequency effects $\Delta f_D(t)$, caused by neutral gas of variable density (where the effect of higher ionized layers is not significant), one can calculate the relationship between the index of refraction and radial distance r, that is, $n(r) = n_n(r)$ with $n_e(r) \approx 0$, and conversely for regions of the ionosphere. To do this, the integral equation for optical refraction is usually used; this is a special case of the well-known Abel equation, widely used in various physical applications. This equation enables us to relate the quantity $n(r)$ to the refraction angle α as a function of the shortest distance from the planet's center of mass to the line of sight. Sometimes a simpler, though no less arduous technique is used for this purpose, where the entire atmosphere, assumed to be spherically symmetric, is broken up into a series of layers, each of which corresponds to a single point on an experimental curve $\Delta f_D(t)$. For each layer the $n(r)$ values, corresponding to the distortion of the radio beam, specified by the angle of refraction α, are computed sequentially.

From the altitude profile of the index of refraction $I(r) = (n - 1) \cdot 10^6$, it is easy to

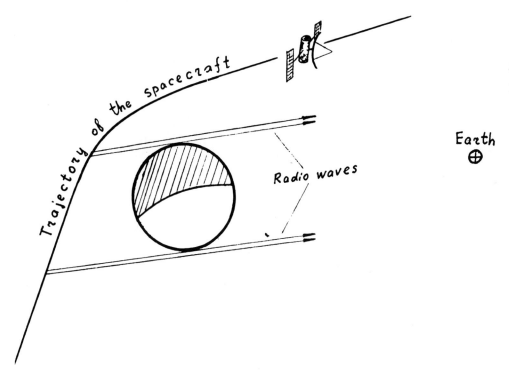

Fig. 5.5. The fly-by trajectory of a spacecraft projected onto the equatorial plane.

calculate the characteristics of the atmosphere, if one takes into account the condition of hydrostatic equilibrium (Eq. [5.1]) and the fact that the index of refraction is proportional to atmospheric density. The density gradients are linked to the gradients of the index of refraction by the following simple dependence:

$$\frac{dn}{dr} = C \frac{d\rho}{dr} \tag{5.10}$$

The coefficient C depends on the chemical composition of the atmosphere (for example, for carbon dioxide $C = 0.23 \ cm^3/g$). Further, because the scale height, with respect to density (see Eq. [5.4]), can be represented in the form $H_\rho = -\rho/(d\rho/dr)$, knowledge of ρ and $d\rho/dr$ enables us to determine the quantity $H_\rho = (RT)/(\bar{\mu} g_{\venus})$ for all conditionally isothermic layers, corresponding to the number of subdivisions within which the scale height is assumed to be constant (for $T, g_{\venus} = const$). Subsequently, the temperature $T(r)$ can be easily calculated from H_ρ for a given $\bar{\mu}$, while the pressure $P(r)$ is determined from the gas equation of state (Eq. [5.2]).

So, essentially, the radio occultation experiment measures the index of refraction and its change with altitude. The altitude range for Venus' atmosphere, specified earlier, is pre-

Fig. 5.6. Geometry of radiowave propagation and the angular relationships, which provide the altitude correlation in the radio occultation experiment (according to Fjeldbo et al., 1971; Kuzmin and Marov, 1974). α is the angle of refraction, $\alpha = \beta_r + \delta_r$; the remaining angles are secondary in nature and enable us to describe the change in frequency of the electromagnetic wave Δf_D in real geometry, by accounting for the relative motion of the spacecraft (r_S, z_S) and the receiving station on Earth $(0, Z_\oplus)$. It allows us to define angle Φ in a more complex form, compared with Eq. (5.9).

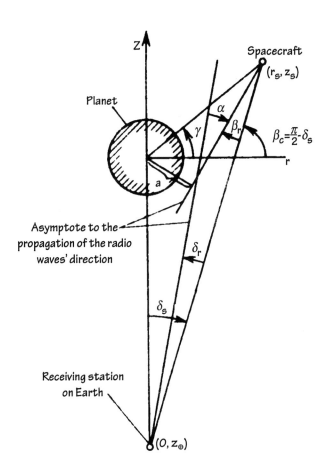

determined by the fact that at an altitude of ~35 km, the critical refraction is reached in Venus' atmosphere (the radius of curvature of the beams becomes equal to the radius of the planet), whereas above ~80 km, the density of the neutral atmosphere is insufficient for the index of refraction to exert a noticeable effect on the passage of radiowaves. In this technique, the altitude Z correlation is achieved via the trajectory parameters, which are measured relative to the gravitational center of the planet, that is, the radial distance r is determined, then the radius R_φ is subtracted. Note that in this method the limiting resolution of the measurable atmospheric parameters with respect to altitude is specified by the size of the first Fresnel zone. Owing to refraction, the curvature of the wave front increases and the Fresnel zones become elliptical instead of circular, with the minor axis of the ellipse oriented in the vertical direction. The altitude resolution is somewhat greater in regions of neutral atmosphere, which defocus the radiowaves, compared with the ionospheric layers whose presence, conversely, focuses the radio beam.

It is obvious that the greater the number of layers into which the $n(r)$ curve is broken, the higher the accuracy of the approximation to the actual T and P profiles is, either in the adjustment method or when numerically solving the Abel equation. Naturally, the accuracy of the obtained results depends on the errors in the technique itself, primarily on the stability of the master oscillator of the onboard radio transmitter, and on the precision in determining the spacecraft's trajectory. Usually, the error in computing T and P does not

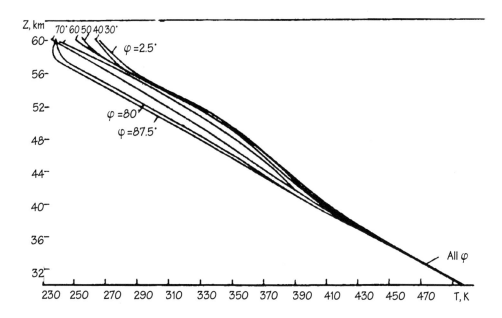

Fig. 5.7. Averaged temperature profiles, which reveal latitudinal temperature variations in the upper portion of the troposphere, from results of radio occultation experiments on the Pioneer-Venus orbital spacecraft (according to Kliore et al., 1985).

exceed $\pm 1\%$ (see Kuzmin and Marov, 1974; Yakovlev, 1974; Fjeldbo and Eshleman, 1969; Kliore and Patel, 1980).

The Pioneer-Venus radio occultation measurements provided the most complete coverage of the middle atmosphere at various latitudes φ (in more than 100 locations). These results are shown in Fig. 5.7. They are in good agreement with similar measurements conducted by the Venera-9 and Venera-10 spacecraft (Yakovlev et al., 1978). Systematic discrepancies in the absolute temperature values that arise when using the radio occultation method, aside from the technical errors already specified, are linked to inaccuracies in the altitude correlation. These systematic errors have been eliminated in the curves presented in Figs. 5.8 and 5.9, by comparing the $T(Z)$ and $P(Z)$ dependences specified by this method with the more accurate direct measurements in regions where the corresponding measurements overlap (at altitudes of $\sim 35-65$ km).

Remote measurements using an infrared spectrometer on the Pioneer-Venus spacecraft cover the altitude range from 50 km to 120 km and apply mainly to the northern hemisphere of Venus. They exhibit good agreement with radio measurement results, especially at altitudes of $60-80$ km; below 60 km the divergence does not exceed ~ 5 K. Intrinsic errors in the method lie within the ± 2.5 K range (at an altitude of 100 km), increasing to ± 12 K at 110 km (Schofield and Taylor, 1982). The altitude resolution inherent in this method is large, however—on the order of $10-15$ km.

Infrared spectrometer data, to altitudes above 100 km, are also in keeping with accelerometer measurements (within $\pm$ 10 K) by the Pioneer-Venus probes and the Venera-12 and Venera-13 landers (Seiff et al., 1985). It is significant to note that remote infrared measurements give a mean temperature with limited altitude resolution, whereas accelerometer measurements correspond to instantaneous values reflecting the actual space-time variations in the middle atmosphere.

In addition to these vertical "slices" of the atmosphere, limited temperature and pres-

Fig. 5.8. Temperature variations in the middle atmo-
sphere at $\varphi = 30°$ latitude,
from a set of measurements
by various techniques:
(1) infrared radiometry
from onboard the Pioneer-
Venus orbital spacecraft;
(2) the Pioneer-Venus
accelerometer, Day probe;
(3) the Venera-11 accelerometer; (4) the Pioneer-
Venus accelerometer, Night
probe; (5) the Pioneer-
Venus Day probe, direct
measurements; (6) the
Pioneer-Venus Night probe,
direct measurements; and
(7) the radio occultation experiment on Pioneer-Venus
(according to Kliore et al.,
1985).

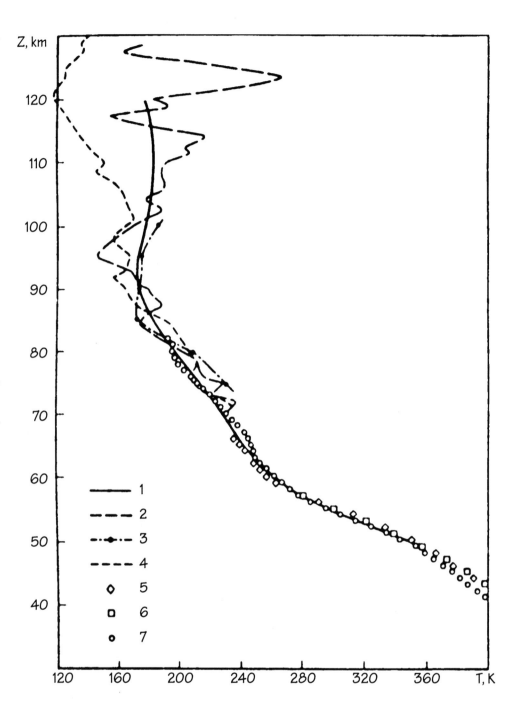

sure measurements along the horizontal have become available in the latitude direction
close to the equator, at an altitude of 53–54 km. These measurements were conducted by
the Vega-1 and Vega-2 balloon probes along both sides of the terminator, and they cover
1/3 of the planet's circumference (Sagdeev et al., 1986). The range of measured parameters, on average, was found to be within 535–620 Mbar in pressure and 300–320 K in
temperature. In these experiments the altitude was correlated to the measured T and P values, based on the existing atmospheric model. The temperature, pressure, and their variations during the drift phase proved to be highly correlated, with minimal variations in

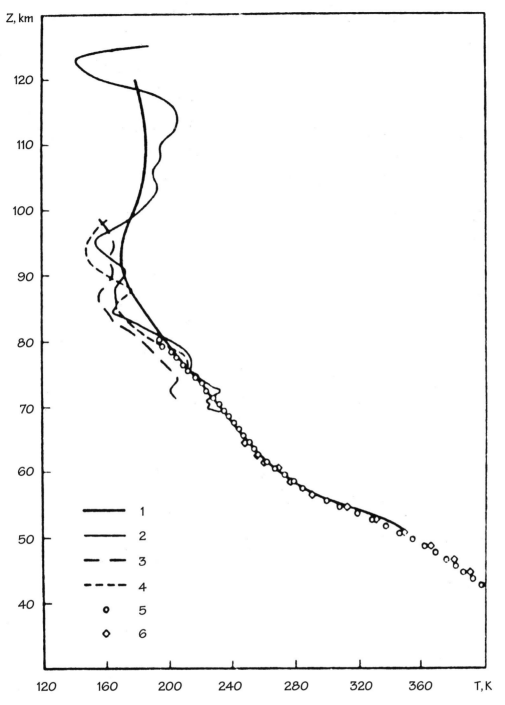

Fig. 5.9. Temperature variations in the middle atmosphere at $\varphi = 5°$ latitude, from a set of measurements by various techniques: (1) infrared radiometry, (2) the Pioneer-Venus accelerometer, (3) the Venera-12 accelerometer, (4) the Venera-13 accelerometer, (5) the radio occultation experiment on Pioneer-Venus, and (6) direct measurements from the Pioneer-Venus Large probe (according to Kliore et al., 1985).

$\Delta T = 0.1$K and $\Delta P = 0.1$ Mbar. From the magnitude of the variations, the deviations from an equilibrium state, brought about by vertical motions in the atmosphere, were estimated.

Structure and Properties of the Atmosphere

We shall now consider the results of measurements of the vertical structure of the atmosphere, and from these results we shall estimate the thermodynamic state of the atmospheric gas. These estimates will serve as the basis for judging the vertical stability of the atmosphere.

In contrast to Earth, where diurnal-latitudinal and seasonal contrasts in temperature in the troposphere and on the surface exceed 100 K, on Venus temperature variations are within several degrees at the surface (see Table 5.2) and less than 10 K in the lower and middle portions of the troposphere. Moreover, the troposphere of Venus is approximately 5 times greater in height than Earth's. Because of the lack of seasonal changes, relatively small differences in the temperature-altitude relations (see Figs. 5.2 and 5.3) measured by the landers in various regions should evidently be attributed to diurnal-latitudinal variations. Unfortunately, because of the limited nature of direct measurements, it is difficult to separate diurnal from latitudinal variations clearly, although an attempt to do this was undertaken, based on measurements from the Pioneer-Venus small probes (Fig. 5.10; see Seiff et al., 1983). In comparison with other measurements (for example, a comparison of the Pioneer-Venus main probe with Venera-10, as shown in the bottom graph of Fig. 5.10), other effects, associated with differences in instrument calibration and altitude correlation of the measurements, may be superimposed. Nevertheless, from a comparison of data from the Pioneer-Venus Day and Night probes, it follows that diurnal temperature oscillations in the subcloud atmosphere (below ~40 km) are small, on the order of 1 K (possibly somewhat greater below ~20 km). On average they remain small even at the altitude of the clouds, where a noticeable periodic component, whose nature is most likely linked to thermal tides and internal gravitational waves, is superimposed.

A latitudinal dependence in the temperature trend with altitude is more clearly detected. From the comparison presented in Fig. 5.10, one could conclude that latitudinal variations in the subcloud atmosphere are less than 10 K, increasing to 10–20 K at cloud level. Thus, latitudinal variation, extending to lower altitudes, is the dominant type of variation in the parameters of Venus' troposphere.

This conclusion follows even more definitively from an analysis of results of radio occultation measurements, over a large time interval, with almost global coverage (see Fig. 5.7). Latitudinal temperature variations increase progressively, starting from an altitude of approximately 40 km and amounting to almost 40 K near 60 km (Seiff et al., 1985). At 40 km these variations do not exceed 10–15 K, which conforms well with direct measurements. On the whole, one can conclude that in the 40–60 km altitude range, the atmosphere becomes colder as one moves from the equator to the pole; whereas initially, the temperature change is small up to ~50° latitude, it subsequently rises substantially. A temperature inversion, observable above 60 km at low latitudes (see Fig. 5.9), develops at an altitude of 58 km at high latitudes (see Fig. 5.8); this reflects the general lowering of the tropospheric boundary, the tropopause, with an increase in latitude (Kliore and Patel, 1982; Seiff et al., 1985).

The possibility of detecting temperature variations in Venus' lower atmosphere that are linked with atmospheric dynamics is of obvious interest, so we shall mention an as-yet-unexplained feature that appeared in measurements of the meteorological parameters by the Vega balloon probes. The probes found a 6.5 K atmospheric temperature difference for

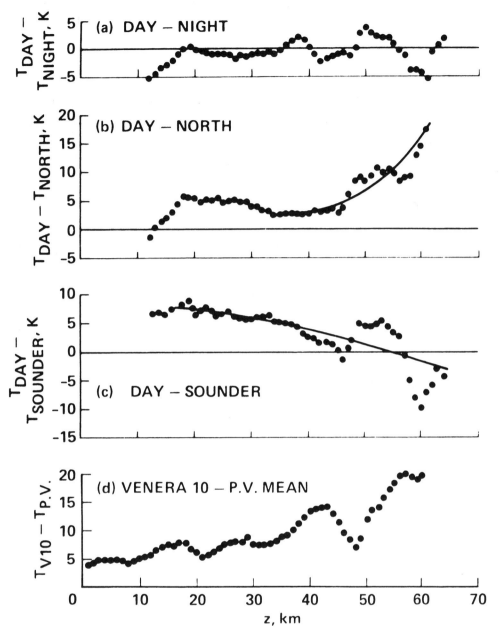

Fig. 5.10. Diurnal-latitudinal temperature variations from Pioneer-Venus probe measurements in comparison with Venera-10 data, in the 4°–60° latitude band. Data from the North probe above 45 km (where latitudinal variations are substantial) were excluded from the upper graph, in order to bring out diurnal temperature contrasts more clearly (according to Seiff et al., 1985).

the same pressure, that is, for essentially the same altitude (Linkin et al., 1986). The temperature in the drift band of the Vega-1 balloon turned out to be even higher, although the balloons were placed nearly symmetrically relative to the equator (7.3° N and 6.6° S). In this case the difference between the P and T values did not change appreciably with longitude (in the drift range from 70° to 180° E longitude). This evidence may be linked not with actual atmospheric variations but with the specific nature of the balloon measurements, resulting from the tendency of balloons to fly along with the same air mass, which in these two regions had different thermal histories. The effect of time variability in the atmosphere or an asymmetry between the northern and southern hemispheres is less likely.

A series of interesting features is observed in the structure of Venus' above-cloud atmosphere (see Figs. 5.7, 5.8, and 5.9). Diurnal and latitudinal variations up to 30° latitude are small, as suggested by the radio occultation, infrared (thermal) radiometry, and accelerometer measurements (Yakovlev et al., 1978; Kliore and Patel, 1980; Courtin et al., 1979; Ksanfomality, 1980; Schofield and Taylor, 1983; Seiff, 1980; Seiff et al., 1985). Temperature oscillations in the near-equatorial middle atmosphere depend directly on the solar longitude; in other words, they are directly linked to the elevation of the Sun in Venus' sky. Here, the oscillations are more noticeable than in the troposphere; however, they do not exceed ~ 10 K. Latitudinal variations have a more complex nature. In this regard, the presence of areas of inversion, the deepest of which is located directly above the upper cloud boundary at high latitudes, draws attention to itself. It is here, in the approximate latitude range of $50°–80°$, that the cold region known as the polar collar, first discovered by infrared spectrometry (F. Taylor et al., 1979, 1980), is located. The polar collar, which surrounds the pole, is centered at an altitude of 65 km and 65° latitude. It divides the middle atmosphere into two regions along the vertical: with an increase in latitude the troposphere ($Z \leq 70$ km) becomes colder, while the overlying stratomesosphere ($Z \geq 70$ km to an altitude of ~ 90 km) becomes warmer. So, near the poles at these altitudes, the warmest regions of the atmosphere are found.

The nature of the temperature variation on Venus near the upper boundary of the clouds is shown by the stereographic projection in Fig. 5.11 (F. Taylor et al., 1980). The image shows Venus' northern hemisphere (up to 50° latitude). This is an example of a series of similar images obtained using the Vortex infrared spectrometer on the Pioneer-Venus spacecraft, in the small window of transparency for CO_2 at a wavelength of 11.5 μm. The difference in brightness temperatures between the collar and dipole regions is up to $\sim 30°$. Such a temperature trend, reflected in corresponding changes in the pressure and density at various altitudes as a function of latitude, results in unique middle atmosphere dynamics. This latitudinal trend is associated with the maintenance of a cyclostrophic balance of zonal winds in the deep atmosphere at middle latitudes (Seiff, 1983; Seiff et al., 1982b; Schubert et al., 1980; Seiff et al., 1985).[3]

This same effect is linked with the presence of jet streams in the zonal wind at altitudes of ~ 65 km and high latitudes ($\sim 65°$) and meridional circulation at altitudes of 85–100 km (Kerzhanovich and Limaye, 1985; Newman et al., 1984). Here the presence of temperature inversions is possibly connected with a change in the regime of radiative equilibrium, in the transitional region from a cloudy to a "clean" atmosphere. The temperature minimum near 60–65 km could be explained by thermal emissions in the CO_2 rotational bands, whose efficiency is suppressed by radiative heating at higher levels (Taylor et al., 1979; Seiff, 1980). It is possible that other dynamic processes, or mechanisms linked with heating from lower-lying regions of the atmosphere, exist. This would make it easier to explain, in particular, the warmer polar regions (as compared to equatorial) in the middle atmosphere above the polar collar (see Chapter 7).

Another interesting set of features of the middle atmosphere is the distinct oscillations superimposed on the mean temperature profile at all latitudes (see Figs. 5.8 and 5.9). Their amplitude increases with altitude, from several degrees at 60 km to ~ 10 K at 80 km, and simultaneously increases with latitude. Aside from a semidiurnal thermal tidal wave

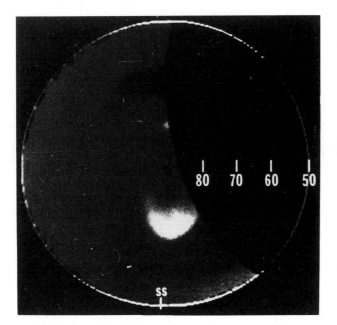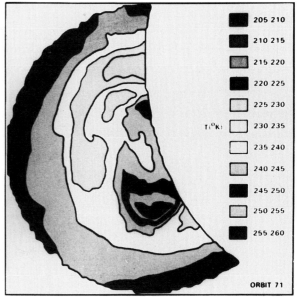

Fig. 5.11. An example of a polar stereographic image of Venus at a wavelength of 11.5 μm, from the results of Pioneer-Venus onboard infrared spectrometer measurements (left) and the corresponding brightness temperature contours of the planet's northern polar region (right). The outer circular contour corresponds to 50° latitude, and the cross in the center to the North Pole. Intermediate latitudes are shown on the image at the left. The curve inside the circle separates a black zone, where data were not obtained due to the satellite's orbital geometry. The white band below (a bright polar dipole above ~70° N latitude) is the region of highest temperature; the black band, at 60°–70° latitude, is the polar collar. The letters SS denote the longitude of the subsolar point. The temperature values, corresponding to the temperature contours on the image at right, are designated there in the form of a series of shadings (according to Taylor et al., 1980).

(which, conversely, decreases with increased latitude) superimposed on the mean temperature profile of the equatorial atmosphere, a periodicity of 5.9 days shows up in these oscillations and, near the pole, a 2.9-day period coincides with the rotational period of the polar dipole (Apt and Leung, 1982). Such periodic perturbations may indicate the propagation of internal gravitational waves, or of other types of waves, outward from the lower-lying atmosphere. In turn, the waves induce temperature variations of ~5 K in the clouds at altitudes of 50–65 km (Seiff et al., 1985). Unfortunately, it is not yet possible to trace the link between these waves and similar global or temporal changes in the troposphere, because only a few instantaneous slices of the atmosphere were obtained by the landers. The errors in determining the atmospheric parameters in the troposphere and middle atmosphere are also large, which may mask the actual diurnal temperature variations caused by thermal tides, as well as the nature of the oscillations, some of which may be local. For example, perturbations reflected in temperature variations on the order of several degrees and associated with large-scale vortices in the deep atmosphere are quite likely. A study of these peculiarities presents an interesting and promising task for the further understanding of Venus.

Fig. 5.12. An analysis of direct measurements of Venus' atmospheric temperature and pressure in lg T–lg P coordinates. This analysis makes it possible to determine the polytropic index ξ, which characterizes the thermodynamic state of the atmospheric gas: (1) Venera-4, (2) Venera-5 and Venera-6, (3) Venera-8, (4) Venera-9, (5) Venera-10, and (6) Venera-12.

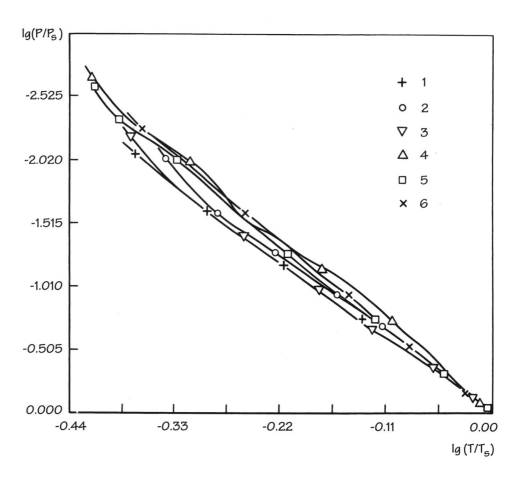

Studying the temperature and pressure values measured by the Venera and Pioneer-Venus spacecraft in P–T coordinates and comparing the corresponding curves with an adiabatic profile for a CO_2 atmosphere at the appropriate P and T (taking into account deviation of the equation of state from the ideal gas equation, incorporating compressibility, in this case added to the right-hand side of Eq. [5.2]), makes it possible to estimate the stability of Venus' atmosphere. In order to do this, it is necessary to determine the polytropic index from the relation $\xi = (1 - d\lg T/d\lg P)^{-1}$ and compare it to the adiabatic index ξ_{ad}. The adiabatic index varies in Venus' lower atmosphere from $\xi_{ad} = 1.21$ at the surface to $\xi_{ad} = 1.28$ at the $P = 1$ atm level ($Z \approx 55$ km); the relationship between ξ and ξ_{ad} follows from an analysis of the curves in lg P–lg T coordinates normalized to lg P_s and lg T_s, respectively (Fig. 5.12).

The condition for stable stratification of the atmosphere is the inequality

$$\frac{dT}{dZ} < \Gamma$$

where Γ is the adiabatic temperature gradient, given by

$$\Gamma = -\frac{g_{\female}}{C_P}\left[1 - \rho\left(\frac{di}{dP}\right)_T\right]$$

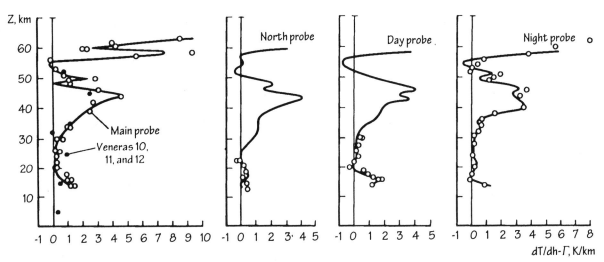

Fig. 5.13. Stability characteristics of Venus' lower atmosphere as a function ($K = dT/dZ - \Gamma$) of altitude in four regions of the planet, from Pioneer-Venus data, compared with Venera 10–12 data (according to Seiff, 1983; Seiff et al., 1985).

where C_P is the specific heat at a constant pressure and i is the enthalpy. In meteorology the concept of potential temperature is often used to determine the static stability, $\Lambda = T(P_0/P)^{R/\mu C_P} \simeq T - \Delta Z\Gamma$, where the pressure P_0 usually refers to the surface and P to the altitude ΔZ. In this case the stability of stratification is characterized by the static stability parameter $K = d\Lambda/dZ$, which may be positive (an unstable atmosphere), negative (a stable atmosphere), or equal to 0 (a neutrally stable atmosphere). It can also be written in the form $K = dT/dZ - \Gamma$, and from the magnitude of K (in K/km) one can trace the degree of deviation of the measured temperatures from an adiabatic profile.

Preliminary analysis of the first Venera data (Avduevsky et al., 1968b, 1970) led to the conclusion that, in the measured altitude range (within the error range of T and P measurements), $\xi < \xi_{ad}$; this suggested an essentially stable atmosphere ($K < 0$). Further, more accurate study of the measured temperature profiles, including so-called entropy diagrams, which enabled us to determine how the entropy of a gaseous medium behaves upon a change in P and T (see Kuzmin and Marov, 1974; Marov, 1978, 1979), indicated a probable deviation from a condition of stability (that is, $\xi \geq \xi_{ad}$, for which a change in entropy $dS \geq 0$, and therefore $K > 0$) in the altitude ranges of 15–25 km (according to Venera-8 data) and 25–33 km (according to Venera 5–6 data). Measurements by the Pioneer-Venus main probe and Venera 11–12 spacecraft contain clearer evidence of the existence of unstable regions in the troposphere, at altitudes of 20–30 km and 50–55 km, as is obvious from the $K(Z)$ graphs in Fig. 5.13. This instability is suggested by temperature measurement data from the Pioneer-Venus small probes (Seiff, 1983; Seiff et al., 1985). Based on analysis of Venera-10 data, a region of minor instability below $\sim$10 km is inferred; this conforms to ideas about the convective state of the near-surface atmosphere (Marov, 1978; Seiff et al., 1985).

So Venus' atmosphere has several regions of instability, which are associated with the presence of convective motions and turbulence. The bulk of Venus' troposphere, however, is statically stable, with a temperature gradient of 7.7 K/km, which is much less than $\Gamma = 8.9$ K/km. This stability has important consequences for atmospheric dynamics, from the

Table 5.3   Model of Venus' lower atmosphere ($Z = 0 - 30$ km)

Z, km	T, K	P, bars	ρ, kg$\times$cm^{-3}	Z, km	T, K	P, bars	ρ, kg$\times$cm^{-3}
0	735.3	92.10	64.79	12	643.2	41.12	33.54
1	727.7	86.45	61.56	14	628.1	35.57	29.74
2	720.2	81.09	58.45	16	613.3	30.66	26.27
3	712.4	76.01	55.47	18	597.1	26.33	23.18
4	704.6	71.20	52.62	20	580.7	22.52	20.39
5	696.8	66.65	49.87	22	564.3	19.17	17.88
6	688.8	62.35	47.24	24	547.5	16.25	15.62
7	681.1	58.28	44.71	26	530.7	13.70	13.59
8	673.6	54.44	42.26	28	513.8	11.49	11.77
9	665.8	50.81	39.95	30	496.9	9.581	10.15
10	658.2	47.39	37.72				

Note: The mean radius of the planet was taken to be $R_\male = 6052$ km; the mean molecular weight of the atmosphere was taken to be 43.44; the gas constant was taken to be R $= 191.4$ J $\times$ kg^{-1} $\times$ mole^{-1} $\times$ K^{-1}.

Table 5.4   Model of Venus' lower (30-65 km) and middle (66-100 km) atmosphere

Z, km	T, K	P, bars	ρ, kg $\times$ m^{-3}	Z, km	T, K	P, bars	ρ, kg $\times$ m^{-3}
		latitude 0–30°				latitude 75°	
33	471.7	7.211	8.041	33	471.7	7.211	8.041
36	448.0	5.346	6.274	36	446.5	5.345	6.294
39	425.1	3.903	4.823	39	420.5	3.894	4.865
42	403.5	2.802	3.646	42	394.5	2.780	3.701
45	385.4	1.979	2.693	45	368.7	1.941	2.762
48	366.4	1.375	1.967	48	343.5	1.321	2.016
51	342.0	0.9347	1.432	51	318.5	0.8741	1.438
54	312.8	0.6160	1.032	54	290.0	0.5582	1.009
57	282.5	0.3891	0.7212	57	258.2	0.3392	0.6878
60	262.8	0.2357	0.4694	60	237.5	0.1948	0.4293
64	245.4	0.1156	0.2443	64	234.3	8.950×10^{-2}	0.1998
68	235.4	5.447×10^{-2}	0.1210	68	242.0	4.180	9.030×10^{-2}
72	224.1	2.476×10^{-2}	5.775×10^{-2}	72	243.6	1.987	4.263×10^{-2}
76	212.1	1.081×10^{-2}	2.663×10^{-2}	76	234.9	9.345×10^{-3}	2.079×10^{-3}
80	197.1	4.476×10^{-3}	1.156×10^{-2}	80	214.7	4.176×10^{-3}	1.016×10^{-3}
84	183.8	1.733×10^{-3}	4.926×10^{-3}	84	196.1	1.734×10^{-4}	4.618×10^{-3}
88	173.6	6.312×10^{-4}	1.898×10^{-3}	88	182.4	6.665×10^{-4}	1.908×10^{-3}
92	167.2	2.191×10^{-4}	6.836×10^{-4}	92	176.1	2.433×10^{-4}	7.250×10^{-4}
96	169.2	7.519×10^{-5}	2.314×10^{-4}	96	169.6	8.586×10^{-5}	2.638×10^{-4}
100	175.4	2.660×10^{-5}	7.890×10^{-5}	100	166.1	2.054×10^{-5}	9.250×10^{-5}

point of view of choosing between radiative-convective equilibrium and planetary circulation models, within the scope of which it should be possible to explain the high atmospheric temperature near Venus' surface and a series of structural features and tropospheric variations. In turn, the middle atmosphere above the cloud level (60–100 km) is distinguished by its even higher stability; here dT/dZ is even less than 3–4 K/km, approaching zero at altitudes of 85–100 km at low and middle latitudes and essentially right above the clouds near the poles, where $\Gamma \geq 11$K/km. The stable, stratified middle atmosphere is, apparently, close to a state of radiative equilibrium (Seiff, 1983).

We therefore can conclude that according to measurements by the Venera and Pioneer-Venus spacecraft, Venus' lower and middle atmosphere turns out to be close to adiabatic. Measurements made by the Vega 1–2 balloon probes confirmed this finding in a major portion of the middle cloud layer. The minor deviation from the adiabat (about 0.3K/km) suggests slightly stable stratification (Linkin et al., 1986). It is interesting to note that as the balloon probe traversed the terminator, the measured temperatures formed into two adiabats, separated by 0.24K relative to one another. This minor deviation from a single adiabat suggests the existence of different air masses, involving rising and descending motions, which flowed around the balloon as it drifted.

Current ideas about the structure and fundamental properties of the lower atmosphere to an altitude of 100 km, including upper-level stratification and spatial-temporal variations, as well as likely unstable regions at several altitude ranges, are summarized in the COSPAR reference model VIRA-85 (Seiff et al., 1985). Detailed tables are presented showing atmospheric parameters at various latitudes, graphs of probable deviations from the mean model values, and thermodynamic gas constants (heat capacity, thermal conductivity, viscosity, the speed of sound) at specific altitudes in the atmosphere. Temperature, pressure, and density values from this model are presented in Tables 5.3 and 5.4.

Measurements and Estimates of Chemical Composition

We shall first consider the data on the chemical composition of Venus' atmosphere that we actually have available, and we shall discuss current estimates of the upper concentration limits for a series of components whose existence can be assumed based on geochemical and other considerations.

In the three decades since the flights of the Venera 4–6 spacecraft, a battery of physical and chemical analysis techniques have been used not only to refine the results of CO_2, N_2, and H_2O analysis but also to obtain extensive information on nearly 20 minor components in various regions of Venus' lower atmosphere. A detailed description of the techniques used to study the chemical composition of the atmosphere by spacecraft can be found in various surveys and monographs (Kuzmin and Marov, 1974; Marov, 1972, 1978, 1979, 1989a; Colin and Hunten, 1977; Surkov et al., 1977; Moroz, 1981; von Zahn et al., 1983). The results obtained enable us to formulate ideas about the chemical processes in the planet's lower atmosphere.

In 1984 a discovery was made that opened up the composition of the lower, sub-cloud atmosphere to ground-based study. Observations of the night side of Venus at near-infrared wavelengths showed anomalously high brightness temperatures at two narrow wavelength intervals centered around 2.3 and 1.7 μm (Allen and Crawford, 1984). This finding was interpreted as thermal radiation from the hot, lower atmosphere escaping through narrow spectral "window" regions, where the dominant opacity of CO_2 and H_2O is not sufficiently great to completely attenuate radiation from these depths. Many observations at higher spatial and spectral resolutions were made following this discovery. Successful interpretation of these spectra had to await new advances in the theoretical understanding of the details of the CO_2 opacity structure under the high T and P conditions found in the lower atmosphere, including numerous "hot bands" that are not found in spectral databases designed for terrestrial applications (Pollack et al., 1993). Observations in these window regions, and in a number of new narrow windows discovered around 1.2 μm, have allowed improved estimates of the abundance of many important species in the lower atmosphere of Venus, including H_2O, HDO, CO, SO_2, OCS, HCl, and HF (Bézard et al., 1990, 1991, 1993; Crisp et al., 1991; De Bergh et al., 1991, 1995; Pollack et al., 1993). Observations with the Near Infrared Mapping Spectrometer (NIMS) instrument on the Galileo spacecraft were made in these spectral windows during the fly-by of Venus en route to Jupiter. These data were especially useful in studying spatial variability of H_2O and CO (Carlson et al. 1991, 1993; Drossart et al., 1993).

The results of measuring component concentrations are summarized in Table 5.5. This table is not exhaustive but does enable the reader to evaluate the present situation concerning the fundamental groups of elements and their compounds, and to review the techniques on which these concentrations are based (more detailed information is contained in the VIRA-85 model; see von Zahn and Moroz, 1985). Aside from carbon dioxide and nitrogen, we have reliably recorded water vapor and carbon monoxide, a series of sulfur- and halogen-bearing compounds, and noble gases. Data on other atmospheric components are based on upper detection limits. We shall comment in detail on the scope of this table, following the works of Marov et al. (1989a) and von Zahn et al. (1983).

Carbon Dioxide

CO_2, the dominant component of Venus' atmosphere, was first determined by the Venera-4 spacecraft, using gas analyzers with a chemical absorber. Using the same technique, Venera 5–6 obtained a relative concentration of this gas (by volume) to a high accuracy: $0.97^{+0.03}_{-0.04}$ (Vinogradov et al., 1968, 1970; Marov, 1972). Measurements conducted using gas chromatographs on the Pioneer-Venus Large probe (Oyama et al., 1980) provided the most accurate value: 0.964 ± 0.010 at an altitude of 22 km, and a considerably lower accuracy at higher altitudes (see Table 5.5). In accordance with these data, the CO_2 content in the lower atmosphere was taken to be equal to $96.5 \pm 1.0\%$ by volume. The CO_2 content over the whole lower atmosphere is taken to be constant, although there is no instrumental analysis of CO_2 in the stratomesosphere (65–110 km). At an altitude of

Table 5.5 Chemical composition of Venus' lower atmosphere according to experimental data

Gas	Altitude (or altitude range of analysis), km	Content[a]	Device[b]	Method	Reference
CO_2	53	90 ± 10	Venera-4 (descent module)	Gas analyzer with a chemical absorber (KOH)	Vinogradov et al., 1968
	~54	97^{+3}_{-4}	Venera-5 (descent module)	Gas analyzer with a chemical absorber (KOH)	Vinogradov et al., 1970
	52	$95.4^{+14.6}_{-20.1}$	Pioneer-Venus (Large probe)	Gas chromatography	Oyama et al., 1980
	42	$95.9^{+4.1}_{-5.8}$	Pioneer-Venus (Large probe)	Gas chromatography	Oyama et al., 1980
	22	96.4 ± 1.0	Pioneer-Venus (Large probe)	Gas chromatography	Oyama et al., 1980
N_2	53	≤7	Venera-4 (descent module)	Gas analyzer with a chemical absorber (Zr)	Vinogradov et al., 1968
	~54	≤3.5	Venera 5–6 (descent module)	Gas analyzer with a chemical absorber (Zr)	Vinogradov et al., 1970
	63–34	~1.8	Venera 9–10 (descent module)	Mass spectrometry	Surkov et al., 1978
	23–1	4.0 ± 0.3	Venera 11–12 (descent module)	Mass spectrometry	Istomin et al., 1979
	62–0	4.0 ± 2.0	Pioneer-Venus (Large probe)	Mass spectrometry	Hoffman et al., 1980a, 1980b
	42–0	2.5 ± 0.3	Venera-12	Gas chromatography	Gelman et al., 1979
	52	4.6 ± 0.14	Pioneer-Venus (Large probe)	Gas chromatography	Oyama et al., 1979
	42	3.54 ± 0.04	Pioneer-Venus (Large probe)	Gas chromatography	Oyama et al., 1980
	22	3.41 ± 0.01	Pioneer-Venus (Large probe)	Gas chromatography	Oyama et al., 1980
CO	~90	180 ± 90	Ground-based astronomy	IR spectroscopy	Wilson et al., 1981
	~70	51 ± 1	Ground-based astronomy	IR spectroscopy	L. Young, 1972, 1974
	52	$32.2^{+61.7}_{-22.2}$	Pioneer-Venus (Large probe)	Gas chromatography	Oyama et al., 1980
	42	30.2 ± 18.0	Pioneer-Venus (Large probe)	Gas chromatography	Oyama et al., 1980
	22	19.9 ± 3.1	Pioneer-Venus	Gas chromatography	Oyama et al., 1980
	12	16.8 ± 1.4	Venera-12 (descent module)	Gas chromatography	Gelman et al., 1979
	42–36	38 ± 7	Venera-12	Gas chromatography	Gelman et al., 1979
	36	23 ± 5	Ground-based astronomy	Near-IR spectroscopy	Pollack et al., 1993

[a] For CO_2 and N_2 the relative concentrations are given in percent by volume; for the remaining components the concentrations are given in ppm.

[b] The spacecraft are either landers or satellites; astronomy is either ground-based or from an artificial Earth satellite.

Table 5.5 (*continued*)

Gas	Altitude (or altitude range of analysis), km	Content[a]	Device[b]	Method	Reference
SO_2	>70	<0.035	Ground-based astronomy	IR spectroscopy	Cruikshank and Kuiper, 1967
	>70		Ground-based astronomy	UV spectroscopy	Barker, 1979
	>70	<0.3	Ground-based astronomy	IR spectroscopy	Anderson et al., 1968
	>70	<0.01	Astronomical Observer	IR spectroscopy	Owen and Sagan, 1972
	>70	<0.15–0.30	Venera 9–10 (satellites)	Spectroscopy of scattered radiation	Krasnopolsky, 1980
	70–80	0.02–0.8	IUE	UV spectroscopy	Conway et al., 1979
	~70	0.01–0.1	Pioneer-Venus (satellite)	UV spectroscopy	Esposito, 1981
	~66	0.7–6.0	Venera 15–16 (satellites)	IR (Fourier) spectroscopy	Moroz et al., 1985a
	52	<10	Pioneer-Venus (Large probe)	Gas chromatography	Oyama et al., 1980
	42	>176	Pioneer-Venus (Large probe)	Gas chromatography	Oyama et al., 1980
	22	185±43	Pioneer-Venus (Large probe)	Gas chromatography	Oyama et al., 1980
	42–0	130–35	Venera-12 (descent module)	Gas chromatography	Gelman et al., 1979
	42	180±70	Ground-based astronomy	Near-IR spectroscopy	Pollack et al., 1993
	35–45	130±40	Ground-based astronomy	Near-IR spectroscopy	Bézard et al., 1993
H_2S	>70	<200	Ground-based astronomy	IR spectroscopy	Cruikshank and Kuiper, 1967
	>70	<0.3	Ground-based astronomy	IR spectroscopy	Anderson et al., 1969
	>70	<0.1	Astronomical Observer	UV spectroscopy	Owen and Sagan, 1972
	37–29	80±40	Venera 13–14 (descent module)	Gas chromatography	Mukhin et al., 1983
	22–0	3±2	Pioneer-Venus (Large probe)	Mass spectrometry	Hoffman et al., 1980a
COS	>70	<0.2	Ground-based astronomy	IR spectroscopy	Anderson et al., 1969
	>70	<0.01	High-altitude aircraft	IR spectroscopy	Kuiper and Forbes, 1967
	>70	<1	Ground-based astronomy	IR spectroscopy	Cruikshank, 1967
	>70	<0.1	Astronomical Observer	UV spectroscopy	Owen and Sagan, 1972
	37–29	40±20	Venera 13–14 (descent module)	Gas chromatography	Mukhim et al., 1983
	22–0	<3	Pioneer-Venus (Large probe)	Mass spectrometry	Hoffman et al., 1980a
	33	4.4±1	Ground-based astronomy	Near-IR spectroscopy	Pollack et al., 1993

[a]For CO_2 and N_2 the relative concentrations are given in percent by volume; for the remaining components the concentrations are given in ppm.

[b]The spacecraft are either landers or satellites; astronomy is either ground-based or from an artificial Earth satellite.

Table 5.5 (continued)

Gas	Altitude (or altitude range of analysis), km	Content[a]	Device[b]	Method	Reference
S$_2$	52–0	0.08–0.2[c]	Venera 11–12 (descent module)	Spectrophotometry	Sanko, 1980
HCl	~70	0.42–0.61	Ground-based astronomy	IR spectroscopy	L. Young, 1972, 1974
	62–0	<10	Pioneer-Venus (Large probe)	Mass spectrometry	Hoffman et al., 1980a
	23.5	0.48±0.12	Ground-based astronomy	Near-IR spectroscopy	Pollack et al., 1993
HF	~70	0.001	Ground-based astronomy	IR spectroscopy	Connes et al., 1967
	~70	0.01	Ground-based astronomy	IR spectroscopy	L. Young, 1972, 1974
	33.5	0.001–0.005	Ground-based astronomy	Near-IR spectroscopy	Pollack et al., 1993
H$_2$O	~70	0.5–40	Ground-based astronomy	IR spectroscopy	Barker, 1975a, 1975b
	60–0	<1000	Radioastronomy	Microwave spectroscopy	Kuzmin and Marov, 1974
	~65	6–100	Pioneer-Venus (satellite)	IR radiometry	Schofield et al., 1982
	~50	30^{+60}_{-20}	Venera 15–16 (satellites)	IR spectrometry	Ertel et al., 1984
	54–50	60–10,000	Venera 4 (descent module)	Gas analyzer	Vinogradov et al., 1968, 1980
		60–10,000	Venera 5–6 (descent module)	Gas analyzer	Vinogradov et al., 1968, 1970
	52	<600	Pioneer-Venus (Large probe)	Gas chromatography	Oyama et al., 1980
	42	5200±700	Pioneer-Venus (Large probe)	Gas chromatography	Oyama et al., 1980
	22	1350±150	Pioneer-Venus (Large probe)	Gas chromatography	Oyama et al., 1980
	58–49	700±300	Venera 13–14 (descent module)	Gas chromatography	Mukhin et al., 1983
	54–50	500±200	Venera 13–14 (descent module)	Spectrophotometry	Moroz et al., 1983b
	50–40	200	Venera 13–14 (descent module)	Spectrophotometry	Moroz et al., 1983b
	40–20	100	Venera 13–14 (descent module)	Spectrophotometry	Moroz et al., 1983b
	10–0	20	Venera 13–14 (descent module)	Spectrophotometry	Moroz et al., 1983b
	60–25	3000–200	Vega-1 (descent module)	Moisture gauge[d]	Surkov et al., 1987c
	60–25	800–150	Vega-2 (descent module)	Moisture gauge[d]	Surkov et al., 1987c
	33	30±6	Ground-based astronomy	Near-IR spectroscopy	Pollack et al., 1993
	23.5	30±7.5	Ground-based astronomy	Near-IR spectroscopy	Pollack et al., 1993

[a] For CO$_2$ and N$_2$ the relative concentrations are given in percent by volume; for the remaining components the concentrations are given in ppm.

[b] The spacecraft are either landers or satellites; astronomy is either ground-based or from an artificial Earth satellite.

[c] The range of concentrations is associated with the uncertainty of the nature of the absorber in the λ <550 nm range (S$_3$ or S$_4$).

[d] Data on H$_2$O analysis from the Vega 1–2 landers are also presented graphically in Fig. 5.16.

Table 5.5 (*continued*)

Gas	Altitude (or altitude range of analysis), km	Content[a]	Device[b]	Method	Reference
H$_2$O (*cont.*)	12	30±10	Ground-based astronomy	Near-IR spectroscopy	Pollack et al., 1993
	0–40	30±15	Ground-based astronomy	Near-IR spectroscopy	De Bergh et al., 1995
H$_2$	58–49	25±10	Venera 13–14 (descent module)	Gas chromatography	Mukhin et al., 1983
O$_2$	~70	<1	Ground-based astronomy	IR spectroscopy	Traub and Carleton, 1973
	52	43.6±25.2	Pioneer-Venus (Large probe)	Gas chromatography	Oyama et al., 1980
	42	16.0±7.4	Pioneer-Venus (Large probe)	Gas chromatography	Oyama et al., 1980
	58–35	18±4	Venera 13–14 (descent module)	Gas chromatography	Mukhin et al., 1983
NH$_3$	44–32	100–1000	Venera-8 (descent module)	Colorimetry	Surkov et al., 1973
	60–0	<8–16	Radioastronomy	Microwave spectroscopy	Kuzmin and Marov, 1974
	>70	<0.1	Astronomical Observer	UV spectroscopy	Owen and Sagan, 1972
CH$_4$	>70	<0.03	Ground-based astronomy	IR spectroscopy	Kuiper and Forbes, 1967
	>70	<1	Ground-based astronomy	IR spectrometry	Connes et al., 1967
	52	<10	Pioneer-Venus (Large probe)	Gas chromatography	Oyama et al., 1980
	42	<3	Pioneer-Venus (Large probe)	Gas chromatography	Oyama et al., 1980
	22	<0.6	Pioneer-Venus (Large probe)	Gas chromatography	Oyama et al., 1980
	30	<0.1	Ground-based astronomy	Near-IR spectrometry	Pollack et al., 1993
	24	<2.0	Ground-based astronomy	Near-IR spectrometry	Pollack et al., 1993
C$_2$H$_2$[e]	70	1	Ground-based astronomy	IR spectrometry	Connes et al., 1967
^{20}Ne	<23	8.6±4.0	Venera 11–12 (descent module)	Mass spectrometry	Istomin et al., 1979
	22	4.3±0.7	Pioneer-Venus (Large probe)	Gas chromatography	Oyama et al., 1979, 1980
	<25	10±7	Pioneer-Venus (Large probe)	Mass spectrometry	Hoffman et al., 1979, 1980b
	<26	~7	Venera 13–14 (descent module)	Mass spectrometry	Istomin et al., 1979, 1983
		7±3[f]			von Zahn et al., 1983
$\Sigma(^{40}$Ar + ^{36}Ar)	<42	50±10	Venera-12 (descent module)	Gas chromatography	Gelman et al., 1979
	<23	110±20	Venera 11–12 (descent module)	Mass spectrometry	Istomin et al., 1979, 1983
	22	67.2±2.3	Pioneer-Venus (Large probe)	Gas chromatography	Oyama et al., 1979, 1980

[a] For CO$_2$ and N$_2$ the relative concentrations are given in percent by volume; for the remaining components the concentrations are given in ppm.

[b] The spacecraft are either landers or satellites; astronomy is either ground-based or from an artificial Earth satellite.

[e] For the remaining hydrocarbons (C$_2$H$_4$, C$_2$H$_6$, C$_3$H$_8$), estimates of the concentration in the 52–22 km altitude range, according to gas chromatography data from the Pioneer-Venus Large probe, lie in 20–1 ppm range, respectively.

[f] Recommended value.

Table 5.5 (*continued*)

Gas	Altitude (or altitude range of analysis), km	Content[a]	Device[b]	Method	Reference
$\Sigma(^{40}\text{Ar} + {}^{36}\text{Ar})$ (*cont.*)	<25	67	Pioneer-Venus (Large probe)	Mass spectrometry	Hoffman et al., 1980a
	<26	~100	Venera 13–14 (descent module)	Mass spectrometry	Istomin et al., 1983
		70±25[f]			von Zahn et al., 1983
^{84}Kr	<23	0.6±0.2	Venera 11–12 (descent module)	Mass spectrometry	Istomin et al.,1979
	<25	0.007–0.0028	Pioneer-Venus (Large probe)	Mass spectrometry	Donahue et al., 1981
	<26	0.018–0.023	Venera 13–14 (descent module)	Mass spectrometry	Istomin et al., 1983
	49–37	0.7±0.3	Venera 13–14 (descent module)	Mass spectrometry	Mukhin, 1983
		0.025[f]			von Zahn et al., 1983
ΣKr	<25	0.0523	Pioneer-Venus (Large probe)	Mass spectrometry	von Zahn et al., 1983
Xe	25	0.087	Pioneer-Venus (Large probe)	Mass spectrometry	Donahue et al., 1981
	26	0.001–0.002	Venera 13–14 (descent module)	Mass spectrometry	Istomin et al., 1983
Hg	<24	<5	Pioneer-Venus (Large probe)	Mass spectrometry	Hoffman et al., 1980a
NO[f]	>70	<1	Astronomical Observer	UV spectrometry	Owen and Sagan, 1972
$\text{NO}_2{}^{g}$	>70	<0.01	Astronomical Observer	UV spectrometry	Owen and Sagan, 1972
	>70	<0.08	Ground-based astronomy	IR spectrometry	Anderson et al., 1969
	>70	<(0.2–0.3)	Venera 9–10	Spectrometry of scattered radiation	Krasnopolsky, 1980
$\text{N}_2\text{O}_4{}^{g}$	>70	<0.04	Astronomical Observer	UV spectrometry	Owen and Sagan, 1972
$\text{CS}_2{}^{g}$	>70	<(40–50)	Venera 9–10 (satellites)	Spectrometry of scattered radiation	Krasnopolsky, 1980
$\text{Br}_2{}^{g}$	>70	<1.5	Venera 9–10 (satellites)	Spectrometry of scattered radiation	Krasnopolsky, 1980
HCN^{g}		<10^{-4}	Venera 11–12	Spectrophotometry	Moroz et al., 1979
	>70	<1	Ground-based astronomy	IR spectrometry	Connes et al., 1967
$\text{Cl}_2{}^{g}$	>70	<0.1	Venera 11–12	Spectrophotometry	Moroz et al., 1979

Source: Data from Marov et al., 1989; von Zahn et al., 1983.

[a] For CO_2 and N_2 the relative concentrations are given in percent by volume; for the remaining components the concentrations are given in ppm.

[b] The spacecraft are either landers or satellites; astronomy is either ground-based or from an artificial Earth satellite.

[f] Recommended value.

[g] Threshold values. Furthermore, the following upper limits of the relative concentrations (by volume) have been established: C_3O_2 (<1.0–0.1), O_3 (<0.01–0.001), CH_3Cl, CH_3F, HCHO, aldehydes, and ketones (<1 ppm); see Moroz and von Zahn, 1985.

~ 100 km ($P = 0.02$ Mbar, $T = 178$ K), the CO_2 number density was estimated at $6 \cdot 10^{14} cm^{-3}$, and the time to establish diffusive equilibrium is ~ 13 years (von Zahn et al., 1983). The level of diffusive-gravitational separation of gases (the turbopause) is higher. Therefore, one can confidently assume that turbulent mixing affects the composition of the middle atmosphere much more intensely than diffusion, and CO_2 in the stratomesosphere is not subject to diffusive separation.

Carbon Monoxide and Other Carbon Compounds

Carbon monoxide (CO) plays an important role in Venus' atmosphere. Data on its concentration have been obtained by various methods and provide quite a complete picture. Aside from spectroscopic analyses, starting in 1968 (Connes et al., 1968), ground-based measurements in the microwave range near $\lambda = 2.6$ mm have been carried out over a wide range of phase angles. Early studies using this method found that the percentage of this gas by volume (mixing ratio X_{CO}) increases from 10^{-4} at 85 km to 10^{-3} at 105 km (Schloerb et al., 1980; Wilson, 1981). More recent microwave measurements found $X_{CO} = 10^{-5}$ at 80 km (Clancy and Muhleman, 1985). Diurnal variations in the CO content have also been detected; this is especially pronounced in the 75–85 km interval, where the daytime concentration turned out to be more than 10 times greater than the nighttime. Above 95 km an inversion is observed: maximum concentrations were recorded at midnight and minimum at noon.

Gas chromatographic analyses on the Pioneer-Venus Large probe (Oyama et al., 1980) indicated a regular decrease in CO content as one approached the planet's surface (see Table 5.5). The same trend was noted in interpreting gas chromatographs from the Venera-12 lander: in the 36–12 km altitude range, the CO content decreases by 40% (Gelman et al., 1979). This decrease could be explained by the presence of a CO source in the stratomesosphere, as a product of CO_2 photodissociation, and a sink in the subcloud troposphere as a result of chemical reactions causing oxidation of CO into CO_2.

Until recently, CO was the only carbon-bearing minor species that had been positively identified in the Venusian atmosphere. The presence of carbonyl sulfide (COS), which had been assumed based on theoretical estimates (Lewis, 1969, 1970), had not yet been confirmed and no trace was found (Cruikshank and Sill, 1967). The measurements, including gas chromatographic measurements by the Venera 13–14 spacecraft, were contradictory (see Table 5.5). Equilibrium calculations (Volkov, 1983) suggested that the near-surface atmosphere might contain up to 20 ppm COS. Observations in the 2.3-μm near-infrared window, however, have more conclusively determined a COS mixing ratio of $4.4 \pm 1.1 \times 10^{-6}$.

Synthetic spectra were fit to observations in the 2.3-μm near-infrared window to derive a CO mixing ratio of 2.3×10^{-5} at an altitude of 36 km (Pollack et al., 1993). Evidence for a gradient of CO, with abundance decreasing with decreasing altitude, was also found in this analysis. This same work also found an increase of COS with altitude in the same region and attributed this correlation to buffering reactions by surface minerals in which CO acts as one of the source gases for COS (Fegley and Treiman, 1992).

Experimental data on the chemical composition of the near-surface atmosphere, including the CO content, are lacking. Assuming quasi-equilibrium of the atmospheric gases with the surface rocks (see below), the CO content was estimated from equilibrium calculations of the C–H–O–S system. For $T = 735$ K, $P = 90$ atm ($Z = 0$ km), it is $\sim$7 ppm (Zolotov and Khodakovsky, 1985). A qualitative estimate of the reduction-oxidation parameters at the surface was obtained by the Contrast experiment on the Venera 13–14 landers. The working principle behind the chemical indicator, mounted on the lander support ring, is based on a rapid color reduction reaction of sodium pyrovanadate with the formation of vanadium oxides (Florensky et al., 1983a):

$$Na_4V_2O_7 + CO_2 + CO \rightleftharpoons V_2O_4 + 2Na_2CO_3 \qquad (a)$$

$$Na_4V_2O_7 + CO \rightleftharpoons V_2O_3 + 2Na_2CO_3 \qquad (b)$$

Such reactions occur for a low partial pressure of oxygen at the temperature of the Venusian surface. Research on the albedo characteristics of details on the lander support ring has indicated that the coefficient of reflection of the indicator plate is extremely low; this definitely indicates a sodium pyrovanadate reduction reaction and suggests the presence of a reducing agent, for example, CO. The equilibrium concentration of CO, computed according to equilibrium constants for the reactions presented above, corresponds to a relative concentration of $X_{CO} \approx 10^{-5}(\sim 10$ ppm) for $X_{O_2} \approx 10^{-23}$; this conforms well with the results of model thermodynamic calculations (in this case we likewise estimate $X_{O_2} \approx 10^{-23}$).

The estimates of the concentration of hydrocarbons are more contradictory. The mass spectrometer experiment on the Pioneer-Venus Large probe detected $\sim$2 ppm C_2H_6 and a large upper limit of methane. According to calculations, however, the hydrocarbon content in an equilibrium atmosphere is negligible ($X_{CH_4} = 5 \cdot 10^{-20}$), and the near-infrared window observations have set a CH_4 upper limit at 10^{-7} in the lower atmosphere.

Stability of CO_2

The CO content, and the stability of the dominant constituent in Venus' atmosphere, CO_2, are linked with problems of current interest in atmospheric chemistry. Primarily we are talking about a mechanism that impedes the accumulation of CO (as well as oxygen) under conditions of CO_2 photodissociation by intense ultraviolet solar radiation. Many works (see, for example, Kuzmin and Marov, 1974; Krasnopolsky, 1986; Prinn, 1971; Sze and McElroy, 1975; Yung and DeMore, 1982; von Zahn et al., 1983) have been devoted to the study of suitable photochemical models for the cyclic transformations of CO_2, CO, O, and O_2 in Venus' stratomesosphere and troposphere, with the participation of other components. We will now briefly touch upon several likely processes and will return to these problems during subsequent analysis of atmospheric components.

Ultraviolet radiation capable of ensuring CO_2 photolysis ($\lambda < 2240$ Å) penetrates to an altitude of $Z \approx 65$ km:

$$CO_2 + h\nu \rightarrow CO + O \qquad (5.11)$$

whereas the CO_2 recombination reaction with the assistance of neutral particles

$$CO + O + M \rightarrow CO_2 + M \qquad (5.12)$$

is, as is said, spin-forbidden and occurs very slowly. At the same time, molecular oxygen forms from atomic oxygen at a rate exceeding that of Eq. (5.12) by five orders of magnitude:

$$O + O + M \rightarrow O_2 + M \qquad (5.13)$$

As a consequence, the disappearance of all CO_2 from the atmosphere would take only several thousand years, and the atmosphere would have to consist of a mixture of CO and O_2. This means that there is actually some efficient mechanism for recombination:

$$CO + \frac{1}{2} O_2 \rightarrow CO_2 \qquad (5.14)$$

One of the mechanisms responsible for preserving a predominantly CO_2 atmosphere on Venus and for ensuring photochemical equilibrium in the stratomesosphere could be photolysis of hydrochloric acid (HCl):[4]

$$HCl + h\nu \rightarrow \dot{H} + Cl \qquad (5.15)$$

with recombination in the reversible process:

$$Cl + H_2 \rightleftharpoons HCl + \dot{H} \qquad (5.16)$$

Calculations showed that only several ppm of HCl is sufficient to accomplish this (Krasnopolsky, 1986; Yung and DeMore, 1982). In this case, hydrogen or its radicals, which arise only in the constant presence of HCl, serve as catalyzing agents during oxidation of CO to CO_2, for example:

$$HCl + O \rightarrow \dot{O}H + Cl \qquad (5.17)$$

In this case, the overall hydrogen content in the atmosphere plays a decisive role in the photochemical oxidation of CO. According to Yung and DeMore (1982), if $X_{H_2} \approx 10^{-6}$, then a sufficient quantity of $\dot{O}H_X$ radicals are formed, and CO_2 recombination proceeds according to:

$$CO + \dot{O}H \rightarrow CO_2 + \dot{H} \qquad (5.18)$$

If, however, $X_{H_2} \approx 10^{-7}$, then the necessary additional amount of $\dot{O}H_X$ for subsequent oxidation of CO, according to reaction (5.18), may be produced with the assistance of nitrogen compounds of the form NO_X:

$$NO + HO_2 \rightarrow NO_2 + \dot{O}H \qquad\qquad (5.19)$$

Finally, for $X_{H_2} < 10^{-7}$, when the concentration of OH_X radicals is negligible, another scheme is possible, with the participation of chlorine-bearing radicals of the form ClO_X:

$$Cl + CO + M \rightarrow Cl\dot{C}O + M \qquad\qquad (5.20)$$

$$ClCO + O_2 \rightarrow ClO + CO_2 \qquad\qquad (5.21)$$

Unfortunately, data on the hydrogen content in the lower atmosphere and strato-mesosphere are not precise enough (see Table 5.5) to decide whether this mechanism is realistic or to choose between any particular photochemical processes. Thermochemical reactions with the assistance of sulfur oxides, SO_2 and SO_3 (see reactions [5.31] and [5.32]), are more realistic mechanisms for the sink of CO, given sulfur compounds in Venus' atmosphere.

Nitrogen and Its Compounds

The second most abundant component of Venus' atmosphere is nitrogen. Molecular nitrogen was first discovered in Venus' lower atmosphere using a gas analyzer with a chemical absorber installed on the Venera-4 descent vehicle. A more accurate upper limit of nitrogen's relative concentration was obtained by the Venera 5–6 spacecraft, 3.5% by volume (Vinogradov et al., 1968, 1974; Marov, 1972). Mass spectroscopic measurements by the Venera 9–10 landers gave a lower value, ~1.8% (Surkov et al., 1978).

During subsequent spacecraft experiments, in which gas chromatographs and more sophisticated mass spectrometers were used, it was shown that the N_2 content is actually much closer to the value provided by the gas analyzers (see Table 5.5). Here preference is given to the gas chromatograph data, because the masking effect of CO (which has a mass number of 28, the same as N_2) creates difficulties in interpreting the mass spectra.

The vertical N_2 concentration gradient, detected by the gas chromatograph experiment on the Pioneer-Venus Large probe and exceeding possible analysis errors (Oyama et al., 1980), draws our attention. Such a result does not conform to the simple model of a well-mixed lower atmosphere and currently has no realistic explanation. Nitrogen bonding in cloud condensates, with the formation of $NOHSO_4$ crystals, is theoretically feasible. Such a hypothetical mechanism of N_2 chemical loss, however, is not supported by experimental research data on the composition of cloud particles (see Chapter 6). At present, it is safe to take $3.5 \pm 1\%$ by volume (von Zahn et al., 1983) as the mean concentration of molecular nitrogen in Venus' lower atmosphere.

In spite of the great abundance of nitrogen, the concentration of nitrogen oxides in Venus' atmosphere is negligible. NO absorption lines were first discovered using the International Ultraviolet Explorer (IUE) satellite, during nightglow research of Venus (Feldman et al., 1979). The maximum luminescence caused by NO was attributed to an altitude of ~115 km. This result, however, needs additional confirmation. At the same time, the upper spectroscopic limit for NO, according to astronomical satellite data, is < 1 ppm

(Owen and Sagan, 1972). For other oxides (NO_2 and N_2O_4, the upper limits are < 0.01–0.8 ppm (see Table 5.5). From spectral analysis of scattered radiation at the limb of the planet, obtained using ultraviolet spectrometers aboard the Venera 9–10 spacecraft ($Z = 77$–85 km), an upper limit for $NO_2 < 0.2$–0.3 ppm (Krasnopolsky, 1980, 1983) was established.

It is most likely that NO generation in the middle atmosphere satisfies

$$N + O \rightarrow NO + h\nu \qquad (5.22)$$

where the N and O atoms are drawn from the daylight side, owing to planetary circulation of the atmosphere. Electrical discharges could be the source of NO in the lower atmosphere. According to Ksanfomality (1979, 1983b) and Scarf and Russell (1983), if lightning occurs on Venus as often as on Earth, then up to 1 ppb NO may accumulate in Venus' stratosphere. At higher NO_X concentrations (up to 30 ppb), NO could serve as a catalyst in oxidation cycles of CO by molecular oxygen (see reaction [5.14]) or SO_2 in photochemical processes during CO_2 recombination and H_2SO_4 generation (Yung and DeMore, 1982).

In 1973, the Venera-8 lander measured ammonia (NH_3) content using a gas analyzer that was based on the linear colorimetric technique for identifying a gas by the color change of a chemical reagent recorded by photoconductive cells. Identical results were obtained at two altitudes in the atmosphere, 44 and 32 km: 100–1000 ppm (Surkov et al., 1973). This result, however, was subjected to serious doubt by modeling of the absorption spectrum of microwave radiation in Venus' atmosphere, which produced estimates for the upper limit of NH_3 of 8 ppm, given approximately 0.1% water vapor, and 16 ppm in a dry atmosphere (Kuzmin and Marov, 1974). The spectroscopic limit is lower yet, < 0.1 ppm (Owen and Sagan, 1972). Mass spectrometers on the Venera 9–10 spacecraft fixed only the upper NH_3 limit (500 ppm) in the same altitude range (Surkov et al., 1978), while in experiments on the Pioneer-Venus probes and Venera 11–14 landers, ammonia was not identified at all, with either mass spectrometers or gas chromatography (see Table 5.5).

In 1977, A. Young showed that the linear colorimetric method used in Venera-8 for identifying NH_3 was ineffective. In his opinion, the color change in the acid-base indicator used, given NH_4^+ ions in an acid medium (if H_2SO_4 aerosols entered the intake), would not be sufficiently pronounced to be registered. Therefore, indications by the Venera-8 gas analyzer should be considered indirect proof of the presence of sulfuric acid, not ammonia.

Thermodynamic calculations of the interaction between the near-surface atmosphere and surface rock (Mueller, 1964; Khodakovsky et al., 1979a, 1979b) indicated that there is essentially no bonding of nitrogen in minerals, and the equilibrium concentration of NH_3 is negligible: $X_{NH_3} = 10^{-11}$–10^{-14}. NO and HCN may be found in even smaller quantities.

Water Vapor

Venus' atmosphere contains a small amount of water vapor, whose altitude distribution is extremely irregular, as a consequence of condensation and chemical interaction (see Table 5.5).

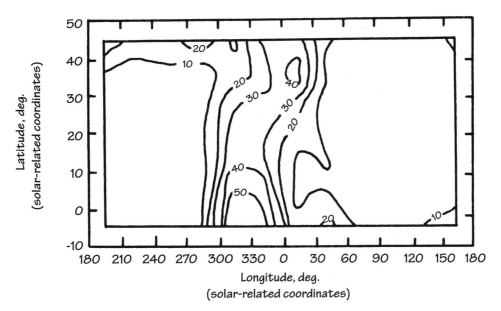

Fig. 5.14. A horizontal distribution of water vapor content (in microns of precipitated water, corresponding to the numbers in the isolines) at an altitude of 67–68 km in solar-related coordinates (according to Schofield et al., 1982). The specific altitude corresponds to $\tau = 1$ at $\lambda = 11.5\ \mu m$; in Venus' equatorial zone, 1 micron of precipitated water in a column with optical depth $\tau \approx 1$ is equivalent to a relative concentration $X_{H_2O} \approx 2$ ppm by volume.

Weak H_2O absorption bands in Venus' infrared spectrum had already been recorded in 1964 (see Chapter 2). Because the formation of these bands is attributed to an altitude approximately coinciding with the upper boundary of the cloud layer (~ 70 km), they characterize only the above-cloud atmosphere. The water vapor content proved to be negligible, less than in the dry atmosphere of Mars; explaining this phenomenon posed a complex problem for researchers. Subsequent measurements on the whole confirmed the initial estimate. The most accurate astronomical measurements in the near-infrared spectral region, obtained as a result of long-term observations (Fink et al., 1972; Barker et al., 1975b), provided quite a wide range of X_{H_2O} values: from 0.5 to 40 ppm. Interpretations of infrared spectra recorded by the Kuiper Airborne Observatory, and results of infrared radiometry from the Mariner-10 fly-by (Kuiper et al., 1969; Taylor, 1975) led to similar results.

Long-term, detailed research on the water vapor content in Venus' middle atmosphere, using remote sensing techniques, was carried out by the Pioneer-Venus spacecraft at altitudes of 55–85 km, using infrared radiometry (Schofield and Taylor, 1982). A region of maximum H_2O concentration (up to 100 ppm) was detected, encompassing a sector from the equator to $\sim 50°$ N, and confined to the period of 14–16 hours local solar time (Fig. 5.14). On the night side of the planet, an extremely low H_2O content was recorded (~ 6 ppm). A monotonic increase was observed in the vertical distribution of X_{H_2O} with decreased altitude (Fig. 5.15). At the same time, a deep minimum (5 ppm) was noted on the daylight side at an altitude of 60 km, with a sharp increase in moisture content near the lower measurement boundary (up to 1000 ppm at 55 km). The authors of the experiment attributed these significant diurnal variations in water vapor content either to photochemical formation of sulfuric acid aerosol on the day side of the planet or to periodic water vapor transport from lower tiers of the cloud layer, in Hadley cells.

Fig. 5.15. A vertical distribution of water vapor content (according to Schofield et al., 1982): (1) night side, 9 P.M., $\varphi = 25° N$, $\theta = 225°$; (2) day side, 1 A.M., $\varphi = 25° N$, $\theta = 345°$ (local standard time).

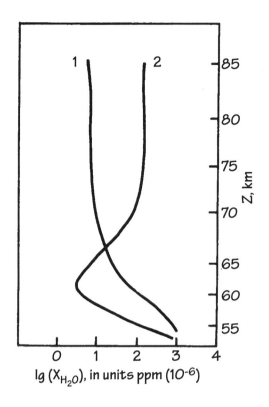

lg (X_{H_2O}), in units ppm (10^{-6})

Data from the Venera 15–16 infrared experiments, conducted using Fourier spectrometers with the goal of revealing the spatial-temporal distribution of water vapor (Moroz et al., 1985a), do not agree with the results. On a background of an overall irregular water vapor distribution, the depletion of water vapor in the night atmosphere cannot be confirmed. At the same time, the X_{H_2O} (30^{+60}_{-20} ppm) attributed to the 50 km level is much lower. In connection with this, note that remote infrared radiometric and spectroscopic techniques for determining H_2O concentration depend strongly on the assumed atmospheric temperature profile in the altitude range under study, on the model optical properties of the cloud layer, and also on the actual level at which H_2O absorption bands form. These considerations are especially significant when interpreting measurements conducted over a rather broad spectral window, as was the case with the Pioneer-Venus radiometric experiment (33–55 μm). Therefore, it is difficult to isolate real variations from methodological effects, as was pointed out by Zasova et al. (1985) when analyzing discrepancies in measurements from the Venera 15–16 and Pioneer-Venus spacecraft. Accordingly, one should approach the variations in Figs. 5.14 and 5.15 (from the quantitative point of view) with particular care. Nevertheless, on the whole one can conclude that, on average, there are 30–40 ppm H_2O in Venus' stratosphere, and spatial-temporal variations along the horizontal and vertical are common and can possibly reach two orders of magnitude.

The situation is no less complex with data on the water vapor content of Venus' troposphere. The upper limit for the H_2O saturation point over the whole thickness of the atmosphere (specified by the densest, lowest-lying layers) was first imposed by the results of radio astronomy and radar measurements. A study of the thermal emission in the microwave region (at the H_2O absorption line $\lambda = 1.35$ cm) furnished an estimate $\leq 0.1\%$ (Pollack and Morrison, 1970; Kuzmin and Marov, 1974; Rossow and Sagan, 1975). Later, Janssen and Klein (1981) obtained a value of ~ 3000 ppm (and additionally, an upper limit for SO_2 of 180 ppm), assuming the presence of both these components in a uniformly mixed atmosphere below 50 km.

The first in situ analyses of the water vapor content in the Venusian atmosphere were conducted by the Venera 4–6 descent vehicles, using gas analyzers with a chemical water vapor absorber (Vinogradov et al., 1968, 1970). The absorption was recorded using manometric and electrolytic techniques; the instruments were designed to work in the 54–30 km altitude range. These measurements provided threshold concentrations in the range of 600 ppm $< X_{H_2O} < 10,000$ ppm. Later, according to the results of a more detailed analysis of the information obtained by the Venera 5–6 spacecraft, despite the relatively low measurement accuracy, a tendency toward a decrease in H_2O content was detected in the direction of the planet's surface from several thousand ppm in the cloud layer to 30–50 ppm at the ~ 20 km level (Marov, 1972; Andreychikov, 1978).

On a series of Venera landers and the Pioneer-Venus Large probe, a set of measurements was conducted using mass spectrometers and gas chromatographs. Unfortunately, mass spectrometer studies did not help significantly in solving the problem of the H_2O altitude distribution in Venus' troposphere. From the Pioneer-Venus measurements in the 60–50 km range, only an upper limit on the H_2O content (< 1000 ppm) was established; starting at an altitude of $Z = 50$ km, the flow of information ceased. The instrument resumed working only at 30 km, where another peak was recorded with a mass of 18 atomic mass units (amus), probably corresponding to H_2O. The authors of the experiment (Hoffman et al., 1979, 1980b) explain this strange instrumental behavior by the fact that drops of sulfuric acid aerosol may have become stuck in the mass spectrometer inlet system as the lander passed through the cloud layer. Subsequently, with increased temperature, they evaporated, and at a lower altitude the instrument resumed working.

The gas chromatograph installed on the Venera-12 spacecraft (Gelman et al., 1979) provided a rough upper limit for the H_2O content in the subcloud atmosphere ($X_{H_2O} < 50^{+50}_{-25}$ ppm). Using the same technique, the Venera 13–14 spacecraft estimated the water vapor content within the cloud layer (58–49 km) to be 700 ± 300 ppm and confirmed the general rule that the maximum H_2O concentration falls within the cloud layer. At the same time the gas chromatograph on the Pioneer-Venus Large probe measured a significantly higher water vapor content in samples of atmospheric gas taken below the cloud layer (42 and 22 km): 5200 ppm and 1350 ppm, respectively (Oyama et al., 1980). It is likely, however, that these are overestimates, in particular because in the working altitude range of the gas chromatograph, the temperature gradient value ($dT/dZ \approx 8.0$ K/km), given in the previous section, conforms better to an H_2O concentration in the 10–300 ppm range (Pollack et al., 1980b).

The water vapor content was also measured by the intensity of scattered solar radiation in H_2O absorption bands ($\lambda = 0.72, 0.82, 0.84, 1.13\mu$m) over almost the entire troposphere, starting from 55 km to the surface, during the descent phase of the Venera landers (Moroz et al., 1979, 1983b). The results are integral values for a specific layer of the atmosphere. According to the initial analysis of these results, the mean H_2O content in the cloud layer is 500 ± 200 ppm and decreases monotonically in the subcloud atmosphere, from 200 ppm immediately below the clouds to 10–50 ppm in the near-surface layer (0–10 km).

It is important to note that interpretation of the measurement data in the 0.94 μm leads to these conclusions, whereas analysis of another strong band (1.13 μm), carried out by L. Young and A. Young, indicates a more uniform distribution over the whole subcloud atmosphere, at a level of ~20–30 ppm H_2O (see von Zahn et al., 1983). Lewis and Grinspoon (1990) suggested that a contribution by H_2SO_4 to absorption in the 0.94 μm band could resolve this discrepancy. Recently, considerable progress has been made concerning the opacity in CO_2 gas bands in these spectral regions, and this promises more accurate analyses of these spectra (Pollack et al., 1993). At present, one can say only that an X_{H_2O} content at a level of ~100 ppm satisfies these models better than a value an order of magnitude greater.

Counterposed to the spectrophotometric data and in support of the gas chromatograph results from the Pioneer-Venus spacecraft, estimates of the H_2O content were obtained by the Venera 13–14 spacecraft, using moisture gauges. These gas analysis instruments for measuring moisture content, with an electrical sensor using a saturated LiCl salt solution as a base,[5] recorded an extremely high concentration of water vapor (2000 $\pm$ 400 ppm) at the lower boundary of the clouds, 50–46 km. More sophisticated instruments of this type were used on the Vega 1–2 spacecraft (Surkov et al., 1987). At the beginning of the descent phase into the atmosphere (60–47 km), a coulometer was turned on, in which the water vapor, contained in the analyzed gas, was absorbed by phosphoric anhydride (P_2O_5) and underwent electrolysis. By measuring the thermal conductivity, which increases due to the inflow of hydrogen formed during electrolysis, one can determine the H_2O concentration. Upon reaching an atmospheric pressure exceeding 1.5 bar (~47 km) to the lower measurement boundary (~25 km), the moisture content was determined, in parallel, using a thermoelectric sensor, based on calculation of the equilibrium temperature dependence, established in the measurement chamber above the saturated LiCl solution, versus H_2O content in the atmospheric gas being analyzed. To safeguard against aerosol droplets falling into the intake system, a special filter was installed. The relative measurement error was 40% at altitudes of 60–45 km, decreasing to 20% at altitudes of 45–25 km.

Using these moisture gauges, the first continuous vertical profiles of the water vapor content in the cloud layer and subcloud atmosphere, in the 60–25 km range, were obtained (Surkov et al., 1987c). The Vega-1 moisture gauge indicated an extended maximum H_2O concentration, 2500^{+2000}_{-1000} ppm, in the 60–52 km altitude range (where the middle cloud layer lies). A similar instrument on Vega-2 recorded a maximum (~900 ppm) at a lower altitude (~48 km), and the H_2O content during descent increased monotonically, starting from 500 ppm at 60 km. In the subcloud atmosphere, a sharp decrease in moisture content is at first observed, then there is a gradual decrease to concentrations of 100 $\pm$

50 ppm in the 25–45 km range; this conforms with the altitude profile of water vapor content constructed from spectrophotometric data.

As with the data obtained by the first Venera spacecraft, these data suggest a maximum moisture content in a region where, as we now know, the cloud base is located, or somewhat higher. Contradictions still remain, however. The existence of such extended water vapor maxima should be reflected in the vertical temperature profile above $\sim$50 km; this is similar to the situation that drew our attention when analyzing the Pioneer-Venus gas chromatograph data at altitudes below 42 km. Here, too, the quantity ξ matches ξ_{ad} better for a distinctly lower X_{H_2O} value.

As far as the water vapor content in the near-surface troposphere (below 20 km) is concerned, unfortunately there are no direct in situ measurements. The importance of obtaining such information is difficult to overemphasize, as it plays a decisive role in evaluating the conditions and results of lithospheric-atmospheric interaction on Venus (Barsukov et al., 1980; Fegley and Treiman, 1992; Fegley et al., 1993). The results obtained using the moisture gauges on the Vega 1–2 spacecraft do not enable us to confirm or refute the data from the Venera 11–14 onboard spectrophotometers concerning the monotonic decrease in water vapor content as one approaches the planet's surface. It is, however, important to emphasize that the Vega data indicate the absence of an H_2O concentration gradient over a broad measurement interval from 40 to 25 km (Surkov et al., 1987c).

Observations of medium- and high-resolution spectra in the near-infrared window regions, begun in the late 1980s, created a new opportunity for remote sensing of water at several altitude regions in the lower atmosphere. Because the different window regions have contribution functions that peak at different altitudes, profiles can be examined. Specifically, the windows at 2.3, 1.7, and 1.2 microns are most sensitive to radiation originating at altitudes of approximately 35, 23, and 12 km, respectively. Repeated observations over long time periods allow temporal variations to be examined, and high-resolution, or spacecraft, observations, as with Galileo NIMS, in these windows can be used to look for spatial variations. The results of many observations in these windows have generally confirmed the lower range of abundance estimates from the various experiments described above. In all three window regions, the spectra can be well fit with an abundance of 30 ppm H_2O, with an uncertainty of about 10 ppm (Pollack et al., 1993, Bézard et al., 1991). This estimate strongly supports the existence of a very dry lower atmosphere with a mixing ratio that is relatively constant with altitude. There is a slight hint, from attempts to fit one of the weaker windows around 1.2 microns, of a falloff toward lower water abundance in the lowest scale height toward the surface, but this observation is tentative at best. Ground-based and spacecraft (Galileo) observations in these windows also suggest that spatial and temporal deviations from a lower-atmosphere H_2O abundance of 30 ppm are both minor (Drossart et al., 1993; De Bergh et al., 1995).

Thus, we can sum up as follows (see Table 5.5 and Fig. 5.16). A strong altitude dependence for the water vapor content in the Venusian troposphere, which was strongly suggested by previous data, is now called into question by near-infrared observations. Significant spatial-temporal variations are characteristic of its upper portion and the

Fig. 5.16. A set of fundamental measurements on the water vapor content in Venus' troposphere: (1) Vega-1 (Surkov et al., 1986); (2) Vega-2 (Surkov et al., 1986); (3) Venera 15–16 (Moroz et al., 1985); (4) ground-based spectroscopy (Barker, 1975a); (5) Venera 11–14, spectrophotometry (Moroz et al., 1979, 1983); (6) Pioneer-Venus Large probe, gas chromatography (Oyama et al., 1980); (7) Venera-12, gas chromatography (Gelman et al., 1979); (8) Venera 13–14, gas chromatography (Mukhin et al., 1983); (9) Venera 13–14, moisture gauge (Surkov et al., 1983); (10) Venera 4–6, gas analyzers (Vinogradov et al., 1970); and (11) ground-based spectroscopy (Fink et al., 1972).

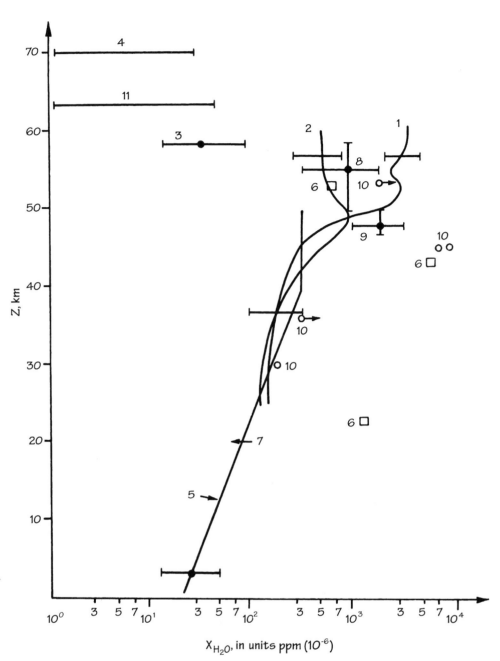

stratosphere. The mean water vapor content, at the level where H_2O absorption bands form (70–65 km), varies by two orders of magnitude, from approximately 1 to 100 ppm, as a function of the time over the Venusian day. The mean H_2O content in the cloud layer is estimated to be 500 ± 200 ppm. Water content maxima, up to several times greater, may appear in the lower or middle cloud layers (58–46 km). These maxima are probably caused by the physical-chemical properties of concentrated sulfuric acid, whose vapor and aerosols experience thermal breakdown in the hot troposphere of the planet. In the subcloud troposphere (below 45–47 km), the water vapor content is relatively constant with altitude at 30 ± 10 ppm and may decrease to still lower values in the near-surface layer (below 20 km). The mechanism for the sink of water vapor in the subcloud troposphere (if such a sink actually exists) is not yet clear. It has been proposed that a gas phase reaction between H_2O and CO_2 could be producing gaseous carbonic acid (H_2CO_3) near the surface (Lewis and Grinspoon, 1990).

Sulfur Compounds and the Sulfur Cycle

Sulfur-bearing gases play an important role in Venus' atmospheric chemistry, although until recently this had been less than obvious. For many years ground-based spectroscopic measurements could not determine the concentration of sulfur-bearing gases; only an upper detection limit could be given (0.01–1 ppm). The first success in this area is associated with Conway et al. (1979) in an experiment on the IUE, which found, by analyzing the ultraviolet spectrum of Venus, that the sulfur dioxide (SO_2) concentration is 0.02–0.8 ppm at altitudes of 70–80 km (a pressure interval of 3–26 Mbar) (see Table 5.5). These data were confirmed by the results of ultraviolet spectrometry on the Venera 9–10 and Pioneer-Venus spacecraft (Krasnopolsky, 1980; Esposito, 1981). Moreover, the Pioneer-Venus data indicated that over an eight-year observational period (1978–86) a gradual decrease in SO_2 content occurred (from 0.1 ppm to 0.01 ppm), correlated with a drop in the density of aerosol polar haze above the clouds (which was most dense in 1978). This effect cannot be explained by instrumental errors in the observations. Esposito (1979, 1984) and Esposito et al. (1988) proposed a connection between the indicated variations in SO_2 content and a corresponding injection of SO_2 from recent volcanic activity on Venus. This idea was advocated by Ksanfomality (1985). Yet it is also possible that a change in the intensity of the SO_2 band may occur without an actual change in the relative SO_2 content, caused by changes, for example, in the cloud structure, affecting band formation (von Zahn and Moroz, 1985). This possibility is indicated, in particular, by the fact that infrared measurement results from the Venera-15 Fourier spectrometer do not match the findings with respect to SO_2 variations. There is also a significant divergence between the Pioneer-Venus data and SO_2 estimates in Venus' above-cloud atmosphere obtained by the IUE satellite from 1979 through 1988; this divergence was pointed out by the authors of the theory (Esposito et al., 1988). Moreover, all measurements steadily converge to an estimate of a very low scale height for SO_2 at the upper boundary of the clouds ($\sim$2 km); this, evidently, suggests intense photochemical destruction of SO_2 at this altitude.

Gas chromatograph analyses of SO_2 in Venus' troposphere at the Venera-12 landing site and by the Pioneer-Venus Large probe (Gelman et al., 1979; Oyama et al., 1980) provide convincing indications that this gas is the dominant form of sulfur in the atmosphere. The vertical SO_2 concentration profile correlates with the H_2O concentration and conforms to notions about sulfuric acid aerosol as the fundamental component of the cloud layer.

High-resolution spectra of the night side of Venus in the 2.3 μm infrared window have been used to model SO_2 absorption at 2.45 μm, allowing an accurate determination of its abundance (Bézard et al., 1993). This work derived an abundance of 130 ± 40 ppm at several locations, pertaining to an altitude region at 35–45 km. This technique will allow remote monitoring of the SO_2 abundance below the Venusian clouds and thus may allow monitoring of volcanic activity. Constraints on SO_2 abundance in the lower atmosphere are also placed by interferometric technique in the millimeter wavelength (Good and Schloerb, 1983).

The situation is much more uncertain for other sulfur-bearing gases. Detections of COS, discussed in the section on carbon compounds, were considered until recently to be contradictory and unreliable. Observations in the near-infrared windows have found an abundance of 4.4 ppm for this species. Much uncertainty exists with respect to hydrogen sulfide (H_2S). The Pioneer-Venus mass spectrometers indicate its content to be 3 ± 2 ppm below 20 km and 1 ppm at the upper boundary of the cloud layer (Hoffman et al., 1980b), whereas the gas chromatograph experiment on the same spacecraft gave an upper limit of $X_{H_2S} < 2$ ppm (Oyama et al., 1980). Within tolerances, these data could somehow be reconciled. But at the same time, H_2S was reportedly found in amounts of 80 ± 40 ppm by Venera 13–14 gas chromatograph analysis of the atmosphere (Mukhin et al., 1983). Simultaneous identification of molecular oxygen ($\sim$20 ppm) excludes the possibility of consistently interpreting these data (see Andreeva et al., 1985, for further discussion).

Gaseous sulfur may play an important role in the chemistry of sulfur compounds on Venus. In spectrophotometric measurements by the Venera 11–12 spacecraft, strong absorption in the blue region of the spectrum from 450 to 600 nm was discovered at altitudes between 30 and 10 km (Ekonomov et al., 1980; Moroz, 1981). Allotropes of gaseous sulfur (either S_2 or S_4) are most likely the source of the absorption. The curves in Fig. 5.17 characterize the change in transmission of solar scattered radiation in various portions of the visible and near-infrared spectral regions, as the Venera 13–14 spacecraft descended into the atmosphere, starting at an altitude of 62 km and ending at the surface. Essentially, the curves represent spectral "slices" of the atmosphere (though with low resolution) at several levels. The degree of absorption increases with depth, caused by the presence of particular components, as a consequence of an increase in their integrated abundances. To identify these components, synthetic spectra were used to fit the measurement data. Below $Z \approx$ 30 km, S_2 dimers were found to dominate, whereas larger polymerized molecules S_{5-8} apparently have a higher abundance within the cloud layer, where condensation of S_8 from the gaseous phase is possible. In constructing the S_n distribution curves with respect to altitude, it was assumed that the chemical equilibrium restoration rate between the allotropes is higher than the rate of chemical reactions with the participation of $S_{n(g)}$.[6] According to Prinn (1979), in particular, the following type of photochemical reactions may occur in the subcloud troposphere, owing to solar radiation in the visible and near-infrared region:

$$S_2 + COS \rightleftharpoons CO + S_3 \qquad (5.23)$$

This possibility, however, is insufficiently supported by experimental data on the kinetics of the respective processes.

Summing up, we have briefly touched upon the general problem of cyclical transformations of sulfur and its compounds in Venus' lower atmosphere. Apparently, the behavior of sulfur-bearing gases in the stratomesosphere and troposphere is controlled by the com-

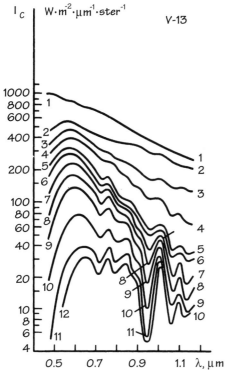

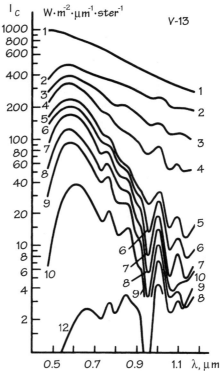

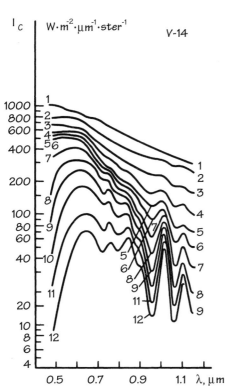

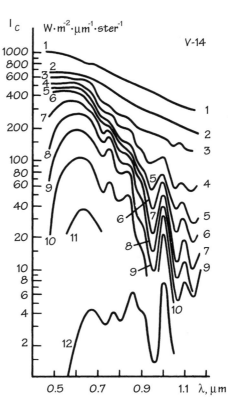

Fig. 5.17. Spectra of scattered solar radiation in Venus' atmosphere at various altitude levels, from the results of measurements by the Venera 13–14 spacecraft (according to Moshkin et al., 1983): (a) radiation incident from the zenith (upper hemisphere); (b) radiation arriving from a direction that makes an angle of 159° with the zenith (lower hemisphere). The numbers next to the curves correspond to the following altitudes: (1) outside the atmosphere (calculation), (2) 62 km, (3) 55 km, (4) 52 km, (5) 49 km, (6) 40 km, (7) 25 km, (8) 16.5 km, (9) 10 km, (10) 4.5 km, (11) 1 km, and (12) 0 km. Computational (synthetic) spectra are compared with these spectra, making it possible to identify polymerized sulfur molecules S_n.

petition between photochemical ($Z > 58$ km) and thermochemical gas-phase reactions, where the principal form of sulfur is undoubtedly found in oxides—SO_2, SO_3, and probably SO radicals. Chemical processes, which determine the formation and retention of a solid cloud cover of H_2SO_4 aerosols, are intimately linked with CO_2 photolysis reactions, which most likely take place by catalysis of hydrogen- and chlorine-bearing radicals. As a function of the rate of gas-phase reactions, one can distinguish rapid (associated with oxidized forms of sulfur) and slow (including reduced forms) atmospheric cycles. Closing the reaction chain with the assistance of sulfur compounds, according to current thinking, occurs through a very slow geological cycle (see von Zahn et al., 1983; Marov et al., 1989a).

The *rapid atmospheric cycle* encompasses a series of photochemical reactions in the above-cloud atmosphere and cloud layer and may be represented by various alternative schemes—(5.24)–(5.26) or (5.27)–(5.30):

$$CO_2 + h\nu \to CO + O \qquad (5.24)$$

$$SO_2 + O + M \to SO_3 + M \qquad (5.25)$$

$$SO_3 + H_2O \to H_2SO_{4(solu.)} \qquad (5.26)$$

$$SO_2 + h\nu \to \dot{S}O + O \qquad (5.27)$$

$$\dot{S}O + O + M \to SO_2 + M \qquad (5.28)$$

$$SO_2 + O + M \to SO_3 + M \qquad (5.29)$$

$$SO_3 + H_2O \to H_2SO_{4(solu.)} \qquad (5.30)$$

The principal mechanism for producing oxygen, by the first scheme, is CO_2 photolysis; the catalyzing agents for reaction (5.25) are OH and HO_2 radicals (hydroxyl and perhydroxyl), which in turn result from the photodissociation of HCl. According to the second scheme, the source of the oxygen necessary for oxidation of sulfur-bearing gases into SO_3 may be SO_2, which undergoes photolysis in the above-cloud atmosphere; here elemental sulfur is formed as a by-product. Both schemes, involving formation of SO_2 followed by its oxidation to SO_3 with H_2SO_4 acid as the final product, are presented in Fig. 5.18a and conform to the observational data—that is, to a sharp decrease in the SO_2 and H_2O content in the stratosphere, compared with the troposphere (Krasnopolsky and Parshev, 1979, 1981a, 1981b, 1983; Winick and Stewart, 1980). Although according to (5.24)–(5.26) the concentration of O_2 and CO exceeds known experimental estimates by 6 and 100 times, respectively, this discrepancy can be eliminated by introducing a hypothetical mechanism of CO oxidation by ClO_X radicals (see reactions [5.20] and [5.21]), based on calculated constants of a series of gas-phase reactions (Yung and DeMore, 1982).

Condensates of concentrated sulfuric acid, which form within the cloud layer, undergo thermal dissociation at an altitude of 47–49 km, at a temperature of 365–380 K; this thermal dissociation specifies the position of the lower boundary of the cloud layer:

$$H_2SO_4 \to H_2O + SO_3 \qquad (5.31)$$

The rapid cycle is probably closed within the subcloud atmosphere by reducing SO_3 to SO_2 through the following reaction:

$$SO_3 + CO \rightleftharpoons SO_2 + CO_2 \qquad (5.31')$$

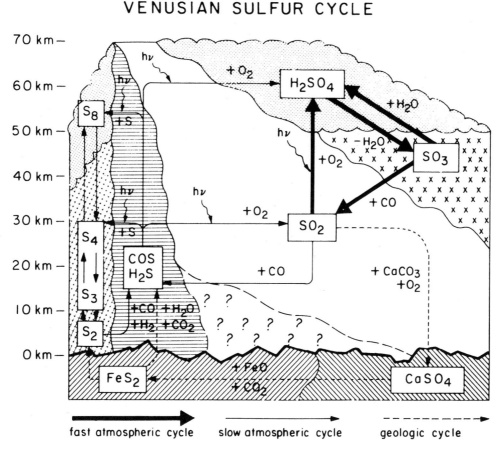

VENUSIAN SULFUR CYCLE

fast atmospheric cycle   slow atmospheric cycle   geologic cycle

Fig. 5.18a. The sulfur cycle on Venus. Zones of stability and relative concentrations of the main sulfur-bearing components are shown qualitatively by the position and size of the dashed regions corresponding to each such component. The sequence of transformations in the rapid atmospheric, slow atmospheric, and geological cycles is indicated by the bold solid arrows, the thin solid arrows, and the dashed arrows, respectively (according to Lewis and Prinn, 1984).

Here the presence of CO makes this reaction possible and even necessary, so that SO_2 regeneration may occur. At these altitudes, the scheme (Prinn, 1979) is also possible:

$$2CO + SO_2 \rightleftharpoons \frac{1}{2} S_2 + 2CO_2 \tag{5.32}$$

The maximum residence time of an SO_2 molecule in the rapid cycle is estimated to be several years (Rossow, 1978).

The *slow atmospheric cycle* takes place in the lower atmosphere and includes photochemical and thermochemical processes associated with the existence of reduced forms of sulfur-bearing gases (H_2S, COS) and of elemental sulfur in a gaseous and condensed form (see Fig. 5.18a). In this case, the above-cloud atmosphere and cloud layer are considered regions of H_2S and COS outflow, which are oxidized by molecular oxygen into SO_3 or undergo photodissociation. The series of processes, aside from (5.24) and (5.26), thus include the following reactions:

$$H_2S + \frac{3}{2} O_2 \rightarrow SO_3 + H_2 \tag{5.33}$$

$$COS + \frac{3}{2} O_2 \rightarrow SO_3 + CO \tag{5.34}$$

Elemental sulfur is brought about either upon H_2S and COS photolysis or upon oxidation of H_2S in the presence of CO and SO_2 with the assistance of HS and H radicals (Prinn, 1975; Volkov, 1983). Its production is associated with the presence of strong ultraviolet absorption in the upper portion of the clouds (see Chapter 6). Closing the cycle can occur in the subcloud troposphere, owing to anticipated thermochemical reactions (Lewis and Prinn, 1984):

$$COS + 3SO_2 \rightleftharpoons SO_3 + 4CO \tag{5.35}$$

$$SO_3 + H_2 + 3CO \rightleftharpoons H_2S + 3CO_2 \tag{5.36}$$

$$H_2S \rightleftharpoons \frac{1}{n}S_n + H_2 \tag{5.37}$$

$$\frac{1}{n}S_n + CO \rightleftharpoons COS \tag{5.38}$$

Surface rock minerals may serve as the catalysts of similar reactions. Note, however, that reactions (5.36) and (5.37) are difficult to reconcile with the observational data, because reliable analyses of H_2 concentrations in the lower atmosphere are lacking. The residence time of a sulfur atom in the slow atmospheric cycle is estimated to be not less than 10 years. It is precisely this value that is computed as the maximum lifetime of sulfur in an oxidized form (H_2SO_4 aerosols) in the rapid atmospheric cycle.

Both of the atmospheric cycles proceed from the assumption that the ratios of all fundamental sulfur compounds, for sufficiently rapid gas-phase reactions, remain approximately constant. In other words, gas phase chemical equilibrium is maintained in the atmosphere, and the interaction of atmospheric gases with the surface does not substantially affect this equilibrium. In fact, the situation is more complex; the surface actually supplies these gases to the atmosphere, and this process may be continuous (from chemical interaction and weathering) as well as cyclical or sporadic (from volcanism, for example). As a simple variation, it would be interesting to consider likely reactions with surface rock minerals, which could enable us to close the whole sequence of processes through a third, *geologic cycle*.

In this hypothetical cycle (see Fig. 5.18a), our initial assumption is the presence of the mineral pyrite (FeS_2) in the surface rock and primordial de-gassing of reduced sulfur gas compounds H_2S and COS with subsequent transformation into SO_2, as follows:

$$FeS_2 + 2H_2O \rightarrow FeO + 2H_2S + \frac{1}{2}O_2 \tag{5.39}$$

$$FeS_2 + 2CO_2 \rightarrow FeO + 2COS + \frac{1}{2}O_2 \tag{5.40}$$

$$2H_2S + 3O_2 \rightarrow 2H_2O + 2SO_2 \tag{5.41}$$

$$2COS + 3O_2 \rightarrow 2CO_2 + 2SO_2 \tag{5.42}$$

Further, a whole complex of reactions, (5.24)–(5.38), may occur in the rapid and slow atmospheric cycles, including the formation and decomposition of sulfuric acid cloud aerosol, again with the formation of SO_2. However, existing theoretical estimates (Khodakovsky, 1982; Khodakovsky et al., 1979; Barsukov et al., 1980, 1982; Lewis and Kreimendahl, 1980) suggest that under conditions of chemical equilibrium between Venus' surface and atmosphere, SO_2 cannot be the dominant sulfur-bearing component. Therefore, owing to thermochemical reactions at the surface, either SO_2 bonds in anhydrite ($CaSO_4$), which enters into the composition of the rock, or the reverse transformation process of SO_2 into other, reduced forms—H_2S and COS—must occur.

Unfortunately, it is impossible to choose between these processes at present, if we keep in mind the lack of reliable data on the reaction rates of the corresponding processes. Therefore, setting aside for the moment the possible violation of thermochemical equilibrium in the lower troposphere (or the cyclical production of components), we shall point out the possibility of bringing the system into equilibrium by successive transformation of SO_2 into anhydrite and pyrite according to the scheme adopted by von Zahn et al. (1983) and Volkov (1987) in the geologic cycle:

$$4SO_2 + 2O_2 + 4CaCO_3 \rightarrow 4CaSO_4 + 4CO_2 \qquad (5.43)$$

$$4CaSO_4 + 2FeO + 4CO_2 \rightarrow 2FeS_2 + 4CaCO_3 + 7O_2 \qquad (5.44)$$

Theoretically, the system of sulfur compound sources and sinks, in this case, turns out to be closed.

The Halogens and Their Compounds

Halogens and their compounds present in Venus' atmosphere are linked with possible important processes in atmospheric chemistry.

Minor amounts of gaseous *chlorine* and *fluorine* compounds were first identified in Venus' above-cloud atmosphere from ground-based astronomical data in the near-infrared spectral range (Connes and Connes, 1967). The HCl content was estimated at 0.6 ppm, and HF at 5 ppb. Later these estimates were corrected somewhat (see Table 5.5). Unfortunately, the concentration of halogen-bearing gases is below the sensitivity threshold of instruments for measuring chemical composition that were installed on the Venera and Pioneer-Venus spacecraft. Only an upper limit for the HCl content in the subcloud troposphere could be estimated (< 10 ppm) using the mass spectrometer experiment on the Pioneer-Venus Large probe (Hoffman et al., 1979, 1980b).

Observations in the near-infrared windows at 1.7 and 2.3 μm have been used to determine the lower atmosphere abundances of HCl and HF, respectively. At an altitude of 23.5 km, HCl is found to have an abundance of 0.48 ± 0.12 ppm. The abundance of HF is determined, for an altitude of 33.5 km, to be in the range of 0.001–0.005 ppm (Pollack et al., 1993).

Chlorine-bearing radicals, ClO_X, which form owing to HCl photolysis in Venus' stratosphere, may play the role of catalyst in the CO_2 recombination process, along with HO_X radicals (see reactions [5.20] and [5.21]). In the Krasnopolsky and Parshev model (1981b), the role of chlorine-bearing radicals is assumed to be dominant, whereas the Yung and DeMore model (1982) derives a possible Cl_2 content in the atmosphere of 1 ppm; in this case one could assume that it is an as-yet-unidentified ultraviolet-absorber above the clouds. The somewhat lower upper limit was obtained from spectrometric measurements of scattered radiation at the planet's limb by the Venera 9–10 spacecraft, according to which the concentration of molecular $Cl_2 \lesssim 0.4$ ppm (Krasnopolsky, 1980).

Within the framework of ideas developed by Yung and DeMore (1982), the appearance of high concentrations of Cl_2 is unavoidable in the above-cloud atmosphere as a product of HCl photolysis, lacking a supply of molecular oxygen from the troposphere, and ClCO and $ClCO_3$ radicals, along with ClO_X, also participate in CO_2 recombination processes, for example, according to reaction (5.21). Subsequent development of these hypotheses led to ideas about the possible direction of Venus' atmospheric evolution, toward an irreversible accumulation of gaseous Cl_2 and $COCl_2$ with a gradual depletion of all hydrogen-bearing gas reserves.

The role of fluorine compounds in Venus' stratomesosphere is completely different. Photochemical equilibrium calculations, conducted by Parisot and Moreels (1984), indicated that HF photolysis is essentially absent, because the shortwave solar radiation in the $\lambda < 190$ nm range, necessary for HF dissociation, is nearly completely absorbed over the thickness of Venus' CO_2 atmosphere and does not reach the cloud layer. Moreover, HF does not enter into reactions with OH and O, so fluorine radicals may be present only in vanishingly small amounts and can hardly play a significant role in atmospheric chemistry. Consequently, the irreversible outflow of fluorine in the stratomesosphere occurs in the form of HF, and the maximum atomic F concentration (90–100 km) does not exceed 10 cm^{-3} ($X_F \approx 10^{-15}$).

From the results of the Venera-13 gas chromatograph experiment, the possible presence of gaseous sulfur hexafluoride (SF_6) was inferred, in amounts of 0.2 ± 0.1 ppm in the 58–35 km altitude range (Mukhin et al., 1983). If volcanic gases on Venus are depleted in water vapor ($X_{H_2O} = 10^{-5}$–10^{-4}), then, as Zolotov and Khodakovsky (1985) showed, sulfur fluorides could have been their prime fluorine-bearing components ($SF_4 \gtrsim HF > SF_6 > F_2$). If, however, in the near-surface troposphere chemical equilibrium exists between the gases, then the SF_6 concentration must be negligibly small ($X_{SF_6} < 10^{-50}$). This second estimate is difficult to reconcile with notions about the possible accumulation of this gas in Venus' atmosphere, even taking into account its chemical inertia.

Chlorine and fluorine compounds may exert an influence on the chemistry of the clouds as well as on the atmosphere itself. Chlorine is likely found in the composition of cloud condensate as a minor component (see Chapter 6). It is difficult to be specific concerning its form, however. The presence of fluorine in cloud particles is less likely, although the reaction

$$HF_{(gas)} + H_2SO_{4(solution)} = H_2O_{(gas)} + HSO_3F_{(solution)} \tag{5.45}$$

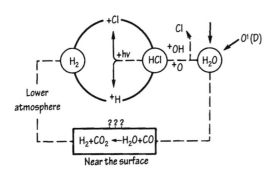

Fig. 5.18b. The hydrogen cycle in Venus' lower and middle atmosphere, without taking into account sulfur-bearing gases (according to the model of von Zahn et al., 1983).

was put forward as a possibility by Young (1975). We do not exclude the possibility of Cl and F bonding in the cloud condensate, based on estimates of chemical quasi-equilibrium between the atmosphere and surface rocks of Venus, within the framework of the lithospheric-atmospheric interaction model (Volkov and Khodakovsky, 1984).

Hydrogen and Oxygen

In the processes considered above, in Venus' atmosphere and clouds, a considerable role is given to *hydrogen* and *oxygen*. The fundamental chemical form of hydrogen in Venus' atmosphere is water vapor, whose content and altitude distribution, as well as that of the fundamental H_2O radicals, halogen and sulfur compounds, were discussed above in detail (see reactions [5.15]–[5.19], [5.30], [5.31], and [5.33]–[5.38]). At present there is no generally accepted model of the hydrogen cycle; one possible model is shown in Fig. 5.18b.

The only direct analysis of molecular hydrogen in Venus' lower atmosphere, at altitudes of 58–49 km (the cloud layer), was carried out using gas chromatography on the Venera-13 lander; a value of 25 ppm was obtained. Moreover, using the same technique, the Pioneer-Venus Large probe established upper limits for detecting H_2 at 52 km ($<$ 200 ppm), 42 km ($<$ 70 ppm), and 22 km ($<$ 10 ppm). The Venera-13 measurement data, however, are difficult to reconcile with theoretical estimates for the tropospheric composition, based on thermodynamic calculations within the scope of chemical models, because, as follows from these calculations, the H_2 equilibrium concentration must not exceed several ppb.

The H_2 content in the stratomesosphere (60–110 km), where photochemical processes dominate, is the main limitation to the hypothesis that allows the production of HO_X or ClO_X radicals as components determining catalytic oxidation of CO to CO_2—one of the basic mechanisms studied earlier for maintaining CO_2 as the main component in Venus' atmosphere. Unfortunately, reliable estimates of the hydrogen content at these altitudes are lacking. A peak, corresponding to 2 amu, was recorded using an ion mass spectrometer on the Pioneer-Venus spacecraft (H. Taylor et al., 1980). It was hypothesized that this peak corresponds to H_2^+ ions, produced in the daytime ionosphere through photodissociation of H_2 and destroyed upon interaction with atomic oxygen (Kumar et al., 1978, 1981):

$$H_2 + h\nu \rightarrow H_2^+ + e \qquad (5.46)$$

$$H_2^+ + O \rightarrow OH^+ + H \qquad (5.47)$$

Hence, knowing direct measurement data of the H_2^+ and O density, Kumar et al. (1981) obtained an H_2 concentration in Venus' stratomesosphere on the order of 10 ppm. According to this model, in the thermosphere (above ~140 km), where atomic oxygen is extremely prevalent, the H_2 and H content falls to 10^5–$10^6 cm^{-3}$.

The compositional measurement of the stratomesosphere using ion mass spectrometry has another possible interpretation: the 2 amu peak can be explained by the presence of deuterium ions (Hartle and Taylor, 1983; McElroy et al., 1982). Then the ratio D/H $\approx$ $(2.2 \pm 0.6) \cdot 10^{-2}$, which is 100 times greater than observed in Earth's atmosphere. An explanation for this can be found in selective non-thermal escape of hydrogen. The source of such non-thermal energy could be the collision of thermal hydrogen atoms with fast oxygen atoms (McElroy et al., 1982):[7]

$$O_{fast} + H \rightarrow O_{fast} + H_{fast} \; (v \lesssim 10^6 \, cm \cdot sec^{-1}) \qquad (5.48)$$

In turn, fast O atoms may be produced through dissociative recombinations of O_2^+ (Rodriguez et al., 1984; Gurwell and Yung, 1993):

$$O_2^+ + e \rightarrow O(^1D) + O(^3P) \qquad (5.49)$$

Based on reaction (5.48), McElroy et al. (1982) estimated the current escape rate of H atoms at 10^7 atoms $cm^{-2}sec^{-1}$; in this case the escape rate of deuterium is negligibly small. If one assumes that the adopted hydrogen escape rate has been characteristic over the whole geological history of Venus, then escape at this rate easily provides the necessary deuterium concentration in Venus' present-day atmosphere.

Another source of nonthermal hydrogen dissipation was proposed by Kumar et al. (1985), who modeled charge exchange between "cold" H atoms in the exosphere and high-temperature H^+ ions in the ionosphere. This mechanism also impedes the escape of deuterium and results in an increase in D/H by 1000 times over geological time. The actual total escape flux is model dependent and is not completely understood, but current estimates range from 0.4×10^7 to 3.7×10^7 hydrogen atoms $cm^{-2}sec^{-1}$. The fractionation factor, f, which defines the relative escape efficiency of deuterium to that of hydrogen (so that $f = 1$ if D escapes as easily as H), is estimated at 0.13 (Grinspoon, 1993; Donahue and Hartle, 1992).

An estimate of D/H was also made in interpreting mass spectrometer measurements from the Pioneer-Venus Large probe in the 50–27 km range, when the atmospheric gas intake system was probably shut down by H_2SO_4 droplets from the cloud layer (Donahue et al., 1982). Identification of the 19 amu peak with the HDO^+ ion enabled HDO/H_2O to be estimated at $(3.2 \pm 0.4) \cdot 10^{-2}$, that is, D/H $= (1.6 \pm 0.2) \cdot 10^{-2}$ (Table 5.6).

Table 5.6   Isotopic ratios of hydrogen and inert gases in the atmospheres of Venus and Earth

Isotopic ratio	Venus	Earth
D/H^a	$2.4 \pm 0.3 \times 10^{-2}$	1.6×10^{-4}
^{3}He/^{4}He	$<3 \times 10^{-4}$	1.38×10^{-6}
^{22}Ne/^{20}Ne	8.5×10^{-2}	0.102
^{40}Ar/^{36}Ar	1.1	295.5
^{38}Ar/^{36}Ar	0.19	0.19
^{86}Kr/^{84}Kr	0.17	0.30
$(^{80}$Kr $+$ ^{82}Kr $+$ ^{83}Kr$)/^{84}$Kr	0.92	0.44

Source: The isotopic content in Venus' atmosphere is from von Zahn and Moroz, 1985 (VIRA-85); the isotopic content in Earth's atmosphere is from Ozima and Podosek (1983).
^{a} D/H in Venus' atmosphere is from Donahue et al., 1998.

More recently, De Bergh et al. (1991) made a spectroscopic determination of the lower atmosphere D/H ratio in the near-infrared window regions. They found a value, consistent with the uncertainty in the PV measurement, of 120 ± 40 times the terrestrial D/H ratio. The interpretation of the concentration of deuterium in the Venusian atmosphere could have far-reaching consequences for the development of ideas about Venus' geological past and the evolution of its atmosphere (see Chapter 9).

Analysis of the *oxygen* content, from the first direct measurements of the chemical composition of Venus' troposphere by the Venera 4–6 spacecraft, was derived by the thermochemical method. This method consisted of passing an electrical current through a spiral tungsten filament in a gas analysis cell; with free oxygen in the gas sample, the filament must burn out. This is how the upper limit of $O_2 < 1000$ ppm was set (Vinogradov et al., 1968, 1970). At the same time spectroscopic estimates, attributed to the upper boundary of the clouds, indicated extremely low concentrations (not more than 1–20 ppm; see Table 5.5).

Using the gas chromatograph on the Pioneer-Venus Large probe, the O_2 content was determined in the 52–22 km interval (Oyama et al., 1980). It turned out that the oxygen content at an altitude of 52 km is 43.6 ± 25.2 ppm, and 16.0 ± 7.4 ppm at 42 km, while at 22 km the O_2 content was so small that it became difficult to estimate. Gas chromatographs on the Venera 13–14 landers, which were designed to analyze oxygen and argon in Venus' atmosphere without separating them in a chromatograph column and with higher sensitivity, measured the molecular oxygen content in the 58–35 altitude range, at 18 ± 4 ppm (Mukhin et al., 1983). This value coincides with the data of Oyama et al. within an order of magnitude; however, no O_2 concentration gradient was detected in the vertical profile over 23 kilometers, in contrast to the Pioneer-Venus data.

In discussing the results of these measurements, we are faced with the question, To what extent do they agree with our ideas about the photochemical processes in Venus' lower atmosphere? This question is far from incidental because according to all currently available theoretical models, the existence of molecular oxygen at altitudes below 60–70 km is, in general, considered unrealistic. Actually, a study of the photochemical transformation of CO_2 into CO in Venus' stratomesosphere indicates the possibility of generating molecular oxygen, with a ratio $CO/O_2 \approx 2$; however, the actual ratio is ≈ 45 (see Table 5.5). Probably such a deficiency in O_2 can be explained by its loss during CO oxidation into CO_2, or upon oxidation of SO_2 and other sulfur-bearing gases in the rapid atmospheric cycle of sulfur, leading to the appearance of sulfuric acid aerosol (Winick and Stewart, 1980). Another path of oxygen escape

$$SO_2 + \frac{1}{2}O_2 + H_2O \rightarrow H_2SO_4 \qquad (5.50)$$

is not sufficiently effective, and the O_2 concentration in the stratosphere would have to be 50 times higher than is actually observed (Yung and DeMore, 1982). In principle, this deficit could be eliminated by transforming atomic oxygen into molecular oxygen, with the assistance of chlorine- and nitrogen-bearing radicals—for example, in the following chlorine cycle:

$$O + ClO \rightarrow O_2 + Cl \qquad (5.51)$$

But one way or another, all photochemical calculations indicate that the stratosphere cannot serve as the oxygen source in the cloud layer. Therefore, not one of the photochemical atmospheric models takes into account data from gas chromatograph analysis of O_2 in Venus' troposphere.

In discussing possible sources of oxygen, a hypothesis was put forward about the existence of sulfur trioxide (SO_3), a product of the thermal decay of H_2SO_4 aerosols, as the dominant gas in the subcloud troposphere (Craig et al., 1983). Thermodynamic calculation of the composition in the near-surface atmosphere, given inert behavior by CO and assuming a relationship between the chemical composition of the remaining components and altitude, indicated that, if $X_{CO} = 16$ ppm, then $X_{SO_2} = 176$ ppm and $X_{H_2O} = 4$–40 ppm (as a function of the H_2SO_4 concentration in the aerosol droplets). The SO_3 concentration near the surface would then have to be approximately 1000 ppm. It is not possible to verify this assumption using these analysis techniques, because gas chromatographs were not designed to analyze SO_3. Even with this approach, however, it appears impossible to explain the lack of O_2 at an altitude of 22 km in the Pioneer-Venus experiment, or its constant concentration in the 58–35 km range in the gas chromatograph experiments on the Venera 13–14 spacecraft. Thus, the identification of molecular oxygen in Venus' troposphere and its concentration require confirmation.

Existing experimental data concerning the altitude distribution of chemically active components in Venus' atmosphere are summarized in Fig. 5.19. In particular, the measured concentrations of trace components are indicated. This figure provides a graphic representation of the change in concentration with altitude, caused by a complex set of chemical and physical processes in the atmosphere, taking into account phase changes at the level of the clouds.

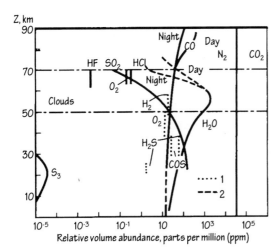

Fig. 5.19. A generalized scheme of the altitude distribution of macro- and microcomponents in Venus' troposphere, according to experimental data: (1) measurement intervals of the gas concentration, which require confirmation; (2) an extrapolation. The dotted-dashed lines distinguish the cloud layer.

Noble Gases

To solve the problem of the origin and evolution of Venus' atmosphere, it is extremely important to have data on the abundance of *noble gases*, which do not enter into compounds with other atmospheric components or surface rock material, owing to their chemically inert qualities.

Direct analysis of the *helium* (He) content in the lower atmosphere has not been conducted. In the upper atmosphere the He concentration was first measured during nightglow studies using ultraviolet spectrometers, installed on the Mariner-10 and Venera 11–12 fly-by spacecraft (Kumar and Broadfoot, 1975; Berteaux et al., 1985). Later, a detailed profile of the helium concentration in the 130–140 km altitude range was obtained during mass spectrometer studies from the Pioneer-Venus satellite and Large probe bus (von Zahn et al., 1980, 1983). The mean helium content in Venus' upper atmosphere is taken to be 450 ppm, which is 150 times greater than on Earth; extrapolation to the lower atmosphere depends on the assumed temperature profile of the exosphere and the eddy diffusion coefficient, defining turbulent mixing. Accordingly, the He content at altitudes ≤ 130 km was estimated to be 12^{+24}_{-6} ppm.

Determining the He isotopic composition using mass spectrometers involves considerable difficulties, because the peak corresponding to an amu = 3 contains a background contribution of HD^+ and H_3^+, produced by the instrument's ion source (Hoffman et al., 1980a). Interpretation of the mass spectra leads to the conclusion of a vanishingly small ^3He content in estimating the upper limit $^3He/^4He < 3 \cdot 10^{-4}$ (Hoffman et al., 1980a). At present one can say only that the ratio $^3He/^4He$ in Venus' atmosphere most likely does not exceed the known value for Earth's atmosphere ($\sim 10^{-6}$) (see Table 5.6).

Based on data analysis of potassium, uranium, and thorium in Venusian rocks, and the well-known correlation between the amounts of ^4He and ^{40}Ar in Earth's atmosphere, an attempt was undertaken to estimate the overall amount of helium that could have entered Venus' atmosphere upon radioactive decay of uranium and thorium in the planet's crust (Pollack and Black, 1982). If the escape of ^4He is negligibly small, then the amount of helium, even corresponding to the upper estimate of 450 ppm, could have accumulated only after 10.5 billion years, which is more than twice the age of the solar system! Consequently, if the measurement data are correct, then we are forced to infer that the bulk of the helium in Venus' present-day atmosphere is of primordial origin.

Analyses of *neon* were carried out using mass spectrometers on the four Venera 11–14 landers and the Pioneer-Venus Large probe (Istomin et al., 1979, 1983; Hoffman et al., 1980a,b). The concentration of neon was estimated from the ratio $^{20}Ne/^{36}Ar$. The results of Soviet and American measurements of the ^{36}Ar content differ quite significantly (46.1 ppm and 31 ppm, respectively); this discrepancy is taken into consideration when comparing the data on neon. Neon was also detected by Pioneer-Venus using the gas chromatograph. If we keep in mind that measurements of the Ne content have significant margins of error, then the results referred to may be considered consistent (see Table 5.5).

Determining the ^{22}Ne isotopic content, from mass spectrometer data, is complicated owing to masking of the mass 22 peak by the CO_2^{++} ion. It is possible that the isotopic composition of neon ($^{22}Ne/^{20}Ne$) in Venus' atmosphere differs from Earth's (see Table 5.6), as well as from solar composition (Istomin et al., 1983). This may indicate significant differences in the evolution of Venus' atmosphere from that of the Earth (see Chapter 9).

Independent analyses of *argon* were conducted five times using mass spectrometers (Venera 11–14 and the Pioneer-Venus Large probe) and twice using gas chromatographs (Venera-12 and the Pioneer-Venus Large probe; see Table 5.5). The results obtained by the various methods differ so much that the analysis error bars do not overlap. Von Zahn et al. (1983) believe that the gas chromatograph technique has definite advantages, whereas from a methodological point of view, the data of Istomin et al. (1979, 1983) favor mass spectrometer analysis. The criteria for calculating the errors in these analysis methods are different, however, therefore the value $\Sigma Ar = (70 \pm 25$ ppm), given by these methods, is essentially an educated guess. Berttaux et al. (1985), for example, accept a broader range of values, 70^{+50}_{-30} ppm.

It was also suggested (Istomin et al., 1979, 1983) that the divergence in the argon data in the Venera and Pioneer-Venus mass spectroscopy experiments could be explained by a change in the sensitivity of the PV instrument during flight, connected with the presence of $\sim 14\%$ He near the ion source. This phenomenon, judging by post-flight laboratory experiments, can cause an apparent decrease in the argon concentration by 15–20%. The corresponding corrected argon content, $^{36}Ar \approx 50$ ppm, would be closer to the Venera data.

The isotopic composition of argon reveals sharp differences from the terrestrial value (see Table 5.6), expressed mainly as an almost 300-fold depletion of radiogenic ^{40}Ar in Venus' atmosphere. A discussion of this most important indicator, reflecting the de-gassing history of planetary material, can be found in Chapter 9, along with an analysis of the isotopic ratios of other noble gases, normalized with respect to ^{36}Ar.

Krypton was first identified by the mass spectrometer experiments on the Venera 11–12 landers. After completing the measurement analysis, Istomin et al. (1979) estimated the

Kr concentration to be 0.6 ± 0.2 ppm (see Table 5.5). Upon interpreting the mass spectra for the Pioneer-Venus Large probe, the upper limits of Kr content were, subsequently, lowered from < 1 ppm to < 0.2 ppm, and finally, an estimate of ~ 0.025 ppm was given (Hoffman et al., 1979, 1980b; Donahue et al., 1981). Von Zahn et al. (1983) put forward a series of reasons for favoring the Venera 11–12 results over the Pioneer-Venus data. In fact, processing the new mass spectrometer data, obtained by Venera 13–14, led Istomin et al. (1983) to conclude that a krypton concentration of $0.018-0.023$ ppm is closest to reality; this essentially coincides with the estimate by Donahue et al. (1981). Gas chromatographs on the Venera 13–14 landers measured essentially the same krypton concentration, previously determined by the Venera 11–12 mass spectrometers (Mukhin et al., 1983).

The ratios of various Kr isotopes (see Table 5.6) were determined on a semiquantitative basis. It was assumed that there are no significant differences between the Kr isotopic ratios in Venus' atmosphere and those in Earth's.

Mass spectrometer measurements of *xenon* were carried out by the Venera 13–14 landers and the Pioneer-Venus Large probe (see Table 5.5). The two data sets differ by almost an order of magnitude and provide only an upper limit for the Xe concentration. Istomin et al. (1983), in discussing how to interpret these mass spectra, noted that there is a danger of understating the measured Kr and Xe concentrations, owing to sorption of these gases by components of the mass spectrometer intake system. Moreover, it is unclear to what extent one can extrapolate the sensitivity values for the instrument to concentrations in the 1–10 ppb range. Therefore, information on the xenon content of Venus' atmosphere must be considered particularly tentative.

Models of Atmospheric Chemistry

The chemical composition of Venus' atmosphere raises a series of complex questions. These questions focus on how and why a massive carbon dioxide atmosphere formed on the planet; what Venus' place is in the general (at least hypothetical) formation of planetary atmospheres; what can explain the presence of various minor admixtures in Venus' atmosphere that are essentially absent in Earth's atmosphere; what governs the equilibrium state of the existing atmospheric chemical composition, and where it differs from an equilibrium state; and, finally, what the content of noble gases and their isotopic ratios tell us, if we compare them with corresponding data for other planets of the terrestrial group. These problems are associated with attempts to obtain the necessary criteria and limits for the concentrations of minor components (making up less than 0.1% but playing a fundamental role in the chemical processes of the atmosphere), based on computational (cosmochemical) thermodynamic models, and thereby to independently, critically interpret the results of measurements or threshold estimates. We shall attempt, based on principles of atmospheric chemistry, to answer at least some of these questions, which have a direct relation to the study of general problems in atmospheric evolution, comparative meteorology, and the paleoclimate of planets. Chapter 9 is devoted to the cosmogonic and evolutionary aspects of these problems.

If we examine the concentrations in the Sun and in the atmospheres of the terrestrial planets of the most characteristic representatives of the atmophile elements, that is, the noble gases, which are not subjected to chemical interactions and therefore maintain the greatest stability during evolution, then their marked scarcity on the planets is revealed (see Fig. 4.1). A similar depletion in atmophile elements is observed in the material of chondritic meteorites. Taking into account ideas about the sequential nature of condensation of the protoplanetary nebula (see Chapter 4), as a model for the formation of various meteorite classes, this leads us to a point of departure concerning planetary atmospheres. Although the timing of major events in planetary evolution involving early global differentiation essentially at the accretion phase, accompanied by formation of crust and atmospheric outgassing, is difficult to establish, it is not likely that all volatiles outgassed from the interiors, and produced by impact de-gassing, could be retained on the initially hot planet. We thus pursue the assumption that, in contrast to the giant planets, the primordial atmospheres of the terrestrial planets were largely lost during accretion. The present-day secondary atmospheres then formed during subsequent thermal evolution. They formed from condensed material, in which a portion of the volatiles was chemically bonded or absorbed, through de-gassing of the interior, primarily as a result of volcanic eruptions, and from impact de-gassing of the volatiles in the impacting objects of a late-accreting veneer (Zahnle et al., 1988). It is possible that the ultimate loss of the original atmospheres occurred somewhat later, as the young Sun passed through the T Tauri stage. In this case the gaseous envelopes of the nearest planets were swept out, whereas the giant planets by this time were already sufficiently massive to withstand the sweeping out and the escape of even the most highly volatile elements (see Urey, 1952; Sagan, 1967; Marov, 1986).

On Earth, water vapor and carbon dioxide gas make up the bulk of volcanic gases, whose composition depends on the conditions of eruption (primarily temperature) and the participating mantle material. Water amounts to approximately 20% of the volume of outflowing terrestrial basalts. The relative ratio of water vapor to carbon dioxide gas is about 5 to 1. At higher temperatures sulfur compounds and halogens are present in the volcanic gases, and at relatively lower temperatures nitrogen compounds are present; hydrogen is present as an admixture. Usually there is no free nitrogen or oxygen in volcanic emissions, and their presence in the atmosphere is a result of secondary reaction products (Vinogradov, 1967).

Thus, the migration of these gases into the atmosphere should be considered a natural evolutionary stage, which accompanies the heating of planetary interiors through gravitational compression and radiogenic heat sources. One can discern a definite analogy between the de-gassing products of the mantle and the amount of volatiles in material of chondritic composition. Here the degree of de-gassing depends on the stage through which the planet is passing in its thermal evolution: according to current concepts, degassing must be least extensive on Earth and Venus. This is confirmed for Earth by current estimates of the concentration of volatiles in Earth's mantle, relative to which gases and vapors concentrated in the hydrosphere and atmosphere make up several or merely fractions of a percent in all. The remainder are found in a bound state as oxides and hydroxides of metals and silicon, and as metal compounds with nitrogen, carbon, and sulfur (nitrides, carbides, and sulfides). Based on the corresponding acid salts, these volatiles form a broad class of natural minerals in Earth's crust: nitrates, carbonates, and sulfates. In addition, the concentration of fundamental volatiles in the present-day hydrosphere and atmosphere exceeds by many times that which could be liberated from minerals of volcanic, crustal rock; this is the so-called volatile excess, established in the early 1950s by Rubey (1951). The volatile excess is closest to the composition of volcanic gases and once more suggests the decisive role that interior, terrestrial material played in de-gassing.

Fundamental data on the chemical composition of structural parameters of Venus' lower atmosphere, compared with similar characteristics of Earth's atmosphere, and with other planets of the terrestrial group and Jupiter (as the most characteristic representative of the giant planets) are presented in Table 5.7. Earth, Venus, and Mars are of paramount interest. Mercury, by virtue of its small mass and proximity to the Sun, has essentially lost its atmosphere. In contrast, Jupiter has preserved a composition in the upper portion of its gaseous envelope (atmosphere) that is close to the original material composition which formed the planets far from the Sun.

From the point of view of interior de-gassing and volatile excess, the composition of the Venusian and Martian atmospheres, with a dominant concentration of carbon dioxide, is easier to understand than the nitrogen-oxygen composition of Earth's atmosphere. It is usually assumed that photosynthesis, in response to solar radiation, exerted a crucial influence on the evolution of Earth's primitive atmosphere and led to the appearance of free oxygen. In turn, this brought about the oxidation of ammonia and ammonium compounds in the volcanic gases and the liberation into the atmosphere of free nitrogen, a chemically non-active gas, which does not directly interact with the surface rocks and also possesses a high threshold of dissociation (separation of molecules into atoms), which is especially significant in a low-density atmosphere.

The photosynthetic genesis of oxygen on Earth, supported by the discovery of organic remains in ancient rock formations, has given rise to the suggestion that the initial step along the path of biological evolution was the so-called autotrophs, which are one variety of anaerobe. In this case, oxygen could have been liberated through biochemical splitting of water molecules and could have gradually stimulated the transition from primitive autotrophs to more developed forms of photosynthesis. Geological data suggest that the beginning of an oxidizing metabolism dates back to the lower Proterozoic (about 2–2.5 billion years ago); the discovered remains of the simplest single-celled organisms — organelles in the rocks of eastern California, on the African continent, and other Precambrian shields — date to this time. They were probably the predecessors of blue-green algae that developed in the upper layers of the ocean and brought about an intense inflow of oxygen into the atmosphere through oxidation-reduction reactions, which produce carbohydrates and oxygen from carbon dioxide and water (Goldsmith and Owen, 1992).

The photosynthetic nature of oxygen on Venus (and, apparently, on Mars) can be ruled out. In Venus' atmosphere, the oxygen content hardly exceeds thousandths of a percent, and even this estimate is subject to doubt. In the Martian atmosphere oxygen is

Table 5.7 Fundamental parameters of the atmosphere of Venus and other planets

Parameter	Earth	Mercury	Venus	Mars	Jupiter
Chemical composition (percent by volume)[a]	N_2 78.1	$He \sim 20$	CO_2 96.5	CO_2 95	H_2 87
	O_2 20.9	$H_2 \sim 18$	N_2 3.5	N_2 2–3	He 12.8
	Ar 0.93	$Ne \sim 40–60$	Ar 0.01	Ar 1–2	H_2O 1×10^{-4}
	H_2O 0.1–1	$Ar \sim 2$	H_2O 3×10^{-5}	H_2O $10^{-3}–10^{-1}$	CH_4 7×10^{-2}
	CO_2 0.03	$CO_2 \sim 2$	CO 2.3×10^{-5}	CO 4×10^{-3}	NH_3 2×10^{-2}
	CO 10^{-5}		HCl 4×10^{-7}	O_2 0.1–0.4	HCN 10^{-5}
	CH_4 10^{-4}		HF 10^{-6}	Ne $<10^{-3}$	C_2H_6 4×10^{-2}
	H_2 5×10^{-5}		O_2 $<2 \times 10^{-4}$	Kr $<2 \times 10^{-3}$	C_2H_2 8×10^{-3}
	Ne 2×10^{-3}		SO_2 2×10^{-4}	Xe $<5 \times 10^{-3}$	PH_3 4×10^{-5}
	He 10^{-4}		H_2S 8×10^{-3}		CO 2×10^{-7}
			Ne 10^{-3}		CH_3D 2×10^{-3}
	Kr 10^{-4}		Kr 4×10^{-5}		
	Xe 10^{-6}		Xe $10^{-6}–10^{-5}$		
Mean molecular mass	28.97		43.2	43.5	2.25
Temperature at the surface $T_{\max}$ (K)	320	500	740	270	135
$T_{\min}$ (K)	220	100		150	
Mean pressure at the surface P (atm)	1	$<2 \times 10^{-14}$	95	6×10^{-3}	0.5
Mean density at the surface ρ (g/cm^3)	1.27×10^{-3}	$<10^{-17}$	61×10^{-3}	1.2×10^{-5}	10^{-4}

[a] Mean relative concentrations of basic components are given, corresponding to the most reliable data from Table 5.5.

0.1–0.4% of the main constituent, CO_2, or, in absolute concentration, four to five orders of magnitude less than in Earth's atmosphere. As far as nitrogen is concerned, its measured content in Venus' atmosphere, about 3.5%, means that the absolute amount of nitrogen in Venus' atmosphere exceeds that of Earth by about five times. Because nitrogen only weakly enters sedimentary rock and because denitrification (the reduction of nitrates into molecular nitrogen, brought about by bacteria) can be completely ruled out, juvenile fixed nitrogen evidently accumulated in Venus' atmosphere owing to more effective de-gassing from the mantle. On Mars, where the atmospheric nitrogen mixing ratio is 2.7%, corresponding to a partial pressure of $\sim 10^{-4}$ bar, such a process was apparently ineffective.

Taking into account the composition of volcanic gases, we can explain the presence of halogens in Venus' atmosphere in the form of hydrogen compounds with chlorine and fluorine that are given off liberally under terrestrial conditions, along with sulfur dioxide and hydrogen sulfide from volcanic craters, fissures on the slopes of volcanoes (fumaroles), and cooling lavas. Subsequently, they are washed out of Earth's atmosphere by rains, dissolved in the oceans, or combined in reactions with solid material in Earth's crust. The huge volume of these gases also form hydrothermal solutions in underground waters on Earth, which is impossible under current conditions on Venus. As far as sulfurous compounds are concerned, processes responsible for the formation and maintenance of an equilibrium state of the Venusian clouds play a fundamental role in their cyclical transformation.

For Venus' subcloud atmosphere (as well as the atmosphere of Mars), on the whole, a low relative water vapor content is typical, at the level of thousandths of a percent. On Mars approximately 80% of the water vapor is concentrated in the near-surface layer of the atmosphere, several kilometers thick, and in contrast to Venus, the water vapor content fluctuates by two orders of magnitude, as a function of season, latitude, and time of day.

Turning now to measurement data on the relative amounts of minor components at various altitudes (see Table 5.5), we arrive at the conclusion that Venus' troposphere may be considered a homogeneous, well-mixed gaseous envelope only with respect to carbon dioxide, nitrogen, and inert gases. Apparently, there is a complex interrelation between physical and chemical processes in the atmosphere. Among the physical processes are turbulent mixing and planetary circulation along the horizontal and vertical; among the chemical processes are condensation and evaporation of cloud particles, gas phase reactions, and interactions of gases with minerals. These processes give rise to vertical concentration gradients of the minor components (H_2O, SO_2, CO, COS). One cannot completely exclude the presence of horizontal (latitudinal) gradients, or of temporal variations for some minor species (see Fig. 5.19).

The high surface temperature of Venus should certainly be considered a factor contributing to the active chemical interaction of atmospheric gases with the rocks. As a consequence, the composition of the atmosphere depends on the results of heterogeneous chemical processes at the atmosphere-surface interface.

As early as 1964 it was proposed (Mueller, 1964) that one can distinguish zones in the vertical profile of Venus' atmosphere as a function of the prevalence of a particular chemical process: a zone of thermochemical reactions, in which the composition of the atmosphere is buffered by minerals of the surface rocks, and a zone of "frozen" chemical equilibrium, where the composition of the gases corresponds to their equilibrium ratios in the near-surface layer of the troposphere. In this case it was emphasized that chemical reactions on the surface, at the atmosphere-crust interface, will take place given a constant supply of reactants, that is, under conditions of geological-tectonic activity in the crust.

The principle of global chemical quasi-equilibrium makes it possible to apply thermodynamic calculations for computing the equilibrium concentration of many gases whose measurements are uncertain or absent altogether. Such an approach (the principle of buffering the atmospheric composition by surface rock) was used to calculate the composition of the near-surface atmosphere soon after obtaining the results of atmospheric analysis from the Venera 4–6 spacecraft (Lewis, 1970; Khodakovsky et al., 1978; Florensky et al., 1978). In addition to these data, spectroscopic data on the CO, HCl, and HF concentrations were used, and the assumption was introduced that chemical equilibrium has been established over the whole troposphere up to the upper cloud boundary (the level of the spectroscopic measurements). The results of later calculations (Zolotov, 1985) are presented in Table 5.8, compared with the measurement data.

For many years a hypothesis by Urey (1952) concerning wollastonite equilibrium as a mechanism for buffering the CO_2 content in a global atmosphere-crust equilibrium system, according to the inverse reaction

$$CaCO_3 + SiO_2 \rightleftharpoons CaSiO_3 + CO_2, \tag{5.52}$$

has been discussed in the literature. Thermodynamic calculations carried out in works by Mueller (1963), Vinogradov and Volkov (1971), and Lewis and Kreimendahl (1980) indicated that the calcite–quartz–wollastonite mineral association on Venus' surface could buffer $P \approx 90$ bar CO_2 at a temperature of ~ 740K, which corresponds to conditions at the surface. Interpretation of the multisystem calculations (Volkov, 1983; Volkov et al., 1986; Khodakovsky et al., 1979), however, has led to the conclusion that carbonates are unstable and that one of the determining factors in this process is the relatively high SO_2 content in the troposphere. It is questionable whether wollastonite equilibrium, in pure form, can be considered the basis for a chemical model of Venus' atmosphere (Yung and DeMore, 1982; Zolotov, 1985).

As the idea of possible chemical equilibrium in the subcloud atmosphere has evolved, three zones have been distinguished over the whole lower atmosphere of Venus: (1) a stratosphere with an upper tier of clouds—a zone of photochemical processes; (2) a cloud zone, where photochemical (above) and thermochemical (below) processes compete; and (3) a subcloud troposphere (below ~ 48 km)—a zone where thermochemical equilibrium processes dominate. Calculations of the chemical composition of the subcloud troposphere, within the scope of the H–O–N–C–S system, with respect to the equilibrium constants of gas phase reactions, have indicated that this zone is in a state of disequilibrium, at least with respect to certain microcomponents (SO_2, H_2O), and that disequilibrium conditions undoubtedly characterize the second zone, the clouds. Under

Table 5.8 Chemical composition of Venus' subcloud troposphere from experimental data, compared with its model equilibrium composition

Gas	$X_{i(experimental)}$[a]	$X_{i(calculated\ equilibrium)}$[b]
CO_2	0.965	
N_2	0.35	
H_2O	3×10^{-5}	2×10^{-5}
SO_2	$\sim 1.8 \times 10^{-4}$	1.3×10^{-4}
S_2	1.3×10^{-8}	1.3×10^{-8}
CO	2.3×10^{-5}	7.2×10^{-6}
H_2S	8×10^{-5}	8×10^{-9}
COS	4×10^{-6}	3×10^{-6}
SO_3	—	8×10^{-13}
O_2	1.8×10^{-5}	10^{-23}
H_2	2.5×10^{-5}	10^{-9}

[a] According to data from Table 5.5.
[b] According to estimates by Zolotov (1985), obtained based on thermodynamic calculations using the first five components, $CO_2 \dots S_2$, as a starting point.

conditions of chemical equilibrium, a vertical gradient of microcomponent composition must exist, governed by the temperature and pressure profiles of the atmosphere and the buffering action of the surface rocks. So, the SO_2 content at an altitude of 26 km must decrease by 8–9 orders of magnitude compared to the surface level, if one of the assumptions is to accept a uniform distribution of CO_2, N_2, H_2O, and CO over the entire subcloud atmosphere. The existence of vertical concentration gradients of sulfur-bearing gases (and, in the actual circumstances, H_2O) could in principle be explained by their participation in the formation processes of the cloud condensate. Calculations have likewise led to the conclusion that the near-surface atmosphere is characterized by an extremely low oxygen content ($X_{O_2} \sim 10^{-19}$), which corresponds to the obtained partial pressure of $O_2 \sim 10^{-21}$ bar.

To develop the concept of wollastonite equilibrium, a pyrite anhydride buffer was taken as the starting point of the processes responsible for the scheme in Fig. 5.18a. In new modern models of atmospheric chemical composition:

$$FeS_2 + 2CaCO_3 + \frac{7}{2}O_2 \rightleftharpoons 2CaSO_4 + FeO + 2CO_2 \tag{5.53}$$

This buffer, as Barsukov et al. (1980) and Lewis and Kreimendahl (1980) proposed, may regulate the oxygen content in the atmosphere. Moreover, as calculations showed, the overall amount of sulfur-bearing gases ($SO_2 + H_2S + COS$) must decrease rapidly with increased O_2 content and hence, according to reaction (5.53), with an increase in FeO and $CaSO_4$ content in the crust. Thus even a moderately oxidized crust on Venus is inconsis-

Fig. 5.20. Gas concentrations versus altitude at the cloud level and in Venus' above-cloud atmosphere, expressed as the number density versus altitude $\mathcal{N}(Z)$. (It is derived from the partial pressure $P^* = nkT$, where T is the temperature and k is the Boltzmann constant. $\mathcal{N}$ is also easily determined from the density ρ, which enters into the equation of state [5.2], because $\rho = m_H \mu \mathcal{N}$, where μ is the molecular weight of the corresponding component, and m_H is the mass of the hydrogen atom, $m_H = 1.66 \cdot 10^{-24}$ g.) The altitude profiles presented are according to the model computed by Krasnopolsky and Parshev (1981). S and H_2SO_4 concentrations are given for the aerosol phase.

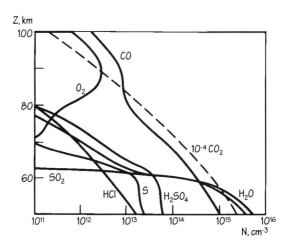

tent with the presence of significant amounts of sulfur compounds (at levels ≥ 100 ppm) in the troposphere. In this case H_2S and COS must predominate, and the SO_2 content at the surface must be ~ 0.1 of that amount, which conforms to the measurements higher in the atmosphere.

In models where the rates of chemical and physical processes are compared, the troposphere (with the exception of the several-kilometers-thick near-surface layer with the highest temperature) is considered to be in a state of disequilibrium (Khodakovsky et al., 1974; Krasnopolsky and Parshev, 1979). Heterogeneous catalysis at the atmosphere-surface interface, however, may be conducive to the establishment of chemical equilibrium with respect to certain components. Moreover, in this computational model, shown as an altitude distribution $\mathcal{N}(Z)$ in Fig. 5.20, it is assumed that the chemical composition of microcomponents in a vertical cross section of the subcloud troposphere does not change and corresponds to frozen equilibrium at the atmosphere-surface interface for T = 735 K and $P = 90$ bar. This principle of frozen equilibrium is also used for theoretical estimates of the composition of cloud condensates.

Under these assumptions, model compositions for the subcloud troposphere were calculated and were improved as new experimental data became available (Zolotov, 1985). The last variant, presented in Table 5.8, was computed for the following initial values: $X_{CO_2} = 0.965$; $X_{H_2O} = 2 \cdot 10^{-5}$; $X_{SO_2} = 1.3 \cdot 10^{-4}$; $X_{S_2} = 1.3 \cdot 10^{-8}$; and $X_{CO} = 7.2 \cdot 10^{-6}$ (the last value was computed with respect to the equilibrium constant for reaction [5.32] at $P = 90$ atm and $T = 735$ K). The resulting model composition of the near-surface troposphere is clearly characterized by the prevalence of oxidized forms of sulfur-bearing gases ($SO_2 > H_2S + COS$) and the extremely low concentration of H_2 and O_2. In reference to the O_2 problem, the model clearly indicates the impossibility of reconciling the gas chromatograph detection of oxygen on the Venera 13–14 with the simultaneous presence of H_2S and COS atmospheric gases in these samples (80 ppm and 40 ppm, respectively). In fact, catalytic oxidation of H_2S into oxides at 573 K is completed in 10–30 minutes, with the formation of elemental sulfur as the final product (Andreeva et al., 1985). Therefore, the coexistence of H_2S and O_2 under Venusian surface conditions is essentially impossible. The experimental data on H_2S, COS, and O_2 concentrations are obviously incompatible with the model compositions, from a comparison of them with the right-hand column in Table 5.8. The results of the Contrast experiment also support the low partial pressure of oxygen, which corresponds to the relation $X_{SO_2} > X_{H_2S} + X_{COS}$ in the quasi-equilibrium, atmosphere-surface system of the planet.

Thus, today one can speak about Venus' atmosphere as a naturally occurring system in which gas phase photochemical and thermochemical reactions, and heterogeneous reactions of gases with surface minerals, compete. We can formulate certain consequences from the model we have considered (Marov et al., 1989):

- On the whole, the troposphere of Venus is in a state of disequilibrium.
- The near-surface layer of the troposphere may exist in a state of chemical quasi-equilibrium with respect to certain microcomponents, in particular, sulfur-bearing gases.
- Chemical interaction of the troposphere with minerals results in the loss of sulfur in solid phases. Here it is assumed that SO_2 is extremely dominant over reduced forms of sulfur-bearing gases (H_2S and COS).
- The possible existence of a vertical H_2O gradient in the subcloud atmosphere within the scope of the model is inexplicable. Moreover, the model composition is not consistent with the measurements of free oxygen from the Venera 13–14 and Pioneer-Venus spacecraft.

Further research on the planet, primarily highly accurate measurements of minor atmospheric constituents in the lower atmosphere near the surface along with measurements of the surface rock composition, is needed to answer the remaining questions about the chemical processes in Venus' atmosphere and to support or refute these models.

Clouds

....

6 Of the numerous unique physical properties of Venus, one of the most vexing and interesting is the nature of its clouds. The clouds, located high in the atmosphere, shroud the entire planet and hinder optical observations of its surface (Fig. 6.1). We have only recently been able to study the properties of the atmosphere from information on the upper cloud boundary visible from Earth (which approximately corresponds to a pressure level of $P \approx 0.1$ bar, corresponding to an optical thickness of $\tau \leq 2$ and an altitude of 65–70 km), and from certain narrow spectral windows that transmit radiation from beneath the clouds. Questions about the physical structure of the clouds and their altitude stratification have long remained unanswered. Huge progress has been made through the penetration of spacecraft into the atmosphere and through a series of direct measurements that have made it possible to obtain information about the altitude, spatial distribution, and properties of the aerosol component.

Initial Facts and Hypotheses

Research on the clouds of Venus has a very interesting and instructive history. Several important stages in the study of the structure, properties, and composition of the clouds were covered in Chapter 2. Here we shall discuss this subject more comprehensively, in the context of progressively broadening perspectives and improvements in measurement techniques.

For a long time it was assumed that Venusian clouds were very dense—similar, perhaps, to the well-known cumulus clouds on Earth. It was also proposed that these clouds were not condensed but made of dust, which extended to the surface (recall the eolospheric model of Venus). The most diverse hypotheses were proposed concerning the nature of the condensate, although the theory of water-ice particles had, for a long time, the greatest number of adherents. Hydrochloric acid droplets, ammonium chloride (NH_4Cl), iron oxide, carbon suboxide (C_3O_2), hydrocarbons, formaldehyde polymers, silicate dust, and a series of other materials were also named (see, e.g., Kuiper, 1971; Kuzmin and Marov, 1974). None of these compounds, however, could satisfy all the observational data.

A fortunate exception was the hypothesis that the cloud particles were composed of a concentrated solution of sulfuric acid (H_2SO_4), independently proposed by G. Sill (1972) and by L. and A. Young (1973). Despite its unconventional nature, this hypothesis, supported by a series of strong arguments, subsequently gained additional confirmation. It is based on an analysis of data obtained by ground-based polarimetry and spectroscopy. The direct identification of the index of refraction for cloud aerosol particles at the cloud tops (Coffeen, 1968; Coffeen and Gehrels, 1969; Hansen and Hovenier, 1974) was especially

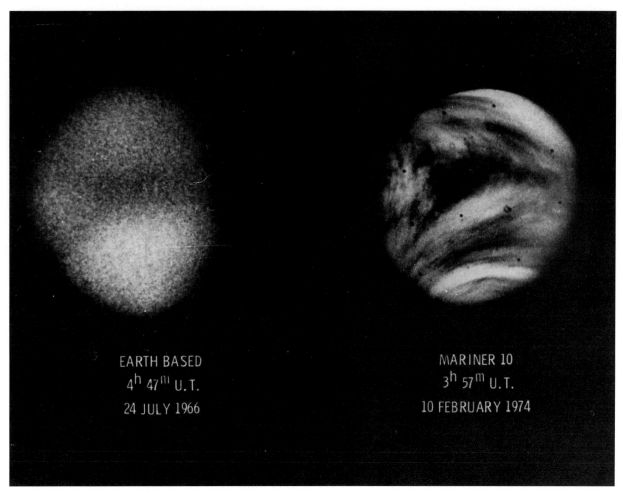

EARTH BASED
$4^h 47^m$ U.T.
24 JULY 1966

MARINER 10
$3^h 57^m$ U.T.
10 FEBRUARY 1974

Fig. 6.1. Venus' cloud cover from ground-based observations and Mariner-10. The photographs were obtained in the near-ultraviolet spectral region; in visible light, irregularities on the disk are much less pronounced. Only the upper boundary of the clouds is observed here; this is also true of remote sensing data in other spectral regions. We thus have only extremely limited information on the structure and properties of the clouds, and on the characteristics of the underlying atmosphere.

convincing. Based on analysis of the polarimetric measurements (Fig. 6.2a and b), the index of refraction was estimated to be $n = 1.44 \pm 0.02$, whereas $n = 1.33$ for water and $n = 1.31$ for water ice. This value for the index of refraction, as well as its dependence on sulfuric acid concentration (Fig. 6.2c) and certain spectral features of reflected light in the near-infrared region of the spectrum ($\lambda = 1-4$ μm), conform well with the n value and laboratory spectra of a 75–85% solution of H_2SO_4 (A. Young, 1973, 1977; Pollack et al., 1978).

A spectrum of reflected light from Venus in the $1-4$ μm range is shown in Fig. 6.3a; this is compared with the spectral characteristics of the cloud particle candidates mentioned above: water, ice, ammonium chloride, hydrochloric acid, hydrates of ferrous chloride, and even mercury, with various particle sizes $\bar{r}$. As we can see, at wavelength $\lambda >$ 3μm, Venus' spectrum is very dark. In other words, here the clouds reflect sunlight very weakly. Yet all the materials referred to above are highly reflective in this spectral region; this rules out their presence. Sulfuric acid in highly concentrated aqueous solution fits Venus' reflection spectrum best. It was found that the lower concentration threshold, of about 70%, is defined by the lack of typical water features in the infrared spectrum, and the upper limit is constrained by the relative concentration of SO_2 in the above-cloud atmo-

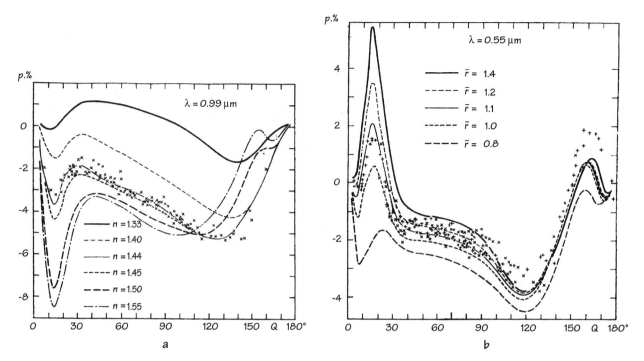

a

b

Fig. 6.2a. The relation between the degree of polarization and phase angle. The theoretical curves shown (Hansen, 1972) approximate the results of Coffeen and Gehrels (1969) at $\lambda = 0.99$ μm (crosses) for various values of the index of refraction n. In the calculations, particles with a mean radius $\bar{r}$ were used, which provide the best agreement with the measurements at all wavelengths: $\bar{r} = 0.7$, 0.8, 1.1, 1.1, 1.2, and 1.2 μm, respectively, starting from $n = 1.33$. The adopted albedo was $A = 0.90$ (at $\lambda = 0.99$ μm).

Fig. 6.2b. A theoretical approximation of observations from Lyot, 1926 (plus signs), and from Coffeen and Gehrels, 1969 (crosses), at $\lambda = 0.55$ μm. Calculated curves by Hansen (1972) correspond to $n = 1.45$ for various $\bar{r}$ values.

Fig. 6.2c. The relation between the index of refraction and the concentration of a sulfuric acid solution at 250 K. The arrow indicates the range of values corresponding to the polarimetric measurements (according to Young, 1973).

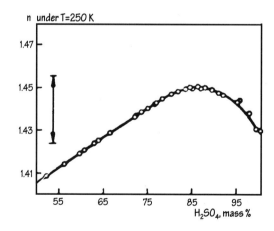

sphere, no higher than 10^{-7} (Barker, 1979). The closest agreement with the observational data was obtained for a sulfuric acid concentration of $84 \pm 2\%$. The well-known broad depression at $\lambda = 11.2$ μm in Venus' emission spectrum has long been identified with H_2SO_4 aerosols, as is obvious from Fig. 1.10. A feature near $\lambda = 20$ μm is also identifiable with H_2SO_4.

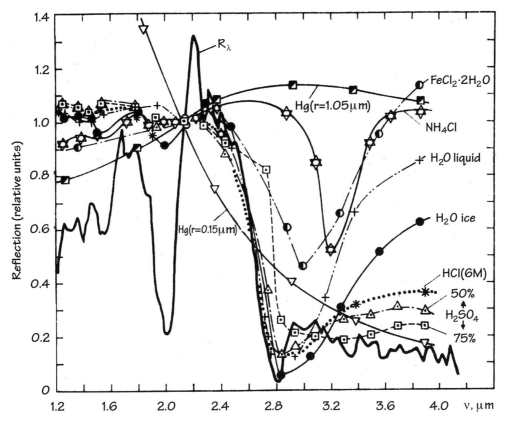

Fig. 6.3a. The spectrum of reflected radiation from Venus in the 1–4 μm range (bold solid line). Its characteristic feature is the extremely low reflection at wavelengths $\lambda > 3$ μm, which rules out the presence of various materials identified as cloud particle candidates (curves corresponding to these materials are shown in the figure), except for a concentrated aqueous solution of sulfuric acid (according to Pollack et al., 1975).

From polarimetric measurements at $\lambda = 0.55$ μm, estimates of the mean particle size at the upper cloud boundary were obtained (the cross section weighted mean radius $r \approx 1.05 \pm 0.10$ μm) upon comparison with a calculated model of the cloud. In this model the degree of polarization of reflected solar light p was determined by taking into account multiple scattering in a Rayleigh medium containing homogeneously distributed spherical particles (Hansen and Arking, 1971; Hansen and Hovenier, 1974). The estimated values of n and r from the results of such a comparison, which best satisfy the observational data, are shown in Fig. 6.2b. The excellent agreement between the early observational data by Lyot (1929) and later measurement by Coffeen and Gehrels (1969) deserves mentioning. Note also that the value of the complex index of refraction $n = 1.44$ corresponds to the middle portion of visible wavelengths (the so-called maximum visibility green line at $\lambda = 0.55$μm). At the same time an obvious trend toward a weak decrease in n between the ultraviolet and infrared spectral regions was discovered (from $n = 1.46$ at $\lambda = 0.37$ μm to $n = 1.43$ at $\lambda = 0.99$ μm). In turn, because the imaginary part of the complex index of refraction turned out to be negligibly small, one could conclude that the cloud aerosols essentially do not absorb light and that their reflective properties are entirely characterized by scattering. On the whole, the polarimetry results conform to H_2SO_4 concentrations starting from approximately 65% and higher (see Fig. 6.2c).

Observational data also provided some indirect evidence in favor of such an exotic aerosol chemical composition as sulfuric acid. In 1973, the temperature of the upper cloud

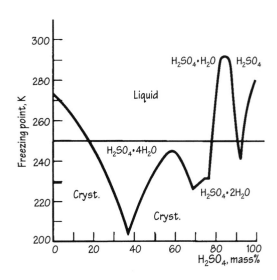

Fig. 6.3b. Freezing curves for sulfuric acid solutions of various concentrations (according to Young, 1973). The horizontal curve corresponds to the temperature of the upper boundary of the clouds, which was inferred from ground-based measurements; the actual boundary is approximately 10–30 K lower.

boundary was estimated at 250 ± 10 K, based on ground-based observations and measurements from the Venera 4–6 and Mariner-5 spacecraft. Later, taking into account measurement data from the Pioneer-Venus probes, it was established that the actual upper cloud boundary corresponds to a temperature interval of 220–243 K (Seiff et al., 1985). The values for the index of refraction of 75–80% H_2SO_4 solutions seem to be in good agreement with observational data at precisely these temperatures. Moreover, the low temperatures prevailing in the zone of the liquid-droplet aerosol (considerably below the freezing point of water) must be in line with the existence of "antifreeze." Sulfuric acid is quite capable of acting as an antifreeze, because at 250 K its solutions maintain a liquid state over a broad concentration range (Fig. 6.3b). Finally, the amazing dryness of the Venusian above-cloud atmosphere ($X_{H_2O} \leq 10^{-5}$) compared with Earth and even Mars (suggested by the lack of H_2O features in the infrared spectrum) enables us to assume the presence of an extremely hygroscopic aerosol. This, again, is consistent with the properties of sulfuric acid, which transforms H_2O ions into hydronium ions:

$$H_2O + H_2SO_4 \rightleftharpoons H_3O^+ + HSO_4^-$$

Note also that a series of geochemical arguments also suggest a sulfuric acid composition for the Venusian clouds. It is well known that H_2SO_4 vapor is present in the composition of volcanic fumes, and the cosmic abundance of sulfur and the degree of de-gassing from Venus' interior do not contradict mass estimates of H_2SO_4 in the cloud layer. The discovery that minute droplets of sulfuric acid are a main component of Earth's stratospheric aerosols is also important.

The observational data presented above pertain only to a relatively thin layer at the upper boundary of the clouds and do not provide information on their structure and properties in the lower-lying atmosphere. The first information on the altitude distribution of aerosols in Venus' lower atmosphere was obtained by photometric luminosity measurements from the Venera-8 spacecraft (Avduevsky et al., 1973a; Marov et al., 1973c). Three zones can be distinguished, with various degrees of solar radiation attenuation, in terms of

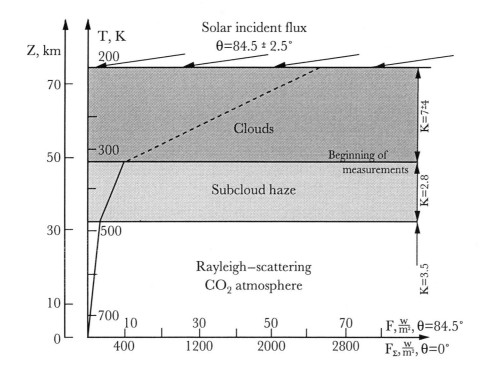

Fig. 6.4. Attenuation of solar energy in the atmosphere F (watts/m^2) according to measurements from Venera-8 for a solar zenith angle $\theta =$ 84.5° $\pm$ 2.5°. The thick solid curve denotes measurements; the thick broken line is an extrapolation from the start of F measurements at $Z =$ 49 km to the conditional upper boundary of the atmosphere $Z \approx$ 75 km, where F is the amount of energy supplied to Venus from the Sun at a given θ. k $=$ F_n/F_{n+1} is the index of attenuation in the $Z_{n+1} - Z_n$ layer. Along the x-axis F values are indicated for $\theta =$ 84.5° and scaled F_Σ values for $\theta = 0°$. Regions of the atmosphere are shown below 32 km, where attenuation occurs according to an approximate molecular (Rayleigh) scattering law, and above 32 km, where attenuation, in addition to Rayleigh attenuation, is most likely associated with the presence of aerosols, although their density is different. This is the first experimental analysis of the altitude trend in solar radiation attenuation, its value at Venus' surface, and the position of the clouds and subcloud haze.

amount of solar energy per unit area F (watts/m^2), calculated from the measured illumination for the corresponding solar zenith angle θ (Fig. 6.4).

The greatest attenuation occurs down to an altitude of 49 km, whereas at altitudes between 49 km and 32 km it is significantly less, and below 32 km the atmosphere is essentially transparent—attenuation of solar light is insignificant and completely obeys the law of molecular (Rayleigh) scattering in carbon dioxide gas, the main component of Venus' atmosphere. So, as one goes deeper into the atmosphere the molecular density increases, whereas the optical density falls. This can be explained if one assumes that the primary contribution to the decrease in solar radiation intensity in the two overlying regions is not Rayleigh scattering, whose effectiveness drops with altitude, but scattering by cloud aerosols, which are most dense above 49 km. Accordingly, the separate zones are identified with the densest clouds, located above 49 km, the lower-lying less dense aerosol layer (later named the subcloud haze), and the lower transparent atmosphere below 32 km (see Fig. 6.4). A cloud model with an overall aerosol optical thickness of $\tau = \sigma H = 50$ and a volume coefficient of aerosol scattering $\sigma \approx 3 \cdot 10^{-5}$cm^{-1} was derived from these observations (see Kuzmin and Marov, 1974).

Methods of Measurement

The conclusions made, in general, proved to be correct. They were supported by more comprehensive measurements, using broad- and narrow-band spectrophotometry (the relationship between the solar radiant flux in several wavelength intervals in the visible and near-infrared spectral regions, and altitude), first by the Venera 9–10 spacecraft (the corresponding instruments are shown in Fig. 2.39) and then by Venera 11–14 and the Pioneer-

Venus Large probe (Avduevsky et al., 1976b; Moroz et al., 1979, 1983a,b; Tomasko, 1980a; Economov et al., 1983). Such measurements, however, provide only mean integral characteristics and variations in the optical density of the gaseous and aerosol medium. More delicate techniques are necessary in order to study the cloud structure, the microphysical properties of the aerosols (size, index of refraction, particle concentration), and their chemical composition. Among these techniques, optical methods to measure the physical properties of aerosols stand out because of their efficiency and relative simplicity. These methods can be subdivided into two fundamental groups: nephelometric, which use devices that have their own intrinsic radiation sources and working volumes from several to tens of cm^3 and that hold tens or hundreds of aerosol particles during measurement; and photoelectric (using particle spectrometers), based on the study of individual aerosol particles, also using intrinsic radiation sources.

To determine the integral properties of a medium, the photometric technique still remains important, but on the whole it is less efficient than methods using nephelometers and particle spectrometers. As far as direct analysis of the aerosols' physical properties is concerned, a collection of particles is used, gathered on various types of collectors (filters and separators). The appropriate devices, however, are more complex to realize, compared with optical methods. They are nevertheless extremely effective in conjunction with techniques for analyzing the chemical composition of aerosols, such as mass spectrometry (with preliminary pyrolysis), gas chromatography, and X-ray radiometry.

To illustrate the progress made in studying Venus' clouds, we shall talk about the techniques used on spacecraft in somewhat more detail. Nephelometry (from the Greek word *nephelē*, meaning cloud) is widely used in terrestrial meteorology, and in chemistry and biology, to analyze dispersive (turbid) media, that is, media containing various suspended particles. In planetary exploration using spacecraft, the nephelometric method was first used in 1975 by Venera 9–10 (Marov et al., 1976a, 1980, 1983).

This method is based on measuring the intensity of light created by an artificial source (an incandescent lamp or laser), scattered at various angles. From the nature of the scatter at a given wavelength, which depends on the size and nature of the particles, one can accurately determine the properties of the scattering aerosol medium and can study in detail the characteristics of the cloud particles (particle size, size distribution spectrum, index of refraction). In turn, the index of refraction is very sensitive to the aerosol chemical composition. More strictly speaking, the nephelometric method is based on establishing the microphysical properties of aerosols from the derived characteristics of the radiation scattered by the medium (the scattering indicatrix, or phase function) with careful regard made to possible measurement errors. From a mathematical point of view, such problems belong to the class of ill-posed inverse problems (see Tikhonov and Arsenin, 1979).

Detailed study of the characteristics of scattered light, obtained in many numerical experiments (Shari and Lukashevich, 1977; Marov et al., 1980), has indicated that, assuming a definite distribution law of aerosol particles with respect to size, one can obtain reliable estimates of the microphysical parameters: these are the effective particle size r_{eff}, a parameter that determines the width of the particle distribution with respect to size α, the refractive n, and absorption k indices. The complex refractive index n is represented as a sum of its real and imaginary parts: $n = m + ik$. It turned out that to classify the experimental data in a certain rather broad class of atmospheric aerosols, it is sufficient to investigate light scattering at only a few predetermined angles, in order to reproduce the form of the phase function. In the developed technique, angles of $4°$, $15°$, $45°$, and $180°$ were chosen to be most informative; this technique was first realized in nephelometric experiments on the Venera 9–10 landers.

We shall briefly touch upon the theoretical foundations of this method. Mie scattering theory (see, for example, Deirmendjian, 1969) can serve as a convenient approximation for describing light scattering as an ensemble of particles of size r at wavelength λ (the diffraction parameter $x = 2\pi r/\lambda$). Although this theory strictly holds only for spherical particles, one can under certain assumptions expand it to the case of non-spherical particles (Pollack and Cuzzi, 1979).

We will assume that the particle distribution with respect to the radii corresponds to the following gamma distribution:

$$\mathrm{u}(r) = r^\alpha \exp\left(-\alpha \frac{r}{r_m}\right), \qquad 0 \le r \le \infty \tag{6.1}$$

This function, having an absolute maximum at the modal radius r_m, satisfactorily approximates the distribution with respect to particle size in terrestrial clouds, fog, haze, and atmospheric aerosols (Zuev, 1966). Actually, numerical modeling (Hansen and Travis, 1974; Lukashevich and Shari, 1982) indicated that the form of the distribution function (gamma, log-normal, and bimodal distributions were considered) has little effect on the light scattering characteristics, given that the effective distribution parameters (the effective radius and width) are identical.

The effective radius r_{eff} is specified by the relation

$$r_{\text{eff}} = \frac{1}{G} \int_0^\infty r\pi r^2 \mathrm{u}(r)\, dr \tag{6.2}$$

where $G = \int_0^\infty \pi r^2 \mathrm{u}(r)\, dr$ is the geometric cross section of all particles in a unit volume. The value of r_{eff} is a good approximation of the typical size, which determines the scattering cross section of the particles $\pi r^2 Q_{sc}(r)$, depending on the efficiency factor $Q_{sc}(r)$; the latter is also calculated according to Mie theory.

The dimensionless width of the distribution V_{eff} with respect to r_{eff} is, in turn, determined by the following:

$$V_{\text{eff}} = \frac{1}{Gr_{\text{eff}}^2} \int_0^\infty (r - r_{\text{eff}})^2 \pi r^2 \mathrm{u}(r)\, dr \tag{6.3}$$

In the case where $\mathrm{u}(r)$ corresponds to a gamma distribution ($0 \le r \le \infty$), we have:

$$r_{\text{eff}} = r_m\left(1 + \frac{3}{\alpha}\right) \tag{6.4}$$

$$V_{\text{eff}} = \left(\frac{1}{\alpha + 3}\right) \tag{6.4'}$$

From Eq. (6.4) for a given α the modal radius of the particles r_{m} is simply determined.

The results of nephelometric measurements can be conveniently represented by the index of directed scattering, defined by the expression

$$\bar{\sigma}(\theta_{\text{j}}) = \int_{\lambda_1}^{\lambda_2} \int_{\theta_{\text{j}_1}}^{\theta_{\text{j}_2}} \sigma(\lambda)\, i\,(\lambda, \theta) F(\lambda) f_{\text{j}}\,(\theta)\, \mathrm{d}\theta\, \mathrm{d}\lambda \tag{6.5}$$

where $\sigma(\lambda)$ is the volume coefficient of scattering at wavelength λ; $i(\lambda, \theta)$ is the scattering indicatrix, normalized to unity; $F(\lambda)$ is the spectral characteristic of the transceiving optics of the emitter and light detector in the measurement device; and $f_{\text{j}}(\theta)$ is the intensity distribution over the scattering angle θ for each channel measured. For convenience, $F(\lambda)$ and $f_{\text{j}}(\theta)$ are normalized by introducing a weighting function, provided

$$\int_{\lambda_1}^{\lambda_2} F(\lambda)\, \mathrm{d}\lambda = 1; \qquad \int_{\theta_{\text{j}_1}}^{\theta_{\text{j}_2}} f_{\text{j}}(\theta)\, \mathrm{d}\theta = 1$$

where λ_1 and λ_2 denote the width of the radiation spectrum and $\theta_{\text{j}1}$ and $\theta_{\text{j}2}$ are the limiting scattering angles in each measured channel.

Further, adopting $\bar{\lambda}$ as the effective wavelength ($\bar{\lambda} = 0.92 \; \mu\text{m}$ was used for Venera 9–14), the previous expression can easily be transformed into

$$\bar{\sigma}(\theta_{\text{j}}) = \int_{\theta_{\text{j}_1}}^{\theta_{\text{j}_2}} \sigma(\bar{\lambda}) i(\bar{\lambda}) \theta f_{\text{j}}(\theta)\, \mathrm{d}\theta = \sigma \int_{\theta_{\text{j}_1}}^{\theta_{\text{j}_2}} i(\theta) f_{\text{j}}(\theta)\, \mathrm{d}\theta = \sigma \bar{i}(\theta_{\text{j}}) \tag{6.6}$$

where $\bar{i}(\theta_{\text{j}})$ and $\bar{\sigma}(\theta_{\text{j}})$ are the effective values of the indicatrix and volume coefficient of scattering. The weighting function $f_{\text{j}}(\theta)$ is computed for all nephelometric channels, taking into account analysis of the instruments' optical design parameters. Errors in the differences between $\bar{\sigma}(\theta_{\text{j}})$ and $\sigma(\theta_{\text{j}})$ can thus be minimized by calculating the actual values of the $f_{\text{j}}(\theta)$ function during numerical modeling of the measurement results. Further, by solving the inverse problem or, in other words, by comparing the values obtained in experiment $\bar{\sigma}(\theta_{\text{j}})$ with the computed values $i(\theta_{\text{j}})$, the microphysical characteristics of polydispersive systems of spherical particles $r_{\text{eff}}, \alpha, n$ can be found, as can the integral optical characteristics of the medium, the coefficient of volume scattering σ, the particle concentration N, and the optical thickness τ of each layer, correlated with altitude. To do this, accepted models of the aerosol under study are used:

$$\sigma(z) = \frac{\bar{\sigma}(\theta_{\text{j}}, Z)}{\overline{i(\theta_{\text{j}}, Z)}}; \qquad \tau = \int_{Z_0}^{Z} \sigma(Z')\, \mathrm{d}Z' \tag{6.7}$$

where Z_0 is the altitude at the start of measurements. In turn, the particle concentration N is related to σ by

$$\sigma = N \bar{Q}_{sc}(n, r_{\text{eff}}, \alpha) \pi \left[\frac{(\alpha + 1)(\alpha + 2)}{(\alpha + 3)^2}\right] r_{\text{eff}}^2 \tag{6.8}$$

where $\overline{Q}_{sc}$ is the mean scattering factor, given by the following expression:

$$\overline{Q}_{sc}(n, r_{\mathrm{eff}}, \alpha) = \frac{\displaystyle\int_0^\infty Q_{sc}(r)\pi r^2 \mathrm{u}(r)\,\mathrm{d}r}{\displaystyle\int_0^\infty \pi r^2 \mathrm{u}(r)\,\mathrm{d}r}$$

Nephelometric measurements, using multiangular nephelometers on the Venera 9–10 spacecraft, resulted in the discovery of the three-layer Venusian cloud structure and sub-cloud haze, and made it possible to determine the fundamental microphysical characteristics of the aerosol component (Marov et al., 1976a,b; 1980). Later, cloud structure and the properties of cloud particles were studied by this method using only backscattering nephelometers. Such measurements were conducted by the Pioneer-Venus Large and small probes (Ragent and Blamont, 1980) and by Venera-11, Venera-13, and Venera-14 (Marov et al., 1979c, 1983). They confirmed the basic characteristics of the vertical cloud and haze structure, and the microphysical properties of aerosols, obtained by Venera 9–10, at the same time revealing some additional features of their spatial distribution and the optical density of the medium.

Photoelectric measurements of the aerosols in Venus' atmosphere were achieved on only one occasion, at the landing area of the Pioneer-Venus Large probe. These measurements made it possible to obtain additional information on the sizes, degree of dispersion, and local concentration of cloud particles as a function of altitude (Knollenberg and Hunten, 1980; Knollenberg et al., 1980). Measurements were carried out using a projection particle spectrometer. Like nephelometers, these spectrometers are analogous to devices widely used in terrestrial meteorology. The working principle consists of recording the "image" of a particle on a photodiode array, which serves as the detector; at a given moment the particle is located along the path of a laser beam. From the number of photodiode elements shaded by the projection of each individual particle, one can obtain an idea of its size, and based on the movement of the projections, one can estimate the number of particles per unit volume. Note that in this method, the effect of the index of refraction on the measurement accuracy is minor; in principle this makes it possible to decrease systematic errors, even with non-spherical particles in the medium. In the case of a very forward-elongated indicatrix or particles with complex morphology, however, noticeable systematic errors may emerge, which must be kept in mind when interpreting the measurement data.

The Venera-12 and Venera-14 landers were equipped with instruments for analyzing the cloud composition using X-ray radiometry (Surkov et al., 1981, 1982). Aerosol particles, as well as atmospheric gases, were pumped through the instrument and precipitated onto a special filter (Fig. 6.5). The composition of the particles was determined from the characteristic fluorescence emission excited by exposing the samples to radioisotope sources of ^{55}Fe and ^{109}Cd. The characteristic emission was recorded using proportional counters in energy ranges from 1.1 to 10 keV, which made it possible to identify elements from aluminum to mercury. During transit of the cloud layer, over ~9 minutes, about 1 m^3 of atmosphere was pumped through the instrument and a set of spectra was produced dur-

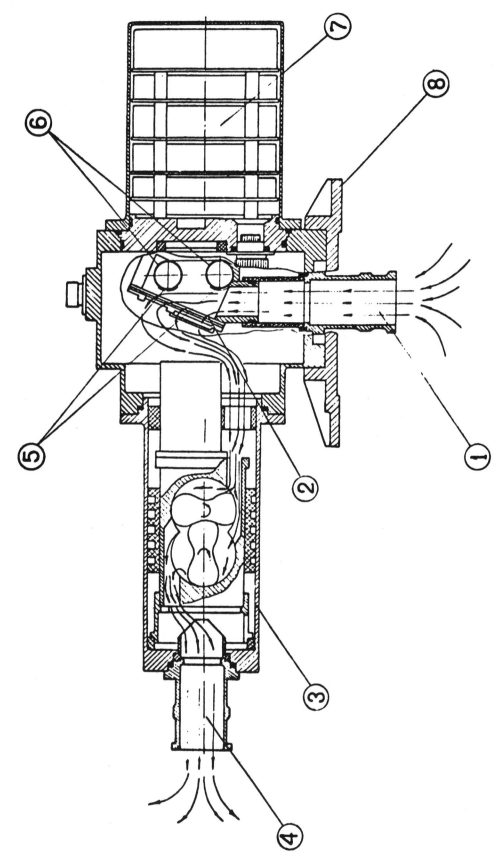

Fig. 6.5. The X-ray fluorescence system for analyzing aerosols that operated on the Venera-12 and Venera-14: (1) the intake for obtaining aerosols from the atmosphere, (2) a micropump, (3) the output with aspiration system, (4) a filter, (5) the ^{55}Fe radioisotope source, (6) proportional counters, (7) the electronic unit, and (8) the support (mounting) (according to Surkov et al., 1981).

ing the entire time the aerosols were collected on the filter. Aside from this, the change in aerosol density with altitude was estimated from the overall intensity of the excited fluorescence emission.

Another instrument for analyzing the composition of the cloud aerosols was installed on the Venus landing module of the Vega-1 spacecraft (Surkov et al., 1987d). It was created through Soviet-French cooperation and consisted of a collector-pyrolyzer and mass spectrometer, which collected the particles and analyzed the gas products of their pyrolytic decomposition. The atmosphere was pumped through an intake and an inertial separator, which separated the particles into two fractions, with sizes more or less than 3 μm. Each fraction entered pyrolytic chambers, where the aerosols were precipitated onto filters. After pyrolysis, the gaseous products passed through an injection system, consisting of microvalves and a capillary tube, into a hyberboloid-type mass analyzer: electrodes, shaped as hyperboloids of revolution, form a closed volume, in which the gas molecules (atoms) are ionized. The ions were classified and subsequently separated by mass, by varying the frequency of the high-frequency voltage. Two-stage selection and analysis of the aerosol sample was conducted using this technique (with parallel multistage analysis of the atmospheric gas) in the 63–52 km and 52–45 km altitude ranges.

The gas chromatograph method for analyzing the chemical composition of aerosols was realized on the Vega 1–2 spacecraft (Gelman et al., 1986). The instrument consisted of various types of detectors, three sets of thermoreactive cells, and two independent analysis channels in which nitrogen and helium were used as carrier gases. The aerosols gradually accumulated on a special fiberglass filter, covered with carbon, which was then heated to 575 K. This experiment was oriented toward a quantitative analysis of sulfuric acid. This was accomplished by analyzing the volumes given off as a result of the following reaction:

$$C + 2H_2SO_4 \rightarrow CO_2 + 2H_2O + 2SO_2$$

Because the products of this reaction are present in the atmosphere, a special working cyclogram was incorporated, which enabled us to isolate them from atmospheric gases.

In addition to the experiments mentioned, remote methods have contributed significantly to the study of the clouds and overlying atmosphere. Measurements using photopolarimeters and infrared radiometers on the Pioneer-Venus spacecraft (Travis et al., 1979a; F. Taylor et al., 1979, 1980), infrared radiometers on Venera 9–10 (Ksanfomality, 1985), and Fourier spectrometers on the Venera 15–16 spacecraft (Ertel et al., 1984) made it possible to detect the presence and nature of spatial-temporal variations in the upper boundary of the clouds and above-cloud haze.

The anomalously bright night-side radiation in the near-infrared window regions discovered in 1984 (see Chapter 5) was found to be highly localized in "hot spots" that moved with the characteristic four-day superrotation. Careful analysis of zonal wind speeds derived from extended observation of these motions placed the origin of these modulations in the middle (50–57 km) or lower (48–50 km) cloud regions (Crisp et al., 1991; Carlson et al., 1991). This evidence supports the interpretation that the spatial variations in this radiation are caused by localized changes in the optical depth of the lower cloud deck. This is the same region of clouds that shows the greatest variation in density between nephelometer traces from different spacecraft. Analysis of the multispectral mosaics of the night side provided by the NIMS instrument on the Galileo spacecraft, taken during its brief fly-by of Venus in late 1990, have added new constraints on cloud particle sizes and vertical and latitudinal structure (Carlson et al., 1993; Grinspoon et al., 1993).

Comparison and complex analysis of all currently available results have enabled us to develop a series of theoretical models of the structure, chemical composition, and formation of the clouds, and have augmented our understanding of the physical-chemical characteristics of the aerosol component in Venus' atmosphere.

Altitude and Spatial Distribution of Aerosols

Aerosols occupy an extended region in Venus' atmosphere, approximately 60 km in altitude, principally located within the altitude range of 30 to 90 km. This distribution is not uniform in optical density or particle properties, and within the indicated altitudes, separate, characteristic zones can be distinguished (Table 6.1). Contemporary views on particulate matter in Venus' atmosphere are summarized by Ragent et al. (1985) and Volkov et al. (1989).

The highest density falls within the interval of 70 to 48 km; this is the main cloud deck. Between 70 and 90 km is the upper (above-cloud) haze; between 48 and 32 km we find the lower (subcloud) haze. Below 32 km the atmosphere contains essentially no aerosols, and its optical properties are almost entirely determined by molecular scattering in CO_2. There are, however, isolated indications that negligible amounts of aerosols may be present even in this region of the atmosphere, starting from an altitude of ~ 10 km, in the form of a light haze or individual thin layers (Marov et al., 1980, 1983; Golovin and Ustinov, 1982; Ekonomov et al., 1983). It is possible that such layers or separate regional formations may occasionally appear because of geological-geochemical processes on Venus' surface. According to current notions, the observed significant spatial-temporal variations in the structure of the upper and lower haze are governed by such processes and atmospheric dynamics.

The Above-Cloud Haze

The apparent homogeneity of the cloud cover, as observed from Earth in the visible spectrum, creates an impression of stability between the upper boundary of the primary clouds and the above-cloud haze. Early aerosol models with a narrow particle size distribution, based on the results of ground-based polarimetry (Hansen and Hovenier, 1974), contributed to this overall impression. However, later polarimetric measurements from the Pioneer-Venus satellite, which were compared with brightness details on the disk, indicated that the aerosol density in the above-cloud haze is subject to considerable spatial-temporal variations, especially as a function of latitude (Kawabata et al., 1980).

The upper haze extends to an altitude of ~ 90 km and is formed by submicron-sized

Table 6.1 Structure of Venus' cloud cover

Region	Altitude range Z, km	Temperature range T, K	Pressure range P, atm
Above-cloud haze	90–68	172–238	3.6×10^{-4}–3.1×10^{-2}
Upper layer I	68–58	235–276	0.06–0.31
Intermediate layer I′	58–57	276–281	0.31–0.36
Middle layer II	57–52	281–330	0.36–0.76
Intermediate layer II′	52–51	330–340	0.76–0.92
Lower layer III[a]	51–48	340–365	0.92–1.3
Subcloud haze	48–32	365–475	1.3–7.6

[a] Including indications of the thickness in the 0.2–0.5 km range, located ~ 0.5 km below the boundary of the layer.

particles (~0.5 μm), which are apparently mixed with particles from the upper cloud layer. These particles are most likely close to mode 1 cloud aerosols in their properties. The haze is very rarefied; the particle concentration is estimated to be ~500 cm^3 at the tops of the clouds, and it decreases with altitude, as does the size of its constituent particles. Most likely, it consists of minute droplets of sulfuric acid, producing an overall optical thickness $\tau \leq 1$ along the vertical. Under this assumption, Krasnopolsky (1983) estimated the radius of the particles at $r \approx 0.2$ μm, using spectrophotometric data at Venus' limb from the Venera 9–10 spacecraft in the visible range, and he found that the number of particles decreases by about three orders of magnitude between 70 and 100 km.

The results of analysis of polarization variations, along Venus' equator and perpendicular to it, from ground-based observations at four wavelengths in the 0.365–0.630 μm range (Starodubtseva, 1989), do not contradict these estimates. These variations (with a period of ~4.5 days) and systematic sign differences in the polarization, in the ultraviolet and red, are consistent with the periodic appearance of a layer of above-cloud haze consisting of particles of 0.2–1 μm in size.

Observations from Pioneer-Venus in the ultraviolet, visible, and infrared spectral regions accordingly suggest that the above-cloud haze is an almost global formation and extends from the equator up to the polar regions (Esposito et al., 1983). Small regions near the pole within ~10° latitude are an exception. With increasing latitude the optical density grows, so that in the near-polar zones it is $\tau \approx 0.6–0.8$, which is almost an order of magnitude greater than at the equator. Consequently, these zones appear much brighter on television images of Venus, although their temperature is several tens of degrees lower. In particular, the haze in the 60–80° latitude range is associated with the formation of a cold collar (see Chapter 5, particularly Fig. 5.11).

In the longitudinal direction the variations are much more local. The densest haze forms near the morning and evening terminators, mainly at equatorial latitudes. In extent, it varies from ~3000 km at the upper boundary of the clouds to ~1000 km at an altitude of 85 km, standing out more intensely in the infrared, than in the ultraviolet spectral region. On this basis, another hypothesis was formulated about the formation of these zones of higher-than-normal optical density and about their chemical composition, which differs from that of the global haze (Ragent et al., 1985). If sulfur dioxide (SO_2) is identified as the most likely source of global haze, consisting of H_2SO_4, then local zones of aerosols near the terminator may possibly consist of SO_2 crystals.

As it turned out, a change in SO_2 content in the above-cloud atmosphere correlates well with temporal variations in the density of the global haze, on a scale of several weeks to several years (Esposito, 1980, 1984, 1988; Zasova et al., 1981). Recall that in Chapter 5, we discussed a possible connection between these variations (which caused a significant increase in global haze density starting in the early 1970s, and then a decrease) and current volcanic processes on Venus. This intriguing idea, however, requires more convincing evidence.

The boundary separating the haze from the upper layer of clouds, at an altitude of 65–70 km, is quite well defined at night, whereas during the day it appears more diffuse. At high latitudes (in the "polar vortex" region), the upper cloud boundary is 5–7 km lower and less homogeneous than it is at the equator; in the longitudinal direction a trend toward a decrease in altitude is observed in the morning hours, and in the evening hours the altitude is at a maximum. The interrelation between the thermal structure and atmospheric dynamics and the processes for forming the cloud and haze condensate is apparently reflected in the similar nature of the variations in the position of the cloud boundaries and haze density.

The most remarkable phenomena in this region of the atmosphere on Venus are designated ultraviolet formations or ultraviolet clouds (see Chapter 2, particularly Figs. 2.6 and 2.7). Such formations last from several hours to weeks; their sizes and forms exhibit great variety, and as the size increases the degree of ultraviolet contrast also increases. Along with the largest structures, reminiscent of a letter Y placed on its side, there are bands inclined at various angles to the equator, spirals, streamers, and even honeycombed structures, which suggest the presence of widespread convection, as revealed by television pictures in ultraviolet light from the Mariner-10 spacecraft (Fig. 6.6). These formations migrate across the disk with a velocity corresponding to the zonal wind velocity at these altitudes and reflecting the planetary circulation system in the troposphere and stratosphere. Wave motions are superimposed on the clouds, with various apparent velocities within individual latitude zones, influencing the observed ultraviolet contrast structures and their motions (Belton et al., 1976a, 1976b).

The morphology of these zones was studied using a large number of images obtained by the Pioneer-Venus photopolarimeter (Travis et al., 1979a, 1979b) (Fig. 6.7). Rossow et al. (1980) distinguished three fundamental zones: a polar zone at latitudes above 45°–50° in both hemispheres; a middle latitude zone between 20° and 45°–50° latitude; and an equatorial zone within a ±20° latitude band. The polar zone, on the whole, is characterized by considerably greater brightness relative to the remaining portion of the disk, even though by virtue of its positional geometry it should appear darker, given an isotropically

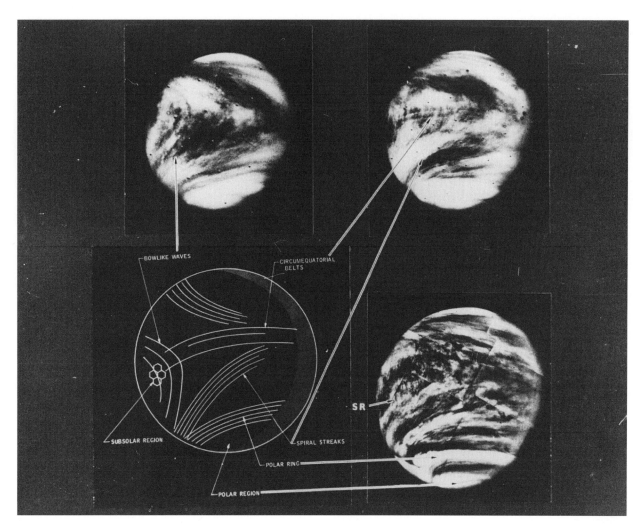

Fig. 6.6. Images of Venus' upper cloud layer in the ultraviolet, obtained by the Mariner-10 photographic experiment. Various forms, observable in the morphology of the cloud cover and above-cloud haze, are summarized in the diagram at the lower left (according to Murray, 1974).

scattering sphere. A dark polar cap is sometimes formed (see Fig. 6.7a–d). The northern and southern polar regions differ little in their morphology. As a rule, the persistence of individual details is observed over an interval of less than a day and essentially never is observed over 4–5 days (Schofield and Diner, 1983).

In contrast to the polar zone, the middle latitude regime and near-equatorial band show a great variety of structures in the form of relatively darker formations, reflecting considerable albedo variations. Along with large-scale structures, extending between 50° N and 50° S latitude, are more or less ordered configurations of smaller size. Their classification, essentially indistinguishable from that presented in Fig. 6.6, is shown in Fig. 6.8. Dark middle latitude bands are clearly visible, for example, in images 6.7b and 6.7c; the dark equatorial band is seen in Figs. 6.7a and 6.7d. One can distinguish the near-equatorial band in all the images in Fig. 6.7. Arched wave processes stand out especially well in Fig. 6.7a. The classical Y-form appears quasi-periodically every four or five days but sometimes completely disappears and is absent for several cycles. All albedo irregularities and their configurations appear to undergo a regular, as though coordinated, oscillating evolution in their orientation relative to the latitude circles, on a global scale, independent of

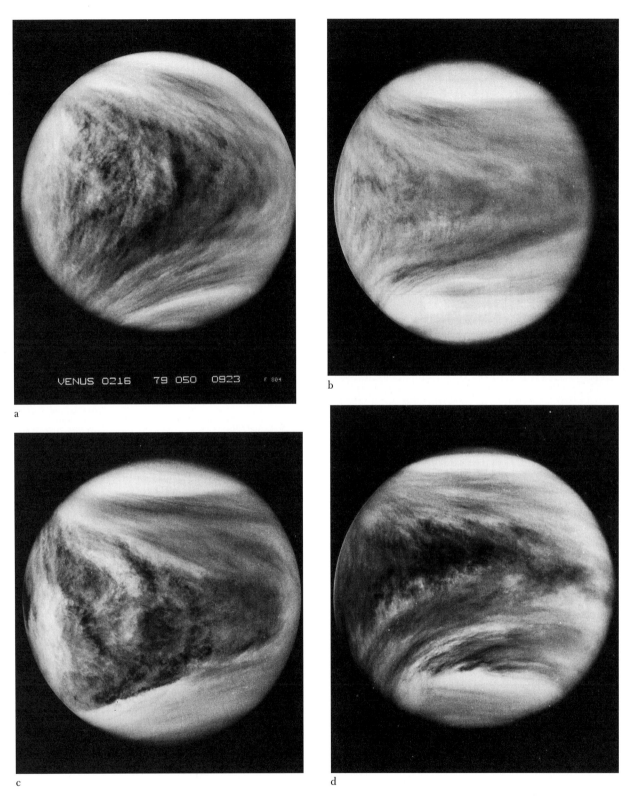

VENUS 0216   79 050   0923   F 804

a

b

c   d

Fig. 6.7a–d. Examples of images of the upper portion of the cloud layer and above-cloud haze in the ultraviolet, obtained using a photo-polarimeter on board the Pioneer-Venus spacecraft. All four images correspond to illumination at a phase angle close to 0°, and they cover a time interval of about 1.5 months. Standing out in the cloud morphology are diverse structural formations, which undergo strong dynamic variations; certain forms repeat periodically and maintain their configuration over various time intervals. See the text for more details (reprinted with permission of L. D. Travis and S. S. Limaye).

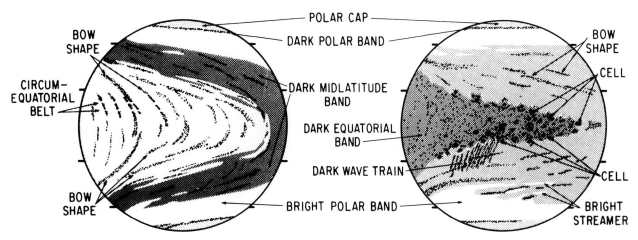

Fig. 6.8. A diagram indicating the fundamental types of structural formations on images of Venus in ultraviolet rays, examples of which are shown in Fig. 6.7. The two views presented here correspond to typical configurations on Venus' disk, separated in time by two days, and are examples of forms exhibiting maximum (left) and minimum (right) inclinations relative to latitude (according to Rossow et al., 1980).

whether the Y-form is present. The arched forms and convective cells are mainly concentrated toward the subsolar point in the equatorial zone but are also sometimes observed near the morning terminator. Together these data reflect complex processes in atmospheric dynamics, as well as in cloud and atmospheric chemistry, exhibited in the morphology of ultraviolet formations and their evolution (Rossow et al., 1980).

What is the nature of these formations and how are they linked with processes occurring in the upper cloud layer and above-cloud haze? Unfortunately, we cannot answer this question yet. It is obvious that regions of contrasts correspond to the greatest light absorption in the ultraviolet spectral region; in other words, there must be a suitable absorber here (Sill, 1975; Travis, 1975). Based on available experimental data, one can assert that in the spectral range shorter than 3200 Å the primary absorber is sulfur dioxide, whose concentration correlates well with the degree of ultraviolet contrast. Moreover, in the $\lambda > 3200$ Å range, one should assume the existence of at least one more absorber (Pollack et al., 1980a). It is highly improbable that the haze itself is the absorber, because the required optical thickness $\tau \approx 0.5{-}1$ would lead to appreciable light polarization in regions of corresponding UV contrasts; such polarization is not observed. Also, the appearance and disappearance of contrast features, on the order of hours and days, does not conform to notions about processes of haze formation, which require a timescale at least an order of magnitude greater (Esposito, 1980; Esposito and Travis, 1982; Ragent et al., 1985). There is also no noticeable correlation in regions of UV contrasts with the temperature of the surrounding atmosphere, which was measured from the broadening of the CO_2 line.

The UV formations on Venus' disk enable us to visualize horizontal as well as vertical transport, and to build up an interesting picture of the dynamic processes in this region. Among these are turbulent mixing, upwelling, and wave motions including internal gravitational waves propagating from the lower-lying atmosphere, which are suggested by the temperature stratification in the stratosphere (see Chapter 5). One could imagine that because of vertical drift, either a direct absorber of solar shortwave radiation, or components resulting in its formation in a chain of chemical (including photochemical) reactions at these altitudes, may be supplied to regions of observed UV contrasts. So, the explanations for the formation of UV contrast features should be sought in the interaction of dynamic and chemical processes. It should be emphasized that the required vertical motions, leading to a change in sulfur dioxide content by several times, are not great; they are in the 1–2 km range, because this is precisely the scale height for SO_2 in the above-cloud atmosphere (Ragent et al., 1985).

Gaseous chlorine has been suggested as another potential absorber, based on its reasonable identification in Venus' reflection spectrum at $\lambda > 3200$ Å. Pollack et al. (1980a) found that the chlorine content in this case must be not very great (~ 1 ppm); however, this estimate is not supported by the possibility of Cl_2 generation due to HCl photodissociation. In fact, there are no reasonable chemical arguments for the possible coexistence of SO_2 and Cl_2.

Very fine particles of polymorphic amorphous sulfur have also been proposed (Toon and Turco, 1982). This substance also matches the spectral absorption features at $\lambda > 3200$ Å, and it naturally fits into existing ideas concerning the sulfur cycle in Venus' atmosphere (see Fig. 5.18a), including the presence of SO_2 and the bimodal nature of the particle size distribution in the clouds. This theory has been associated with the hypothesis about the existence of yet another particle fraction at the upper cloud boundary, provisionally named mode 0 by Suomi et al. (1980) and Tomasko et al. (1980a). This hypothesis was put forward in an attempt to interpret the nature of absorption in the infrared spectral region (at $\lambda \approx 10$ μm) from radiometry data. The existence of this mode, however, was found to be unnecessary in light of the more complete picture of the radiative effects of Venus' main cloud deck provided by the NIMS measurements during the Galileo fly-by of Venus (Grinspoon et al, 1993; Roos et al., 1993). Moreover, one cannot rule out the possibility that, along with SO_2, some unidentified absorber in a gaseous phase, which is an intermediate product of the sulfur cycle at high altitudes, may be responsible for the formation of UV contrasts and for the trend in Venus' spectral albedo. In any case, independent of the nature of the absorber, we have evidence of significant instability in the distribution of gas components and haze aerosols above the clouds; this instability is associated with variations in brightness and UV formations on Venus' disk.

The Main Cloud Deck

According to the classification introduced as a result of the Venera-9 and Venera-10 nephelometer experiments (Marov et al., 1976a, 1976b, 1980), three layers stand out within the main clouds, at the boundaries of which there are changes in aerosol characteristics (see Table 6.1): the upper layer (layer I) from 68 to 58 km; the middle layer (layer II) from 57 to 52 km; and the lower layer (layer III) from 51 to 48 km. No sharp boundary exists between the layers. The regions of variation in particle characteristics between layers I–II and II–III are further classified as transitional layers I′ (58–57 km) and II′ (52–51 km). A similar structure with three layers within the main clouds was found from the results of the Pioneer-Venus aerosol studies (Ragent and Blamont, 1980; Knollenberg and Hunten, 1980); these layers were designated B (lower), C (middle), and D (upper).

The upper boundary of the main clouds tends to drop to lower altitudes to the north and south of 55° latitude: if at the equator it is located at the 30 Mbar pressure level, then at 65° it is at the 40 Mbar level—lower by 5 km. The lower cloud boundary (LCB) also does not remain at the same level in various probe regions; it is found within the 47–49 km altitude range. The Night and North Pioneer-Venus probes found it approximately 2 km closer to the surface. From analysis of radio attenuation measurements, Cimino et al. (1980) concluded that at high latitudes a stable depression is observed in the position of the LCB, compared with low and middle latitudes. In the polar vortex region it was recorded at an altitude of 40 km. Below the LCB there is sometimes a narrow transitional layer, 0.5–1 km lower and 0.1–0.2 km thick. This region, distinguished by Pioneer-Venus data as a separate A layer with great variability (often it disappeared completely), may be classified as a zone of cloud precursors (Marov et al., 1983).

At first approximation the cloud density (or turbidity of the medium) can be characterized by the integral index of backscattering $\bar{\sigma}$ (180°), which is the product of the volume scattering coefficient σ and the effective value of the scattering indicatrix $\bar{i}$ (180°). A more rigorous estimate of the optical density is provided directly by the volume scattering coefficient σ for a medium with a known phase function $\bar{i}$. Determining $\bar{i}$, however, is linked with the necessity of measuring the light scattering intensity at several additional angles at least. Initial results of such measurements, taken with a four-channel nephelometer (see Fig. 2.40) on the Venera-9 and Venera-10 landers, are shown in Fig. 6.9. Analysis of these data by the technique described in the previous section provided an interpretation of the subsequent $\bar{\sigma}$ measurements. Inside the basic clouds $\bar{\sigma}$ (180°), according to all spacecraft measurements, varies within the limits from $0.5 \cdot 10^{-4}$ to $2.0 \cdot 10^{-4} m^{-1} sterad^{-1}$, and σ varies from 0.2 to 6 km^{-1}.

The structure of the basic clouds (as a vertical profile of the coefficient σ), in five regions of the atmosphere probed by the Venera spacecraft and four regions probed by the Pioneer-Venus probes, is presented in Fig. 6.10 (Marov et al., 1983; Esposito et al., 1983; Ragent et al., 1985). The results suggest stability in the altitude stratification of the clouds, and a well-preserved spatial homogeneity in the equatorial and middle latitudes. Over approximately 20 km, no breaks are detected in the clouds, although the optical density in various regions undergoes noticeable changes. Such quasi-homogeneity in the vertical and horizontal directions is apparently explained to a considerable extent by the existence of stable mass transfer in Venus' four-day atmospheric circulation system, and by associated widespread turbulence in the cloud zone (Kerzhanovich and Marov, 1974, 1983).

Substantial changes in the optical density are observed in layer I, while the structure of layer II generally remains most constant over all regions studied. Layer III (near the LCB), which sometimes completely disappears, combining with layer II, exhibits the greatest variability. An investigation of the spatial variations in the cloud deck at the time of the Galileo fly-by, using the NIMS instrument, showed a variation in total cloud optical depth from 25 to 40, with essentially all the variation occurring because of changes in mode 3 abundance in the lower cloud deck. This is consistent with a spatially variable LCB resulting from condensation and evaporation of H_2SO_4 cloud droplets in regions of

Fig. 6.9. The relation between the scattering indices $\bar{\sigma}$ and altitude for four nephelometer channels $\bar{\sigma}$ (4°), $\bar{\sigma}$(15°), $\bar{\sigma}$(45°), and $\bar{\sigma}$ (180°) in the area of the Venera-10 descent. Analysis of these data enabled us to obtain information on the structure and microphysical characteristics of clouds, according to type indicated in Fig. 6.11 from similar Venera-9 measurements. (1) is $\bar{\sigma}$(4°) · 10²; (2) is $\bar{\sigma}$(15°) · 10³; (3) is $\bar{\sigma}$(45°) · 10⁴; and (4) is $\bar{\sigma}$(180°) · 10⁴.

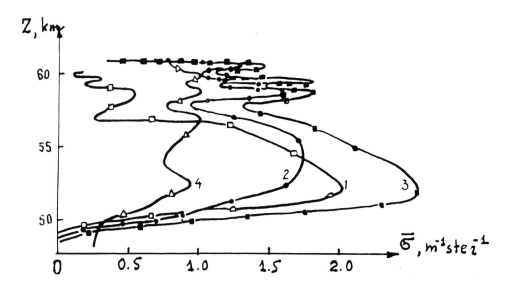

downwelling and upwelling in the turbulent atmospheric layer, which apparently directly underlies the main cloud deck (Grinspoon et al., 1993). The presence of narrow transitional layers below the LCB is an additional indication of stratification instability, relative to the small subcloud region, where the H_2O content changes rapidly with altitude and the potential temperature gradient is not equal to zero.

We shall now turn to the microphysical properties of the clouds and shall analyze their constituent particles, based on available experimental data. The aerosol parameters, which were calculated using the technique described earlier and which satisfy the nephelometer data from Veneras 9–14, are presented in Table 6.2. Taking into account the initial assumption that the particle spectrum is represented by a gamma size distribution, the cloud aerosols, from the point of view of light scattering, are characterized by the effective radius r_{eff} and effective distribution width V_{eff}. For layer I, the measured characteristics may be satisfactorily represented by a single-mode particle size distribution curve, whereas starting at layer I′ and over all layer II, a deviation from a single-mode distribution is observed. Hence, it follows that there are at least two particle fractions with different r_{eff} in the middle cloud layer. In layer III, where the measured angular light scattering characteristics undergo most vigorous changes over a broad range of values, it was difficult to interpret the nephelometer data by assuming anything other than a single-mode distribution, although there were indications of deviation from a single-mode state. It proved possible to interpret the measurement data by assuming a broader particle size distribution u(r) in layer III than in layers I and II, and this gave a more acceptable approximation. Such a distribution corresponds to a low α value ($\sim$2–3) in Eq. (6.1). To sum up, from nephelometer results on Venera-9 and Venera-10, it was concluded that cloud layer I is predominantly populated by particles with effective radii of 1 μm, layer II by particles with effective radii of 0.3 μm (mode 1) and 2–3 μm (mode 2), and layer III by particles with effective radii of 1–1.5 μm (Marov et al., 1976a, 1976b, 1980).

With a particle spectrometer used during the descent of the Pioneer-Venus Large probe (Knollenberg and Hunten, 1980), three particle modes were distinguished, and

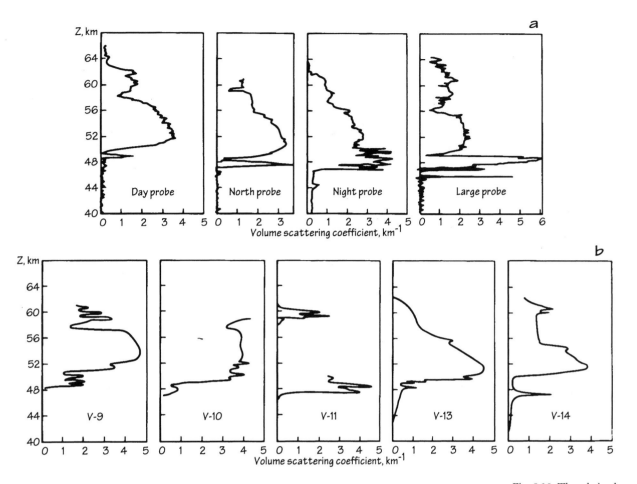

Fig. 6.10. The relation between the volume scattering coefficient and altitude in nine descent regions (the Pioneer-Venus Day, North, Night, and Large probes, and Venera 9–11, 13, and 14). In each region a basic three-layer cloud structure is detected (at times the lower layer III combines with the middle layer II); below layer III a very narrow transitional layer of cloud precursors is sometimes detected.

the concentration of each mode was determined in the upper, middle, and lower layers (Table 6.3). According to these data the fundamental mode is mode 2, which essentially determines the microphysical properties of all three cloud layers; its modal radius is 1–1.5 μm, which is in good agreement with the modal radius r_m, corresponding to r_{eff} from the Venera nephelometer data (see Eq. [6.4]). The narrowness of this mode's distribution also corresponds to the small V_{eff} value according to nephelometric data (Eq. [6.4′]), although it is somewhat masked by the presence of mode 1.

Mode 1, represented by submicron-sized particles with mean radii of 0.30–0.40 μm, like mode 2, enters into the composition of each of the three cloud layers, as well as the composition of the above-cloud haze and possibly the subcloud haze. Its presence in layer I is not identified directly from the nephelometric data, whereas its presence in layer II apparently causes the bimodal state mentioned earlier. It seems likely that mode 2 particles grow from finer mode 1 particles because sulfuric acid vapor forms during photochemical and diffusion processes, and parent molecules are transported from below. The maximum diffusion increase occurs, according to estimates, at an altitude of 60 km, leading to a very narrow size distribution of mode 2 particles (α up to ~20). A study of the mechanism for the formation and destruction of such particles (Young, 1977) suggests very long lifetimes at the cloud levels in Venus' atmosphere.

The strong broadening of the distribution in layer III may be governed by an increase

Table 6.2 Microphysical characteristics of aerosols in Venus' cloud cover

Layer	Effective radius r_{eff}, μm	α	Effective width of distribution[a]	Number of particles N, cm^{-3}	Index of refraction n	Volume scattering coefficient σ, km^{-1} [b]	Optical depth τ^b	Mass concentration ρ_{a}, mg $\times$ m^{-3}
Upper I	1.0–1.4	8–20	0.091–0.043	200–350	1.46±0.01	2–7	7–20 (0.1–4.4)	1.0–2.8 (2–16)
Middle II	1.7–2.5	8–20	0.091–0.043	250–350	1.35±0.02	3–5 (0.8–4.7)	26–27 (10–23)	5.5–11
Lower III	1.7–0.9	2–4	0.2–0.143	50–150	1.41±0.01	1–5 (0.2–4.7)	6–12 (5–7)	2.0–0.45
Subcloud haze	0.15	2–12	0.2–0.067	2	1.7	0.1–0.5 (0.03–0.2)	3 (<3)	—

Note: Data are from nephelometric measurements from the Venera 9–14 spacecraft.

[a] Assuming a gamma particle distribution with respect to size

$$\mathrm{u}(r) = r^{\alpha}\exp\left(-\alpha\,\frac{r}{r_{\mathrm{m}}}\right); \qquad r_{\mathrm{eff}} = \frac{1}{G}\int_{0}^{\infty} r\pi r^{2}\mathrm{u}(r)\,dr; \qquad V_{\mathrm{eff}} = \frac{1}{Gr_{\mathrm{eff}}^{2}}\,(r - r_{\mathrm{eff}})^{2}\pi r^{2}\cdot\mathrm{u}(r)\,dr,$$

where r is the radius of the particles; r_{m} is the modal radius, $r_{\mathrm{m}} = r_{\mathrm{eff}}/\left(\left(1 + \dfrac{3}{\alpha}\right); V_{\mathrm{eff}} = \dfrac{1}{\alpha + 3}\right.$

[b] The σ and τ values from Venera 13–14 data are presented in parentheses.

Table 6.3   Microphysical characteristics of Venusian cloud particles

Layer	Mean radius, μm	Number of particles N, cm^{-3}	Index of refraction n	Volume scattering coefficient σ, cm^{-1}	Optical depth τ	Mass concentration ρ_a, mg $\times$ m^{-3}	Single scattering albedo ω
Upper I	Mode 1: 0.2	1500	1.55–1.45	0.2–1.6	6.0–8.0	1.0	0.990
	Mode 2: 1.0	50					0.997
Middle II	Mode 1: 0.5	300					0.9992
	Mode 2: 1.25	30	1.42–1.38	0.3–4.2	8.0–10.0	6.2	0.9997
	Mode 3: 3.5	10					
Lower III	Mode 1: 0.2	1200					
	Mode 2: 1.0	50	1.33–1.44	0.2–13.5	6.0–12.0	20.0	0.9995
	Mode 3: 4.0	50					
Subcloud haze	0.1	220	—	—	0.1–0.2	—	0.998

Note: Data are from measurements made with the particle spectrometer on the Pioneer-Venus Large probe.

in the relative contribution of large particles and may also serve as qualitative evidence of a deviation from a single-mode distribution. The presence of a relatively small amount of particles with a modal radius as large as 7–8 μm was discovered in a particle spectrometer experiment on the Pioneer-Venus Large probe. In layer III they form a noticeable peak, whereas in layer II and the intermediate layer II' there are no such particles, which suggests that their appearance may be associated with condensation and destruction conditions at the lower cloud boundary, at temperatures of 340–365 K. In fact the situation is not so simple, because a peak of mode 2 particles clearly stands out independent of this peak, and no correlation is found between them. Taking this and other considerations into account, these particles were classified separately as mode 3 (Knollenberg and Hunten, 1980). Nevertheless, one should not exclude the possibility that the most significant fraction of cloud particles could still be considered a tail of the primary mode 2 distribution, and not a separate mode 3, with its own distribution. Such a proposition was put forward, based on well-founded considerations of possible errors in the calibration technique of the particle spectrometer, by Toon et al. (1982). Yet, later results of additional calibration appear to exclude this possibility (Knollenberg et al., 1984).

The index of refraction n of the particles and their change between layers, obtained from Venera-9 and Venera-10 measurements, are presented in Table 6.2. In layer I, the best model proved to be the one with $n = 1.46 \pm 0.01$. Within the limits of intermediate layer I' it was found that the index of refraction decreases to 1.42 ± 0.01 and then drops sharply to 1.33–1.35 in layer II. This result was surprising because water has the same index of refraction, although the presence of water as a candidate for the cloud aerosols has been completely ruled out. In layer III, the index of refraction again proved to be high, equal to $n = 1.41 \pm 0.01$ (see Fig. 6.2c).

Nephelometric measurements on the Pioneer-Venus Large and small probes were interpreted in conjunction with aerosol measurements using the particle spectrometer on the Large probe. The n values from the measurements were, on the whole, in good agreement with the results presented in Table 6.2. Moreover, a higher upper limit, $n = 1.45$–1.55, was obtained for the upper layer; this is associated with the possible influence of mode 1 particles, where sulfur, possessing a high index of refraction ($n \approx 1.9$), enters into the composition. It was also independently shown that the Venera result, concerning the low index of refraction in the middle cloud layer, was not erroneous: Pioneer-Venus data were interpreted independently to show a sharp decrease in n to a value ≤ 1.33 (Ragent and Blamont, 1980).

The question of why such a decrease in n occurs has brought about lively discussion. A model was proposed as one possible variant for this layer, according to which $n = 1.33 \pm 0.02$ for large particle modes and $n = 1.42 \pm 0.01$ for fine particle modes. It was difficult, however, to support such a model with any independent estimates or to find an acceptable "companion" for H_2SO_4 upon which to base it. In another model, absorbing particles were assumed to be present in layer II, with an imaginary part to the complex index of refraction k no greater than 10^{-3}. Here the assumption of a low n value is not required, and $n = 1.42$ could be accepted for both modes of cloud particles. However, this contradicted spectrometry data (Moroz et al., 1979, 1983a), which rule out the presence of any appreciable absorption in the region near 0.9 μm, which corresponds to the working wavelength of the nephelometer radiation source. A hypothesis concerning the presence of crystal particles (mode 3), in the form of thin flakes, has also been widely discussed; in this case the particles would possess the requisite n value. At the same time, if there were such particles, the mass concentration of particles in the clouds would prove to be unrealistically high (Marov et al., 1980; Knollenberg et al., 1980). It was also suggested that if drops of sulfuric acid are completely coated with sulfur, then their optical scattering properties will correspond to those observed and will not require us to assume low n values to interpret the nephelometer data (Young, 1983).

Simulations of the spatially varying emission at 1.7 and 2.3 μm on the night side observed with the NIMS experiment during the Galileo fly-by have been used to investigate the composition of the mode 3 particles responsible for the opacity variations. These simulations resulted in the conclusion that mode 3 particles cannot be composed of pure sulfur or any substance that absorbs less strongly than H_2SO_4. It was also found, however, that coated particles with a non-absorbing core material composing up to 50% of the volume of mode 3 particles cannot be ruled out by this data set (Grinspoon et al., 1993).

What is the concentration of each particle mode in the upper, middle, and lower layers? The nephelometric method provides an estimate of some mean concentration of particles N with effective dimensions r_{eff}, which govern the optical properties of the medium; however, in the event of a multi-modal distribution $n(r)$, it is difficult to indicate how many particles of an individual mode contribute to light scattering. Mean N values, based on Venera-9 and Venera-10 measurements, are presented in Table 6.2. Approximately analogous estimates were obtained from measurements with the Venera-13 and Venera-14 spacecraft, with a somewhat greater concentration range in layer III (Marov et al., 1983).

The fundamental mean cloud characteristics in the area of the Venera-9 descent and their variation with altitude (r_{eff}, n, σ, N, ρ_a, τ) are shown in Table 6.2 and in Fig. 6.11. These values on the whole satisfactorily characterize the properties of the clouds and the variations in their optical density, and may be considered an approximation for a more detailed model, which includes concentrations of particles from each mode. As a basis for the model, presented in Table 6.3, values are set, which were specified in the descent area of the Pioneer-Venus Large probe (Knollenberg et al., 1980; Simino, 1982; Esposito et al., 1983) and which, generally speaking, represent local cloud characteristics. They are basically in accord with the data summarized in Table 6.2. The actual N values are variable within distinct limits, as follows from variations in $\sigma(Z)$ in various explored regions of the atmosphere (see Fig. 6.10). The size distributions of cloud particles in individual tiers, from Pioneer-Venus data, is shown in histogram form in Fig. 6.12, and their distribution with altitude is shown in Fig. 6.13.

For a mean particle concentration in the clouds $N \approx 250$ cm^{-3}, corresponding to nephelometric measurements, the mean distance between particles is 1.5 mm, and its ratio to particle size (analogous to the Knudsen number) turns out to be $\sim 10^{-3}$ (Marov et al., 1976b). Thus, the clouds on Venus are quite a rarefied formation; the visibility is on the or-

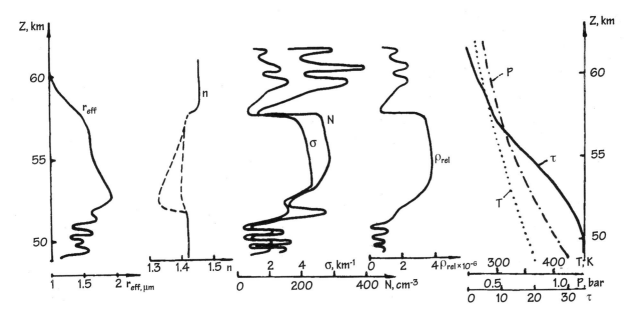

Fig. 6.11. Fundamental mean cloud characteristics in the region of the Venera-9 descent and their change with altitude: the effective particle radius r_{eff}, the index of refraction n, the volume coefficient of scattering σ, the number of particles N, and the mass concentration ρ_a. The change in the optical thickness τ along with the temperature T and pressure P profiles in Venus' atmosphere are shown in the rightmost graph.

der of 1–2 km in the clouds. Minor concentrations with a small characteristic particle size provide a small mass concentration of condensate in the clouds. Assuming that particles in all layers are homogeneous and are a concentrated (80%) solution of sulfuric acid with a density of $\rho^* = 1.6$ g/cm^3, we can obtain values for the mass concentration $\rho_a = 2 \cdot 10^{-9}$ g/cm^3 in the upper layer, $3 \cdot 10^{-9}$ g/cm^3 in the middle layer, and $5 \cdot 10^{-9}$ g/cm^3 in the lower layer. The mean cloud concentration is thus about $3.0 \cdot 10^{-9}$ g/cm^3, or 3 mg/m^3 (see Tables 6.2 and 6.3). This corresponds to the concentration $n_{H_2SO_4}$ and altitude trend in accordance with the model in Fig. 5.20. The relative concentration of condensate in the middle portion of the clouds ($Z = 55$ km), that is, with respect to the mass of the surrounding gaseous medium, turns out to be $\rho_{rel} = 4 \cdot 10^{-6}$ (see Fig. 6.11).

Essentially analogous values were obtained by Surkov et al. (1981, 1982) from the results of X-ray radiometric analysis and measurements of the vertical structure of cloud aerosol using the instrument shown in Fig. 6.5, during Venera-12 and Venera-14 experiments. These results are shown in Fig. 6.14.

Values for the volume scattering coefficient σ, along with the optical depth of the atmosphere for aerosols τ in each layer and in the subcloud haze, are presented in Tables 6.2 and 6.3. According to Venera-9 and Venera-10 nephelometric measurements, and taking into account actual oscillations in the optical density, the σ values are (2–7) km^{-1} in the upper layer, (3–5) km^{-1} in the middle layer, and (0.05–5) km^{-1} in the lower layer. According to the Venera-13 and Venera-14 backscattering data (for the same model of the aerosol microphysical properties), the ranges of variation in σ in these layers are different: the minimum σ values were obtained in the upper layer and maximum values in the lower layer. According to data from the Pioneer-Venus Large probe, σ is 0.2–1.6 km^{-1} in the upper layer, 0.3–4.2 km^{-1} in the middle layer, and 0.2–13.5 km^{-1} in the lower layer.

The mean value τ of the clouds is 25–50 according to Venera data and 25–35 according to Pioneer-Venus data (see Tables 6.2 and 6.3). At the time of the Galileo fly-by of

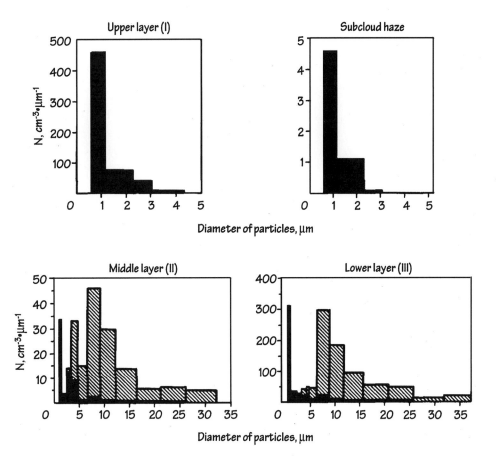

Fig. 6.12. Histograms of the particle size distribution in the upper, middle, and lower cloud layers and in the subcloud haze. The particle concentration per unit volume is shown along the y-axis. The bimodal nature of the particle distribution is most clearly seen in the middle and lower layers; however, the presence of mode 3 (larger-scale) particles is also assumed. The upper layer is represented by modes 1 and 2, although the presence of mode 1 (the finest particles) and, consequently, the layer's bimodal nature are not clearly detected at all altitudes. The subcloud haze is represented by a single-mode particle distribution, with average sizes less than the size of mode 2 particles (according to Knollenberg and Hunten, 1980).

Venus, the total cloud optical depth varied spatially from 25 to 40 (Grinspoon et al., 1993). In Table 6.3, single scattering albedo values ω in the clouds are also presented, which are in good agreement with the Venera spectrophotometry data (Golovin and Ustinov, 1982; Ekonomov et al., 1983). This quantity, which is the ratio of the scattering coefficient to the extinction coefficient $\sigma/(\sigma + k)$, characterizes the degree of absorption in the medium, specified by the absorption coefficient k. In a physical sense ω is the probability of quantum survival. In the clouds of Venus $\omega \rightarrow 1$; that is, absorption is negligible and the clouds' optical properties are almost entirely determined by scattering. Note also, that at $Z > 48$ km τ is essentially independent of λ, whereas in the haze above the clouds it increases with a decrease in λ.

The Subcloud Haze

The subcloud haze begins immediately below the LCB and is characterized by a backscattering index $\bar{\sigma}$ of $(0.05–0.1) \times 10^{-4}$ m^{-1}sterad^{-1}, which suggests that it is highly rarefied. The extent of the subcloud haze, according to Venera 8–10 and Pioneer-Venus probe data, is approximately 15 km, so that its lower boundary is found at an altitude of 31–35 km. Yet in Venera-13 and Venera-14 measurements, the lower haze boundary was recorded at an altitude greater than 41 km (at a signal level corresponding to a backscattering index of 0.05×10^{-4} m^{-1}sterad^{-1}). This corresponds to a volume scattering coeffi-

Fig. 6.13. A distribution of concentrations with respect to particle mass in the clouds as a function of altitude, according to measurements using particle spectrometers on the Pioneer-Venus Large probe. In addition to the particle concentrations of the two fundamental modes (1 and 2), mode 3 particles, whose existence remains uncertain, are also detected (according to Knollenberg and Hunten, 1980).

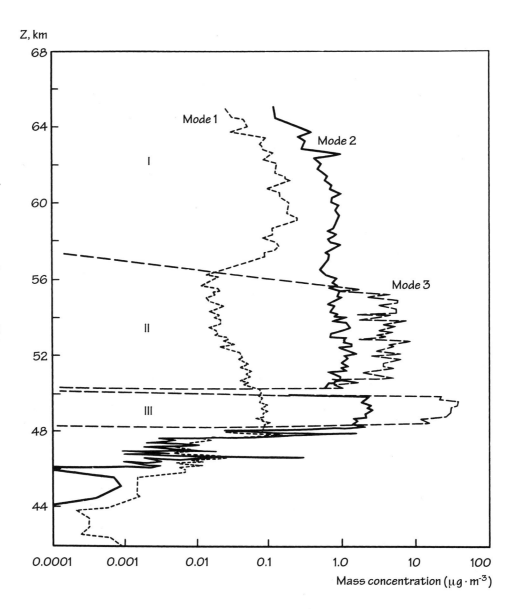

cient $\sigma = (0.03 - 0.2)$ km^{-1} and apparently reflects actual temporal or spatial variations in the position of the lower haze boundary. The τ value for the subcloud haze was determined to be ≤ 3.

The light scattering characteristics in the subcloud haze, according to nephelometric measurement data, are best described based on a particle model with effective radii of approximately 0.15 μm (Marov et al., 1980, 1983). The results of spectrophotometry (Golovin and Ustinov, 1982; Ekonomov et al., 1983) do not contradict these estimates. The spectrophotometry results contain indications that the haze sometimes extends to a depth of 10 km and even to 5–7 km. The possible contribution to scattering from aerosols is estimated in the lower atmosphere by a τ value in the 0–2.5 range. Note that at individual probe landing sites, the presence of large dust particles ($\sim$10–100 μm in diameter) is observed in the near-surface layer of the atmosphere. These particles probably cover the

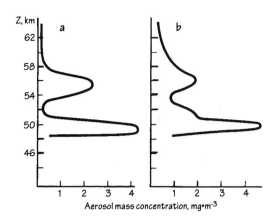

Fig. 6.14. An altitude distribution of aerosols in Venus' clouds according to the results of measurements using an X-ray radiometric analyzer on the Venera-12 (a) and Venera-14 (b) landers (according to Surkov et al., 1982).

surface in these areas and were most likely thrown up into the atmosphere by the impact of the spacecraft on the surface.

The Formation and Phase State of the Clouds

Study of the formation and maintenance of the unique stability in Venusian clouds is linked with problems in meteorology, geochemistry, and planetary evolution. Cloud kinetics are characterized by a complex set of photochemical, phase-transition, diffusion, and dynamic processes. An indirect reflection of these processes are short- and long-period variations in the visible cloud boundary, as recorded from Earth, including contrast changes in the ultraviolet spectral region on a scale of several hours to several months, variations in the composition of the above-cloud haze over several years, and the complex variability of structures in the polar regions. In all probability cyclical processes for producing particles dominate within the clouds, through photochemical oxidation of SO_2 and hydration of SO_3 at the upper boundary, with subsequent condensation of H_2SO_4 down to the cloud base, below which there is decomposition of the condensate and SO_2 regeneration.

Mode 2 is most closely identified with a concentrated H_2SO_4 solution. According to estimates (Turco et al., 1982), droplets of this mode exist for the longest period of time (up to several months) in the upper cloud layer, whereas their lifetime in the lower layer apparently does not exceed several hours. As a result, H_2SO_4 vapor at high altitudes turns out to be in a state of severe supersaturation, promoting the formation of condensation nuclei and consequently a bimodal distribution, preserved over the entire thickness of the clouds. Condensation occurs most actively near 60 km, where the maximum concentration of mode 1 particles is observed. These particles have been identified with H_2SO_4 droplets and with solid sulfur particles, and they possibly also enter the composition of the above-cloud haze. Theoretical models indicate that condensation may continue to the lower boundary of the clouds, partially compensating for the destruction of aerosol droplets due to evaporation. The lower haze is probably also represented by nuclei that remain after evaporation, which are similar to mode 1 particles and are transported downward owing to vertical diffusion processes; the extent of these nuclei below the LCB is determined by the intensity of these processes.

It is natural to assume that the size stability of basic mode 2 particles, and consequently the narrowness of their size distribution, are caused by competing processes, which vary the saturation of vapor above the surfaces of the difference curve (the Kelvin effect), and by an increase in the concentration of the acid solution. The increase in acid concentration evidently prevents the growth of droplets through coagulation, as occurs in terrestrial clouds (Rossow, 1978; Esposito et al., 1983). Given the general sedimentation stability of the clouds (precipitation is compensated by convective transport upward), one could imagine that drops of condensate gradually consolidate within the middle and lower layers. Their intense evaporation occurs at the lower cloud boundary (Marov et al., 1980).

Stratiform clouds apparently dominate the optical and mass structure of cloud aerosol over the extent of Venus; this differentiates them from terrestrial clouds. Another important difference is that heat exchange processes are mainly controlled not by the liberation of heat of condensation but by radiative transfer. Moreover, clouds are involved in zonal circulation at altitudes corresponding to the maximum wind velocity and the maximum vertical component of atmospheric motions. The existence of H_2SO_4 aerosols conforms to chemical condensation laws: concentrated drops of H_2SO_4 adapt to the new equilibrium state. The unbroken planetary cloud covering is thus maintained by the rapid circulation and low volatility of its constituent particles.

The kinetics for the formation of aerosols, consisting of an H_2SO_4 solution at various concentrations, are directly connected with the problem of the water vapor content and its variation with altitude within the existing cloud structures; this is predetermined by the high chemical affinity of H_2SO_4 for water. Physical-chemical analysis enables us to give theoretical estimates of the change in sulfuric acid concentration in the droplets and the corresponding mole fraction of water vapor in vertical profile, and to compare them with the measurement data (see Figs. 5.19 and 5.20).

Young (1975) obtained an estimate for the mean relative H_2SO_4 concentration by mass of $5 \cdot 10^{-6}$, based on a cloud model consisting of sulfuric acid drops of 75% concentration, $r = 1.1$ μm in size at the upper boundary of the clouds ($P = 50$ Mbar, $T = 250 \pm 10$ K), and considering the optical characteristics of Venus at the reflection level. If, however, we take into account the considerable scatter (by more than two orders of magnitude; see Chapter 5) in the instrumental data for determining the water vapor content, then two cloud condensate variants are possible: given $X_{H_2O} = 4 \cdot 10^{-3}$ the relative mass content of H_2SO_4 is $5 \cdot 10^{-3}$ (the humid model), and given $X_{H_2O} = 10^{-5}$, it is $3 \cdot 10^{-6}$ (the dry model). In both variants the lower boundary of the clouds was defined as the evaporation level of the sulfuric acid drops. For the humid model, this boundary corresponded to an altitude of 26 km, and for the dry model it was 44 km. Hence H_2SO_4 rain droplets, if they existed, could not reach the surface in either case. Note that on this basis, an attempt was made to link the start of evaporation of drops of the appropriate radius with an unusual inversion in atmospheric density near the $T = 400$ K level (as appeared in measurements using an ionization density gauge on the Venera-4 descent module; see Avduevsky et al., 1968a), and with the nature of absorption in the microwave wavelength range ($\lambda = 1$–3 cm).

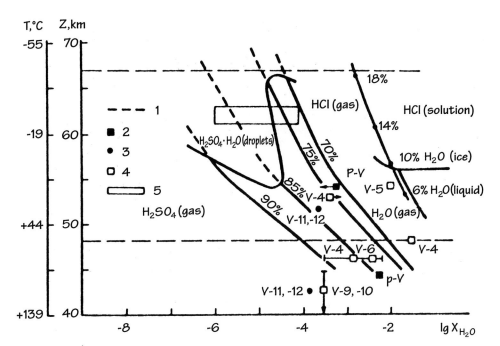

Fig. 6.15. A diagram of phase ratios for sulfuric acid, hydrochloric acid, and water for Venus' troposphere in $\lg X_{H_2O}\text{-}T(Z)$ coordinates. Solid lines represent a family of H_2SO_4 curves at various concentrations (70–90%); the dashed lines are their continuation above the freezing curve. The solid curve with dots at the right corresponds to the diagram of state for HCl. Horizontal dashed lines (1) limit the altitude range within which the three fundamental cloud layers are located. The various designations identified at left denote results of measuring the H_2O content, correlated to the appropriate axes on the graph: (2) Oyama et al., 1979; (3) Moroz et al., 1979; (4) Vinogradov et al., 1970; (5) Barker, 1975b. V-4 . . . V-12 are the Venera spacecraft; P-V is Pioneer-Venus (according to Volkov et al., 1979, 1989).

Based on physical-chemical analysis of the H_2SO_4–H_2O diagram of state, Pollack et al. (1978) showed that under conditions in Venus' atmosphere almost all the water is present in a gaseous phase, not in the composition of aerosols. At the upper boundary of the clouds the relative H_2O content by volume was estimated to be $2 \cdot 10^{-6}$, in good agreement with Young's model, and $X_{H_2O} = X_{H_2SO_4}$. It was also found that at all cloud levels the H_2SO_4 concentration in droplets is controlled by the concentration of water vapor, with the exception of the upper boundary ($Z = 67$–70 km), where there is photochemical generation of H_2SO_4 from sulfur-bearing gases.

Volkov et al. (1979; 1989) conducted a detailed analysis of phase ratios in the H_2SO_4–H_2O system and compared them with available data for determining the water vapor in the atmosphere. A diagram in $\lg X_{H_2O}$-$T(Z)$ coordinates is shown in Fig. 6.15, where X_{H_2O} is the mole fraction of $H_2O_{(gas)}$, T is the temperature, and Z is the altitude above the planet's surface. Here $T(Z)$ and $P(Z)$ profiles are adopted from the atmospheric model of Marov and Ryabov (1972). The values of the mole fraction X_{H_2O} satisfy the relation $X_{H_2O} = P_{H_2O}/P$ for ideal gases, and the partial pressures P_{H_2O} satisfy the equilibrium condition with a concentrated H_2SO_4 solution (70–90%). The family of curves, obtained from this data, of the form $H_2SO_{4(solution)}$–$H_2SO_{4(gas)}$ intersects the freezing curve. Using this phase diagram it is easy to determine X_{H_2O} in the gaseous phase under equilibrium conditions with H_2SO_4 droplets of various concentrations; the corresponding experimental data on the water vapor content at various altitudes serve as a limitation for the sulfuric acid aerosol models.

The high $H_2O_{(gas)}$ abundances in the cloud layer, measured by the Venera 4–6 and Pioneer-Venus spacecraft, are well in line with partial pressure values for saturated water vapor over 75–85% H_2SO_4 solutions. The upper threshold for the existence of an H_2SO_4

Fig. 6.16. Altitude profiles of H_2SO_4, SO_3, and H_2O concentrations in Venus' troposphere, constructed according to laboratory studies of the H_2SO_4–H_2O system (according to Ayers et al., 1980; Esposito et al., 1983).

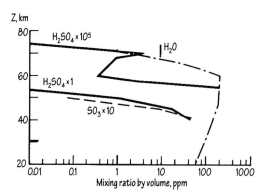

solution at 70–75% concentrations coincides with the upper boundary of the basic cloud layer (~68 km), corresponding to the freezing curve. At the same time, H_2SO_4 solutions with concentrations exceeding 75% already freeze at 58 km. At these and higher altitudes, however, supercooled solutions may exist (at temperatures approximately 50 K lower), corresponding to the dashed curves.

The maintenance of supercooled sulfuric acid droplets in a liquid state at concentrations up to 84–86% is supported by spectral and polarimetric observations. According to Young (1975), interactions of H_2SO_4 with HCl and HF in gaseous phases could promote supercooling processes, for example, in accordance with reaction (5.45). Incidentally, such a process is capable of explaining the marked prevalence of chlorine over fluorine in the Venusian troposphere (see Table 5.5), despite the approximate parity in the cosmic abundance of these elements.

The position of the lower boundary of the cloud layer in the 47–50 km altitude interval is also in good agreement with the physical-chemical properties of the H_2SO_4–H_2O system. The evaporation level of H_2SO_4 droplets is determined by the $H_2O_{(gas)}$, $H_2SO_{4(gas)}$, and $SO_{3(gas)}$ concentrations. The H_2SO_4 and SO_3 vapor concentrations in Venus' troposphere are shown in Fig. 6.16, calculated by Ayers et al. (1980), for the H_2O system, proceeding from the assumption that at the 46 km level ($T = 379$ K, $P = 1.71$ bar), the concentration of H_2SO_4 solution reaches 96% (the H_2O concentration profile corresponds to the experimental data). It is obvious that at 40 km ($T = 416$ K) the $H_2O_{(gas)}$ and $H_2SO_{4(gas)}$ partial pressures essentially coincide, whereas the H_2SO_4 vapor concentration reaches extremely high values (~150 mg $\cdot$ m^{-3}) in the zone of thermal dissociation. Unfortunately, we have not been able to determine the concentration of products from the thermal dissociation of sulfuric acid ($H_2SO_{4(gas)}$, $HSO_{3(gas)}$) by mass spectrometer and gas chromatograph. An estimate of the mass concentration of H_2SO_4 condensate at the 48 km level gives a value of ~10 mg $\cdot$ m^{-3}, which, generally speaking, does not contradict the value (~5 mg/m^3) obtained from an analysis of microphysical characteristics in the lower cloud layer (see Tables 6.2 and 6.3).

Summing up, we arrive at the most likely picture for phase processes in Venus' atmosphere and for the formation of its clouds. At altitudes $Z > 60$ km, mode 1 aerosol ($r = 0.2$ μm) forms from H_2SO_4 vapor. In deeper cloud levels (to $Z = 51$ km) there is a preferred increase in mode 2 droplets ($r = 1–1.2$ μm). The concentration of the H_2SO_4 droplets decreases from $\sim$90% at 68 km to 82% at 51 km; in the lower level ($Z = 47–51$ km) evaporation begins (the H_2SO_4 concentration in the drops increases), and the droplet size starts to decrease. The $H_2SO_{4(gas)}$ concentration in the 31–48 km interval, however, is sufficient for recondensation of the sulfuric acid in the subcloud haze down to an altitude of $Z = 31$ km. The lifetime of H_2SO_4 droplets in the upper level of clouds (58–70 km) is estimated at several months ($\sim$10^7 sec) and in the lower level, several hours ($\sim$10^4 sec).

The results of direct analysis of the cloud aerosol composition by the method of X-ray fluorescence analysis, undertaken on the Venera-12, Venera-14, and Vega-1 landers, proved to be contradictory. According to Venera-12 data, the mass concentration of sulfur in the aerosol is 0.1 mg $\cdot$ m^{-3} in the 47–54 km altitude interval (and possibly even less), whereas chlorine is an order of magnitude greater, 0.43–2.1 mg $\cdot$ m^{-3} (Table 6.4). Therefore the sulfur concentration turned out to be at least an order of magnitude less than the mass concentration of aerosols in the middle and lower cloud levels, determined by nephelometer and particle spectrometer data from Venera and Pioneer-Venus. Subsequent experiments on the Venera-14 and Vega-1 spacecraft furnished much more encouraging results: the mass concentration of sulfur is 1.1 and 2.0 mg $\cdot$ m^{-3}, and chlorine is 0.16 and 0.3 mg $\cdot$ m^{-3}, respectively, that is, in inverse correlation (see Table 6.4).

The Vega-1 and Vega-2 experiments, using gas chromatographs and mass spectrometers, were the most convincing. From the gas chromatograph data we succeeded in establishing that the mean H_2SO_4 content is 1 mg $\cdot$ m^{-3} in the 63–48 km range. Analysis of mass spectra of gaseous pyrolysis products from the large-scale fraction of cloud particles ($r \geq 1.5$ μm) indicated the presence of a distinct peak at 64 amu, identified as SO_2^+. In other words, the aerosol contains a condensate that breaks down under heating, with the release of SO_2; this serves as additional evidence of the presence of sulfuric acid.

Genesis of the Sulfuric Acid Aerosol and Possible Contaminants

The genesis of H_2SO_4 and elemental sulfur is most likely a single cyclical atmospheric process and can be described by a set of photochemical and thermochemical reactions, which we considered in detail in the previous chapter. We arrived at a scheme for the genesis of H_2SO_4 droplets (Fig. 6.17). Recall that in the above-cloud atmosphere and upper layer of clouds, according to this scheme, oxidation of SO_2 into SO_3 occurs with the assistance of atomic oxygen, produced either upon CO_2 photolysis with the participation of OH and HO_2 catalysts-radicals (which, in turn, are brought about by HCl photolysis) or directly during SO_2 photolysis (Winick and Stewart, 1980). Essentially all $H_2SO_{4(gas)}$ results from gas phase processes, such as $SO_3 + H_2O$, at altitudes of $Z > 60$ km, that is, where SO_2 photolysis is possible. Further hydration of SO_3 occurs with a subsequent diffusion increase in H_2SO_4 droplets. In the hot subcloud atmosphere, thermal dissociation of H_2SO_4 aerosols occurs, and CO reduces SO_3 to SO_2, in this way closing the rapid atmospheric

Table 6.4   Condensed particles in Venus' clouds

Component	Measurement intervals, km	Condensation (theoretical estimate), km	Mass concentration of aerosol, mg × m^{-3}	Spacecraft	Reference
H$_2$SO$_4$ (solution)	62.0–48.0	70.0–48.0	3.0	Venera 9–10	Marov et al., 1976b, 1980
	70.0–56.5	—	1.0	Pioneer-Venus	Knollenberg & Hunten, 1980
	56.5–50.5	—	5.0	Pioneer-Venus	Knollenberg & Hunten, 1980
	62.0–54.0	—	1.0–2.0	Vega 1	Surkov et al., 1987d
	63–48.0	—	1.0	Vega 1–2	Gelman et al., 1986
S-bearing component	54.0–47.0	—	0.10±0.03	Venera-12	Surkov et al., 1981
	54.0–47.0	—	<0.1	Venera-12	Petryakov et al., 1981
	63.0–47.0	—	1.10±0.13	Venera-14	Surkov et al., 1982
			2.0	Vega-1	Surkov et al., 1987d
S$_{(c,\,l)}$[a]		~48	~0.6		Volkov et al., 1982
Cl-bearing component	54.0–47.0	—	2.10±0.06	Venera-12	Surkov et al., 1981
	54.0–47.0	—	0.43±0.06	Venera-12	Petryakov et al., 1981
	63.0–47.0	—	0.16±0.04	Venera-14	Surkov et al., 1982
			~0.3	Vega-1	Surkov et al., 1987d

[a] c is crystalline; l is liquid

sulfur cycle (reactions [5.24–5.26] or [5.27–5.30]) in accordance with reactions (5.31) and (5.31′).

Thus, high-temperature thermochemical reactions with the assistance of products of chemical decomposition of H$_2$SO$_4$ into H$_2$O and SO$_3$ are the primary source of SO$_2$. However, there may also be an influx of SO$_2$ from the near-surface atmosphere in turbulent gas flows.

Along with reactions (5.24)–(5.30), Sill (1983) proposed an alternative mechanism for generating H$_2$SO$_4$ aerosols. According to this scheme, SO$_2$ may be oxidized by nitrogen compounds, as was used in an old industrial technique for producing H$_2$SO$_4$:

$$2NO + O_2 \rightarrow 2NO_2 \tag{6.9}$$

$$SO_2 + NO_2 \rightarrow SO_3 + NO \tag{6.10}$$

$$SO_3 + H_2O \rightarrow H_2SO_4 \tag{6.11}$$

Fig. 6.17. Cyclical processes in Venus' atmosphere at the cloud levels with the participation of sulfur compounds. The diagram illustrates the processes of genesis and destruction of sulfuric acid drops. The dashed line indicates a hypothetical process with the participation of mode 3 cloud particles, whose existence is subject to doubt (according to Knollenberg and Hunten, 1980).

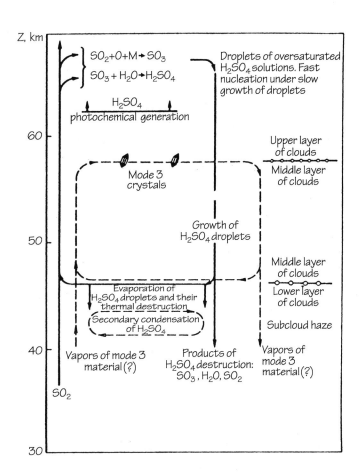

The mechanism for forming nitrogen oxides (for example, through electrical discharges in the atmosphere) is quite hypothetical; moreover, neither NO_2 nor NO was discovered in the above-cloud or the subcloud atmosphere, at upper limits of tenths and hundredths ppm (see Table 5.5).

The situation is approximately analogous with another theoretically possible mechanism for the generation of H_2SO_4 (and S_n) condensates from COS and H_2S during the slow atmospheric cycle of photochemical (in the middle and lower cloud layers) and thermochemical (in the subcloud atmosphere) reactions (5.33)–(5.38), along with (5.24) and (5.26). This mechanism is based on the idea of cyclical transformations of sulfur between compounds with reduced (H_2S, COS, S_n) and oxidized (SO_2, SO_3, H_2SO_4) properties.

The different variants for producing S_n as a by-product of H_2SO_4 formation for SO_2 and COS, and for producing condensation of aerosols from gaseous sulfur vapor, were studied by a series of researchers (see, for example, Prinn, 1979; Young, 1979; Dorofeeva et al., 1981; Volkov, 1983; Esposito et al., 1983). On the whole, these ideas are consistent concerning the fact that condensates of elemental sulfur are formed in regions of the upper layer of clouds, depleted in oxygen, in ascending turbulent currents during photochemical reactions according to:

$$3SO_2 + 2H_2O \xrightarrow{h\nu} S + 2H_2SO_4 \qquad (6.12)$$

In this case the rate of SO_2 transformation into S and H_2SO_4 depends on the water vapor and oxygen content. Condensation of particles is most likely on H_2SO_4 droplets upon supersaturation of S_8 octamer vapor. In addition to what was said concerning the upper boundary of the clouds and the above-cloud haze, note that the absorption phenomenon in the ultraviolet range and formation of contrasts on Venus' disk may be linked not with condensates but with gaseous allotropes S_3–S_4 (which absorb strongly at wavelengths with centers near 0.4 and 0.53 μm), which do not survive more than several days. The most likely absorber along with SO_2, however, is amorphous sulfur, which forms as a by-product of the whole transformation cycle of sulfur-bearing compounds in the atmosphere and clouds. This occurs even though the sulfur particles themselves are a negligible portion of the overall mass of the upper cloud layer. Liquid or solid mixtures in mode 1 and mode 2 sulfuric acid aerosols are the most likely forms of sulfur.

Because the identity of the UV absorber has not been determined, and because compounds other than those considered here cannot be ruled out, Titov (1983) advanced an exotic hypothesis: ammonium pyrosulfite $(NH_4)_2S_2O_5$ may be formed as a result of SO_2 interaction with traces of NH_3 (for $X_{NH_3} < 10^{-6}$) in the upper cloud layer ($Z > 58$ km). Such an assumption seems quite artificial, if we keep in mind the extremely small probability of even traces of ammonium in Venus' atmosphere (at a level of several ppb as an upper limit; see Table 5.5).

We shall now return to the problem of possible sulfuric acid aerosol contaminants, considering the possibility of coexistence of materials of different chemical nature and the consistency with the available experimental data. The question of possible hydrochloric acid formation in aerosol clouds, based on gas-phase HCl discovered in Venus' atmosphere, has been discussed in the literature for many years. Lewis (1968, 1971), having analyzed the corresponding phase diagrams for the HCl–H_2O system with the measured atmospheric T and P values, found that the spectroscopically determined equilibrium concentration of HCl and H_2O in the gas phase corresponds to a 6-mole solution of HCl (25% concentration by weight). Here the temperature over the solution is about 200 K and the coefficient of refraction of the aerosol particles is $n \approx 1.42$, in satisfactory agreement with polarimetry data. The spectral characteristics of the HCl aqueous solution at the given temperature also turned out to be in good agreement with the Venusian spectrum between 3 and 4.5 μm, although certain features could have been explained by the presence of hydronium ions in the acid, which exhibit a broad absorption band in the $\lambda = 3$–5 μm region and longer-wave range (Kuzmin and Marov, 1974). And although definite discrepancies were discovered in the estimates of thermodynamic parameters, following from analysis of the phase diagrams and those which correspond to the formation level of the corresponding spectral lines, the hypothesis about hydrochloric acid aerosols has not been definitively rejected. Young (1973) also later came to the conclusion that HCl droplets possibly enter into the composition of the main cloud deck.

After the discovery of chlorine in the composition of cloud aerosol by the Venera spacecraft, in amounts of 0.2–2.0 mg $\cdot$ m^{-3}, it became necessary to reconsider the possibility of HCl droplet formation. An $HCl_{(gas)}$–$HCl_{(solution)}$ phase curve is shown in Fig. 6.15

in $T(Z)$–$\lg X_{H_2O}$ coordinates, corresponding to various concentrations of HCl solution for a constant HCl mole fraction in a gaseous phase. To construct this diagram, Volkov et al. (1979) used an initial family of isopleths of $HCl_{(solution)}$ (6–26%) concentrations and isopleths of X_{HCl} (10^{-5}–10^{-7}) in T–$\lg P_{HCl}$ coordinates. From the intersections of these two families of curves, the $\lg P_{H_2O}$–T relations were found for the indicated range of hydrochloric acid concentrations. The various $HCl_{(solution)}$ concentrations are indicated by individual points on the $HCl_{(gas)}$–$HCl_{(solution)}$ phase curve. In Fig. 6.15 only one $HCl_{(gas)}$–$HCl_{(solution)}$ curve is presented, corresponding to the spectroscopic estimate $X_{HCl} = 10^{-6}$. Equilibrium condensation of concentrated ($\sim$10–18%) hydrochloric acid drops in Venus' clouds cannot be ruled out theoretically, starting from an altitude of $\sim$57 km, but could exist only given extremely high water vapor concentrations ($X_{H_2O} \geq 10^{-3}$). Although such concentrations have been recorded (see Table 5.5), more recent measurements suggest that the water vapor content is significantly lower. The long-term survival of HCl droplets in the presence of sulfuric acid aerosols is unlikely: hydrochloric acid aerosols must rapidly dissociate as a consequence of significant differences in the water vapor pressure over the H_2SO_4 and HCl solutions.

Volkov et al. (1979) also estimated the thermodynamic possibility of condensation of a series of liquid sulfur and chlorine compounds. Under terrestrial surface conditions (on Venus, this corresponds to an altitude of 50–52 km), many of these substances are hydrolyzed with an explosion! It has been shown that at $T = 298$ K, $P = 1$ atm, equilibrium mole fractions of the gas phase of six liquid Cl and S compounds ($SOCl_2$, SO_2Cl_2, S_2Cl_2, H_2S_2, $SO_2(OSO_2Cl)_2$, and HSO_3Cl) exceed the mole fraction estimates of H_2SO_4 and H_2O by several orders of magnitude or are commensurate with them. Consequently, liquid droplets of possible chlorine and sulfur compounds cannot coexist with sulfuric acid aerosol. Moreover, approximate estimates of the partial vapor pressure were made for $SOCl_2$ and SO_2Cl_2 according to the equilibrium constants of their hydrolysis and the probable partial pressure values P_x for H_2O, SO_2, SO_3, and HCl in the cloud layer (at $P = 1$ atm, $T = 300$ K); it was shown that $P_{SO_2Cl_2}$ and P_{SOCl_2} are 10–15 orders of magnitude lower than the corresponding partial pressure values for these components in a state of saturation. Thus, liquid chlorine and sulfur compounds are thermodynamically unstable over all altitudes where the presence of chlorine was recorded in the composition of cloud aerosol.

In a similar way, also based on thermodynamic considerations, Volkov et al. (1982, 1989) analyzed the possibility of chlorine condensation in Venus' atmosphere. They proceeded from the assumption that equilibrium condensation of the aerosol is reached at a certain altitude level in the troposphere, the process goes to completion, and the mole fraction of the gas corresponds to "frozen" chemical equilibrium at the planet's atmosphere-surface interface. It was proposed, moreover, that reserves of these elements in the outer shells of Earth and Venus are the same, and that during de-gassing of the crust they completely transform into a gas phase. Calculations were carried out for elements whose mixing ratios in the troposphere are tentatively considered sufficient, so that during condensation, along with Cl, cloud particles form with a mass concentration ≥ 0.1 mg $\cdot$ m^{-3}

Table 6.5   Thermodynamic estimates of the possibility that certain volatile components will condense in the troposphere and on Venus' surface

Gas[a]	Mole fraction of gas[b]	Condensation level in the troposphere, km[c]	Condensed phase on the surface[c]
AS$_4$, AS$_4$O$_6$, <u>AsS</u>, As$_2$, AsCl$_3$, AsO, As	$\leq 10^{-7}$	—	As$_2$S$_{3(l)}$, As$_4$S$_{4(l)}$
<u>SbS</u>, SbCl$_3$, Sb$_4$O$_6$, SbCl, Sb$_4$ Sb$_2$, Sb, SbO, SbCl$_5$	$\leq 3 \times 10^{-8}$	—	Sb$_2$S$_{3(c)}$, Sb$_4$O$_6$
SO$_2$, COS, H$_2$S, <u>S</u>$_n$	$X_{S_2} \simeq 1.8 \times 10^{-7}$	≥ 48, S$_{(c, l)}$	FeS$_2$ (pyrite), CaSO$_4$ (anhydrite)
<u>Se</u>$_2$, Se, SeCl$_2$, SeO$_2$, SeO	$\leq 10^{-5}$	≥ 18, Se$_{(l)}$	Se$_{(l)}$
Te$_2$, TeOCl$_2$, Te, TeCl$_2$, TeO$_2$ TeO, TeCl$_4$	$\leq 10^{-7}$	≥ 18, Te$_{(c)}$	Te$_{(l)}$
<u>Hg</u>, HgCl$_2$, HgCl	$\leq 10^{-8}$	≥ 62, Hg$_{(c)}$	—
<u>PbCl</u>$_2$, PbCl, Pb, PbO	$\leq 1.2 \times 10^{-8}$	≥ 16, PbCl$_{2(c)}$	PbS (galena)
<u>FeCl</u>$_2$, FeCl$_3$	$\leq 10^{-12}$	—	—
<u>Al</u>$_2$<u>Cl</u>$_6$	$\leq 10^{-39}$	—	—

Source: Data from Volkov et al., 1989.

[a] At the start of the series the dominant gas phase is shown; theoretically possible condensates in the clouds are underlined.

[b] The sensitivity limit of modern techniques for analyzing the chemical composition of the atmosphere is $X_i \leq 10^{-7}$.

[c] l is liquid; c is crystalline.

(which must satisfy a mole fraction of condensing gas on the order of 10^{-7}–10^{-6}, corresponding to the sensitivity limit of modern instruments). Compounds of chlorine with 47 chemical elements were considered, and the stability of condensed phases at the surface was studied. The dominant form of gaseous material was established, and its equilibrium mole fraction in the near-surface troposphere was computed; the altitude level of condensation was also estimated according to the vapor saturation curve of the dominant gas. As a result it was concluded that condensation of chlorides of practically all elements can be ruled out either because of thermodynamics or because the equilibrium mole fraction of the gaseous component $X_x \leq 10^{-7}$ cannot be recorded by existing methods. In particular, this conclusion pertains to chlorides of ammonium, iron, aluminum, and mercury, identified earlier as possible cloud layer condensates (see Table 6.5). Only antimony oxychloride (SbOCl, which crystallizes from the gaseous phase of SbCl$_3$) could be a thermodynamically stable compound; however, according to Lewis and Fegley (1982), the dominant Sb-bearing gas in Venus' troposphere should more likely be a sulfide of antimony SbS (which is stable even at the surface), not a chloride. Moreover, calculations provide a strict limit on the concentration of Sb-bearing gases and exclude antimony compounds from the number of possible condensates in Venus' clouds. The same can be said for arsenic.

In connection with the morphological and chemical features of cloud particles, the existence and nature of mode 3 in the lower cloud layer remain problematic. The supposition by Knollenberg and Hunten (1980) concerning highly elongated, crystalline particles, 7–8 μm in size, does not satisfy the limitation on the mass concentration of aerosols in the clouds. The hypothesis concerning chlorides, attractive at first glance, encounters difficulties that result from an analysis of the chemical processes in the atmosphere and clouds. In an attempt to circumvent these difficulties, and because of the need to satisfy the coexistence of hypothetical particles of this mode and a concentrated solution of sulfuric acid, Watson et al. (1979) proposed the possible formation of solid nitrosylsulfuric acid ($NOHSO_4$) particles in the clouds in the presence of nitrogen oxides, which may interact with H_2SO_4 aerosols according to the following reaction:

$$NO + NO_2 + 2H_2SO_4 \rightarrow 2NOHSO_4 + H_2O$$

The vulnerability of the given hypothesis, however, like the possible mechanism for generating sulfuric acid according to reactions (6.9)–(6.11), is the lack of any noticeable concentrations of nitrogen oxides in the atmosphere. From this point of view, Sill's (1983) idea about the formation of solid $NOHSO_4$ particles surrounded by a liquid shell of H_2SO_4 solution also seems unconvincing.

Finally, a number of other possible contaminants in the Venusian clouds, such as mercury, selenium, and tellurium, have at various times been mentioned. A hypothesis concerning mercury clouds was advanced by Lewis (1969), based on complete de-gassing of mercury from the planet's crust into the atmosphere, and also on equilibrium condensation of its vapor with the formation of liquid Hg and crystalline HgS, $HgCl_2$, and so on. It was shown that if the Venusian and terrestrial reserves of this element are equal, then the mole fraction X_{Hg} will be 10^{-5}, while liquid and solid Hg condensates, as follows from thermodynamic analysis of its phase state, would be stable over the whole cloud zone. Direct measurements (Surkov et al., 1981), however, showed that the Hg content in the aerosol does not exceed 0.05 mg $\cdot$ m^{-3}, which corresponds to $X_{Hg} \leq 10^{-8}$. Somewhat earlier an analysis of radiowave absorption on Mariner-10 (Kliore et al., 1979) led to the conclusion that the measurement data are incompatible with the existence of mercury in Venus' cloud layer. Theoretical estimates (Volkov et al., 1982) also showed that the threshold mercury concentration in the clouds is 3.5 orders of magnitude lower than the concentration that would exist in the troposphere, assuming complete Hg de-gassing from the crust.

Liquid selenium and tellurium, according to data from thermodynamic analysis, are stable on Venus' surface (Volkov et al., 1982, 1989). The pressure of saturated vapor over them, within one order of magnitude, corresponds to values obtained assuming that the entire reserve of Se and Te ($X_{Se} = 5 \cdot 10^{-6}$; $X_{Te} = 10^{-7}$) de-gasses from the crust, and that the gas condensation levels for these elements do not exceed 18 km from the surface. Thus, if condensed selenium and tellurium particles actually exist in Venus' atmosphere, then their calculated concentration would be no less than 200 cm^{-3}, given the radii of the latter $r \approx 1$ μm. In fact, below 47 km only individual rarefied layers of aerosol are observed with a calculated concentration of 20 cm^{-3}. Consequently, actual mole fractions of Se_2 and

Te_2 in the gas phase are several orders of magnitude lower than estimated from terrestrial analogy; this again can be explained by incomplete de-gassing of Se and Te from Venus' interior.

The unusual (by our terrestrial standards and persistent perceptions) chemical composition of Venus' clouds, and the possible presence of exotic contaminants, are naturally predetermined by the striking difference in the thermal regime of Venus' atmosphere from that of Earth, and the lack of a water cycle on the planet. At the same time, Venus has clearly been volcanically active and almost certainly remains so. But on Earth, sulfur compounds, halogens, and other mixtures ejected by volcanoes in abundance, which accompany the primary de-gassing products, CO_2 and H_2O, for the most part were washed out of the atmosphere and dissolved in the oceans. On hot Venus, where there are no oceans, these mixtures condensed in the atmosphere at those altitudes where temperature conditions allowed. The flow of gases into the atmosphere continues to the present on Earth and, very likely, on its sibling planet, Venus. Venusian clouds from this point of view are analogous to terrestrial oceans.

In conclusion, we note that the study of Venus' clouds is especially important in connection with the increasing urgency of the purely terrestrial problem of defending the environment from pollution. The discharge of industrial wastes into the atmosphere, leading to the formation of smog, has many serious consequences: air pollution exceeding the health norms in city basins, the destruction of ecological equilibrium, the catastrophic effects on architectural monuments, and so on. One of the reasons for the appearance of stable smogs is the rise in atmospheric sulfur dioxide, its subsequent oxidation to SO_3, and hydration to form sulfuric acid droplets. In contrast to normal fogs, these droplets do not disappear in sunshine; on the contrary, they become more pronounced owing to photochemical transformation. Venusian clouds are, apparently, similar to these types of smog. Therefore, it is necessary to understand the fundamental physical-chemical and dynamic processes in the clouds. This understanding is directly linked to highly accurate analysis of the minor components participating in reactions, the appropriate chemical constants, and the dynamic constants for such important processes as condensation, coagulation, and sedimentation of cloud particles at various levels. This is one of the most important applied problems in the study of Venus.

7

Planets, like all cold bodies (in contrast to hot, emitting stars like our Sun), absorb and reflect radiation. Absorbed solar energy is converted into other forms of energy, but eventually it must all be released from the system through re-radiation back to space, thus maintaining heat balance on the planet. In terms of energy exchange, the two main planetary subsystems—a solid (or liquid) surface and a surrounding gaseous atmosphere—are closely interrelated and therefore can be considered a single unified system driven by solar energy. The energy transmitted through a planetary atmosphere is accompanied by scattering and absorption from the beginning of its arrival at the outer edge of the atmosphere to its ultimate absorption by the planetary surface. Absorption may occur over the whole range of wavelengths, or it may be concentrated in several wavelength bands. Absorbed energy is a source of internal energy of atmospheric gas in the form of molecular motion (of which temperature is an indicator) or kinetic energy, which is a function of wind velocity and is eventually dissipated through friction. These two forms of energy are balanced with gravitational potential energy of gas molecules, which is a function of height over the planetary surface.

We shall introduce the discussion of atmospheric radiative transfer with a reminder of some basic physics. If a body radiates at the greatest possible intensity (radiation from a unit surface, say 1 cm^2) for all wavelengths, it is termed a blackbody, or perfect radiator. This term refers to an ideal body capable of absorbing and emitting equal parts of radiation, independent of color. Radiation from real bodies is usually poorly approximated by a blackbody analog, though a blackbody can sometimes provide a reasonable match over a certain spectral band. The spectral distribution of blackbody radiation is specified by the Planck radiation law. Radiation intensity is strongly dependent on temperature and has a definite maximum at some wavelength: the higher the temperature, the shorter this wavelength. The respective mathematical expressions for these two statements are referred to as the Stefan-Boltzmann and Wien laws for a blackbody.

It is, for example, easy to estimate that the maximum radiation from our Sun, which closely resembles a blackbody at a temperature of about 5700 K, lies in the visible part of the spectrum near 0.55 μm, and the bulk of the solar energy occurs at wavelengths from 0.4 to 0.7 μm. At the same time, much colder planets mainly radiate in the infrared beginning at about 6 μm for Mercury. On the average, the intensity of this longwave planetary radiation is defined by the effective temperature T_E (introduced in Chapter 1) and in particular is strongly affected by the thickness of the gas envelope for those planets possessing atmospheres. For Earth, Venus, and Mars, the strongest emission occupies a part of the spectrum around 10–14 μm (the whole length ranges from 5 to 50 μm) and it shifts to the far infrared for the outer solar system, reaching a maximum at nearly 100 μm for Pluto.

For the terrestrial planets and their atmospheres, radiation from the Sun is essentially the only source of energy. Depletion of the solar beam increases with distance traveled through the atmosphere on the way to the solid surface. The amount of radiation received at a point on the ground (insolation) is strongly affected by the elevation angle of the Sun in the sky and the amount of atmospheric mass through which incident solar radiation must pass. For the planet as a whole over an entire year, however, the disposal of the incoming solar energy remains nearly unchanged (for a given climate stability) and can be evaluated as portions of the energy received by a planet from the Sun. The quantity that Earth receives is the solar constant E_s (see Eq. 1.1). Because the Earth's Bond albedo, within some uncertainties, is $A = 0.31$, the remaining 0.69 of the incoming radiation must be absorbed. Radiation is generally reflected from clouds (0.23) and solid aerosols (0.08), while from the remaining part 0.22 is absorbed in the atmosphere (water vapor, clouds, dust, and ozone) and 0.47 is absorbed on the ground (see, for example, Riehl, 1978).

That solar incident radiation on Earth is mostly absorbed at the surface can be easily explained by the large transparency of its atmosphere in the visible part of the Sun's spectrum, which allows the radiation to pass through nearly unaltered. The directly heated surface acts as the immediate heat source for the lower atmospheric levels. The heat balance of the atmosphere and ground is mainly controlled by water vapor and carbon dioxide, although these are only trace constituents of Earth's atmosphere. These gases selectively absorb longwave radiation in certain bands. Clouds contribute substantially to this process, absorbing thermal radiation at Earth's temperatures very effectively (and thus radiating as blackbodies). Both atmospheric gas absorbers and clouds limit the heat flow from the surface and reduce radiative cooling of the ground and near-surface air, causing the temperature to rise. This trapping of solar radiation is known as the greenhouse effect, which is in particular responsible for an increase of mean temperature at Earth's surface by about 30° C. Indeed, without this built-in energy-storing mechanism in the atmosphere, our planet would have a mean temperature of slightly below freezing.

A secondary source of energy is provided by longwave re-radiation from the atmosphere back to the surface. The Stefan-Boltzmann law states that the amount of radiation from a blackbody is proportional to the fourth power of the absolute temperature, so at the mean surface temperature Earth would radiate to space energy greater than the solar constant by some 15%. But in reality the heat balance is accomplished under T_e at the top of Earth's troposphere and is mainly controlled by the abundances of the main atmospheric absorbers (water vapor and carbon dioxide) and clouds. In other words, with such absorbers and clouds present, the zone of net radiation loss is shifted from the ground upward to the cloud top and to the top of the moisture layer.

It is of great interest to compare the situation for Earth with the attenuation of shortwave and longwave radiation and energy balance on Venus, which has a huge atmosphere mainly composed of the efficient infrared absorber CO_2 and an enormous cloud deck high in the troposphere. Much information on this extreme environment has become available recently and will be discussed in more detail in the following sections of this chapter, with the main goal of explaining the unique greenhouse environment on this planet. Another extreme is Mars, which possesses a very thin atmosphere also composed of CO_2. Its albedo is close to that of Earth, although its atmosphere is even more transparent to incident solar radiation when there are no severe dust storms, and thus most of this solar flux, including shorter ultraviolet wavelengths, reaches the surface. Nonetheless, even the thin CO_2 atmosphere of Mars creates a noticeable greenhouse effect responsible for some increase of the global mean temperature.

In addition to radiation, energy in the atmosphere can be also transferred from one place to another by thermal conduction and dynamic mass motions. The role of conduction is negligible in atmospheric heat exchange, except at a very narrow boundary layer near the surface. The characteristic modes of mass motions are convection, advection, and turbulent diffusion, in which energized molecules transfer bodily from one location to another. Convection is the most important mode of vertical transport, whereas advection is related to horizontal mass exchange due to circulation. Natural convection develops as the result of differences in atmospheric density, whereas circulation and forced convection related to it are precipitated by external forces. Turbulent motions may contribute significantly to heat exchange.

Before we discuss various patterns and specific phenomena of the thermal regime and atmospheric dynamics on Venus, we shall address the general problems of heat balance on a planet in more detail. We have seen that the thermal balance is determined by that portion of solar radiation reaching the surface, and by reflection and re-radiation of solar and thermal radiation by the surface, atmosphere, and clouds (Fig. 7.1a). Because of the radiation penetrating below the clouds, a radiative influx of heat into the atmosphere is ensured; this is a constantly operating non-adiabatic factor. In other words, the absorption of solar energy, as it penetrates deep into the planet's atmosphere, must cause an instability in the thermal flux with altitude, in contrast to stellar atmospheres (see Chandrasekhar, 1950; Goody, 1964; Kondratyev, 1969). The amount of energy release in the medium (or the energy loss), due to the interaction with radiation, is determined by the divergence of the radiation flux, so that comparison of the radiative energy fluxes may characterize the role of radiative heat exchange in the medium.

If a regime of radiative equilibrium is realized in the atmosphere (as is often assumed, for example, in Earth's stratosphere), then the amount of solar radiation flowing in must be compensated for, at every level, by outflowing radiation. More strictly speaking, the rate of solar radiation inflow and the cooling rate of the atmosphere due to infrared radiation must in this case be equal. Mathematically, this is expressed by the divergence of the overall radiation flux being equal to zero or, in other words, by the derivatives of the solar (F) and thermal (W) fluxes at a given level in the atmosphere being equal. Because from the thermal conductivity equation we have

$$\rho C_p \left(\frac{dT}{dt} \right)_{SOL,IR} = \frac{d(F,W)}{dZ}$$

(ρ is the gas density, C_p is the heat capacity given constant pressure), then it follows that under conditions of radiative equilibrium, aside from $dF/dZ = dW/dZ$, $(dT/dt)_{SOL} = (dT/dt)_{IR}$ also holds. If, by contrast, the condition for radiative-convective equilibrium is

Fig. 7.1a. Radiative energy transfer in a planetary atmosphere.

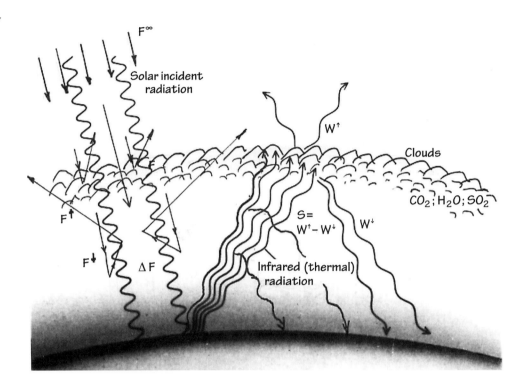

Fig. 7.1b. Quasi-monochromatic solar radiation flow versus altitude in five spectral intervals from Venera-10 lander measurements. Curves $\Phi 1$–$\Phi 5$ correspond to the spectral characteristics of channels $\Phi 1$–$\Phi 9$, shown in Fig. 2.39 (lower right), for maximum values 0.52 μm, 0.59 μm, 0.65 μm, 0.72 μm, and 0.96 μm, respectively.

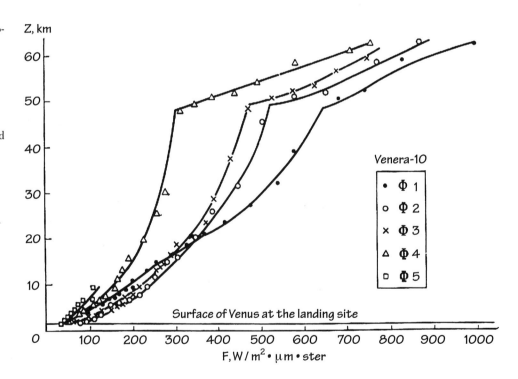

fulfilled (which is usually the case in the troposphere), then the influx of solar energy at each level in the atmosphere is compensated by the overall thermal flux due to outflowing thermal radiation and convective transfer. These alternatives serve as local approximations to the more complete picture of heat exchange on a planetary scale, in which dynamic processes on various spatial scales play a key role. To describe these processes here, it is necessary to draw upon the complete system of hydrodynamic equations, along with the equations for radiative transfer. The contribution of radiative heat exchange (loss or inflow of energy) is accounted for in the energy equation of the hydrodynamic system by the use of an additional term, which represents the divergence of the radiative flux at each level in the atmosphere. In other words, radiative energy flux is added to the overall hydrodynamic energy flux (this is discussed in more detail in Zeldovich and Raizer, 1966; Kuzmin and Marov, 1974; and Marov and Kolesnichenko, 1986).

Planetary dynamics reflect the balance between the rate of generation of potential energy due to solar radiation and the rate of loss of mechanical energy due to dissipation. From this point of view, the planet's atmosphere is often compared to a heat engine, where the equatorial regions serve as the heater and the poles as the cooler. The efficiency of such an engine is small; it does not exceed several percent. So the source of motions on various spatial scales is the inequality between the incoming and outgoing energy in individual regions of the planet, given the strict realization of thermal balance on a global scale, characterized by the effective temperature (see Chapter 1). In other words, the formation of horizontal temperature gradients as a consequence of differential heating must be compensated for by the development of large-scale motions over a broad spectrum of spatial sizes.

These considerations lead us directly to key questions, which must be answered in order to understand the fundamental features of the thermal regime in Venus' atmosphere. These questions concern the shape of the altitude profiles for F and W, how they are formed in the atmosphere, and what the fundamental features of radiative heat exchange are. A comparison of these profiles and their derivatives, in turn, makes it possible to compare the role of radiative heat exchange and dynamic processes, and, from available observational data, to formulate ideas concerning the mechanism of planetary circulation, which plays a most important role in the thermal regime. Ultimately, all this provides an approach for answering the question, What causes Venus' unusually high surface temperature and its peculiar climate?

Before proceeding with an analysis of radiative fluxes and atmospheric dynamics, it is useful to make certain preliminary estimates that characterize the thermal inertia of the atmosphere and the intensity of diurnal temperature variations. It is easy to show that, given the measured parameters in Venus' atmosphere, the difference between temperatures on the day and night sides at the surface must be extremely small. Actually an estimate of the heat reserve in a gas column of unit cross section can be obtained from the following obvious relation:

$$i_{\male}^{\Sigma} = \int_{(T)} C_p(T)\, dT = \frac{5}{2} k \int_0^{\infty} T(Z) \mathcal{N}(Z)\, dZ$$

Here i is enthalpy, C_p is the specific heat capacity at constant pressure, $\mathcal{N}$ is number density of particles, and k is the Boltzmann constant. The actual value turns out to be equal to $i_{\male}^{\Sigma} \approx 4.2 \cdot 10^{14}\, \mathrm{erg} \cdot \mathrm{cm}^{-2} \approx 10^4\, \mathrm{kcal} \cdot \mathrm{cm}^{-2}$ (the same value on Earth is almost three orders of magnitude less). At the same time the heat loss over a Venusian night ($t^n = 58.4$ terrestrial days) is $\sigma T_e^4 \cdot t^n \approx 10^{12}\, \mathrm{erg} \cdot \mathrm{cm}^{-2} \approx 24\, \mathrm{kcal} \cdot \mathrm{cm}^{-2}$; that is, it does not exceed 0.25% of

Table 7.1   Characteristic temperatures and thermal relaxation times

	$m\,C_p,$ $J/(m^2 \cdot$ degree)	T_e, K	T, K	$\tau_{tr},$ sec.	$\tau_{re},$ sec.
Venus	10^9	230	500	10^{10}	2×10^{10}
Earth	10^7	250	$\sim T_E$	10^7	2×10^7
Mars	10^5	200	$\sim T_E$	3×10^5	6×10^5

the energy present in the atmosphere and cannot in any way be significantly reflected in the thermal balance of the planet. The maximum diurnal temperature variations in Venus' lower atmosphere, taking into account the incident solar radiation F absorbed by the planet, accordingly must not exceed

$$\Delta T = \frac{T/2}{i_\varphi^\Sigma/F^\infty \cdot t^n} \approx 1 \text{ K}$$

A similar estimate of the diurnal temperature change at the surface may be obtained from the relation $\Delta T = (g\sigma T_e^4 t^n)/(P_S C_p)$, where P_S is the pressure at the planet's surface. This is completely consistent with available experimental data (see Chapter 5).

The values presented here are directly connected to two parameters used to characterize the thermal regime of a planet's atmosphere: (1) the thermal relaxation time $\tau_{tr} = mC_p\overline{T}/\sigma T_e^4$, which characterizes the timescale for temperature inhomogeneities, with respect to radiation perturbations, to equalize, and (2) the radiative heat exchange time $\tau_{re} = C_p\overline{T}/q(H)$, which can be considered the characteristic timescale for energy transfer upward, ensuring compensation of radiative losses at the upper boundary of the atmosphere (see Gierasch et al., 1970; Golitsyn, 1970; Leovy and Pollack, 1973). Here m is the mass of a unit column of atmosphere, q(H) is the cooling rate per unit mass, from a level on the order of the scale height H (that is, approximately corresponding to 1/3 m), and $\overline{T}$ is the mean atmospheric temperature with respect to depth. If one takes the efficiency of atmospheric radiation to be $\varepsilon \approx 0.2$, then $\tau_{tr} \approx (mC_p\overline{T})/(0.5\sigma T_e^4) \approx 2\tau_{re}$. The values of τ_{tr} and τ_{re} for Venus, in comparison with similar estimates for Earth and Mars, are presented in Table 7.1. For Venus, both parameters exceed analogous values for Earth and Mars by approximately 3 and 5 orders of magnitude, respectively. This suggests that the formation of thermal inhomogeneities and their equalization in Venus' lower atmosphere is extremely inefficient, and also that the process of heat output from depth is extremely long in duration, in the event of a deviation from thermal equilibrium at the emission level for whatever reason.

Incidentally, another curious estimate follows from this. If Venus were suddenly to stop receiving solar energy (let us say that it was somehow screened from the Sun), then its cooling to the heat content of the terrestrial atmosphere $i_\oplus^\Sigma \approx 24.5$ kcal $\cdot$ cm^{-2}, due to radiative heat loss (corresponding to the nighttime heat loss on Venus), would take approximately 65 terrestrial years, all other parameters remaining fixed. This is of course a rough estimate, which does not take into account changes in the opacity of the atmosphere during cooling, changes in planetary circulation, or the disruption in the equilibrium conditions that formed the gaseous shell of the planet, accompanied by changes in its chemical composition.

Solar Radiation

In Chapter 1 the characteristics of the radiation field of Venus, including fundamental features of its thermal balance, were presented. But these integral properties say essentially nothing about how the planet's thermal regime below the radiating layer, located near the observable cloud boundary, is formed, or about the features of radiative heat exchange and atmospheric dynamics.

Naturally, the fundamental question is how that portion of solar energy F^∞ not reflected back into space, specified by Eq. (1.1) given an integral Bond albedo for Venus of $A = 0.76$, is distributed in the clouds and subcloud atmosphere. Venera-8 was the first spacecraft to enable us to answer this question (Avduevsky et al., 1973a, 1973b; Marov et al., 1973c), having provided an estimate of the solar energy flux at the surface and over the extent of the clouds (see Fig. 6.4). The relationships between quasi-monochromatic solar radiation flux and altitude are presented in Fig. 7.1b for five intervals in the visible and near-infrared spectral regions (0.45–1.1 μm) from Venera-10 measurements (Avduevsky et al., 1976b; Marov, 1978, 1979). Quite similar results were obtained by Venera-9. These results significantly broadened our notions concerning the illumination intensity and, moreover, provided information about the relationship between radiation attenuation and wavelength. Much additional information, concerning the contribution of minor components to attenuation and the characteristics of the radiation field, was provided by similar measurements with a higher spectral resolution in the same wavelength range by the Venera 11–14 spacecraft (Moroz et al., 1981, 1983b; Ekonomov et al., 1983) and in one narrow (near 0.63 μm) and two broad (0.4–1.8 μm) spectral channels on the Pioneer-Venus Large probe (Tomasko et al., 1980).

The curves in Fig. 7.1b clearly indicate a break in the monotonic trend (a jump in the derivative of solar intensity with altitude) at an altitude of 49 km, which corresponds to the position of the lower cloud boundary. Note that this position proved to be in amazing coincidence with the start of Venera-8 measurements and, on the whole, was correctly interpreted in Fig. 6.4. The degree of non-monotonicity increases somewhat in the longwave portion of the spectrum, and approaching the surface attenuation in the shortwave range, grows. Hence it follows that near the surface red rays dominate and, apparently, give Venus' sky and its landscape a reddish (or more likely an orange) hue.

Figure 7.2 shows the relationship between the descending solar radiation reaching the planet's surface, averaged by the least squares method using all available measurements, and the solar zenith angle, according to the VIRA model (Moroz et al., 1985a). The flux is averaged over the spectrum and corresponds to the value that would be measured by a

Fig. 7.2. Bolometric flow of solar radiation at the planet's surface as a function of solar zenith angle θ, from the set of all available measurements (according to the VIRA-85 model; Kliore et al., 1985).

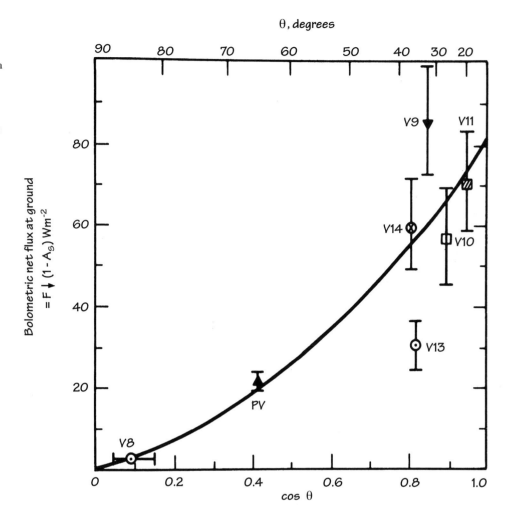

bolometer located on the surface. It is obvious that this is entirely scattered radiation; direct solar radiation is not visible on the surface because scattered radiation dominates, starting at approximately 60 km, owing to the high optical thickness. If one averages this flux over the planet's surface, for example, using Sobolev's (1972) theorem of multiple scattering in the deep layers of the atmosphere, then the amount of solar radiation falling per unit surface on Venus and cited in the VIRA model turns out to be $F_s = 16.8 \pm 2.3$ Watts $\cdot$ m^{-2}. This is somewhat more than 10% of the energy absorbed by Venus and averaged over the entire surface of the planet. Because the estimate of the surface albedo, obtained from analysis of photometric measurements and panoramas (Marov, 1978), in $A_S \approx 0.1$, almost 90% of the given flux is absorbed by the surface and re-radiated by it in the infrared wavelength range.

The VIRA model also provides the integral (over the spectrum) value of the overall solar radiation flux $\Delta F(Z)$, constructed from direct measurements and reproduced in Fig. 7.3.[1] The model continuation corresponds to a condition of thermal balance according to the results of calculations, taking into account remote measurements and the optical properties of atmospheric gases and aerosols. An analysis of this curve (for now ignoring certain features of its behavior in the subcloud atmosphere) provides a rigorous quantita-

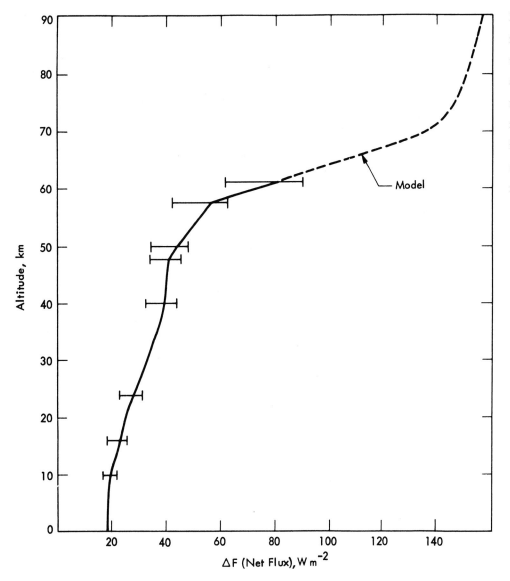

Fig. 7.3. Globally averaged flow of overall solar radiation ΔF as a function of altitude. The solid curve represents direct measurements; the dashed curve is the model extension (according to the VIRA-85 model; Kliore et al., 1985).

tive estimate of how much solar energy is absorbed at various altitudes. In particular, one can conclude that more than 65% of the solar flux is absorbed in the upper portion of the clouds and above-cloud haze (in the 60–90 km altitude range), whereas absorption in the middle and lower cloud layers does not exceed 8%. The remaining portion is absorbed by the subcloud atmosphere (about 15%) and the surface (about 11%).

Transmission Functions of Atmospheric Gases

The study of Venus' atmospheric thermal regime essentially boils down to explaining the high temperature at the planet's surface, based on the likelihood of and conditions for realizing a greenhouse effect and the effect of atmospheric motions at various spatial scales. The first task is linked with the need, first of all, to study the optical properties of the fundamental atmospheric components, which absorb infrared (thermal) radiation, or their transmission functions. The latter serve as a basis for calculating radiative heat exchange models, which, in turn, create the necessary prerequisites for understanding the role of dynamic processes, and for finding a solution to the problem of planetary circulation.

Radiation absorption in a gas is determined by several independent parameters, including the monochromatic coefficient of absorption, temperature, the overall pressure of the medium, the partial pressure of the absorbing gas, the relationship between the spectral line width and mixtures of other gases, the optical path, and the nature of the mass distribution along the optical path. The main opacity in the atmosphere is created by numerous rotation-vibration bands, for which the coefficient of absorption varies rapidly with a change in wavelength λ, or its corresponding wave number ν.[2] Clearly, a large number of individual points must be considered in order to compute the overall energy flux. Actually, for $P \approx 1$ atm the halfwidth of rotational lines is usually less than 10^{-1} cm^{-1}. Thus even for a simplified approach, assuming a constant coefficient of absorption within the splitting interval, it would be necessary to consider intervals 10^{-2} cm^{-1} in width. Consequently, in computing the integral fluxes for a spectral region corresponding to equilibrium (Planck) emission at Venus' surface temperature and encompassing a range of wave numbers from approximately 10^4 to 10^2 cm^{-1} (1–100 μm), it would be necessary to sum no fewer than 10^6 individual points. Apart from the high computational intensity, such an approach is difficult to realize practically, in connection with the considerable modification of the fine rotational band structure and the change in the functional dependence of its effective width with an increase in T and P. In this case, given a large path, absorption becomes extremely efficient at the edges of the bands, because of the influence of the wings of strong lines located in the center of the band. Moreover, at high temperatures and pressures many new features appear in the spectra of the polyatomic molecules, caused by thermally excited states and pressure-induced transitions (see Kuzmin and Marov, 1974; Pollack et al., 1993).

All this illustrates the need not simply to use various distribution models for spectral line within a band (for example, the well-known Goody [1964] statistical distributions) but to look for specialized approaches. The difficulties in calculating CO_2 and H_2O absorption spectra at high temperatures, pressures, and partial component concentrations, and the impossibility of obtaining suitable absorption coefficients under laboratory conditions, have resulted in insufficient accuracy in modeling the effective thermal flux on Venus. Initially these efforts generally reduced to extrapolations of results obtained under normal terrestrial conditions; these are extremely rough approximations. The significant progress attained in this regard over the past three decades was achieved primarily through a series of theoretical and experimental studies of the optical properties of CO_2 gas, and of a number of relatively minor atmospheric components, over a broad pressure and temperature range, by obtaining data on the influx of solar radiation into the atmosphere and the properties of Venus' clouds, and by refining the altitude profiles of atmospheric parameters, taking into account diurnal variations. At the same time, techniques for numerically modeling radiative heat transfer were developed (Marov and Shari, 1973; Kuzmin and Marov, 1974; Pollack et al., 1980, 1993), and measurements of the effective thermal fluxes in several regions of the planet were conducted (Suomi et al., 1980). This enabled us to go beyond relatively simple estimates to a more complete analysis of the thermal regime and to construct a radiative model. Within the scope of such a model, the possibility of directly verifying available data on the concentrations of minor contaminants, which exert the greatest influence on the opacity of the medium, is also ensured, based on the condition of thermal balance in the atmosphere at various levels.

In radiative heat exchange in a planetary atmosphere, it is necessary to take into account those molecules that are able to absorb radiation in the visible and infrared spectral ranges. Essentially all gases found in Venus' troposphere (or in Earth's) absorb IR radiation. The molecules of the dominant component, CO_2, play the fundamental role in radiation absorption in Venus' atmosphere. As for the second most abundant component, nitrogen, like other homoatomic molecules, does not possess intrinsic absorption bands (IR transitions for such molecules are forbidden). Nevertheless, N_2 molecules may form so-called synchronous transitions with triatomic molecules like CO_2 (this is linked with a special case of an induced absorption spectrum); therefore the effect of the indicated transitions cannot be ignored in many cases. In particular, nitrogen on Venus may provide a certain contribution at very high pressure, owing to the appearance of an induced dipole moment, especially near the $\lambda = 4.3$ μm CO_2 band and within the H_2O rotational band ($\lambda = 17$–100 μm).

Taking into account concentration and infrared absorption spectra, the following gases effectively contribute to the overall opacity of Venus' atmosphere: CO_2, H_2O, SO_2, COS, CO, and H_2S. Among the relatively minor components, H_2O and SO_2 play the most important roles. The effect of these gases on the transfer of thermal radiation is largely determined by the fact that their characteristic absorption bands do not coincide with the most intense absorption bands of a purely CO_2 atmosphere. As far as the remaining sulfur compounds are concerned, within the limits of current accuracy for computing radiative heat flow in a CO_2–H_2O–SO_2 atmosphere, their effect at the present stage can be ignored. In this regard, we shall now consider the optical properties of carbon dioxide, water vapor, and sulfur dioxide in more detail.

Carbon Dioxide

The fundamental rotation-vibration bands of CO_2 are given in Table 7.2. Owing to the lack of a constant dipole moment (with the exception of asymmetric isotopes), there is no purely rotational spectrum of normal CO_2 absorption. The bands presented in Table 7.2 possess the greatest intensity. Weaker bands also exist, especially for asymmetric isotopes.

The high CO_2 content along the beam path (on the order of 10^8 atm · cm) and the high temperature mean that Venus' atmosphere is opaque in the region of intense bands, and radiative transfer is accomplished between them in so-called windows of transparency. Absorption in windows of transparency is caused by remote wings of strong bands and by weak bands falling within the windows of transparency. These kinds of absorption complicate theoretical and experimental studies on the opacity of CO_2 gas. In theoretical calculations it is necessary to assign line contours, which are not known for high accuracy in

Table 7.2 CO_2 absorption bands

Frequency of the band center ν, cm^{-1}	9517	8293	7594	6972	6677	6267	4977	
Wavelength of the center λ, microns	1.0	1.2	1.3	1.4	1.5	1.6	2.0	
Frequency of the band center ν, cm^{-1}	3714	2613	2349	2076	1932	1063	960	667
Wavelength of the center λ, microns	2.7	3.8	4.3	4.8	5.2	9.4	10.4	15

remote wings. Weak absorption bands in windows of transparency have also been little studied, because their role in traditionally solvable problems of heat exchange is usually ignored. Major progress in this area has been made with the assembly of a high-temperature spectral database for CO_2 using the direct numerical diagonalization theory (Watson and Rothman, 1986, 1992). This database includes more than 7×10^6 lines that are members of about 6×10^4 vibrational bands. This represents an increase of more than two orders of magnitude over the HITRAN spectral database, which had been used previously. The HITRAN database (Rothman et al., 1987) was developed for terrestrial applications and does not include hot bands, making it inadequate for modeling applications in the Venus atmosphere. The new high-temperature database contains hot bands, bands of other CO_2 isotopes, and weak, ground-state transitions. This database has already been successfully applied to spectroscopic modeling in the near-infrared window regions (Pollack et al., 1993). It has not yet been applied to the problem of the radiative balance of Venus but promises to yield greatly improved results.

In laboratory measurements it is difficult to realize the high CO_2 column density, and the conditions in Venus' lower atmosphere. Therefore, the various models and approximations are of increasing importance. Pollack et al. (1980), Tomasko et al. (1980), and Revercomb et al. (1982, 1985) used a technique for extrapolating the integral transmission functions of the fundamental CO_2 absorption bands, determined by laboratory measurements (I. Howard et al., 1956). But because these measurements were conducted under terrestrial atmospheric conditions, their extrapolation to Venusian atmospheric conditions may contain errors that are difficult to control. In fact, theoretical and experimental studies of the CO_2 absorption spectrum, conducted by Galtsev and Odishariya (1970) and Galtsev and Osipov (1971), indicated that absorption in the center of the bands does not depend on pressure for values greater than about 5 atm. On the contrary, in remote wings absorption is observed to be proportional to pressure, in contrast to the findings of Howard et al. Moreover, because Howard et al. did not account for the temperature dependence, they had understated absorption values; this was shown in particular for an example of hot (9.4–10.4 μm) bands. Finally, measurements of induced absorption spectra indicated the existence of a large number of additional bands located in the near-infrared spectral region, aside from the 7.5 μm and rotational bands determined earlier by Birnbaum et al. (1971) and Burch et al. (1971). Although they are less intense, they definitely contribute to the CO_2 opacity.

All this makes it possible to suppose that the technique mentioned earlier, using integral transmission functions, may lead to understated values for the opacity of Venus' atmosphere. Marov and Shari (1973), Shari (1976), and Marov et al. (1984, 1985, 1989b) used a different approach, according to which the optical properties of the CO_2 spectrum may be found based on an improved semi-empirical method used in the spectroscopy of simple molecules. The authors first calculated the vibrational transitions whose integral intensity exceeds a certain minimal value ($\sim 10^{-8}$ cm^{-2}atm^{-1}), and then, taking into account the selection rules with respect to the rotational quantum numbers, the frequency ν and intensity S of the rotational lines were determined in the 300–800 K temperature range, in 100 K steps. Using this technique, they calculated the spectral transmission of carbon dioxide gas from 0.7 to 100 μm for a set of pressure, temperature, and concentration values, corresponding to conditions in Venus' subcloud atmosphere. Based on these calculations, an improved quasi-statistical model, proposed by Osipov and Galtsev (1971), has been adopted. The normalizing interval was taken to be 10 cm^{-1} in the range of wave numbers $100 \leq \nu \leq 4000$ cm^{-1} and 50 cm^{-1} in the range of $4000 \leq \nu \leq 12{,}000$ cm^{-1}.

Analysis of theoretical and experimental data has shown that the transmission Φ in the $\delta\nu$ interval can be expressed as

$$\Phi^{CO_2} = \exp[-(K_1 + K_2 P)\Psi^{CO_2}]$$

where $\Psi^{CO_2} = PL(1 + 0.005P)^{273/T}$ is the CO_2 content over the length of the optical path L (in cm), taking into account the dependence on the carbon dioxide gas pressure P (in atm) and temperature T (in degrees Kelvin). In a physical sense K_1 represents the mean intensity of the lines located in the $\delta\nu$ interval, and K_2 takes into account the contribution of the lines' wings, located outside the interval, which depends on the line form. Note that both these coefficients are essentially adjustment parameters, therefore K_1 may not coincide with the mean intensity of the lines and K_2 may take on negative values depending on the structure of the vibrational band.

The induced CO_2 absorption spectrum may strongly affect the transfer of thermal radiation in Venus' lower atmosphere. This can be explained by the fact that the induced absorption bands are located in a region of the spectrum where, under normal conditions, the usual absorption bands are absent, with the exception of weak bands of asymmetric CO_2 isotopes. Unfortunately, theoretical calculations of induced CO_2 absorption spectra do not ensure the required accuracy. Therefore, in estimating the optical characteristics, one must rely on available experimental data. Taking into account that the width of the rotational lines in the induced spectrum is much greater than the distance between them, the transmission in the $\delta\nu$ interval for the induced bands can be represented in the following form:

$$\Phi^{CO_2}_{ind} = \exp -(K_{ind}\Psi^{CO_2}P)$$

In this approximation, absorption coefficient values K_{ind} were calculated as a function of wavelength and temperature for three separate spectral regions: in the $18-1000$ μm ($10-550$ cm^{-1}) range, where translational and rotational absorption bands are located; in the $4.78-9.1$ μm ($2090-1100$ cm^{-1}) range, where the fundamental CO_2 band is located (forbidden in the normal spectrum by virtue of symmetry and centered on $\lambda = 7.5$ μm), as well as the 5.0 μm band; and in the $2.14-3.74$ μm ($2675-4660$ cm^{-1}) range, where several bands are located, whose maxima are centered at $\lambda = 3.74$, 3.58, 3.32, 2.46, 2.28, and 2.15 μm. These bands are caused by transitions from the ground state to three groups of interacting Fermi states. Here, because the absorption spectrum is governed by the remote wings of strong normal and induced transition bands, it contains the sum of the K_2 and K_{ind} coefficients.

Further examination of this method would require a more thorough introduction to quantum physics and molecular spectroscopy. Therefore, we shall limit ourselves to the observation that the CO_2 absorption spectrum, calculated using this technique, displayed quite satisfactory correlation with available experimental data. This spectrum was the basis for the calculations of radiative heat flow presented in the following section.

Water Vapor

Among the absorption bands of the nonlinear water vapor molecule, the 6.5 μm and 2.7 μm bands have the highest integral intensity. Because the H_2O molecule, in contrast to CO_2, has a constant dipole moment, the rotational band, located in the spectral region near 50 μm, is also active in absorption. The fundamental H_2O bands, which must be taken into account in calculating radiative heat flow in Venus' atmosphere, are presented in Table 7.3.

Detailed calculations of the water vapor absorption spectrum for various H_2O mixing ratios were conducted by Rothman et al. (1987). They are a poor match for the conditions in Venus' atmosphere, however, because in calculating the H_2O rotational spectrum (the line intensities and halfwidths), they accounted for only a limited number of vibrational states, resulting in errors at higher temperatures. Moreover, there are still no reliable data on the contour of water vapor lines, which may cause errors in calculating absorption at higher pressures and concentrations. With these considerations in mind, data from models by Ferriso et al. (1966) and Roberts et al. (1976) proved to be more suitable. For the characteristic H_2O bands, the transmission in the 25 cm^{-1} spectral intervals can be represented in the form

$$\Phi^{H_2O} = \exp\left[-\frac{K\Psi^{H_2O}}{\left(1+\dfrac{K\Psi^{H_2O}}{4a}\right)^{1/2}}\right]$$

and in the continuous portion of the spectrum (so-called continuum absorption in the $8-30$ μm region)

Table 7.3 H_2O absorption bands

Band frequency ν, cm^{-1}	5331	5276	5234	5180	4666	3755	3736	3657
Wavelength λ, microns	1.88	1.89	1.91	1.93	2.14	2.66	2.67	2.73

Band frequency ν, cm^{-1}	3640	3151	3072	2161	2062	1594	1556
Wavelength λ, microns	2.75	3.17	3.25	4.63	4.85	6.27	6.42

$$\Phi_{\text{cont}}^{H_2O} = \exp[-K_{\text{cont}}B\Psi^{H_2O}P].$$

Here Ψ^{H_2O} is the H_2O content along the line of sight (in atm · cm); K and K_{cont} are absorption coefficients in the band and continuum, respectively, and the coefficients a and B are model parameters, which, in particular, incorporate the relationship between absorption and atmospheric temperature T and pressure P. In this case parametrization in the continuous portion of the spectrum was obtained based on data on the change in the opacity of water vapor in a nitrogen medium. At the same time, it is known (Varanasi, 1972) that broadening of certain H_2O lines by carbon dioxide exceeds by 1.5–2 times the broadening caused by nitrogen. Therefore, continuous absorption by water vapor in a carbon dioxide atmosphere must be higher than in a nitrogen atmosphere but lower than in an atmosphere of pure water vapor. Therefore, calculations by the given technique provide a minimum estimate of the continuous H_2O absorption. The high-temperature spectral database for H_2O being developed as in the case of CO_2, using direct numerical diagonalization (Wattson and Rothman, 1986, 1992), promises to greatly improve models of the thermal effects of absorption by this molecule in the atmosphere of Venus (Pollack et al., 1993).

Sulfur Dioxide

The SO_2 molecule, like the H_2O molecule, is an asymmetric gyroscope. The most important absorption bands for the transfer of thermal radiation are given in Table 7.4. Although data exist on the characteristics of rotational lines for SO_2 (Rothman et al., 1983), it is impossible to use them directly to calculate thermal emission in Venus' lower atmosphere, for the same reasons given for the H_2O molecule. Therefore an approximation method, given in works of Marov et al. (1989b, 1989c), is more acceptable. It is based on the idea that because the asymmetry parameter Ψ of the SO_2 molecule is close to unity ($\Psi = -0.95$), its spectrum can be calculated as for a symmetric gyroscope. Then, in accordance with the selection rules and neglecting centrifugal perturbation and other minor effects, one can determine the frequency of the rotational lines ν and the population density of rotational

Table 7.4   SO$_2$ absorption bands

Band frequency ν, cm^{-1}	2809	2715	2499	2255	1875	1361
Wavelength λ, microns	3.56	3.68	4.00	4.43	5.33	7.35

Band frequency ν, cm^{-1}	1151	844	517
Wavelength λ, microns	8.69	11.8	19.3

levels as a function of ν, as well as the spectral absorption coefficient K for the vibrational band ν_v. Accordingly, the transmission in the $\delta\nu$ interval is expressed as

$$\Phi^{SO_2} = \exp[-K(\nu - \nu_v)\Psi^{SO_2}]$$

where Ψ^{SO_2} is the SO$_2$ content along the line of sight.

The Clouds

Cloud cover can exert a strong effect on the nature of heat exchange in a planet's troposphere. From everyday observations it is well known that during a clear, starry night on Earth, the surface and near-surface atmosphere cool more vigorously than when the sky is covered with clouds. This is because our atmosphere is much less dense than the Venusian atmosphere, and therefore the relative contribution of cloud cover to the formation of temperature (and humidity) fields is significant, owing to the greenhouse effect. In this case the near-infrared region is more sensitive to the appearance of clouds than the far-infrared (see, for example, Kondratyev et al., 1967). This is explained by different Planck functions versus temperature in various spectral regions: this function is most sensitive to temperature variations in the near-infrared spectral regions. At the same time, in the fundamental 15 μm CO$_2$ absorption band (the 12–18 μm region) the integral radiative flux responds least to the appearance of cloud cover.

Hence it follows that with an increase in the planet's surface and atmospheric temperatures (which, according to Wien's law, means a shift in the thermal radiation maximum on the Planck curve toward shorter wavelengths), the influence of cloud cover must increase somewhat. This effect, however, may be almost completely compensated for in the event of effective gaseous absorption in the near-infrared region. This is precisely the situation on Venus. Therefore, given the high opacity of the molecular atmosphere, the contribution of cloud cover to the effective heat flow, over the whole low-lying atmosphere and especially near the planet's surface, can be ignored. It becomes necessary to account for it at high altitudes, starting at approximately 30–40 km.

Calculating the radiative transfer in clouds, specified by the microphysical properties of aerosols, represents quite a complicated, independent problem. In addition, even when the most reliable data presently available on the effective particle sizes (see Table 6.2) are used, the results of such calculations assume substantial inaccuracies. Under these condi-

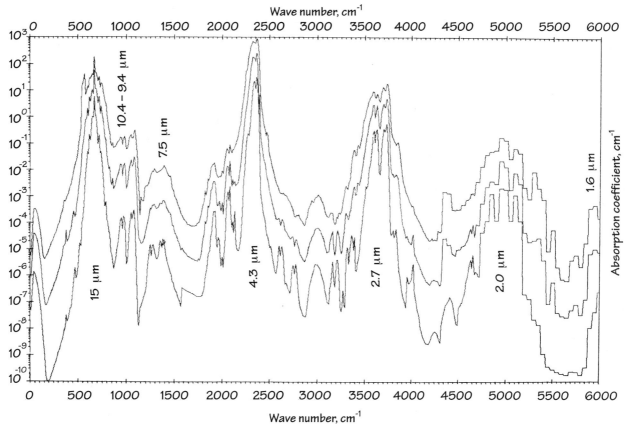

Fig. 7.4. The spectral coefficient of absorption of carbon dioxide $\varkappa$ (cm^{-1}) versus wave number ν for three levels in the atmosphere, corresponding to temperatures of 300 K (lower curve), 500 K (middle curve), and 700 K (upper curve). The most characteristic CO_2 absorption bands are pointed out.

tions an alternative approach may be to parameterize the optical characteristics of the clouds, treated as a surface with a definite degree of thermal radiation reflection (or, as is still said, with a distinct degree of blackness). This parameter can be conditionally attributed to the lower boundary of the cloud cover, located at an altitude of 47–49 km.

Optical Properties of the Atmosphere and Radiative Heat Flow

Synthetic spectra, calculated at temperature and pressure values corresponding to $T(Z)$ and $P(Z)$ altitude profiles from the VIRA-85 model (see Chapter 5), provide a graphical representation of the optical properties of Venus' subcloud atmosphere. Examples are shown as functions of the volume spectral coefficients of absorption $\varkappa$ versus ν for various altitude levels in the atmosphere, corresponding to the appropriate temperature in the 300–800 K limit, in Figs. 7.4–7.7. The transfer of thermal radiation falls within the range of wave numbers from 10 to 6000 cm^{-1}. Here, nearly all the equilibrium radiation of the atmosphere is concentrated. So, for a surface temperature $T_S = 735$ K, the overall unidirectional flow of equilibrium radiation is more than 99.7% of the entire radiation, specified by the Stefan-Boltzmann law $W = \sigma T_S^4$ (for the value of σ see Eq. 1.1), and for lower temperatures at other altitudes the fraction is even larger.

Figure 7.4 depicts the situation for a purely CO_2 atmosphere at levels with temperatures of 300, 500, and 700 K. It serves as a starting point for subsequent analysis. From this figure it is clear which spectral regions have the greatest transmission (where the window of

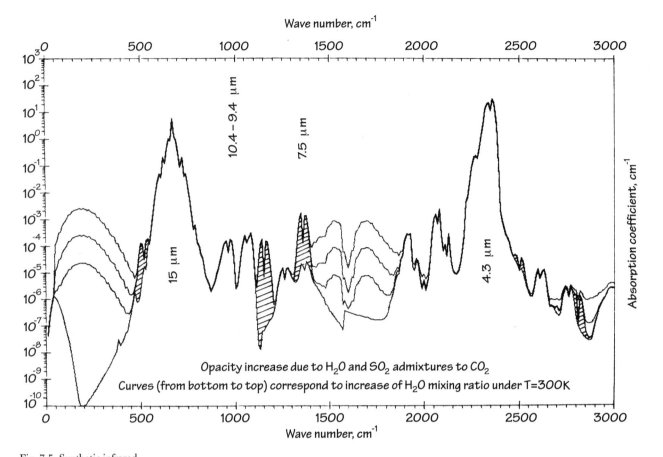

Fig. 7.5. Synthetic infrared spectrum (the spectral coefficient of absorption versus wavelength) for Venus' carbon dioxide atmosphere with water vapor and sulfur dioxide admixtures at the $T = 300$ K temperature level. The SO_2 content is constant and equal to $2 \cdot 10^{-4}$, whereas the three topmost curves are for an H_2O mixing ratio of 10^{-5}, 10^{-4}, and 10^{-3}, respectively; the lowest curve is for pure (97%) CO_2. The superposition of CO_2 windows of transparency is clearly visible, increasing with an increase in X_{H_2O}. The effect of SO_2 on the moist carbon-dioxide atmosphere is indicated by the shaded area.

transparency is located), and how the atmospheric opacity changes for thermal radiation with an increase in temperature and pressure; it increases sharply. Now, turning our attention to Figs. 7.5–7.7, it is easy to trace how the windows of transparency in the CO_2 infrared spectrum overlap in the presence of H_2O for relative water vapor concentrations of 10^{-5}, 10^{-4}, and 10^{-3} (see Table 5.5) at three altitude levels with temperatures of 300, 500, and 700 K. To calculate these spectra, initial data on permitted CO_2 absorption under the indicated conditions, as well as the induced CO_2 absorption spectrum, were used, taking into account the temperature dependence of CO_2 absorption and corresponding data on H_2O absorption, which was discussed in the previous section. The role of H_2O vapor in raising absorption is especially great in the 50–500 cm^{-1} and 1100–2050 cm^{-1} intervals (the second range includes the maximum equilibrium radiation at the surface temperature), and also at 2650–3450, 3750–4300, and 5150–5900 cm^{-1}. The effect of H_2O decreases with an increase in temperature, that is, as one approaches the surface. Finally, the contribution of a minor amount of SO_2 to overall absorption is shown on the graphs, for a modeled sulfur dioxide content in Venus' lower atmosphere of $\sim 2 \cdot 10^{-4}$ (see Chapter 5). The more recent observation of a slightly lower SO_2 abundance in the lower atmosphere of $1.3 \pm 0.4 \cdot 10^{-4}$ (Bézard et al., 1993) will not change the results greatly. All the fundamental SO_2 bands fall within the spectral region up to 3000 cm^{-1}. It is obvious that the influence of SO_2 is significantly less than that of H_2O; nevertheless, it is noticeable. In this case the SO_2 contribution depends inversely on temperature: the minimum effect corresponds to the maximum temperature at the surface (analogous to the effect of H_2O). The

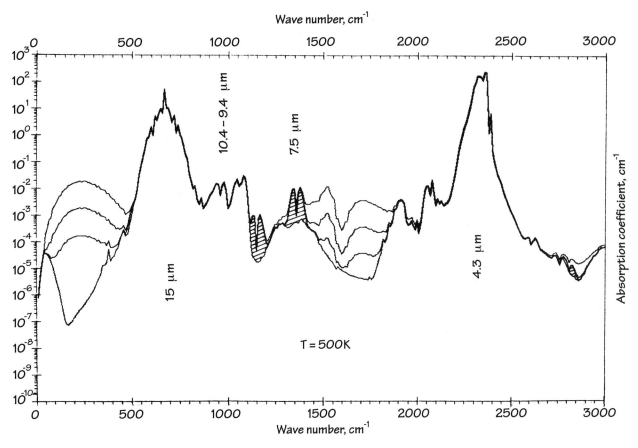

Fig. 7.6. The same as

Fig. 7.5, for $T = 500$ K.

maximum SO_2 contribution to the opacity of the CO_2 atmosphere and to modification of the altitude profiles of radiative heat flow occurs in the region of the spectrum, located between the characteristic CO_2 bands, from 7.5 μm to 9.4 μm (where H_2O bands do not exert a noticeable influence given relative concentrations of no higher than $\sim 10^{-3}$) and within the 7.5 μm induced band itself (in the wave number interval from 1300 to 1400 cm^{-1}), where the maximum equilibrium radiation at the planet's surface temperature is located.

We shall now proceed with a quantitative and qualitative analysis of the altitude distribution of radiative heat flow, based on the given calculated absorption spectra (Marov et al., 1985, 1989b). The entire ν range was broken up into 506 spectral resolution intervals: 10 cm^{-1} in the 10–4280 cm^{-1} region and 50 cm^{-1} in the 4850–6000 cm^{-1} region. The flow of net thermal radiation S is found by summing the flows transported in each of these intervals. It is the algebraic sum of unidirectional flows (to the upper and lower hemispheres), taking into account the angular distribution of the radiation field in each atmospheric altitude layer: $S = W\uparrow - W\downarrow$ (see Fig. 7.1a).

Examples of $S(Z)$ altitude profile calculations are shown in Figs. 7.8 and 7.9. The first thing which draws our attention in Fig. 7.8 is how great the export of thermal energy is in a purely CO_2 atmosphere. At the maximum, it greatly exceeds the inflow of solar energy to Venus $F^\infty = 157 \pm 6$ Watts $\cdot$ m^{-2}. This excess is maintained even if we assume that 70% of the thermal radiation arriving at the lower surface of the clouds is reflected back; this would correspond to a degree of blackness of $\varepsilon_c < 1$ (see the example of $\varepsilon_c = 0.3$ in the

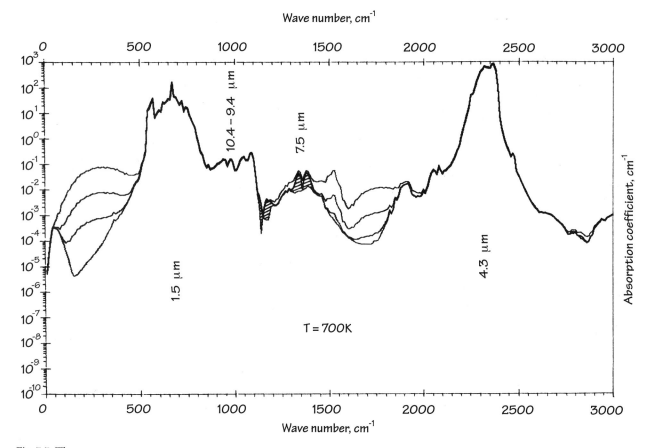

Wave number, cm⁻¹

Fig. 7.7. The same as

Fig. 7.5, for $T = 700$ K.

figure). Such high degrees of reflection, however, are unlikely and are considered here only to show that absorption of thermal radiation in Venus' subcloud atmosphere is also determined by molecular components other than CO_2 (primarily water vapor). This is suggested by the remaining S(Z) curves presented in Fig. 7.8. A more realistic value of $\varepsilon_c = 0.8-0.9$, as it is easy to determine from similar calculations, exerts a minor influence on the heat flow profile only in the vicinity of the lower cloud boundary.

The addition of water vapor most strongly affects the amount and the altitude trend of radiative heat flow. As can be clearly seen in Fig. 7.8, even for an H_2O mixing ratio on the order of 10^{-5}, the situation changes in a fundamental way, and for $X_{H_2O} \approx 10^{-3}$ the absorption of infrared radiation proves to be excessively high. The results of these calculations thus enable us to estimate the maximum X_{H_2O} value necessary to ensure an integral thermal balance on Venus. Varying X_{H_2O} by even hundredths of a percent, especially in the middle portion of the troposphere, appreciably affects the overall heat flow. The best match for thermal balance is ensured for $X_{H_2O} = 10^{-4}$, although additional consideration of SO_2 makes it possible to correct this estimate somewhat. Indication of a possible change in X_{H_2O} with altitude, obtained from experiment (Moroz et al., 1981; Lewis and Grinspoon, 1990; Donahue and Hodges, 1992), is difficult to confirm or refute by this method, taking into account remaining errors in calculating the opacity of atmospheric constituents, primarily CO_2 (Marov et al., 1984). As far as SO_2 is concerned, it most strongly affects the structure and amount of heat flow above about 20 km, especially given small H_2O concentrations. Without adding water vapor, the presence of even greater SO_2 concentrations cannot ensure thermal balance.

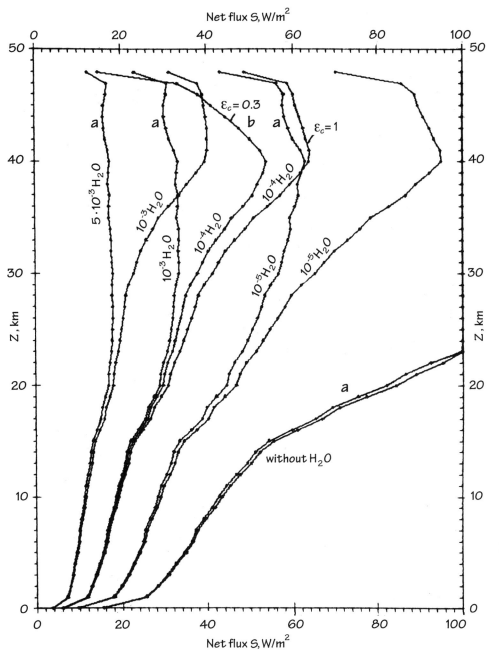

Fig. 7.8. Overall radiative heat flows as a function of altitude ($0°-30°$ latitude) in Venus' subcloud CO_2 atmosphere, with and without H_2O or SO_2 as contaminants. Curves without the index "a" correspond to an atmosphere without SO_2 for various H_2O mixing ratios (from right to left, 0, 10^{-5}, 10^{-4}, and 10^{-3}, respectively). Curves with the "a" index correspond to the same cases but with SO_2 at a $2 \cdot 10^{-4}$ mixing ratio. The strong influence of SO_2 is evident above about 20 km, especially for small H_2O concentrations. All computed curves correspond to the degree of blackness of the lower cloud boundary $\varepsilon_c = 1$. The "b" index denotes the $\varepsilon_c = 0.3$ case with $X_{H_2O} = 10^{-4}$.

The following windows of transparency are responsible for the transfer of thermal radiation in Venus' subcloud atmosphere: (1) $10-550$ cm^{-1}, (2) $1100-2150$ cm^{-1}, (3) $2500-3500$ cm^{-1}, (4) $3750-4800$ cm^{-1}, and (5) $5200-6000$ cm^{-1}. An example of the flow distribution $S(Z)$ over the whole infrared spectral region under study, $10-6000$ cm^{-1}, with respect to individual windows, is shown in Fig. 7.9 for near-equatorial latitudes $0°-30°$ and a model atmospheric composition of 0.97 CO_2, 10^{-4} H_2O, and $2 \cdot 10^{-4}$ SO_2. Analogous computed curves for various H_2O mixing ratios lead to the conclusion that water vapor even in insignificant concentrations, such as $X_{H_2O} = 10^{-5}$, substantially screens the export of energy in the first and second windows, although the same H_2O concentration does not essentially affect the transfer of radiation in the third and

Fig. 7.9. A comparison of the net radiative heat flow, transmitted through the most characteristic windows of transparency in Venus' lower atmosphere. The corresponding detected windows are denoted by the numerals 1–5. Curves without the "a" index refer to a moist CO_2 atmosphere (97% CO_2 and 10^{-4} H_2O), and curves with the "a" index are for an atmosphere containing additional SO_2 at a $2 \cdot 10^{-4}$ mixing ratio. The results of calculations refer to the near-equatorial atmosphere (0°–30° latitude). The selected windows 1–5 are slightly broader than those referred to in the text.

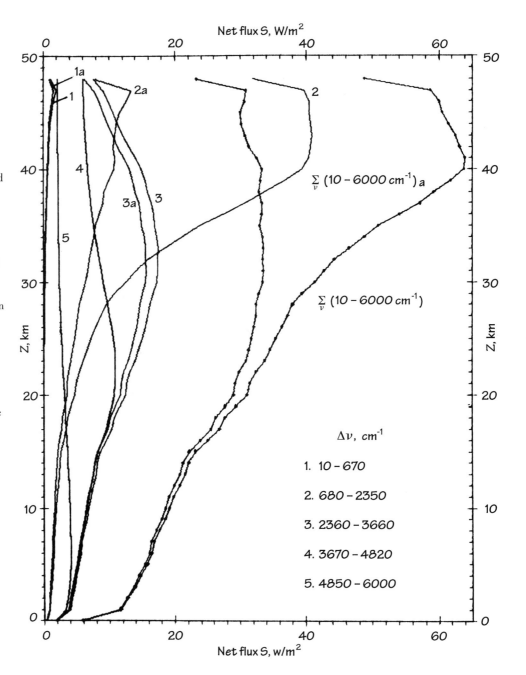

fourth windows. The decisive contribution to flux at altitudes greater than 35 km, for all H_2O concentrations considered up to $5 \cdot 10^{-3}$, is furnished by the second window. This contribution decreases toward the planet's surface and disappears near it at an H_2O concentration near 10^{-3}. The maximum contribution to overall flux in the first window does not exceed 5% in a humid atmosphere, and below 30 km is essentially absent. Below 20 km, flows in the third and fourth windows are similar, and at altitudes of ∼35 km differ the most. The level of the flux in the fifth window decreases in the same way with an increase in the water vapor content in the troposphere. In Fig. 7.10 we present another example of similar calculations, taking into account the contribution from the high-latitude

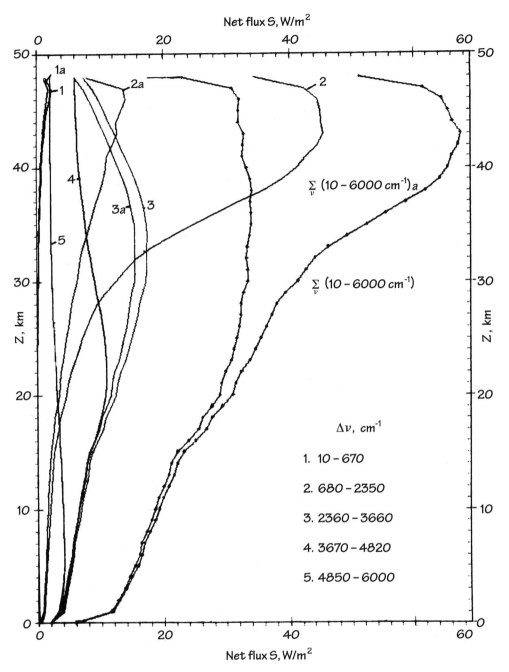

Net flux S, W/m²

Z, km

$\sum\limits_{\nu} (10 - 6000\ cm^{-1})_a$

$\sum\limits_{\nu} (10 - 6000\ cm^{-1})$

$\Delta\nu,\ cm^{-1}$

1. 10 – 670

2. 680 – 2350

3. 2360 – 3660

4. 3670 – 4820

5. 4850 – 6000

Net flux S, W/m²

Fig. 7.10. The same as Fig. 7.9, for latitudes of 0°–60°.

atmosphere (0°–60° latitude), which suggests a minor quantitative difference in the corresponding curves without changing their character.

In Fig. 7.11 we compare the radiative heat flow measured by Pioneer-Venus (Revercomb et al., 1985) with the heat flows calculated by us for the same latitudes and various concentrations of water vapor. To match the theoretical profiles with the flow level obtained by the Night (27° S) and Day (31° S) probes, one must assume an atmospheric X_{H_2O} content of $(1-5) \cdot 10^{-3}$. A comparison with measurements of the North probe (~60° N) provides a concentration of $X_{H_2O} = 5 \cdot 10^{-5}$–$10^{-4}$. Thus, if the measurement data reflect the actual latitude dependence of the altitude profiles of physical parameters and the actual

Fig. 7.11. A comparison of the calculated altitude profiles of the net heat flow with those measured by the Pioneer-Venus Day, Night, and North probes (shaded regions). The notations to the calculated curves, obtained for an atmosphere with 97% CO_2, correspond to $X_{SO_2} = -2 \cdot 10^{-4}$ and various X_{H_2O} values: (1) 10^{-5}, (2) $-2 \cdot 10^{-5}$, (3) $-5 \cdot 10^{-5}$, (4) -10^{-4}, and (5) -10^{-3}.

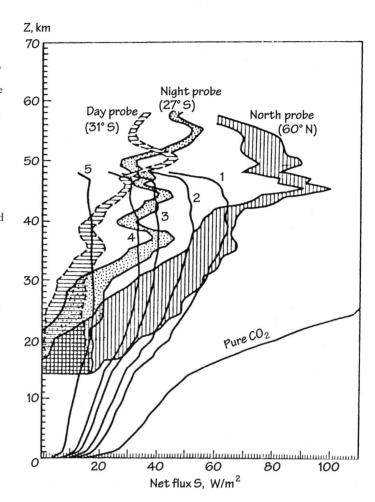

heat flow, then one can expect that Venus' subcloud atmosphere at high latitudes is drier. Later observations in the near-infrared window regions, however, both from the ground and with the Galileo NIMS experiment, strongly suggest that the lower atmosphere water abundance is much lower, ~30 ppm, and does not vary strongly with latitude (Pollack et al., 1993; Drossart et al., 1993).

Certain Characteristics of Radiative Heat Exchange

The results of calculations presented here, which on the whole agree with the results of other authors (Pollack et al., 1980b; Schofield et al., 1982), unfortunately do not enable us to uniquely interpret the quantitative and qualitative features of the $S(Z)$ profiles; understanding the nature of the high temperature on Venus depends on this. The remaining uncertainties in estimating the opacity of the carbon dioxide atmosphere with admixtures of minor components over a broad range of partial pressures, temperatures, and cloud aerosol properties in the transfer of thermal radiation, are the primary limitations.

Unfortunately, the results of heat flow measurements by the Pioneer-Venus Large and small probes have not clarified the situation, owing to large errors in the onboard radiometers, especially in the lower atmosphere (Suomi et al., 1980). To perform an absolute calibration of the data, the authors of the experiment took advantage of the results of heat flow calculations taken as standard values at an altitude of 14 km, where the measurements ceased (Revercomb et al., 1982, 1985). Assuming that measurement errors on all probes were systematic in nature, the authors produced a shift in each measured altitude profile of heat flow by a certain constant value, so that at the moment the measurements stopped, the measured value corresponded to that calculated (based on the idea that closer to the surface the accuracy in the calculations is higher). This calculated value was taken in accordance with data from Pollack et al. (1980b), equal to 16 Watts/m^2, that is, close to the amount of solar energy re-radiated by the planet's surface. But even if we assume that after incorporating this correction the relationship between radiative heat flow and altitude remains valid, then within the limits of measurement error one can, in general, make only qualitative conclusions about the nature of heat exchange in the atmosphere (Marov et al., 1985).

Below the cloud boundary, as follows from the calculations, water vapor provides the main contribution to opacity. The necessary energy balance is ensured even for a mean relative concentration less than 10^{-4}. The contribution of sulfur dioxide is considerably less, therefore the existing scatter in the SO_2 content estimates, by a factor of 2–3 times, does not greatly influence the overall heat flow. At the same time, SO_2 not only quantitatively affects the integral radiative heat flow S but also noticeably modifies its altitude profile, especially in the upper portion of the subcloud atmosphere (see Figs. 7.9 and 7.10). As a result, the altitude where the flow divergence changes sign turns out to be lower, and consequently it turns out that instead of infrared cooling, infrared heating occurs over a substantial portion of the subcloud atmosphere.

Radiation transfer within clouds (which at first approximation can be considered absolutely black, taking into account their sulfuric acid composition) is most strongly affected by large (mode 3) particles. Strong absorption of solar radiation in the clouds and especially in the subcloud (middle) atmosphere, associated with photolysis, thermal mass exchange with numerous chemical (including catalytic) reactions, and multiple scattering (see Chapter 5), poses additional difficulties in calculating radiative heat exchange. The most likely sets of reactions controlling the composition of structure of the cloud zone and subcloud atmosphere were considered in detail in Chapters 5 and 6.

The remaining errors in the measured altitude profiles of the overall heat flow S(Z) in the lower atmosphere, after shifting them by a constant value (after "absolute calibration" with respect to calculated data in the deep atmosphere), proved to be less than the differences between the profiles themselves, measured by various probes. This makes it possible to relate these differences to actual changes in the opacity of the atmosphere at various landing areas and, thus, to conclude that significant latitude variations exist in radiative heat exchange (see Fig. 7.11). The validity of this conclusion is supported by a considerable latitude trend detected in the atmospheric temperature altitude profiles, given their simultaneous latitudinal symmetry (Seiff et al., 1980; Kliore and Patel, 1980; see Chapter 5), and likewise by the predominantly zonal nature of the circulation in almost all altitude intervals under consideration (Counselman et al., 1980). One interpretation of the various S(Z) curves, shown in Fig. 7.11, was made with the assumption that the high-latitude atmosphere is significantly drier than the low or middle latitudes, yet the required difference

Fig. 7.12. Latitudinal variation in absorbed solar energy ΔF (dashed line) and transmitted (net) thermal energy S (solid line) in Venus' atmosphere (according to Revercomb et al., 1985): (1) the Pioneer-Venus Large probe, (2) the Day and Night probes, and (3) the North probe.

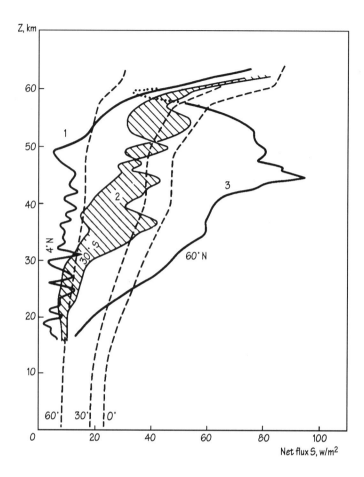

in the water vapor content reaches almost one order of magnitude (the $X_{SO_2} = 2 \cdot 10^{-4}$ concentration is assumed constant in this case), and this is not seen in near-infrared observations (Drossart et al., 1993). At low latitudes the atmospheric moisture content, responsible for the specified opacity, corresponds approximately to results of gas chromatograph measurements but exceeds the limit imposed by spectrophotometry (see Table 5.5) and ground-based near-infrared spectroscopy by more than an order of magnitude.

In order to judge the specific nature of heat exchange and the role of atmospheric dynamics (in particular, circulation), it is important to compare the distributions of the overall solar ΔF and thermal S radiation profiles with respect to altitude (each of which is the difference between the descending and ascending flows) and their derivatives (Fig. 7.12). In addition to our calculations (presented previously), calculations from experimental data were carried out by Tomasko (1983) and Tomasko et al. (1980b) for the 12–94 km altitude range. These authors also consistently point out that solar and radiative heat flows, as a rule, are not balanced out. At certain altitudes in the troposphere, variations in the heat flow profile (and, consequently, in the rate of atmospheric radiative cooling) considerably exceed variations in the altitude profile of solar radiant energy (and, consequently, the rate of heat inflow from the Sun).

To what extent are these differences real? For now it is difficult to answer this question. Even taking into account the errors in the opacity estimates of atmospheric gases and the insufficient reliability in the experimental data on heat flow values, however, one can maintain that these differences are extremely likely to be real. Errors in determining the calculated and experimental S(Z) values evidently do not exceed 10–20% of the effective flow value in the upper troposphere, but even such values are associated with extremely serious consequences. In particular, this leaves open the question of how real the greenhouse effect contribution is in maintaining the high temperature on the planet, and of what role heat transfer, via convection, plays.

Yet another interesting feature is revealed in analyzing the $\Delta F(Z)$ and S(Z) distributions in Fig. 7.12. The inflow of heat from the Sun as a function of latitude turns out to be uncompensated by lost radiative heat flow: at low latitudes with the greater inflow of solar radiation the heat loss is less, whereas at high latitudes it is greater.

The explanation for this phenomenon can naturally be found in the planetary circulation mechanism, which provides for the redistribution of heat and, consequently, for additional heating and cooling of the lower atmosphere. As can be seen from Fig. 7.12, below the clouds the latitude gradient of infrared cooling exceeds even the latitude gradient of solar heating at the same altitudes (Revercomb et al., 1985). Thus, from analysis of the radiative heat flow in the subcloud atmosphere, it follows that a significant temperature gradient exists between the equator and the poles. In turn, this gradient must result in the formation of currents in the meridional direction, accompanied by the export of heat to high latitudes.

Some Comments on Planetary Dynamics

Before we turn to a description of circulation processes on Venus, it seems useful to discuss the most general concepts concerning the dynamics of planetary atmospheres. The system of winds on a planet, created through the unequal distribution of solar heat in space and time, depends on whether the thermal response mechanism has a period greater or less than the intrinsic rotational period of the planet, and also on the characteristic thermal relaxation constants (see Table 7.1). From this point of view, Venus is considerably different from Earth or Mars; this difference is manifested in the specific nature of the mechanisms responsible for the thermal balance and dynamic exchange processes on planetary, intermediate (meso-), and local scales (see, e.g., Golitsyn, 1973; Kondratyev and Hunt, 1982).

We shall begin with a simplified form of how to represent circulation on Earth. As a result of thermal expansion, governed by the relationship between gas density and pressure as well as temperature (this property is called baroclinicity), air that is more vigorously heated, and therefore less dense, rises while cooler and heavier air sinks. It thus seems obvious that pressure differences arising through differences in insolation, and therefore horizontal temperature gradients (barometric gradients), must result in the regular transport of air masses (and, consequently, an excess of heat) from the equator to the poles. In this case, along the meridian a gigantic closed convective cell is formed, in the upper portion of which warm air will be transported from the equator to the pole, and along the surface of which cold air will be transported from the pole to the equator. This is called a

Hadley cell, after the well-known English astronomer who in the first half of the eighteenth century put forward and substantiated the hypothesis that differential heating of equatorial and polar regions by the Sun must be the main reason for the general circulation of Earth's atmosphere.

Actually, such circulation, symmetric relative to the equator, has not been established in either Earth's atmosphere or in the atmospheres of other planets. The reason for this is the presence of the Coriolis force due to a planet's rotation. In the dynamics of the atmosphere (and in oceans on Earth), the horizontal component plays the decisive role; it causes air currents to be deflected to the right from their direction of motion in the northern hemisphere and to the left in the southern hemisphere. As a result, the extent of the meridional circulation is confined; in Earth's atmosphere, the Hadley cell dominates only at very low latitudes, to approximately 30° on both sides of the equator, where the trade winds blow. But at middle and high latitudes two additional cells are formed: the Ferrel cell, which is thermally indirect and occupies the region between 30° and 60° latitude, and a polar cell of rather weak circulation. Circulation at these latitudes takes on a zonal character, that is, motions proceed along the parallels of latitude. Inasmuch as temperature gradients are their primary source, the winds themselves are called thermal. In the troposphere westerly winds blow, directed from west to east, whereas in the stratosphere the winds vary their direction: in winter westerly winds blow, and in summer easterly; here wind velocities of 50–100 m/sec are observed. Note that similar values are observed in Venus' atmosphere, as was mentioned when discussing the drift of ultraviolet clouds, but their direction remains unchanged over the course of a year, coinciding with the direction of the planet's intrinsic rotation.

In determining the wind fields on Earth, the concept of geostrophic flow, or geostrophic wind, corresponding to the situation where the horizontal pressure gradients are balanced by Coriolis forces, provides a convenient approximation, which can be practically realized in the atmosphere of a rapidly rotating planet. In this case, friction of the air mass over the surface can be ignored. The force of such a wind depends on the pressure gradient and is directed along lines of equal pressure, isobars.

The effect of Coriolis forces on motions is usually characterized by the dimensionless Rossby number: $Ro = U/(2l\Omega \sin \varphi)$, where U is the characteristic horizontal velocity, l is the characteristic scale of motions, Ω is the angular velocity of planetary rotation, and φ is latitude. In other words, this is a dimensionless parameter, representing the ratio of terms associated with acceleration (as a consequence of the barometric gradient) and Coriolis forces in one of the important equations in hydrodynamics, the conservation of momentum. This means that Coriolis forces are dominant for $Ro \ll 1$. This is the case for Earth, where for typical values $U \approx 10$ m/sec at middle latitudes and on a scale of $l \approx 10^3$ km, the specified condition ($Ro \approx 0.1$) is satisfied and the flow has a clearly defined zonal nature, stipulated by Coriolis forces. In turn, on Venus, for the same initial parameters, we have $Ro \approx 30$. This means that Coriolis acceleration can be neglected, because it turns out to be approximately one to two orders of magnitude less than the other terms in the equation of motion.

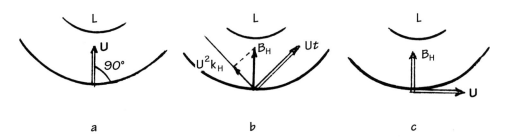

Fig. 7.13. Types of Eulerian motion in examples of tropical cyclones. The horizontal component of the pressure force B_H, which is perpendicular to the isobars and directed toward low pressure L, is equal to acceleration in each case. In case (a), the centripetal acceleration is equal to zero, and the motion begins from a state of rest and is directed along the line to the center of the low pressure L. This is the first type of motion. Type (b) motion corresponds to the case where the tangential and centripetal components of acceleration are approximately identical. These components may be obtained by projecting the pressure force B_H in the indicated directions. These directions correspond to the normal (centripetal) component, in projection onto the horizontal plane $U^2 k_H$, and the tangential component Ut, where k_H is the horizontal projection of spatial trajectory of air parcel, and t is time. This case can be considered a logical extension of type (a) motion, when the Coriolis force is not identically equal to zero, so that it generates a gradual change in wind direction. Finally, in case (c) tangential acceleration is equal to zero, and the motion occurs parallel to the isobars. Such Eulerian motion is called cyclostrophic.

In meteorology, the motion arising in this case is called Eulerian (Haltiner and Martin, 1957). Three types of such motion can be distinguished (Fig. 7.13), as a function of the ratio of acceleration components, tangential and centripetal, acting on a unit volume (parcel) of atmospheric gas. In the limiting cases, one or both of these types of motions may immediately go to zero. If the centripetal component, directed toward the axis of planetary rotation, is negligibly small, then motion develops in the direction perpendicular to the isobars, and the winds blow atmospheric masses from high to low pressure regions. If, by contrast, tangential acceleration, directed along the surfaces of curvature, specified by the figure of the planet and at first approximation coinciding with the curvature of the isobars, is equal to zero, then motion develops parallel to the isobars. This is called cyclostrophic motion (Fig. 7.14). For clarity it is presented in two projections in the figure, in the equatorial and meridional planes. The force caused by centripetal acceleration on a planet of radius a is balanced by the pressure gradient force $dP/dy \approx B_H$ owing to the appearance of zonal currents with velocity U_λ directed orthogonal to these accelerations. At φ latitude the components of these forces, directed toward the equator and pole, respectively, act on an isolated volume of gas (see Fig. 7.14a). The condition for cyclostrophic balance corresponds to the relation

$$\frac{U_\lambda^2 \tan \varphi}{a} = -\frac{1}{\rho}\frac{dP}{dy}$$

This relation can be used to estimate the magnitude of the balanced zonal current from data on the pressure distribution over the geopotential surfaces. In turn, by computing the ver-

Fig. 7.14. The forces acting on a unit volume of atmospheric gas under conditions of cyclostrophic balance in the meridional (a) and equatorial (b) planes with a zonal flow organization U_λ (according to Schubert, 1983). See the text for an explanation.

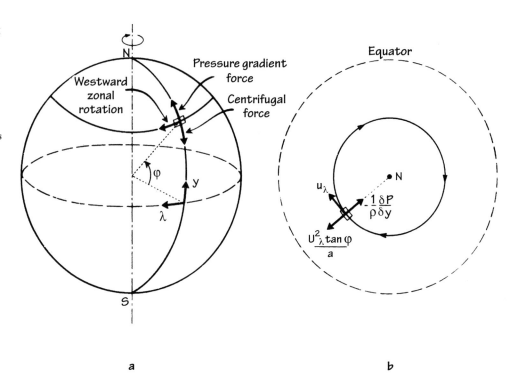

tical gradient of the balanced zonal current specified by this relation, one can obtain an equation for the thermal wind for cyclostrophic balance in the form

$$\frac{dU_\lambda}{dP} = \frac{R}{2\mu U_\lambda \tan \varphi} \left(\frac{dT}{d\varphi}\right)\frac{1}{P}$$

(R is the gas constant and μ is the molecular weight). Then, if one knows the temperature gradient along the meridian $dT/d\varphi$, by integrating the given equation (using the zonal current at a specified level in the atmosphere as the boundary condition) one can obtain the currents at other levels. In this case, however, uncertainty in the balance condition near equatorial latitudes remains.

Finally, the third type of Eulerian motion is an intermediate case, closest to the geostrophic approximation (where the tangential and centripetal accelerations are zero, whereas the Coriolis acceleration is large). In fact, with the appearance of a noticeable Coriolis component, this third type of Eulerian motion converges with geostrophic (Fig. 7.13b).

The various types of Eulerian motion shown in Fig. 7.13 are realized in tropical cyclones on Earth, where the Coriolis parameter $f = 2\Omega \sin \varphi$ is small. Although the geostrophic model describes large-scale extra-tropical circulations well, the best model for motions in tropical cyclones is cyclostrophic motion. It is sometimes linked with the concept of a gradient wind directed along the isobars. From a theoretical analysis it follows that if the temperature decreases from the equator to the poles, then U_λ in Fig. 7.14 increases with altitude, and conversely, $dU_\lambda/\partial Z < 0$, if $dT/\partial\varphi > 0$. Most likely the concept of cyclostrophic balance forms the basis for the mechanism of planetary circulation on Venus. This idea was first expounded by Leovy (1973).

In fact, each of the diagrams in Figs. 7.13 and 7.14 is an idealization. The actual nature of circulation is governed by the superposition of several types of motions, whose degree of disorder depends greatly on the planet's angular rotational velocity. On a rotating planet, wave motions, called Rossby waves, develop. With increasing angular velocity and large temperature differentials along the meridian, such waves become unstable; as they break down, baroclinic vortices arise. In Earth's atmosphere the size of these vortices varies within broad limits, from on the order of millimeters to several thousand kilometers in cross section. Their heat transfer efficiency is even higher than that of meridional currents, which are associated with circulational Hadley cells (Holton, 1979). One should expect strong vorticity, even under conditions of cyclostrophic balance, on a planet with a low rotational velocity, especially in the presence of shear currents.

Small- and medium-scale vortices are elements of atmospheric turbulence, whereas the largest-scale vortices form regions of low and high pressure—cyclones and anticyclones—which are well known on Earth. In cyclones air circulates around the center of low pressure in a counterclockwise direction in the northern hemisphere and in a clockwise direction in the southern hemisphere, whereas in anticyclones the rotational direction about the center of high pressure is reversed. Their lifetime in the atmosphere corresponds, on average, to the estimate for the transformation rate of potential energy into kinetic; for the Earth this is on the order of a week. The instability in Rossby waves associated with large-scale weather systems (or baroclinic instability) is the most efficient mechanism for atmospheric mixing in the meridional direction, for transferring heat from the equator to the poles and for smoothing out the corresponding temperature differences at Earth's surface. Meanwhile, some transient perturbations transmit kinetic energy to the middle zonal flow (mainly to the jet stream, which is found at the altitude of the tropopause), which also contributes to the development of circulation. These transient perturbations must, presumably, also be extremely efficient on Venus.

In spite of huge, multiyear efforts to study the structure of circulation in Earth's atmosphere (which is the basis of dynamic meteorology and which determines the reliability of weather forecasts), many unresolved problems still remain. The main difficulties are associated with the infeasibility of adequately describing the pressure and wind fields as a function of the temperature field, determined by the inflow of solar heat, and the characteristics of the underlying surface. Solving the aforementioned system of hydrodynamic equations using numerical methods, given the limited initial data on meteorological elements and the unavoidable filtering out of a series of harmonic waves in baroclinic models, does not make it possible to account completely for the diversity of interrelated phenomena occurring in the atmosphere.

It is natural to expect that we shall encounter an even more difficult situation when trying to model theoretically the circulation on other planets. The situation is aggravated by the low volume of experimental data and the lack of comprehensive observations. Nevertheless, significant progress has been achieved in theoretical descriptions of observed regularities in motions on Venus (and on Mars and Jupiter). Models that take into account the special conditions on these planets, moreover, help us to better understand many characteristic features of atmospheric dynamics on Earth.

Circulation on Venus

The fact that a significant fraction of the solar energy arriving at Venus in the form of scattered radiation passes through the clouds and reaches the surface removed the main objection of opponents to the greenhouse model. This fact, however, did not completely eliminate the deep circulation model (see Chapter 2). Rather, it led to a certain convergence of these seemingly contradictory mechanisms. As it turns out, first, at the tops of the clouds, almost 70% of the solar radiation is absorbed, and second, hypothetical large-scale circulation cells, analogous to Hadley cells, may in fact not be destroyed as a consequence of hydrodynamic instability, which was pointed out by critics of the model. In principle, they are more stable the deeper solar radiation penetrates the atmosphere, that is, the prerequisites for a superadiabatic initial temperature profile are created, which triggers the convection mechanism. Such a situation could prove to be real, as shown by Avduevsky et al. (1970b, 1971, 1972b), based on an analysis of radiative-convective heat exchange, and Stone (1968) additionally pointed to turbulence as the primary mechanism for vertical diffusion of heat in a deep circulation regime.

Another approach, differing from the deep circulation model, dates back to works of Halley (1686), which were devoted to terrestrial monsoons. The basic idea is that periodically recurrent heating of the atmosphere by the Sun, over the course of a day, may serve as the driving source of circulation. Under the influence of this heating action, large-scale convective Hadley-like cells arise in the plane parallel to the equator, and these cells are slightly inclined as a consequence of phase shift (owing to the thermal inertia of the atmosphere) with respect to the heat source. Such was the nature of motions produced by Schubert and Whitehead (1969) in their experiments with mercury confined in a toroidal vessel and heated from above by a slowly moving gas burner. The gas burner imitated the Sun (and, consequently, the change in insolation during the diurnal-yearly motion of the planet), and the mercury, confined to a limited volume, made it possible to best model the parameters on which Venus' atmospheric dynamics depend (the so-called similarity parameters). The concept was further developed by Schubert and Young (1970); Young et al. (1972); Gierasch (1970); and Gierasch et al. (1973). Thompson (1970) considered a circulation regime with an instability arising in the Hadley cell for some reason, for example as a result of perturbations, imposed in the form of a vertical wind shear profile. It has been shown that this leads to stable zonal flow under a superrotation regime, which is a significant element in the energy transfer of turbulent vortices (so-called Reynolds stresses) from the lower-lying atmosphere upward, with conversion to zonal flow energy. Vasin (1978, 1987) and Vasin and Marov (1977) arrived at similar results through a series of numerical experiments based on a more complete model.

Not one of these models reflects the actual, complex picture. But without doubt, the

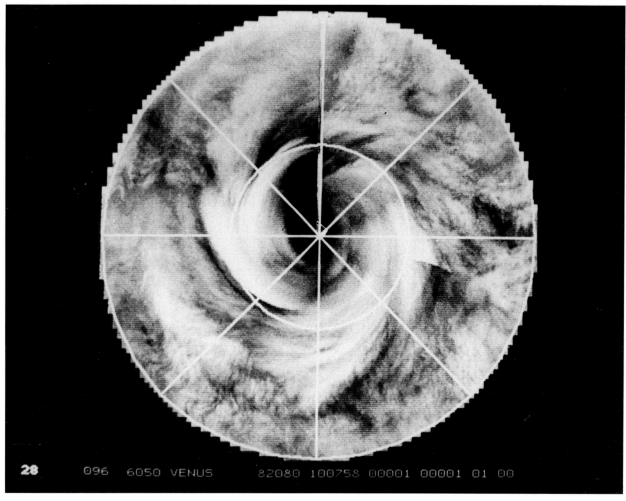

28   096 6050 VENUS   82080 100758 00001 00001 01 00

Fig. 7.15. A composite polar image of Venus' southern hemisphere, taking into account the time lag of individual images from the Pioneer-Venus spacecraft in 1982. Time-lapse composite views were made by displacing the clouds at each latitude using the average measured "wind" at that latitude and accounting for the time interval between images (reprinted by permission of L. D. Travis and S. S. Limaye).

main role in Venus' thermal regime is played by large-scale dynamics, which equalize the temperatures between the equator and poles, and between the day and night hemispheres. What are current notions about the dynamics of Venus' atmosphere? We have seen that circulation on Venus is tracked by the drift of cloud structures seen in the ultraviolet spectral region, and that the fundamental component of the motions has a mean velocity around 100 m/sec in a zonal direction near the top of the troposphere. The meridional component is much weaker and develops from the equator in opposite directions toward the poles; here the flow velocities increase up to $\sim 45°$ latitude in each hemisphere, and then gradually decrease. In television images from Mariner-10 and Pioneer-Venus it is clear from the cloud configurations that the motions are ordered and that, with increasing latitude, the migration acquires a spiral shape, forming a huge vortex around each pole (Figs. 7.15 and 7.16). Consequently, the period of atmospheric rotation decreases with increasing latitude from approximately 4.1 days at the equator to 2.5 days at 45° latitude. Convergence of the flow toward the poles, given such circulation, must presumably lead to descending currents at high latitudes (Limaye and Suomi, 1977, 1981; Suomi and Limaye, 1978; Rossow et al., 1980; Limaye et al., 1982). The general trend toward a warming of the polar regions, compared to middle latitudes, is definitely associated with this phenomenon. As men-

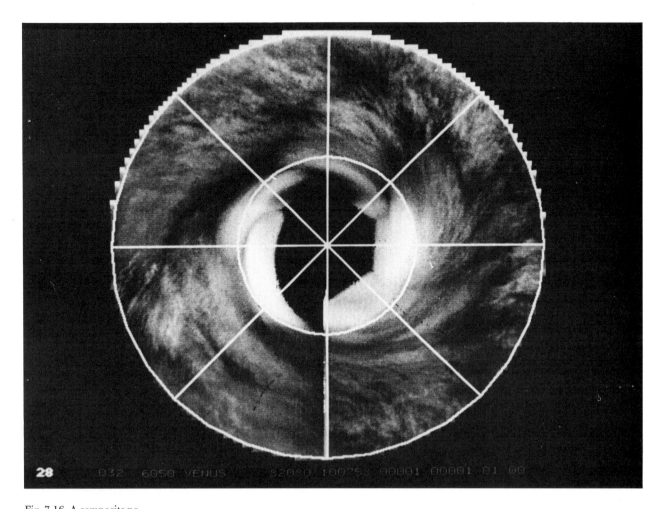

28   032 6050 VENUS   82060 10075 00001 00001 91 00

Fig. 7.16. A composite polar image of Venus' northern hemisphere, made with the same method as Fig. 7.15. Because the spacecraft was located in an orbit with a periapsis (subpoint) primarily in the southern hemisphere, the northern polar region was not observed, which explains the "hole" in the center of the picture (reprinted by permission of L. D. Travis and S. S. Limaye).

tioned in Chapter 6, the observed temperature rise is caused by a lowering of the clouds in the polar zone and is most likely explained by the descending flows in the center of the cyclonic vortex, which is linked with the general system of planetary circulation on Venus.

Where ultraviolet contrasts arise, the velocity of the zonal current probably does not remain constant with altitude. The maximum velocity of the flow (peak zonal flow jet or jet core), according to calculations based on a model of balanced circulational transfer (cyclostrophic approximation), does not coincide with the observed velocity at the cloud tops but is noticeably higher. Periodic changes in the effective level of the cloud boundary are apparently associated with observed variations in the atmosphere's rotational period with an increase in latitude, which, according to Pioneer-Venus data, turned out to be less pronounced than the values obtained from analysis of Mariner-10 television images (Limaye, 1984; Limaye et al., 1988; Del Genio et al., 1986; Newman et al., 1984). Based on an assumption of circulational balance, there are definite reasons to believe that the zonal flow structure, on the whole, is preserved above the cloud boundary. Corresponding calculations on the structure of the thermal field in the stratosphere, measured by various techniques (see Chapter 5), led to this conclusion (Chub and Yakovlev, 1980; Schubert et al., 1980; Seiff, 1982b), at least below an altitude of 80 km. Based on infrared radiometer data

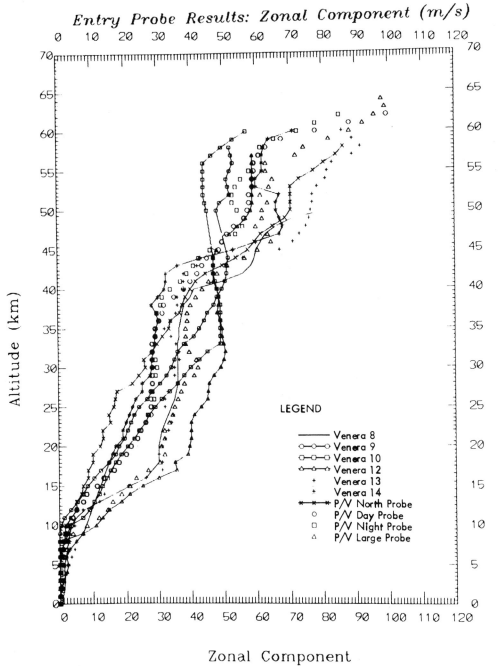

Fig. 7.17. Observed vertical profiles of the zonal wind velocity component on Venus, according to Doppler measurements from the Venera spacecraft and measurements using the DVLBI method with the Pioneer-Venus probes (according to Kerzhanovich and Limaye, 1985).

from Pioneer-Venus, Taylor et al. (1983, 1985) think it possible to raise this level to almost 120 km.

The predominance of a zonal flow as the main characteristic circulation feature on Venus has been established even below the visible cloud boundary, all the way to the surface. This was done using the Doppler method during the descent phase of all Venera landers starting with Venera-4, and was subsequently confirmed by differential very-long-baseline interferometry (DVLBI) using Pioneer-Venus probes. In a series of experiments conducted by Kerzhanovich, Marov, and their colleagues (Kerzhanovich et al., 1972, 1979a, 1979b, 1979c, 1982, 1983; Kerzhanovich and Marov, 1974, 1977, 1983), it was

Fig. 7.18. The structure of the zonal component of wind velocity, measured in various
regions on Venus. Wind velocity height
profiles are plotted on Venus' globe on the
sites of measurements, relative to the Sun's
direction, to take into account possible
influence of diurnal variation.

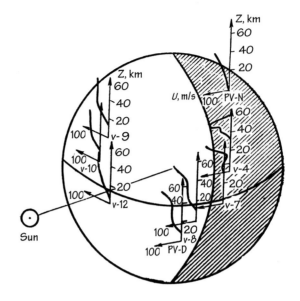

established that the wind velocity is very small near the surface but increases rapidly with
increasing altitude, reaching the drift velocity of the ultraviolet clouds at 60–65 km. The
character of the motions is maintained in various regions of the planet where the Venera
and Pioneer-Venus spacecraft landed, with minor variations in wind velocity (Figs. 7.17
and 7.18). This consistency led to the conclusion that a single circulation system exists on
the planet, encompassing the entire troposphere and stratosphere, with a clearly defined
zonal component reaching a maximum (greater than 100 m/sec) at altitudes above 60 km.

At the same time, the complex, irregular structure of the zonal flow with altitude was
revealed. Its most distinctive features are a very weak increase in wind velocity from the
surface within the first ten kilometers and the presence of three regions with strong vertical
shear flow, between 10 and 18 km, between 40 and 48 km, and above 55 km. Between 18
and 40 km, and between 48 and 55 km, the vertical wind shear profile, as in the portion of
the atmosphere nearest the surface, is minor. The structure of the profiles most likely
undergoes definite latitudinal, and possibly temporal, variations. If we now compare these
regions of shear currents with the deviation in the altitude trend of the temperature gradient dT/dZ from the adiabatic trend Γ shown in Fig. 5.13, which characterizes the degree of
thermal stability in the atmosphere, then some interesting coincidences appear. The regions of minimum shear fall within altitudes where the temperature gradient essentially
does not differ from an adiabatic gradient, and where the atmosphere is neutral to vertical
shifts, whereas regions of maximum shear occur at altitudes where the greatest atmospheric
stability is displayed (Counselman et al., 1980; Schubert et al., 1980). Consequently, in
regions of maximum shear, atmospheric convective activity is evidently suppressed and
static stability increases.

Measurements from the Pioneer-Venus probes have enabled us to clearly distinguish
a meridional component to the currents in the cloud and subcloud atmosphere (Fig. 7.19).
According to Counselman et al. (1980), above 10 km the value of the meridional component is at least an order of magnitude less than the value of the zonal component, whereas in

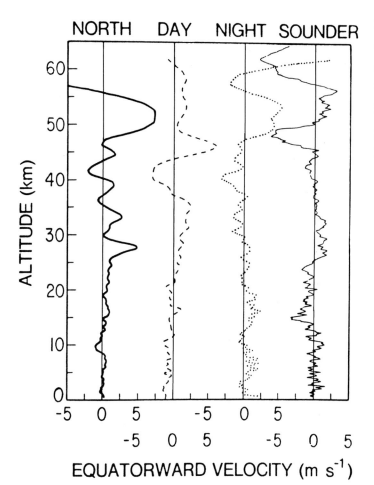

Fig. 7.19. Vertical profiles of the meridional component of wind velocity on Venus, from measurements by the Pioneer-Venus probes (according to Kerzhanovich and Limaye, 1985).

the low-lying atmosphere it becomes comparable to the zonal component or even greater. Earlier, Kerzhanovich et al. (1972) put forward similar ideas based on Venera-4 measurement data. The direction of currents near the surface remains least well determined. According to direct measurements using the cup anemometers shown in Fig. 2.38 (Avduevsky et al., 1976c) and matching Doppler measurements, the wind velocity near the surface varies from 0.3 to 1 m/sec (Fig. 7.20). It is impossible, however, to say anything definite about its direction. From DVLBI data, Schubert et al. (1980) and Schubert (1983) interpreted the observable change in sign of the meridional component, in Fig. 7.19, as an indication of a change in flow direction in a circulational Hadley cell. A change in the direction of the zonal flow in the near-surface atmosphere from retrograde to prograde is also quite likely; this may reflect spatial-temporal variations in atmospheric circulation and may cause a redistribution (taking into account the huge mass of the atmosphere) of the overall angular momentum between the atmosphere and the solid surface of the planet (Hide et al., 1980; Golitsyn, 1982; Waltersheld et al., 1985). In the thousands of wind streaks seen in mapping from Magellan images, there is a clear trend toward an equatorward orientation (Greeley et al., 1992). This is seen as support for a near-surface Hadley cell circulation.

Turning now to the phenomenon of the four-day circulation, we must state that it is not yet completely understood, despite numerous attempts to explain it within the scope of

Fig. 7.20. Results of wind velocity measurements on Venus' surface using cup anemometers (see Fig. 2.38) on the Venera-9 (upper graph) and Venera-10 (lower graph) landers. Each spacecraft was equipped with two detectors; telemetry was sampled at a frequency of 0.4 Hz for 16-second intervals every 2 minutes 8 seconds. On Venera-9 the measurements were carried out over a 49-minute interval by the two instruments, corresponding to the open circles and dots, respectively. On Venera-10 the wind velocity was measured only for the first 90 seconds after landing; telemetry was sampled from only one instrument (open circles). Considerable differences in the wind speed and local variations between the two sites of landing are evident, with an absolute value within 0.5–1 m/sec.

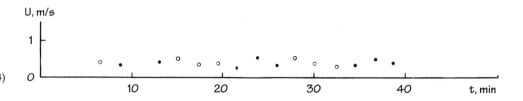

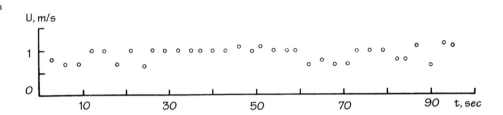

various hydrodynamic models. We refer to the transfer of angular momentum and its connection with large-scale (in the zonal and meridional directions) or small-scale (turbulent) motions. The small-scale motions are associated with Reynolds stresses, or more strictly speaking with the anisotropy arising in the turbulent (Reynolds) stress tensor. Most realistic is the description of large-scale zonal flow based on a model of cyclostrophic balance as one realization of Eulerian motion, which is at least valid in the middle and upper portions of the troposphere, away from the equator. We recall (see Figs. 7.13 and 7.14) that a direct estimate of the cyclostrophic flow can be obtained by making use of observational data on the pressure distribution at a certain level in the atmosphere, considered as a geopotential surface, and that the thermal wind, defined by the temperature gradient in the meridional direction, makes it possible to estimate how the zonal flow varies with altitude (Seiff et al., 1980; Limaye, 1984, 1987; Schubert et al., 1980). It is essential to note that, as follows from the condition for cyclostrophic balance, if the temperature toward the pole drops, the velocity U_λ increases, and vice versa. This is reflected in the gradual increase in the velocity of the zonal current in the subcloud atmosphere and clouds. But the velocity slows above the clouds, where the $\partial T/\partial\varphi$ gradient changes sign (Leovy, 1987).

The corresponding balance relations cannot be applied to the meridional component in this approximation, in contrast to the zonal component. Therefore, it is usually assumed that the mean meridional current is, for the most part, controlled by large- and small-scale vortices (Kerzhanovich and Limaye, 1985). At the same time the transport of angular momentum of vortices to the angular momentum of the mean zonal flow is considered small, although large-scale vortices possibly play an important role in the transfer of angular momentum from low to high latitudes, in this way maintaining the middle latitude jet stream (Suomi, 1974; Gierasch, 1975; Limaye and Suomi, 1981; Limaye et al., 1982).[3]

On the whole, from observations of the structure of ultraviolet contrasts, and from wind turbulence measured by the Doppler method, vortex movements over various spatial and temporal scales and turbulence are important elements of atmospheric dynamics on Venus. They are associated partly with planetary circulation and partly with convective processes (Kerzhanovich and Marov, 1983). At the cloud tops, waves are observed on a planetary scale, superimposed on the primary flow with a velocity amplitude of approximately 15–20 m/sec and a period of 3–6 days. Similar estimates were obtained from analysis of the migration of ultraviolet cloud details. The waves are possibly generated by convective meso-scale cells, which fall within the region of clouds associated with centers of wave activity. Another source may be turbulence as a consequence of small-scale shear instability in the zonal flow (so-called Kelvin-Helmholtz instability) or the disruption of internal gravitational waves propagating from below (for further discussion see Travis, 1978; Del Genio and Rossow, 1982; Limaye et al., 1982; Woo et al., 1980, 1982; Schubert, 1983; Hou and Goody, 1985).

So what we see in television images of Venus represents a complex collection of ordered motions with velocity U_λ, undergoing wave motions with various periods and amplitudes, as well as vortex motions and turbulence on various spatial scales. All these motions are associated, in one way or another, with the system of planetary circulation primarily oriented in the zonal direction whose main component, at the same time, is meridional transport. Through the transfer of angular momentum upward in the retrograde direction, coinciding with the direction of planetary rotation, atmospheric superrotation is maintained. More local processes are linked mainly with contrasts and nascent instabilities. Apparently, such processes can be partially attributed to vertical velocities, usually estimated to be much less than 1 m/sec in the troposphere (Kerzhanovich and Marov, 1983), but sometimes the local motions may appear as intermittent bursts lasting several hours, as shown by Vega balloon measurements in the clouds at an altitude of 54 km, where the horizontal drift velocity, measured using the DVLBI technique, was 70 m/sec (Linkin et al., 1986b).

To sum up, planetary circulation on Venus, whose most characteristic feature is atmospheric superrotation, is generated through the two fundamental mechanisms shown in Figs. 7.21 and 7.22. One mechanism (Fig. 7.21) operates in the equatorial plane, where inclined large-scale convective cells form because of periodic solar heating. Near the subsolar point, where these cells are inclined to the west owing to ascending currents and their associated vortices, retrograde angular momentum is transported upward, whereas in regions where the cells are inclined to the east, prograde angular momentum is transported downward. By correlating the corresponding velocity components and the transfer of momentum, a stable zonal current develops in the equatorial region in a westerly (retrograde) direction. It may be amplified due to initial perturbations and the transport of Reynolds stresses upward (Schubert, 1983).

Another mechanism (Fig. 7.22) operates outside the equatorial plane, meeting the condition for cyclostrophic balance. If, in the first case, the current is initiated by a diurnal variation in atmospheric heating, then the specific mechanism is dictated by the constant difference in insolation between the equator and the pole and the corresponding temperature difference. Therefore it must be based on meridional transport in Hadley cells, which are symmetric relative to the equator; this ensures the heat redistribution responsible for

Fig. 7.21. The motions that explain the mechanism of planetary circulation in the equatorial plane. Superrotation is created by the transfer of angular momentum upward from inclined, large-scale convective cells, which form under the action of a moving heat source — the Sun (according to Schubert, 1983).

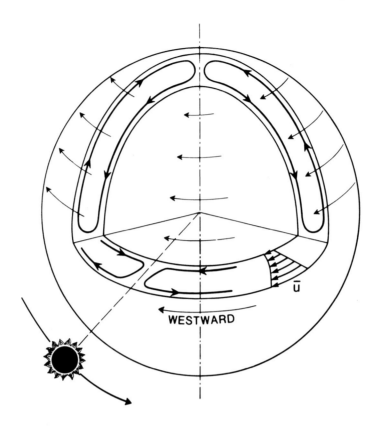

WESTWARD

$\overline{u}$

the difference in the flows ΔF and S in Fig. 7.12. The emerging thermal winds between the equator and poles, under conditions of cyclostrophic balance, acquire a stable zonal component in a westerly direction, which determines the entire nature of the circulation. Naturally, in this case, one would expect ascending currents near the equator and descending currents at the poles, with a flow of gas from high latitudes to the equator near the surface. In fact the situation is apparently much more complicated; more likely several cells form, carrying gas in prograde and retrograde directions (see Fig. 7.22).

The theoretical models by Kalnay de Rivas (1973, 1975), Schubert et al. (1980), and Rossow (1982) illustrate these ideas, although evidence of the presence of such cells is available only from analysis of Mariner-10 and Pioneer-Venus data at the cloud tops. Probably meridional circulation is most strongly developed there, because more than two-thirds of the solar energy reaching the planet is absorbed at those levels. Retrograde meridional cells create additional difficulties in modeling superrotation. But we assume that, despite the complexity of the configuration, such a mechanism (meridional transport) ensures the transfer of retrograde (directed toward the west) angular momentum from equatorial to polar regions and simultaneously upward to the level of the clouds. Vortices, excited by the diurnal high tide, and waves on a planetary scale may contribute to this vertical transport. In particular, the combined effect of thermal and solar gravitational tides is in principle capable of providing the momentum conductive to capturing Venus in resonance (see Chapter 1), as well as increasing the rotational velocity of the atmosphere (Gold and Soter, 1969, 1971, 1979; Covey et al., 1986; Del Genio et al., 1986).

The complex dynamics of Venus' atmosphere could hardly be explained within the

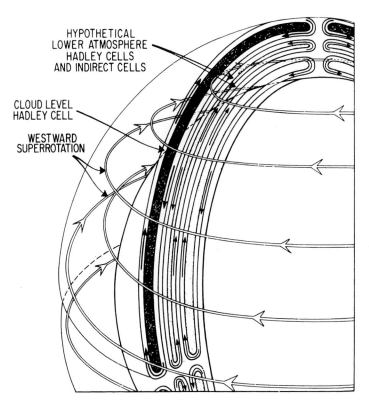

HYPOTHETICAL
LOWER ATMOSPHERE
HADLEY CELLS
AND INDIRECT CELLS

CLOUD LEVEL
HADLEY CELL

WESTWARD
SUPERROTATION

Fig. 7.22. The motions that explain the mechanism of planetary circulation and the formation of superrotation beyond equatorial latitudes. The meridional current in a Hadley cell transmits its excess thermal energy from low to high latitudes, as a consequence of differences in insolation. Below the clouds, including near the surface, several such cells (prograde and retrograde) possibly exist. As a result of cyclostrophic balance (explained in Fig. 7.13), a stable zonal current develops in the retrograde (westerly) direction, with transfer of angular momentum upward along the vertical (according to Schubert, 1983).

scope of any single hydrodynamic mechanism. We have become acquainted with two simplified schemes that enable us to understand, in the most general terms, the still-mysterious phenomenon (from a physical standpoint) of four-day circulation on our neighboring planet. Most researchers currently consider these schemes the most realistic. Moreover, the schemes implicitly embrace many additional processes (several of which we have briefly mentioned) whose effectiveness varies at different levels in the lower and middle atmosphere. In particular, diurnal forcing via tides and other gravitational waves may play an important role in generating superrotation.

   To conclude this chapter we shall make one more important observation. Research conducted using spacecraft has provided a great impetus for studying the basic features of Venus' thermal regime and atmospheric dynamics, and for examining general problems in atmospheric evolution, comparative meteorology, and the paleoclimate of planets, which are directly linked with their formation. First of all, we must discover why Earth and Venus, which display many similar features and are located a relatively short distance from one another, are so different from a climatological point of view. As the closest analog of the Earth, Venus can naturally be considered an extreme "end-member" model of our own planet. Study of its thermal regime will therefore help us to understand the consequences of the continuously expanding anthropogenic effects on Earth's climate.

The Upper Atmosphere

....

8 By the upper atmosphere of a planet we mean regions located above the stratomesosphere, which is also called the middle atmosphere. The structure of Venus' atmosphere and the nomenclature for distinct regions, compared with those of the atmospheres of Earth, Mars, and Jupiter as a typical representative of the giant planets, are shown in Figure 8.1. It is obvious from this figure how temperature behaves in the upper atmosphere, compared with lower-lying regions. Charged particles—electrons and ions—which form the ionosphere, are also shown, as are their concentration with respect to altitude.

The structure and dynamics (the temperature, pressure, compositional distribution, and wind distribution) of the upper atmosphere of Venus, like those of Earth and Mars, are primarily determined by direct absorption of solar electromagnetic radiation in the short-wave spectral range. In the upper portion of the middle atmosphere and, above it, the thermosphere, the primary energy exchange is caused by the absorption of radiation in the ultraviolet and soft X-ray range (from approximately 2300 Å to 10 Å), and by precipitation of energetic particles supplied by the plasma of the solar wind. In lower-lying regions of the middle atmosphere, longer-wave ultraviolet radiation (4000–2000 Å) is absorbed. Below an altitude of approximately 130 km, Venus' atmosphere stays well mixed owing to turbulence, which dominates over molecular diffusion. Beginning at this level, called the homopause (or turbopause), turbulence becomes less effective than molecular diffusion, and each component is distributed in altitude according to its own molecular weight μ, and not the mean molecular weight $\bar{\mu}$, as in the lower-lying troposphere and stratomesosphere. More precisely, above this point each component is distributed according to its scale height $H = RT/\mu g$, where R is the universal gas constant, T is temperature, and g is acceleration due to gravity.

Study of the diverse and complex physical-chemical and dynamic processes occurring in the planet's upper atmosphere is the goal of aeronomy, which uses an extensive arsenal of experimental techniques and theoretical modeling. This modeling is based on techniques used to study the mechanics of multiple-component media, the dynamics of rarified gas and plasma, chemical kinetics, and computational mathematics.

The Neutral Upper Atmosphere

In contrast to processes in Earth's upper atmosphere, where oxygen and nitrogen play a decisive role, processes occurring in the upper atmospheres of Venus and Mars are mainly controlled by the photochemistry of carbon dioxide gas, which remains dominant on Venus to approximately 150–190 km, whereas above that its dissociation products—atomic oxygen and carbon monoxide—begin to dominate. The situation on Mars is simi-

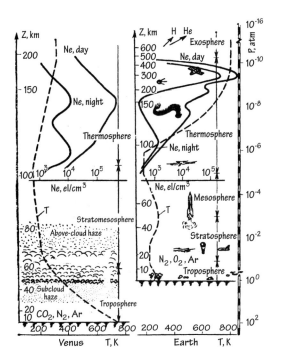

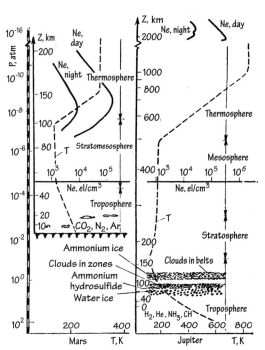

Fig. 8.1. The atmospheric structure of Venus, Earth, Mars, and Jupiter. In the center is the pressure scale (in atm), with respect to which the altitude above the surface of each planet (except for Jupiter) is given. Temperature and electron density profiles, and cloud structures, are shown.

lar, except that O and CO are about an order of magnitude less abundant. Starting at about $Z \geq 300$ km, Venus' atmosphere, like Earth's at exospheric altitudes, becomes mostly helium-hydrogen in composition.

The photochemistry of an atmosphere consisting almost entirely of CO_2 is linked with many problems concerning the structure and dynamics of the upper portion of Venus' gaseous envelope. As a result of CO_2 photolysis, dissociation products form, accompanied by inverse recombination processes under conditions of intense dynamic exchange (turbulent diffusion). Minor atmospheric constituents, primarily hydrogen and nitrogen, play a significant role in Venus' atmospheric chemistry, as do sulfur and chlorine compounds in the lower atmosphere, affecting the altitude distribution of components and the thermal regime of the mesosphere and thermosphere. We shall give special consideration to theoretical modeling and to the unique aeronomy of Venus. But first, we shall discuss the available data.

Venus' upper atmosphere has been studied by various methods using the Venera-4, Venera 9–12, Mariner-5, Mariner-10, and Pioneer-Venus spacecraft. The temperature, density, and concentration of atmospheric components were measured by several different techniques (neutral and ion mass spectrometers, accelerometers, ultraviolet spectrometers, the evolution of a spacecraft's orbit due to drag, and transmission of radio waves through the atmosphere), and generally coherent ideas emerged on the structure and variations of the neutral upper atmosphere and ionosphere of Venus (see Kuzmin and Marov, 1974; Krasnopolsky, 1983; Niemann et al., 1980; Brace et al., 1983; Seiff, 1983; Keating et al., 1985; Marov and Kolesnichenko, 1987; Izakov and Marov, 1989). The set of measurements made in the fall of 1992 during the Pioneer-Venus entry phase resulted in the most reliable data yet obtained. These PV data covered 14 years—more than one full 11-year cycle of solar activity — and were published in a special issue of *Geophysical Research Letters* (20, no. 23, 1993).

According to the results of the first spectral and spectrophotometric measurements, by orbiting spacecraft, of Venus' dayglow in the extreme ultraviolet, a value for the exospheric temperature was obtained, $T_\infty = 375 \pm 100$ K (for HeI 584 Å emission) (Broadfoot et al., 1974). A value of $T_\infty \approx 400$ K on the daylight side was later confirmed by data on drag measurements from Pioneer-Venus. On the night side, the temperature turned out to be ~ 100 K, which is almost 80 K lower than the temperature of Earth's mesopause (Keating et al., 1979, 1980).

The significantly lower temperature of Venus' exosphere, as compared to that of Earth, can be explained by the radiation in CO_2 infrared bands. For this reason the base of the Venusian exosphere lies near 250–300 km, which is approximately 200 km lower than in Earth's atmosphere. And the low temperature of Venus' nightside upper atmosphere, called the cryosphere, can be explained by a very strong non-linearly temperature-dependent CO_2 15 μm band cooling, which is regulated by changing dayside atomic oxygen (Bougher and Roble, 1991), as well as turbulent heat conduction. Other mechanisms have been invoked to explain this low temperature, including the emission of energy in the H_2O rotational band supplementing the CO_2 (Gordiets and Kulikov, 1985), but it is impossible to reconcile this proposition with the available data on the extremely low water vapor content in Venus' mesosphere.

Identification of luminescence spectra on the day and night sides, obtained by various fly-by spacecraft and Venus orbiters, made it possible to obtain the first information on the chemical composition of the neutral upper atmosphere (Kurt et al., 1972; Stewart et al., 1979). The nightglow of Venus' atmosphere in the visible region obtained with the grating spectrometers aboard the Venera-9 and Venera-10 orbiters provided data on oxygen emissions and their altitude profiles (Krasnopolsky, 1983). Specifically, four systems of O_2 bands were found, each forbidden and represented only by zero progression, owing to rapid vibrational relaxation in the atmosphere of CO_2. In the ultraviolet night airglow spectrum, the Cameron CO bands and the γ and δ NO bands were identified (Fig. 8.2). Altitude profiles of upper atmospheric components and their spatial-temporal variations were obtained for altitudes of 150–200 km near solar maximum from measurements using the Pioneer-Venus quadrupole mass spectrometer (Niemann et al., 1980). In addition, mass spectrometric measurements were carried out by the Pioneer-Venus main probe (bus) as it entered the atmosphere (von Zahn et al., 1979). These results were the basis of a semi-empirical model by Hedin et al. (1983) that completely characterizes the composition of the neutral upper atmosphere on the day and night sides of the planet at a high level of solar activity (Fig. 8.3). The collection of data on the concentrations of constituents provides direct information on the overall density and the nature of its variation with altitude (scale height). At the same time temperature values in the upper atmosphere were significantly refined.

The fundamental experimental data, collected before the new set of measurements (made at the entry phase of the Pioneer-Venus mission and during Magellan aerobrak-

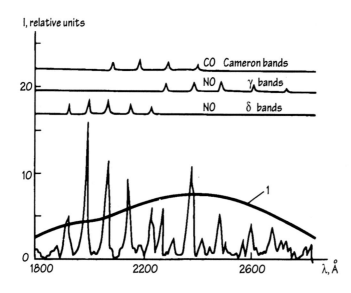

Fig. 8.2. The ultraviolet spectrum of Venus' night luminescence (CO and NO bands) from Pioneer-Venus measurements (Stewart and Barth, 1979). The smooth curve (1) is a graph of the spectrometer sensitivity versus wavelength. The intensity (I) along the vertical axis is in relative units. The absolute intensity of the luminescence, averaged over the hemisphere at $\lambda =$ 1980 Å, is I = 0.8 kR.

ing) became available, were summarized in *The Venus International Reference Atmosphere* (Keating et al., 1985), which made it possible to represent graphically the distribution of parameters in the upper atmosphere and their variations, in a generalized form.

According to this model, the altitude profiles of temperature T, density ρ, and concentration $\mathcal{N}_C$ of neutral components in the upper atmosphere at altitudes $Z \geq 100$ km mainly depend on the solar zenith angle θ; these parameters vary little on the day and night sides, though strong variations exist near the terminator. Variations governed by solar activity are superimposed on this dependence, whereas at altitudes of 100–130 km there are also latitudinal, diurnal, and semi-diurnal variations. Moreover, at all altitudes there are irregular variations, which are several times greater on the night side than on the day side.

The altitude distributions of T, ρ, and $\mathcal{N}_C$ for the fundamental neutral components at noon and midnight near the equator (more precisely, near 16° N, where most of the experimental data was obtained) are shown in Fig. 8.4 for medium solar activity, corresponding to solar radio emission at a wavelength of 10.7 cm, averaged over 81 days, $\overline{F}_{10.7} = 150$ (in units of 10^{-22} Watts $\cdot$ m^{-2} $\cdot$ Hz^{-1}).[1] The temperature of the upper atmosphere increases near midday from $T = 175$ K at $Z = 100$ km to $T = 247$ K at $Z = 150$ km, and then changes to an isothermal distribution: above $Z \approx 210$ km, $T = 283$–284 K. Near midnight the temperature decreases from $T = 175$ K at $Z = 100$ km to an isothermal value $T = 127$ K at $Z \geq 140$ km.

How does the density behave? Recall that the density near Earth's surface is $1.29 \cdot 10^{-3}$ g/cm^3 and the density at 100 km above Earth is approximately a million times less. On Venus, the density near the surface (see Table 5.7) is $61 \cdot 10^{-3}$ g/cm^3; it is lower by almost a thousand times near the upper boundary of the clouds ($\sim$70 km) and, as in Earth's atmosphere, is lower by about a million times at 100 km. Further, atmospheric density continues to decrease exponentially, being dominated by the concentration of CO_2. The concentration of atomic oxygen O begins to predominate over CO_2 at $Z \approx 160$ km during the day and $Z \approx 140$ km at night. At night the density at 100 km is almost the same as during the day, whereas at 250 km the density is substantially lower at night.

Fig. 8.3. Altitude profiles of the density of neutral components in Venus' atmosphere for the 150–250 km altitude range during day and night hours, according to mass spectrometer measurements from the Pioneer-Venus orbital spacecraft. The solid lines (1) are density profiles; the dashed lines (2) are the overall density ρ and temperature T profiles; $\mathcal{N}$ is the particle number density (according to Niemann et al., 1980).

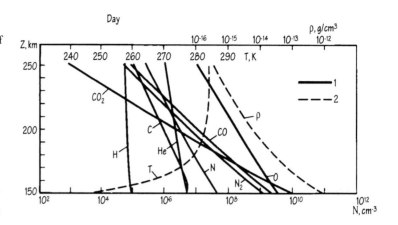

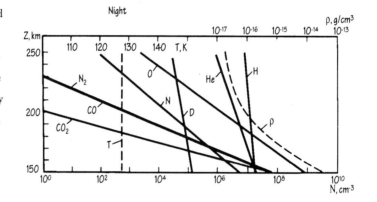

The concentrations of most components at $Z \geq 125$ km increase from night to day, with a simultaneous increase in T and scale height H (see Figs. 8.3 and 8.4). The hydrogen concentration at $Z = 250$ km, by contrast, decreases from $\mathcal{N}_{\mathrm{H}} = (1-3) \cdot 10^7$ cm^{-3} at night to $\mathcal{N}_{\mathrm{H}} \approx 1.0 \cdot 10^5$ cm^{-3} during the day, and the concentration of helium varies from $\mathcal{N}_{\mathrm{He}} = (2-4) \cdot 10^6$ cm^{-3} at night to $\mathcal{N}_{\mathrm{He}} = (1-2) \cdot 10^6$ cm^{-3} during the day. In other words, the diurnal trend in the concentration of these lightest components is inverted; this can be explained by a decrease in their thermal dissipation from the atmosphere with a decrease in temperature.

The first series of Pioneer-Venus in situ measurements was made during the interval from 4 December 1978 to 16 August 1980, when the periapsis was lowered, and covered the range 141–250 km. Thus, only a period of high solar activity ($\overline{F}_{10.7} = 220$) was covered. Nonetheless the major characteristics of Venus' atmospheric behavior were revealed from these measurements and were rather adequately described by the aforementioned models. It was found that variations in the structural parameters of Venus' thermosphere as a function of solar activity are much less than in the thermospheres of Earth and Mars. Models predicted that by increasing $\overline{F}_{10.7}$ from 70 to 200, the temperature of the upper atmosphere (exospheric temperature T_∞) varies from 240 to 300 K, whereas it varies from 500 to 1400 K for Earth and from 200 to 350 K for Mars (Keating et al., 1985; Marov and Kolesnichenko, 1987; Bougher and Dickinson, 1988a).[2]

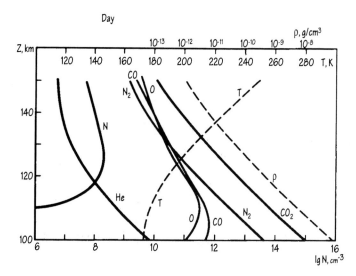

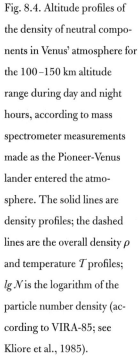

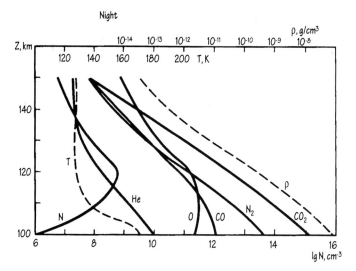

Fig. 8.4. Altitude profiles of the density of neutral components in Venus' atmosphere for the 100–150 km altitude range during day and night hours, according to mass spectrometer measurements made as the Pioneer-Venus lander entered the atmosphere. The solid lines are density profiles; the dashed lines are the overall density ρ and temperature T profiles; $lg \, \mathcal{N}$ is the logarithm of the particle number density (according to VIRA-85; see Kliore et al., 1985).

The comparison of atmospheric and ionospheric conditions for different levels of solar activity was one of the main goals of the Pioneer-Venus in situ measurements during the entry phase, when the level of solar activity ($\overline{F}_{10.7} = 120$) was much lower than when periapsis was at low altitudes in 1979 and 1980 ($\overline{F}_{10.7} = 220$). In general support of the preceding discussion, the neutral atmosphere showed very little variability from solar maximum to solar intermediate conditions, especially on the night side (Strangeway, 1993). The night-side neutral atmosphere appears to be insulated from solar-cycle-dependent changes in the day-side thermosphere, and the relatively small changes in the day-side thermosphere with solar activity have little impact on the night-side thermosphere.

This generalization was clearly manifested by the new results of mass spectrometric and atmospheric drag measurements. A set of neutral composition measurements indicated that in the post-midnight sector of Venus' atmosphere (about one in the morning local time), He was the dominant species above 170 km, O was the dominant species from 140 to 170 km, and CO_2 was the dominant species below 140 km, all the data above

133 km being in a diffusive equilibrium regime with CO_2 (Kasprzak et al., 1993a). Estimated scale-height temperatures for He, O, and CO of about 105–120 K, along with the atmospheric density and morning He bulge position, were essentially similar to those observed in 1978–1980 at higher solar activity. The thorough interpretation of these data in the framework of a theoretical global circulation model of the Venus atmosphere (TGCM) (Bougher et al., 1988a) brought confirmation of a generally consistent picture of the neutral atmospheric parameter distribution and variation.

Drag measurements were carried out at the entry phase of Pioneer-Venus and during the aerobraking maneuver of the Magellan spacecraft. The measurements jointly covered altitudes from 190 km to 130 km near the equator (11° S–11° N) on the night side of Venus at about solar minimum (up to $\overline{F}_{10.7} = 110$). These measurements provided important information on the atmospheric structure (density and temperature height distributions) and their variations in response to solar activity (Hsu et al., 1992; Keating and Hsu, 1993). An example of Magellan drag measurements over one diurnal cycle in 1992–1993 ranging from 190 km to 170 km is shown in Fig. 8.5. The results are compared with the original (asymmetric) empirical VIRA model and its updated (revised) version incorporating more recent observations.

The model supports the aforementioned ideas about compositional changes correlating with the density behavior, that is, the model shows that the ratio of O to CO_2 in the dayside thermosphere strongly responds to the level of solar activity. A negative feedback mechanism appears to be operating, where the increasing rate of CO_2 photodissociation with the increased flux of ultraviolet radiation leads to more efficient $O-CO_2$ cooling and vice versa. Therefore, the thermospheric temperature is only weakly changed. The nightside cryosphere shows even weaker temperature response, although a pronounced decrease of density at the heights where atomic oxygen dominates (above 150 km) was found during lowered solar activity (see Fig. 8.5). Reduced production and transport of O from the day side easily explain this variation, which was taken into account in the revised VIRA model (Keating and Hsu, 1993).

A characteristic feature of Venus' upper atmosphere, which intrinsically distinguishes it from Earth's, is the presence of so-called hot (or non-thermal) components. Measurements in the hydrogen Lyman-alpha line ($\lambda = 1216$ Å) led to the conclusion that, as well as atomic hydrogen distributed with a scale height specified at $T_H = 275$ K (corresponding to the temperature of the thermosphere), there is a minor addition of hydrogen atoms (with a concentration on the order of $\mathcal{N}_H \approx 10^3$ cm^{-3}) at $T_H \approx 1000$–1500 K, which influences hydrogen exosphere distribution and escape from Venus (Rodriguez et al., 1984). In addition to hot hydrogen atoms, hot oxygen, nitrogen, and carbon atoms, with temperatures up to 4000–5000 K, have also been detected. These are probably formed upon dissociative recombination of O_2^+, CO^+, O_2^+, and N_2^+ ions (Nagy et al., 1979; Paxton and Stewart, 1982; Paxton, 1985).

Note also that in Venus' emission spectrum (as in that of Mars), a non-thermal infrared emission was discovered whose source is located in the planet's mesosphere. This emission has been identified with centers of hot 9.4 μm and 10.4 μm bands. In its method

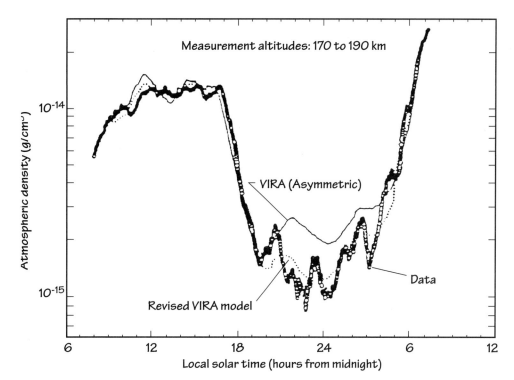

Fig. 8.5. Drag measurements of density of the Venusian atmosphere at heights ranging from 170 to 190 km in 1992–1993 (from Magellan, "cycle 4"). The results are compared to VIRA (asymmetric) and revised VIRA empirical models of Venus' upper atmosphere (according to Keating and Hsu, 1993).

of excitation the emission represents a naturally occurring laser operative in the atmosphere owing to the absorption of solar radiation in the near-infrared region, collisional excitation of the $00°$ I state of CO_2, and its radiative deactivation (Deming et al., 1983). One cannot rule out the possibility that absorption of solar near-infrared radiation by water vapor, with subsequent vibrational excitation of CO_2, may contribute to pumping of the inverse population density, causing this emission.

The Ionosphere

The first direct measurements of the atmosphere, using radio occultation carried out as spacecraft passed behind the planet, indicated that Venus possesses an ionosphere, though one less dense and more closely contracted toward the planet than that of Earth. In Earth's ionosphere the maximum (daytime) electron density (up to 10^6 electrons/cm^3) is observed at an altitude of about 280 km during the day; this is the so-called F_2 layer. Less well defined maxima lie at altitudes of $150–200$ km (the F_1 layer), about 110 km (the E layer), and 80 km (the D layer). Electron densities $\mathcal{N}_e$ in these layers vary from 10^5 electrons/cm^3 in the F_1 layer to 10^3 electrons/cm^3 in the D layer. At night the density of each layer is considerably less. Electron concentrations and the position of maxima are strongly dependent on solar activity. The dominant ion in the F_2 region is atomic oxygen; molecular ions of oxygen, nitrogen, and nitrogen oxide, whose role becomes decisive in the D and E layers, take part in the ionization of the remaining regions. Above the F_2 layer, ion and electron densities (the ionosphere is quasi-neutral) gradually drop and at approximately 1000 km are equal to the neutral gas density (primarily hydrogen). At the same time, neutral com-

Fig. 8.6. Altitude profiles of the electron density in Venus' day and night ionosphere, according to the results of radio occultation experiments from the Venera-9 and Venera-10 spacecraft for various solar zenith angles θ (according to Alexandrov et al., 1976; Ivanov-Kholodny et al., 1979): (1) θ = 14°, 10/2/75; (2) θ = 63°, 12/3/75; (3) θ = 83°, 10/23/75; (4) θ = 144°, 11/2/75; (5) θ = 146°, 11/4/75; and (6) θ = 150°, 10/28/75.

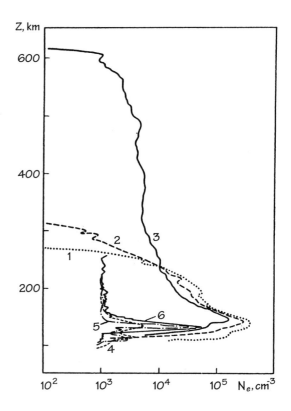

ponents are dominant over the entire lower-lying atmosphere; for example, at the altitude of the F_2 layer, the relative ion concentration does not exceed tenths of a percent.

Examples of electron density profiles $N_e(Z)$ for Venus' day and night ionosphere (Fig. 8.6) were generated from the results of a two-frequency radio occultation experiment conducted by Venera-9 and Venera-10 (Aleksandrov et al., 1976; Ivanov-Kholodny et al., 1979; Savich et al., 1986a, 1986b). A large number of similar measurements were carried out by the Pioneer-Venus orbital spacecraft (Kliore et al., 1979, 1980). A graphical representation of the distribution of Venusian ionospheric parameters, like those of the neutral upper atmosphere, is provided by the appropriate section of the VIRA-85 model (Bauer et al., 1985). It is based on measurements of the characteristics of charged components in the upper atmosphere, by all the methods that were used: radio occultation, Langmuir probes, and ion mass spectrometers. The $N_e(Z)$ profiles averaged over all measurements, in accordance with this model, are shown in Fig. 8.7.

The altitude distributions of electron density $N_e(Z)$ (and, therefore, the overall ion concentration) depend mainly on the solar zenith angle θ (see Fig. 8.7). The daytime maximum, with a density of $(2 - 5) \cdot 10^5$ electrons/cm³, is located at 140 km, in contrast to the terrestrial ionosphere. A sharp drop in electron density is observed at the 250–300 km level; the ionopause is found close to this level (from 290 km at the subsolar point to 1000 km at the terminator). This is the boundary between the thermal ions of the ionosphere (cold plasma) and fluxes of energetic particles from the solar wind plasma (see Fig. 8.6). As the result of heating and shaping of ionospheric plasma by solar wind interactions at the ionopause, an extended zone is formed at the night side—a region of great structure and variability. Some of the ions carried by this nightward flow are heated or accelerated

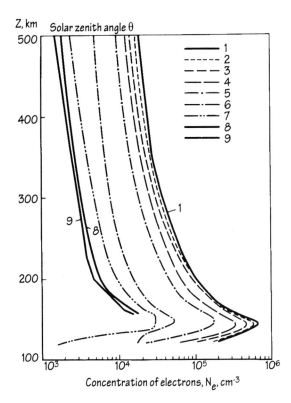

Z, km

Solar zenith angle θ

Concentration of electrons, N_e, cm^{-3}

Fig. 8.7. Altitude profiles of the electron density N_e in Venus' ionosphere, according to the results of radio occultation for various solar zenith angles θ (according to Bauer et al., 1985). (1) 0°–30°, (2) 30°–50°, (3) 50°–70°, (4) 70°–80°, (5) 80°–90°, (6) 90°–100°, (7) 100°–120°, (8) 100°–150°, and (9) 150°–180°.

enough to escape down the Venus wake region, and ionotail rays form in the umbra extending at least a few thousand kilometers downstream (Brace and Kliore, 1991).

Detailed studies of the ion composition, using mass spectrometers and retarding potential analyzers, were conducted by Pioneer-Venus. The mean altitude profiles of fundamental components in Venus' ionosphere during night and morning hours, according to mass spectrometer data collected in the early phase of the mission, are shown in Fig. 8.8 (Taylor et al., 1982; Bauer et al., 1985). Venus' lower ionosphere is mainly formed of O_2^+ ions, as is the Earth's, but with an appreciable admixture of O^+. Ions of O^+, CO^+ (or N_2^+), CO_2^+, and H^+ were also identified, and at higher altitudes ($Z \geq 180$ km) we find C^+, N^+, He^+, and $^{18}O^+$ ions (Taylor et al., 1980). Above ~160 km at night and above ~200 km during the day, O^+ ions dominate, except in predawn conditions, when H^+ has a similar concentration.

The composition of ions in the lower ionosphere and their concentration are subject to considerable variations (see Figs. 8.7 and 8.8). Based on measurements using an ion mass spectrometer, the VIRA model assumes that at the primary maximum the O_2^+ ion absolutely dominates (the O^+ and NO^+ ion density constitutes several percent of the O_2^+ density); at altitudes $Z \geq 170$ km the O^+ ion becomes dominant, whereas on the night side the contribution of H^+ ions increases.

The state of Venus' daytime ionosphere, on the whole, satisfies the condition of photochemical equilibrium; that is, it is controlled mainly by photochemical processes involving the local production and destruction of ions, similar to the ionosphere of Mars and the F_1 layer in Earth's ionosphere. The situation on the night side is more complex. Mainte-

Fig. 8.8. Altitude profiles of the density of ionized components N_i in Venus' ionosphere (according to Taylor et al., 1982).

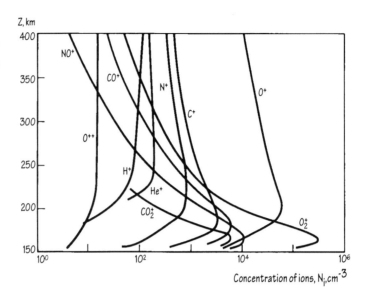

nance of the night-side ionosphere including a narrow maximum N_e, which is formed at approximately the same altitude or somewhat higher than in the daytime ionosphere, is most likely caused by ion transport from the day side, and in its morphology and nature, it more closely corresponds to the F_2 layer in the Earth's ionosphere. The horizontal transport of ions to the night hemisphere is presumably caused by a large diurnal pressure gradient of ionospheric plasma, which does not experience significant damping in the absence of an intrinsic magnetic field (Brace et al., 1983). The precipitation of energetic electrons (Gringauz et al., 1979), which may be attributed to the formation of a secondary N_e peak located below the main peak (see Fig. 8.6), evidently also provides a significant contribution to the ionization of the night-side thermosphere.

Extremely high electron T_e and ion T_i temperatures at altitudes greater than about 150 km are among the additional important features of Venus' ionosphere. They considerably exceed the temperature of neutral particles (Fig. 8.9), whereupon T_e at $Z \geq 200$ km may be as great as $T_e \approx 4000$ K, and T_i at $Z \geq 400$ km converges to T_e. At first glance it is amazing that at zenith angles $\theta = 120°$–$150°$, T_i exceeds T_e, reaching $T_i \approx 9000$ K at $Z \approx$ 1000 km. This again can be explained by the thermalization of O^+ ion flow from the day to night side, with velocities of several kilometers per second recorded near the terminator and on the night side (Knudsen et al., 1980, 1987). Thus a mechanism for supplying energy to the night side of the planet must exist, most likely associated with intense dynamic exchange. At the same time, the invocation of this mechanism also suggests inefficient local temperature relaxation, in contrast to that observed on Earth, where the electron, ion, and neutral particle temperatures do not exhibit any large differences up to an altitude of $Z \approx$ 500 km. Note that high electron and ion temperatures are retained on the night side, on a background of neutral particle temperature $T \approx 100$ K for the cryosphere.

Of special interest is the dependence of N_e, ion composition, and temperature profiles on solar activity, and these effects were studied during the entry phase of the Pioneer-Venus mission. It was found from the earlier measurements that, like the neutral

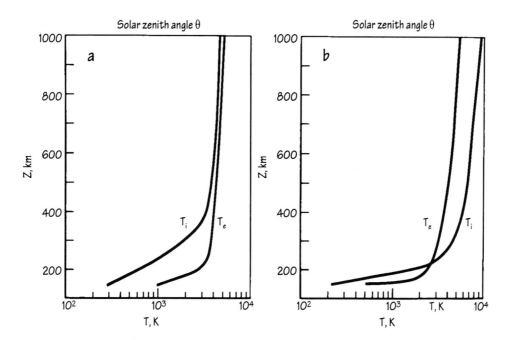

Fig. 8.9. The electron (T_e) and ion (T_i) temperatures versus altitude for two solar zenith angle ranges: $\theta = 0°–30°$ (left) and $\theta = 120°–150°$ (right) (according to Knudsen et al., 1980).

thermosphere, the day-side ionosphere exhibits a rather slight response to the level of solar activity: with an increase in $\overline{F}_{10.7}$ from 75 to 200, the electron density at the primary maximum only increases from $5 \cdot 10^5$ cm^{-3} to $7.5 \cdot 10^5$ cm^{-3}. The $\mathcal{N}_e(Z)$ profile drops off sharply at the ionopause—the area where the ionospheric plasma is compressed by the flow of solar wind plasma; the altitude of the ionopause varies from $250–450$ km for $\theta = 0°$ to $1000–2000$ km for $\theta = 120°$.

Variations in solar activity were expected to have a much stronger effect on the night-side ionosphere. Knudsen et al. (1987) assumed that both sources (ion transport from day side to night side and electron fallout) contribute to ionization of the night-side ionosphere of Venus at solar maximum. Evidently, during solar minimum the day-side $\mathcal{N}_e(Z)$ profile is reduced, causing the ionopause to move planetward and to lower its altitude at the termi-nator, thus reducing the diurnal pressure gradient of ionospheric plasma and cutting off the nightward flow of O$^+$ ions. A similar effect can be expected at some intervals of maxi-mum solar activity owing to high solar wind pressure that is thought to restrict plasma transport and to make this source of ionization less efficient. This effect of reduced trans-port is invoked to explain the observed periods of so-called disappearing ionospheres (Cravens et al., 1982).

The idea that the night-side ionosphere of Venus is depleted at lower levels of solar activity was generally supported by the measurements made during the Pioneer-Venus en-try phase, when pariapsis of the orbiter fell below 185 km to 128 km (Theis and Brace, 1993). It was found that the density of the night-side upper ionosphere experienced a huge solar cycle response, whereas $\mathcal{N}_e(Z)$ in the vicinity of the main peak did not change much between solar maximum and entry (Fig. 8.10), in contrast to the results of radio occultation measurements of Kliore and Mullin (1989) from the previous solar cycle. This fact may have implications for the relative roles of nightward ion transport and local energetic elec-

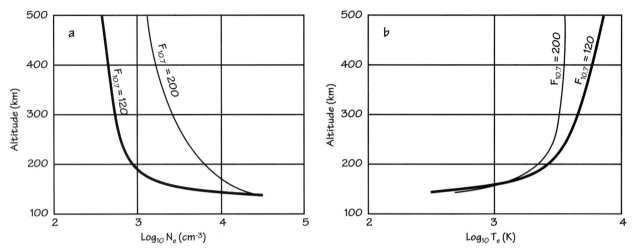

Fig. 8.10. Variations of electron density N_e (a) and electron temperature T_e (b) between maximum $\overline{F}_{10.7} = 200$ and intermediate (closer to minimum, $\overline{F}_{10.7} = 70$) levels of solar activity. N_e at the peak changed little, whereas T_e became much lower. At altitudes above 170 km, N_e was lower and T_e was higher at the lower solar activity at the time of Pioneer-Venus reentry. The explanation of this phenomenon can be found by invoking an idea of partial cutoff of nightward ion flow and heat conduction from the dayside ionosphere (according to Theis and Brace, 1993).

tron precipitation in the $N_e(Z)$ peak formation. One can assume, for example, that energetic electron fallout was much greater at the time of Pioneer-Venus entry than soon after minimum, at the end of the 1980s.

Similarly puzzling was the behavior of electron temperature T_e, which was found to be considerably higher at the heights of $N_e(Z)$ depletion. Indeed, median T_e near the peak (at 140 km) was colder by a factor of 2 than at solar maximum, but exceeded its solar maximum value by a factor of 1.3 at 200 km and by a factor of 2 at 500 km (see Fig. 8.10). These higher temperatures are reminiscent of disappearing ionospheres, and although an actual mechanism has not been identified, lower T_e in the peak region implies a reduction in the night-side heat source at lower levels of solar activity. Note that an intermediate rather than minimum solar activity cycle corresponded to the entry phase, and hence only partial reduction of nightward ion flow might be expected.

Substantial compositional differences were found between the earlier results of the Pioneer-Venus mission and those obtained during its final encounter with the atmosphere. During high solar activity the night-side ionosphere was an extensive formation, with O^+ being the dominant ion at high altitudes, presumably maintained by transport from the day side. By contrast, H^+ was severely depleted relative to O^+. At the intermediate phase of solar cycle during final encounter, a general reduction in concentrations of ions relative to that at solar maximum was observed, the topside O^+ concentrations were much lower, and H^+ dominated at high altitudes on the night side (Grebowsky et al., 1983; Cloutier and al., 1993). In addition, molecular ion species (such as O_2^+, CO^+/N_2^+, and NO^+) composed a greater part of the total ion concentration. It is not clear whether the change reflects a difference in composition of ions transported from the day side, or whether the H^+ is produced by other mechanisms and persists in the absence of O^+ transport. Obviously a distinct change in the overall night-side ionospheric dynamics, including the assumed presence of large-scale drifts, takes place, with decreasing solar activity affecting the process associated with disappearing ionospheres. It was also found that traces of superthermal O^+ ions in the energy range of 30–100 eV in the topside of the ionosphere is one of the properties of this phenomenon, although their source has not been determined. As far as molecular ions are concerned, their formation could be attributed predominantly to energetic electron impact ionization.

In sum, remarkable differences are observed in the structure, composition, and thermal regimes of the terrestrial and Venusian upper neutral atmospheres and ionospheres. Although Venus' orbit is substantially closer to the Sun than Earth's, Venus' thermosphere is much colder ($T_\infty^{\venus} = 120\text{--}300$ K) than the Earth's ($T_\infty^{\oplus} = 500\text{--}1400$ K), and its temperature variations as a function of solar activity ($\Delta T \approx 75$ K) are much less than those on Earth ($\Delta T \approx 900$ K). Venus' ionosphere is appreciably weaker than Earth's, but the diurnal asymmetry in its structure is significantly greater. This can be explained by its direct interaction with the high-energy plasma of the solar wind, because Venus has no intrinsic magnetic field to shield it out. The lack of magnetic field is evidently also linked with a marked difference in the electron-ion and neutral particle temperatures and variations of the ionospheric structure depending on the solar activity cycle. At solar maximum, when solar extreme ultraviolet (EUV) flux was largest, the Venus ionosphere was found to extend to highest altitudes and nightward ion transport was the main source maintaining the nightside ionosphere. At solar minimum, nightward ion transport appeared to be reduced, and electron precipitation was thought to be the main source. Whereas in the upper ionosphere (and in the near planet magnetotail) the effects of solar EUV variations are significant (the electron density decreases about an order of magnitude from high to low solar activity, and the electron temperature at least doubles), in the lower ionosphere lower EUV fluxes are associated with slightly reduced electron density and with higher temperature.

Thus the night-side ionosphere of Venus at low solar activity has many similarities with disappearing ionospheres at solar maximum: both have low plasma densities, drifting plasmas, and $30\text{--}100$ eV superthermal O^+ ions. In an attempt to explain the mechanisms responsible for the structure, composition, and variations of the thermosphere and ionosphere, we shall now turn our attention to some approaches and results of theoretical modeling of aeronomic processes on Venus.

Modeling the Upper Atmosphere

Theoretical models have been constructed concurrently with the accumulation of experimental data, to solve numerically a system of equations that describe physical-chemical processes in the upper atmosphere. By varying the input parameters to the model, we attempt to reconcile the model and experimental data and thereby ultimately to explain the interrelation between these phenomena and to show how they depend on specific factors. Many phenomena of the upper atmosphere are frequently attributed to hypothetical mechanisms that have not been confirmed by measurements. For example, ideas about such parameters as the wind distribution at $Z \geq 100$ km or the distribution of certain minor components are as yet based only on theoretical models.

In each aeronomic model, a planet's upper atmosphere is considered a multicomponent mixture of rarefied gases exposed to shortwave solar photons, which cause molecular photodissociation and photoionization followed by a series of chemical reactions. As a result the gas is heated, its composition changes, and motions arise in the gas; all these processes are closely linked. Because particle concentration N_Σ decreases exponentially with altitude, the mean free transit time of molecules $t_c \approx (N\sigma_g \nu_T)^{-1}$ increases (σ_g is the gas-kinetic collisional cross section, and ν_T is the thermal velocity). To within an order of magnitude, t_c is equal to the equilibration time with respect to the translational degrees of freedom of the molecules. The equilibration time with respect to the vibrational degrees of freedom is typically several orders of magnitude greater, and the characteristic times for a series of chemical reactions is even greater. As a result, at high altitudes there is a deviation from local thermodynamic equilibrium. This is manifested in the difference between the electron T_e, ion T_i, and neutral particle T temperatures (which actually exists on Venus), in the constant presence of certain chemically active components (such as O, O_3, and OH) and particles with excited metastable states, in the contrast between the vibrational and translational temperatures of certain molecules, in the deviation from Kirchhoff's law, and in the difference between the intrinsic emission of the atmosphere at high altitudes and the Planck emission.

Aeronomic Processes

Photochemistry occurs through essentially the entire Venusian atmosphere within and above the clouds. Photolysis and transport play key roles. Ultraviolet and X-ray solar radiation, with a photodissociation threshold at the wavelength $\lambda \leq 2275$ Å, is mainly absorbed by CO_2 molecules; the absorption cross section is very large in the range $\Delta\lambda \approx 150$–$1000$ Å ($\sigma_g \approx 10^{-17}$ cm^2) and remains quite large ($\sigma_g \approx 10^{-19}$ cm^2) up to approximately $\lambda \approx 1600$ Å. Absorbed photons induce CO_2 photodissociation (reaction [5.11]) and ionization, whereas shorter-wave photons excite quantum levels in CO and O.[3] These levels, which are responsible for various excited states, are given in Table 8.1.[4]

We can briefly summarize that solar radiation in the near-ultraviolet spectrum (shorter $\lambda \leq 2275$ Å) is responsible for a weak (spin-forbidden) absorption yielding photodissociation products in ground state CO ($X^1\Sigma^+$) and O (^3P), whereas beginning at 1671 Å permitted absorption yields metastable O (^1D) oxygen atoms and those in even higher exited states O (^1S) at wavelengths below 1288 Å. Triplet metastable CO appears at 1081 Å, and the lowest-lying triplet state is ($a^3\Pi$). Beginning at $\lambda \leq 902$ Å (the threshold for CO_2 ionization) the process of ionization dominates dissociation, although the dissociation channels predominantly yield an ionized fragment below $\lambda \leq 600$ Å.

The energy of excitation is partially quenched in various emissions, and it partially turns into heat, through impact deactivation, as excited particles collide with those adjacent. The discovery of emissions in Venus' upper atmosphere contributed considerably to the study of processes occurring there (see Fig. 8.2). In particular, identification of a transition from the above-mentioned excited state ($a^3\Pi$), which forms at $\lambda > 1081$ Å, with Cameron CO bands well known in molecular spectroscopy (Stewart and Barth, 1979), allowed improved understanding of the energy sink mechanism on Venus.

Table 8.1   Various quantum excited states of CO_2 photodissociation products (CO and O) as a result of reaction (5.11)

CO state	O state		
	^3P	^1D	^1S
X^1	2270	1670	1273
$a^3\Pi$	1080	917	677
$A^1\Pi$	913	798	677

Note: These states correspond to the wavelengths (in angstroms) of ultraviolet radiation.

Although energetic photons are extremely important, the processes in the middle and upper atmosphere are dominated by a relatively weak part of the ultraviolet continuum resulting in ground state O and CO and hence "odd-oxygen" production, because of the larger solar fluxes longward of 1671 Å. In addition, the CO + O (^1D) continuum, which peaks in the mesosphere or lower thermosphere, and the lowest energy continuum of O_2 below this level also contribute to the production rate of odd oxygen, similar to the photochemistry of the atmosphere of Mars (Barth et al., 1992).

A certain fraction of the photons' energy, absorbed during photodissociation, is transformed into heat; the estimates result in a heat release efficiency of $\epsilon \approx 0.2-0.3$ (McElroy et al., 1973). The products of CO_2 photolysis enter into various chemical reactions: in current models about a hundred photochemical and chemical reactions are accounted for (see, for example, Sze and McElroy, 1975; von Zahn et al., 1980; Krasnopolsky, 1983b, 1986; Massey et al., 1983; Shimazaki et al., 1984). Many of these catalyze the chemical recombination of CO and O; this is extremely important, because their direct association through collisions (see Eq. [5.12]) is very slow—the reaction rate is $k = 10^{-37}$ cm^6/sec. This can explain the prevalence of CO_2 over CO and O, up to $Z \approx 160$ km.

In Chapter 5 we showed that processes of atmospheric chemistry in the middle atmosphere are varied and complex. Catalytic reactions with the assistance of hydrogen or its radicals, mainly OH, are most likely the decisive mechanism ensuring the maintenance of a predominantly CO_2 stratomesosphere and lower portion of the thermosphere. As a function of the relative X_{H_2} content, we earlier considered a chain of possible reactions (5.17)–(5.21). It was shown that hydroxyl, which is required for the essential reaction (5.18), is formed through photodissociation of HCl (reactions [5.15] and [5.17]). In addition to these main chemical processes, some additional reactions may be responsible for hydroxyl formation at high altitudes, as follows:

$$H + O_2 + CO_2 \rightarrow HO_2 + CO_2 \qquad (8.1)$$

$$O + HO_2 \rightarrow OH + O_2 \qquad (8.2)$$

$$H + HO_2 \rightarrow 2OH \qquad (8.3)$$

$$HO_2 + HO_2 \rightarrow H_2O_2 + O_2 \qquad (8.4)$$

$$HO_2 + h\nu \rightarrow OH + O \qquad (8.5)$$

$$H + O_3 \rightarrow OH + O_2 \qquad (8.6)$$

$$H_2O_2 + h\nu \rightarrow 2OH \qquad (8.7)$$

In turn, chlorine, obtained as a result of reaction (5.15), also exerts a catalytic effect, and with the participation of chlorine radicals added to reactions (5.20) and (5.21), the following reactions are also possible:

$$Cl + O_2 + CO \rightarrow ClO_2 + CO \qquad (8.8)$$

$$ClO_2 + CO \rightarrow ClO + CO_2 \qquad (8.9)$$

$$ClO + CO \rightarrow Cl + CO_2 \qquad (8.10)$$

Finally, sulfur compounds, which are present at $Z \approx 60\text{--}80$ km, may also affect this general picture, especially reactions associated with the release of free hydrogen (see [5.33], [5.36], and [5.37]). Note that most reaction rates are known from laboratory measurements, although some are only estimates, as summarized by Krasnopolsky and Parshev (1983b). This significantly limits our ability to determine the probabilities of the corresponding processes and, therefore, their actual contributions to the distribution of atmospheric components.

Modeling of the Neutral Atmosphere

It is possible to make a complete description of such a complex system as a planetary upper atmosphere by using a system of modified Boltzmann equations for each atmospheric component, with integrals for elastic and inelastic collisions (see Marov et al., 1996). For certain large-scale problems, however, including a global description of the structure and dynamics of the upper atmosphere, one can employ a system of hydrodynamic equations essentially up to the exosphere (where the mean free path is greater than the scale height): continuum equations for $\mathcal{N}_c$ densities of components with terms that describe the birth and destruction of each type of particle in photochemical and chemical reactions; momentum equations for neutral and charged components; energy balance equations for T, T_i, and T_e with heat sources primarily from the absorption of ultraviolet solar radiation and releases of heat primarily from the infrared emission. This approach and the corresponding theoretical background, used to formulate and solve model problems in aeronomy, have been studied in detail in monographs by Marov and Kolesnichenko (1987), and Chamberlain and Hunten (1987).

We should add that the presence of turbulence up to the homopause, or turbopause (which on Venus is located at an altitude of $Z_T \approx 130$ km and on Mars at $Z_T \approx 135$ km, whereas on Earth it is about 25 km lower), means that for $Z < Z_T$ all components are distributed with the same mean scale height H, whereas at $Z > Z_T$ each component is distributed according to its own scale height H_c. Dynamic processes on all scales involving turbulence are parameterized by introducing the coefficient of turbulent diffusion K_T, which is often used as a parameter for matching the model number density distribution

$\mathcal{N}(Z)$ with experiment.[5] Note that turbulence affects the thermal regime of the upper atmosphere in two ways: on the one hand, dissipation of turbulent energy is an additional source of heat; on the other hand, turbulent heat transfer, which is much greater than molecular, causes the thermosphere to cool. On Venus the second effect is dominant (Marov and Kolesnichenko, 1987).

In an attempt to clarify the composition and atmospheric chemistry, one-dimensional models were developed; the one-dimensionality quite naturally limits the description of the structure and variations in the upper atmosphere. In these models a temperature versus altitude $T(Z)$ dependence is usually introduced, based on experimental data. To accommodate experimental $\mathcal{N}_c(Z)$ profiles, the required K_T values, used as a fitting parameter for the models, turned out to be extremely different, usually lying within 10^6–10^8 cm^2/sec. A more realistic value, $K_T = 5 \cdot 10^6$ cm^2/sec, was obtained in a model, where the composition is calculated along with large-scale motions (Bougher et al., 1988a).

A number of radiative-convective and atmospheric dynamics models that were developed contain certain limitations and simplifications, and some input parameters are varied arbitrarily to reconcile the approximated and experimental data. Nevertheless, these models have enabled us to explain (sometimes semiquantitatively) significant general features of the planet's upper atmosphere, such as thermal structure and seasonal-latitudinal and temporal variations. In particular, low thermospheric temperatures were reproduced with an assumption of low efficiency of conversion of solar EUV radiation to heat on Venus and Mars as compared to Earth (15–20% in contrast to 30–40%), that is, lower than the earlier estimates of McElroy et al. (1973). Such a low heating efficiency is supported by the experimental data on the large intensity of the ultraviolet airglow; the mechanism of non-LTE cooling in the 15-μm of CO_2 molecules has been invoked through their collisions with ambient O atoms and by excitation transfer from metastable atomic oxygen (Stewart, 1972; Bougher and Dickinson, 1988a).

In view of the close connection and interdependence between the thermal regime and the composition and dynamics of the upper atmosphere, the system of three-dimensional continuum, motion, and energy equations must ideally be solved simultaneously in order to provide the most consistent description. Remarkable progress in this approach has been achieved since the first dynamical models of Venus by Dickinson (1972, 1976) and Dickinson and Ridley (1972, 1975, 1977). These models predicted strong night-side convergence and descending winds resulting in substantial adiabatic heating, which was confirmed in later works (Dickinson et al., 1981, 1984; Dickinson and Bougher, 1986). The most comprehensive analysis and calculations of circulation in the Venusian thermosphere (and in that of Mars) were made in the framework of the aforementioned TGCM model, incorporating detailed tidal parameterizations for examining the influence of the lower atmosphere on the thermosphere (Bougher et al., 1988a, 1988b; 1991; Fox and Bougher, 1991). Global distributions of major (CO_2, CO, N_2, and O) and minor (O_2, $N(^2D)$, and $N(^4S)$) species were calculated consistent with the model day-night temperature contrasts and corresponding large-scale winds above 95 km. The model revealed specific patterns in which the calculated densities respond to the wind and temperature structure, with atomic oxygen being subject to global redistribution. It was further suggested that momentum sink terms were needed to match the Pioneer-Venus day-night temperature and density contrasts, thereby reducing the strong night-side heating effect and atomic oxygen bulge. Superrotation was prescribed and added to the mean subsolar-antisolar circulation.

The Venus TGCM also turned out to be an important modeling tool for interpreting the mass spectrometry data of neutral composition measurements at the Pioneer-Venus reentry phase. In particular, the presence of CO, N, and O on the night side implied that these species created only on the day side were transported by day-to-night circulation that was still important at lower solar activity. In other words, upper thermosphere wind speeds are nearly the same for high and medium solar activity levels, in accord with the model's estimates, although some differences in the altitude variation of the wind systems can be anticipated. The abundances of N, N_2, O, CO_2, CO, and He proved to be within about 25% of those predicted by the TGCM model for the respective conditions. A morning helium bulge detected at the earlier data taken at solar maximum (Niemann et al., 1980) was still observed in 1992 and had about the same amplitude.

Superrotating winds, in addition to eddy diffusion and wave drag, are important parameters for simulating the inferred temperatures. Original estimates of the temperature variations as being within 55–70 K between maximum and minimum levels of solar activity conform quite well to the aforementioned idea of an important role for the CO_2–O mechanism in thermospheric cooling. This thermostat seems to effectively buffer the response of the entire upper atmosphere to changes in the solar EUV fluxes, which are the main heating source. The Rayleigh friction, which slows down the pressure-driven winds, can also play an important role in the maintenance of relative isolation of both day-side and night-side thermospheres. The Rayleigh friction is attributed to upward-propagating internal gravity waves, which break at high altitudes, thus modifying the horizontal flow. Wave-like perturbations have been observed in the night-side neutral density data acquired during Pioneer-Venus Orbiter reentry. These data were similar to the earlier findings. It was suggested that CO_2 amplitudes grew in value with altitude up to about 140–170 km and then decreased (Kasprzak et al., 1993b). These perturbations can be interpreted as gravity waves propagating upward from the lower thermosphere (as low as 80 km), in accordance with a model by Mayr et al. (1988), which gives support to the proposed mechanism.

Modeling of the Ionosphere

We now turn to a model description of the ionosphere. Photoionization is produced by photons with $\lambda < 902$ Å:

$$CO_2 + h\nu \rightarrow CO_2^+ + e \qquad (8.11)$$

Photons with λ less than 716 Å, 687 Å, and 640 Å cause ionization with additional excitation of the corresponding quantum levels of the CO_2^+ ion ($A^3\Pi_u$, $B^2\Sigma_u^-$ and $C^2\Sigma_g^+$), with the excitation energy $W^j_{CO_2+}$ of the j-th level of the ion. In turn, photons with wavelengths less than 548 Å and 600 Å produce dissociative photoionization:

$$CO_2 + h\nu \rightarrow CO + O^+ + e \qquad (8.12)$$

$$CO_2 + h\nu \rightarrow CO^+ + O + e \qquad (8.13)$$

Because the energy of all these photons $h\nu \geq 20$ eV, and because the ionization potential $I_{CO_2} = 13.8$ eV, the excess energy $E = h\nu - I_{CO_2} - W_{CO_2^+}^j$ is quite large. The ejected photoelectrons carry away essentially all the energy E and expend it during dissociation, ionization, and excitation of particles through electron collision (see Marov et al., 1996, for more detail).

The ion composition varies significantly as a result of the action of extremely rapid ion-molecular reactions, at rates of $k_{i-m} \approx 10^{-9}\text{–}10^{-10}$ cm^3/sec. As an example of such a process, we shall present the most characteristic reactions:

$$CO_2^+ + O \rightarrow O_2^+ + CO \qquad (8.14)$$

$$CO_2^+ + O \rightarrow O^+ + CO_2 \qquad (8.15)$$

$$CO_2^+ + O_2 \rightarrow O_2^+ + CO_2 \qquad (8.16)$$

$$O^+ + CO_2 \rightarrow O_2^+ + CO \qquad (8.17)$$

In the night-side ionosphere, dissociative photoionization reactions with the participation of light ions (hydrogen and helium) are important, such as the following:

$$H^+ + CO_2 \rightarrow CO + H^+ + O \qquad (8.18)$$

$$He^+ + CO_2 \rightarrow O^+ + CO + He \qquad (8.19)$$

$$He^+ + CO_2 \rightarrow O + CO^+ + He \qquad (8.20)$$

The major source of He^+ is direct photoionization, whereas H^+, as in the terrestrial case, comes from charge exchange with O^+ (Nagy et al., 1983).

In ionospheric chemistry on Venus (as on Earth and Mars), odd nitrogen plays an important role, with NO^+ being the terminal ion in all cases. Whereas the contribution of direct photoionization to the N^+ production rate becomes significant only above $\sim$200 km, this ion can also be produced by dissociative photoionization with the involvement of He^+. An important source at low altitudes is NO^+, which is produced by a variety of reactions. The most efficient is the reaction of O_2^+ with atomic nitrogen. The quenching of produced $N(^2D)$ occurs by O and CO, and this process is sufficient to suppress the formation of NO made and consumed by reaction with $N(^4S)$. Hence the abundance of the surplus $N(^4S)$ is controlled by transport rather than photochemistry (as on Earth). According to Barth et al. (1992), a large fraction of the N is carried to the night side by the thermospheric winds, where it eventually recombines radiatively with atomic oxygen under conditions of higher pressure. Calculations coupling with Venusian odd-nitrogen chemistry and global atmospheric dynamics (TGCM) confirm the role of N-atom transport in producing the observed NO nightglow (Bougher et al., 1990).

The loss of ions and electrons does not occur directly but takes place mainly through so-called dissociative recombination reactions, that is, when the combination of an electron to a molecular ion simultaneously results in its splitting into atoms:

$$O_2^+ + e \rightarrow O + O \tag{8.21}$$

$$CO_2^+ + e \rightarrow CO + O \tag{8.22}$$

Recombination products may be found in various states of excitation; for example, in the CO_2 dissociative recombination reaction, CO molecules may be formed in the ground state ($^1\Sigma$), and O atoms can be formed in either the ground state (^3P) or excited states (^1D), (^1S). Consequently, these exothermic reactions will have various degrees of energy release. In symbolic form this can be written as follows:

$$CO_2^+ + e \rightarrow CO(^1\Sigma) + O(^3P) + 8.35\,eV$$

$$CO_2^+ + e \rightarrow CO(^1\Sigma) + O(^1D) + 6.39\,eV \tag{8.23}$$

$$CO_2^+ + e \rightarrow CO(^1\Sigma) + O(^1S) + 4.18\,eV$$

Moreover, channels of this reaction are possible with the formation of excited CO molecules (in $a^3\Pi$ or $A^1\Pi$ states), although the probabilities of various channels are different (see Izakov and Marov, 1989). Of all these transformation paths, CO formation in the ground state ($^1\Sigma$) and O formation in the (^3P) and (^1D) states are most likely.

The excess energy in the channels presented in reaction (8.23), as in other similar reactions, is thermalized. Moreover, a portion of the energy from excited levels resulting from all photolytic reactions is transformed into heat during impact deactivation. Thermalization channels are clearly numerous, and not all their characteristics are sufficiently well known. Therefore, the efficiency of heat release during photoionization or photodissociation is usually estimated, given certain simplifications. For Venus' ionosphere the quantity turns out to be highly variable: $\epsilon \approx 0.2-0.25$ (Fox and Dalgarno, 1981; Stewart, 1972). The correctness of such estimates is of fundamental importance, because this quantity is often used as a fitting parameter in models of the thermal regime in the upper atmosphere. We must add that with respect to heat, ions with energies higher than the photodissociation threshold ($I_{dis} \geq 6 - 7\,eV$) are apparently especially efficient because, below this threshold, in a carbon dioxide atmosphere energy loss is most likely due to re-radiation from the rotational-vibrational levels of the CO_2 molecule, especially in the fundamental 15 μm band.

In the reactions we have considered, and upon thermalization of photoelectrons, most extreme ultraviolet energy is liberated as heat, because inverse association reactions of CO and O at high altitudes without the participation of tertiary particles as a catalyst are inefficient. Such tertiary particles may play a definite role only in the mesosphere and lower thermosphere, where reactions of photolysis products may occur. The remaining portion of extreme ultraviolet energy is, apparently, scattered and re-radiated by the atmosphere. Emissions of the upper atmosphere (see Fig. 8.2), are characteristic channels for such re-radiation.

In spite of these difficulties and remaining uncertainties, taking into account the processes described earlier, the matching of the ion densities in models of the daytime ionospheric composition with the measurements seems quite satisfactory (Izakov et al., 1981; Bauer et al., 1985; Nagy et al., 1979, 1982). There is also satisfactory agreement between experimental data and modeled electron and ion temperatures.

As an example, in the model by Izakov et al. (1981; see also Marov and Kolesnichenko, 1987), the concentrations of 12 ions (O_2^+, CO_2^+, O^+, He^+, H^+, N^+, CO^+, C^+, CO_2H^+, HCO^+, NO^+, and N_2^+) were calculated at altitudes of 100–220 km, taking into account 64 photochemical and chemical reactions. The calculations were conducted in approximation of photochemical equilibrium, because estimates have shown that at these altitudes the characteristic times for photochemical processes are much shorter than those for diffusion processes. The model ion concentrations, examples of which are shown in Figs. 8.11 and 8.12 for conditions at the subsolar point and near the terminator ($\theta = 80°$), respectively, were quite successfully matched with measured values at essentially all altitudes. To describe the ion composition to an accuracy not worse than 5%, it was sufficient to retain only 48 reactions.

More difficult problems are found in attempts to incorporate into models dynamic phenomena in the ionosphere, and in particular to incorporate its interaction with processes initiated by the oncoming high-energy plasma of the solar wind. Support was found for the idea that a quite dense ionosphere is retained on the night side over the long Venusian night owing to the intense transport of ions from the day side and precipitation of energetic electrons (Knudsen et al., 1980, 1987; Cravens et al., 1983). Transport of ions appears to dominate at high solar activity, and precipitation of electrons appears to dominate at low solar activity. Earlier estimates (Miller et al., 1984; Shimazaki et al., 1984) showed that in order to reconcile the data on the ion density and temperature (obtained using an analyzer with retarding potential on the Pioneer-Venus orbiter), a vertical flow of O^+ ions not less than 10^8 cm^{-2}sec^{-1} is required at the level of the ionopause. To explain such a large flow, an "elongation" of the ionospheric plasma (as a consequence of induced electrical fields) by the onrushing solar wind was assumed to be responsible for the plasma transport to the night side.

More detailed discussion of the chemistry of the Venus night-side ionosphere for the case in which ionization is maintained by plasma transport from the day side was given later by Fox (1992). It was shown, as in other models, that the peak O^+ densities are approximately proportional to the flux of superthermal O^+ ions from the day side, whereas the O_2^+ peak density is roughly proportional to the square root of the O^+ flux. Theoretically the ratio of these ion species at the peak of ionization is valuable for judging the sources of ionization. It was also found that electron precipitation alone was incapable of reproducing the large observed O^+ densities at high altitudes.

Most recently an attempt to find evidence for day-to-night ion transport at low solar activity in the Venus pre-dawn ionosphere was undertaken based on the data collected during the Pioneer-Venus final entry phase (Branon et al., 1993). In parallel to the nightward ion transport model, an auroral model was designed to reproduce the measured O^+ densities through production rates of ions from auroral precipitation. Average ratios of atomic

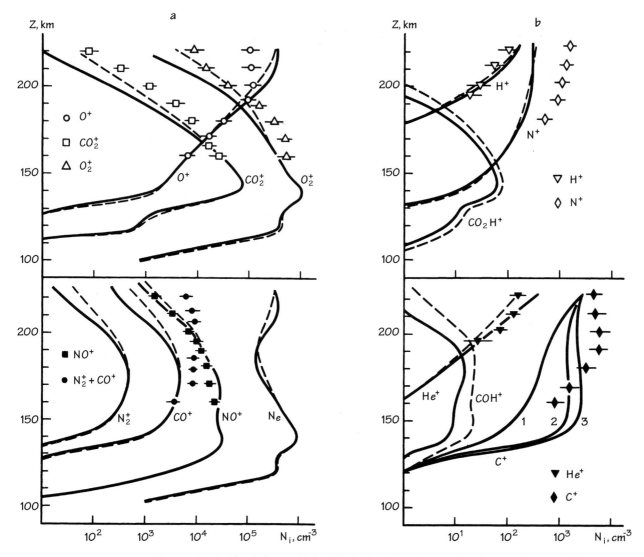

Fig. 8.11. Ion (a, b) and electron (a) densities in Venus' ionosphere at the subsolar point. Graph (a) represents the ions O_2^+, O^+, CO_2^+, N_2^+, CO^+, and NO^+; graph (b) represents the ions H^+, N^+, CO_2H^+, He^+, C^+, and COH^+; curves 1, 2, and 3 relate to C^+ ions for three ratios of the CO_2 ionization cross section, for which C^+ and CO_2^+ arise: $\sigma_{CO_2}^{C^+} / \sigma_{CO_2}^{CO_2^+} = 0, 0.3$, and 0.5, respectively. The solid and dashed lines correspond to two different calculation methods.

ions C^+, N^+, and He^+ for the transport model were computed based on the Pioneer-Venus orbiter neutral mass spectrometer measurements at altitudes of $180-250$ km to determine the ratio of the atomic ion fluxes to impose at the upper boundary of the neutral models. The selection of a neutral model is a potentially significant source of uncertainty for these computations. Ionospheric models were constructed for different solar zenith angles, and for each zenith angle the maximum in the O^+ density profile, as well as the profiles for other ionospheric species, was computed. The downward O^+ flux of 1×10^7 cm^{-2}s^{-1} at the upper boundary of the model was obtained as the average computed from the measured values of the O^+ maxima for the reentry data. A similar fitting procedure in terms of production rates of ions was used to reproduce the measured O^+ maximum in the framework

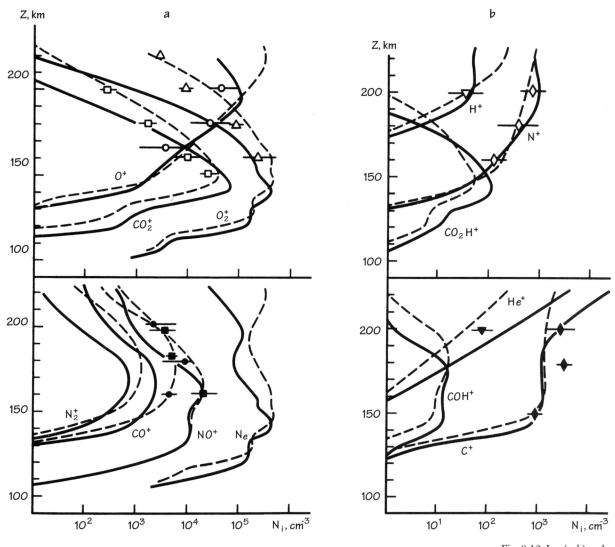

Fig. 8.12. Ion (a, b) and electron (a) densities in Venus' ionosphere for solar zenith angle $\theta = 80°$. Notations are the same as for Fig. 8.11.

of auroral models. It was found, however, that in contrast to the transport model, the auroral model does not fit the other atomic ions at all. This result therefore gave support to the idea that the day-to-night transport of ions is possibly not completely cut off at low solar activity. This important problem needs to be additionally clarified.

To summarize, there is generally satisfactory agreement between the experimental picture and the theoretical models of Venus' upper atmosphere, so we have apparently found an approach for understanding the ongoing processes there. Nevertheless, many details need further study, without which an adequate representation of the aeronomy of our neighboring planet is impossible.

The Near-Planetary Medium

The physics and aeronomy of the upper atmosphere are most intimately linked with the characteristic features of planetary interaction with the oncoming flow of solar plasma. No celestial body within the solar system exists in isolation, unaffected by processes occurring on the Sun. Change in solar activity is accompanied by significant variations in the flow of electromagnetic and corpuscular radiation, which directly interacts with the outermost regions of the medium adjacent to a planet—its upper atmosphere, ionosphere, and magnetosphere. The gas and magnetic "shields" of a planet impede the direct penetration to the surface of the hardest portion of the solar spectrum (ultraviolet and X-ray radiation) and the most energetic charged particles present in the solar plasma flow. Affected regions of the near-planetary medium undergo serious changes: molecules decay into atoms (dissociate); a portion of the atoms and molecules ionize and form the ionosphere. If, moreover, the planet has a magnetic field, then a portion of its force lines are "carried off" to the night side, forming its magnetic tail. In the magnetic field, solar plasma particles are accelerated and focused and, encroaching upon the atmosphere, cause magnificent natural phenomena—the polar aurorae. Particles captured by the force lines of the magnetic field form radiation belts, and other regions of the overall complicated structure of the planetary magnetosphere form. In the absence of a field, other effects arise; in this case the ionosphere plays the main role.

A comparative study of the features of the environments and processes in the boundary regions of planets and in their atmospheres has made it possible to establish some general physical principles and to better understand the solar system as a unified whole. Nature has placed several models at our disposal, whose fundamental differences lie primarily in the presence or absence of an intrinsic magnetic field and gas envelope around the planet. Of the terrestrial planets, only Earth possesses a significant magnetic field. Another extreme case is the Moon, which is deprived of both a magnetic field and an atmosphere. Venus has a thick gaseous envelope but no magnetic field, whereas Mars possibly possesses a weak field (although this has not been confirmed) and a rarefied atmosphere. Jupiter, which has an especially strong magnetic field, in many respects resembles a pulsar (a neutron star possessing an incredibly high magnetic field), or source of pulsating radio radiation; this similarity reflects the common nature of physical mechanisms at work in the universe.

Topology and Processes in Venus' Space Environment

The formation of a transitional ionopause zone on the day side of the planet in a region located beyond the shock wave at altitudes higher than about 300–500 km is the most characteristic feature of solar plasma interaction with Venus. The ionopause (briefly described earlier this chapter) is formed in a zone where the pressure of the solar wind (which works out to be approximately 10^{-14} bar for Venus) is nearly balanced by the kinetic pressure of the atmospheric gas, along with the pressure of the magnetic field induced by ionospheric currents. Thus, Venus develops its own magnetic barrier. In the earlier semi-ideal models of the ionosphere, calculations of the current distribution were based on minimum ohmic dissipation or, in other words, almost infinite conductivity, employing the simplest form of the well-known generalized Ohm's law. Under this assumption, ionospheric currents should be localized in a narrow layer 40–50 km thick at an altitude of 150–200 km, corresponding to maximum conductivity in a partially ionized ionospheric plasma (Cloutier and Daniell, 1973; Spreiter, 1976).

In fact the interactions turned out to be considerably more complicated, as revealed by the results of Venera and Pioneer-Venus plasma experiments. The complex processes in the region of streamline flow, in addition to the formation of an intermediate zone identified with the ionopause, include heating and thermalization of ions and the formation of a rarefied zone behind the shock wave (the ionosheath, approximately analogous to the magnetosheath in Earth's magnetosphere). To this we must add the important role of dynamics and electrical fields within the ionosphere, the formation of a rarefied zone beyond the planet, and so on. Researchers in this field (see, for example, Russell and Vaisberg, 1983; Cloutier et al., 1983; Brace et al., 1983) consider these effects to be fundamentally unlike those observed in solar wind interactions with the magnetosphere of Earth, or with other planets with a magnetic field.

Indeed, the strong compression of the daytime ionosphere, the plasma characteristics in the narrow transitional zone of the shock front, and the lack of any sharp increase in magnetic field intensity are indications of magnetohydrodynamic processes beyond the shock wave, owing to which Venus' streamline flow characteristics differ greatly from the interaction of a plasma flow with a rigid obstacle, as was suggested in earlier models. Therefore, it was reasonably assumed (see Russell and Vaisberg, 1983) that the inflow of hot atmospheric components (mainly O and H) subjected to photoionization and charge exchange, which may subsequently be captured by solar plasma currents, is central to these processes. Study of the nature of streamline flow within the scope of the theoretical model by Cloutier et al. (1983) suggested further that a current system forms under the influence of a solar plasma pressure gradient permeating all regions of the ionosphere, ionopause, and ionosheath. Strong plasma disturbances generated at the obstacles and distributed within the ionosheath, involving turbulent wave fields (instead of being produced locally by such chemical processes as mass loading and charge exchange collisions), modify the local solar wind flow as they travel downstream (Perez-de-Tejada, 1995).

Induced electrical fields and pressure gradients in the ionosphere in turn cause large-scale convection and transport of ionospheric plasma from the day side to the night side. Convective motions are characterized by spatial asymmetry and a maximum velocity U in the direction perpendicular to the magnetic field intensity vector **B** within the ionosphere. This field, which depends on the magnitude and direction of the interplanetary magnetic field, is subject to strong fluctuations. As a result, in regions of the ionosphere with a weakened magnetic field, whose **B** vector makes the largest angle with U, magnetohydrodynamic shear instabilities are likely to develop (like the Kelvin-Helmholtz instability, which is well known in physics).

The Pioneer-Venus final entry phase allowed the detailed examination of the magnetic field and plasma pressures in the low-altitude night-side ionosphere of Venus. These studies showed that above about 160 km the magnetic field generally became stronger when the solar activity decreased. Since the electron density became weaker and the temperature was only slightly greater at that time, the net result was that the magnetic pressure equaled or dominated the thermal pressure above that altitude level, in contrast to the situation at times of high solar activity, when the thermal pressure dominated the magnetic pressure up to at least 300 km (Russell et al., 1993). Thus, when solar activity drops, the magnetic field exercises stronger control over the dynamics of the night-side ionosphere, and probably because of this effect fully developed disappearing ionospheres prevail at solar minimum. These resemble the weaker disappearing ionospheres seen at solar maximum (Knudsen et al., 1987; Zhang et al., 1990).

Even more dramatic variations are attributed to the night-side upper ionosphere, above about 1800 km, where some significant changes to the plasma and magnetic field for different solar EUV conditions were detected (Ho et al., 1993). For the relatively low level of solar activity at the time of Pioneer-Venus entry, the electron density proved to be decreased by about an order of magnitude, as compared with the conditions during high solar activity, when the electron density was more extensive, with larger concentrations extending to higher altitudes. Accompanying this decrease, the electron temperature doubled, and the magnetic field was enhanced by about 2–3 nT. Again, these variations are consistent with the weakening of ion transport from day side to night side owing to the decline of the day-side ionosphere density and ionopause altitude at terminator for lower EUV fluxes. The associated depression in $N_e(Z)$ should be accompanied by an increase of magnetic field and T_e to maintain pressure balance with the magnetosheath. Both are expected to increase because there is likely to be greater magnetic field transport and heat flux from the magnetosheath when the ionosphere is depleted (Theis and Brace, 1993).

A possible explanation for the rather embarrassing situation with $N_e(Z)$ and T_e behavior during the solar cycle (as described in the first section of this chapter) lies in the idea that the depleted upper ionosphere is thermally connected to the hot plasma of the mantle or the ionosheath, rather than to the cool underlying ionosphere. This idea is related to a reduction in the energetic (> 100 eV) electron population ionosheath downstream of Venus that was first detected during Mariner-10 fly-by in 1974 (Bridge et al., 1974) and confirmed in early 1992 by the measurements in the Pioneer-Venus entry phase of the night-side ionosheath by orbiter plasma analyzer (Intriligator et al., 1993). This energetic electron depletion is most probably associated with direct atmospheric scattering, when electrons travel along draped magnetic flux tubes, which thread through Venus' neutral atmosphere, and lose energy from impact ionization with oxygen. The cutoff at about 100 eV is explained by the fact that the cross section for such electron impact ionization of oxygen has a peak near 100 eV and remains high above this energy, so atmospheric loss could provide a natural process for electrons at these energies to be selectively removed. Although an alternative explanation of this phenomenon (incorporating magnetic connection of depleted regions to the weak downstream bow shock and thus resulting in a reduced source strength) is also possible, the described mechanism appears to find confirmation by entry ion mass spectrometer data and therefore supports the idea of ionization sources and maintenance of the night ionosphere (Kar, 1996).

Figure 8.13 pictures the streamline flow by the solar wind around Venus. The figure shows the important role of ionospheric convection and interactions between ionospheric

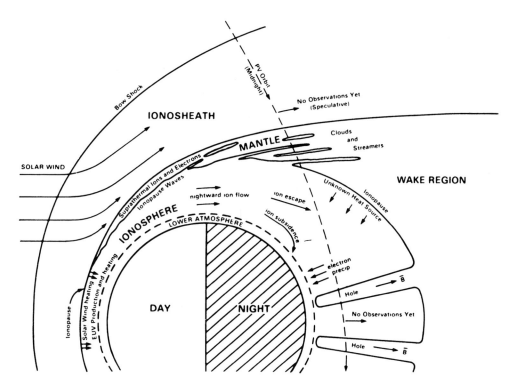

Fig. 8.13. The nature of Venus' interaction with the interplanetary medium as it flows through the solar wind plasma. The most characteristic regions formed by this flow, and a series of accompanying processes, are shown in the figure, constructed from available ionospheric measurements by Brace et al. (1983). (For additional explanations, see the text.) The size of the ionosphere is enlarged by approximately three times compared to other regions of the near-planetary medium.

and interplanetary magnetic fields (Brace et al., 1983). According to this model, the magnetic pressure in the daytime ionosphere is considerably less than the thermal pressure of the plasma and may become comparable only near the terminator, where it presumably exerts its greatest influence on the morphology of motions. The strongest magnetic fields are anticipated in the nighttime ionosphere, especially in regions of large-scale "holes," which evidently reflect interaction between the evacuated zone of the ionotail (the analog of the magnetic tail) and the ionosphere. The magnetic tail is probably formed by the effect of the plasma current flowing around Venus ("accretion" of the interplanetary field by the planet), and especially by the high dynamic pressure of the solar wind, as a consequence of unipolar induction.

The plasma mantle, adjacent to the ionopause at large zenith angles (see Fig. 8.6), is a complex structure governed by asymmetry and fluctuation of the induced magnetic field. According to experimental data, it is thicker at high latitudes than at low, and on the night side it borders a region filled with plasma, which slides into the magnetic tail. In this intermediate zone the concept of an ionopause becomes vague, because there is no well-defined boundary separating the ionosphere, with its high density and weak magnetic field, from the magnetic barrier formed from the inflow of solar plasma, with its strong magnetic field and low particle density. In addition to ions transported by convective movements to the ionosphere, there is apparently partial spillover of electrons from the night side. But on the whole the results of theoretical and experimental studies lead to the conclusion that, despite the lack of an intrinsic magnetic field, essentially no solar wind plasma enters Venus' atmosphere. In the nature of its interaction with the solar wind, Venus is less like Earth and more like a comet with a developed coma.

We shall now address some other peculiar phenomena found in Venus' environment: clusters of whistler-mode (100 Hz) bursts discovered in plasma wave patterns. These whistlers, recorded by the Pioneer-Venus Orbiter, were correlated with very low frequency electromagnetic noises, caused by electrical discharges, detected by Venera landers in the subcloud Venus atmosphere. Lightning in the Venus clouds and even their spatial association with the surface highlands and mountains of possible volcanic origin were suggested as the most plausible interpretation of these phenomena (Ksanfomality et al., 1982, 1983b, 1985). It was further noted that the 100 Hz signals may propagate some distance from the source in the surface-ionosphere waveguide owing to refraction in the atmosphere and irregular structure of the night-side ionosphere where they were recorded (Strangeway, 1991). This interpretation, however, remains controversial (see Taylor et al., 1979; Scarf et al., 1980; Scarf and Russell, 1983). First, known processes for the formation of lightning in clouds require both latent potential instability and the existence of large precipitation particles, which, as was shown in Chapter 6, are not both present on Venus. Second, the assumed planetographic clustering of lightning stimulated by present-day volcanic eruptions was not confirmed by the data. Instead, some evidence suggested that low-frequency electromagnetic noises could be associated with ionospheric disturbances occurring in the vicinity of the Pioneer-Venus Orbiter, and specifically with ion discontinuities, or troughs, resulting from the solar wind interaction with the night-side ionosphere (Taylor and Cloutier, 1992).

Nonetheless, the Pioneer-Venus Orbiter instrument for detecting the electric field (OEFD) measured many 100 Hz plasma wave bursts throughout the low-altitude ionosphere during the final entry phase of the mission, and confirmed that these bursts are whistler-mode phenomena (Strangeway et al., 1993). It was also confirmed that, provided that these signals corresponded to ambient wave noise caused by lightning, they occurred at locations quite remote from the point of observation. Thus the controversy remains, and although additional evidence was found suggesting that the recorded low-frequency whistler-mode noises could be generated by atmospheric lightning, spacecraft-atmosphere interactions and ionospheric plasma instabilities cannot be completely discounted. No optical evidence for lightning was found by Pioneer-Venus during observations of the Beta and Phoebe areas (Borucki et al., 1991).

In sum, owing to several space missions, especially Pioneer-Venus, our knowledge about the neutral upper atmosphere and ionosphere of Venus and about the principal patterns of its interaction with solar plasma has progressed significantly, albeit not always in step with the amount of data available. These data provided a basic framework for the development of theories, placing important constraints on aeronomic processes involved, and contributed to the overall understanding of the formation and evolution of the gaseous envelope of the planet. Some key questions concerning the evolutionary processes occurring on Venus will be addressed in the next chapter.

9 In the course of our exposition, we have attempted not just to recount the features of Venus but also to explain and scrutinize those physical-chemical mechanisms that give rise to the observed features. Nevertheless, the exacting reader may remain unsatisfied if we do not attempt to answer a fundamental question: What made Venus, so long considered Earth's twin, so different from Earth?

When speaking of the striking differences between Venus and Earth, which are somehow linked with evolutionary processes, we have in mind mainly Venus' unique atmosphere, its exotic climate, and the lack of liquid water. The surface morphology, geology, and internal structure are considerably closer to their terrestrial analogs. In any case Venus, like Earth, differs sharply from the other terrestrial planets (the Moon, Mercury, and Mars), which are smaller in size. These smaller bodies are characterized by a globally continuous non-segmented lithosphere, which became stabilized in the early history of the solar system, and their ancient surfaces have preserved numerous impact structures, thick lava flows, and volcanism. In contrast, the distinguishing characteristic of Earth's geology is a segmented, transversely shifting lithosphere, whose individual blocks are involved in global spreading processes, divergence and subduction zones, and convergence zones. Earth loses most of its internal heat through this mechanism, in contrast to single-plate bodies of smaller size, whose outflow of heat occurs mainly through conduction.

The Venusian surface is relatively young and displays abundant volcanism and clearly defined evidence of tectonic activity, which are distinguished by their variety and wide distribution (see Chapter 3). In manifestations of tectonic activity, one can discern indications of regional stress—strain and extensive deformations along both the horizontal and the vertical. It is not yet possible to definitively determine to what extent these signs can be integrated into global tectonic processes, or to quantitatively link them with the removal of heat from Venus' interior (see Chapters 3 and 4). Terrestrial-style plate tectonics is not, strictly speaking, occurring on Venus. However, the question of whether the global tectonic pattern on Venus is some variation on this theme—perhaps with a larger number of smaller, less rigid plates, or even an oscillating system that periodically resembles terrestrial plate tectonics—remains to be answered. The data and models now available indicate that Earth and Venus have many geological features in common. Consequently, from the point of view of its thermal history as a celestial body, Venus may have experienced evolutionary processes not differing greatly from those on Earth since the accretion of the two planets (which apparently occurred almost simultaneously).

The Formation of the Atmosphere

The complex origin of the planets in our solar system was discussed at the beginning of Chapters 4 and 5. Here we shall limit ourselves to the observation that this process occurred through the accretion of cold material in the protoplanetary gaseous-dust nebula (see, for example, Schmidt, 1957; Levin, 1964; Safronov, 1969; Safronov and Vityazev, 1983), followed by intense accretional heating and early differentiation (Schubert et al., 1992). Based on the simplest model of cold condensation (Lewis, 1972), one can assume that the initial composition of this material, throughout the planetary formation zone, corresponded to cosmic elemental abundance (see Fig. 4.1). However, the most abundant elements, hydrogen and helium, were retained only by the massive cold planets, which are composed primarily of these gases and are located at considerable distances from the Sun. In turn, the planets of the terrestrial group were formed by heavier and less abundant elements entering into the composition of an iron-silicate phase (see Chapter 4). The lightest atmophile elements and, in part, other volatiles were dissipated into space owing to the small mass of these planets and the high temperature and shortwave radiation flux in the vicinity of the Sun. Consequently, their original atmospheres, produced through gases released on impact of volatile-containing planetesimals during accretion (Zahnle et al., 1988) and through early outgassing at the differentiation phase, were partially lost. An early phase of rapid hydrodynamic outflow, possibly fueled by an elevated flux of extreme ultraviolet (EUV) radiation from the Sun, may have facilitated early loss and fractionation of the terrestrial planets' atmospheres (Hunten et al., 1989). These primitive atmospheres were augmented by volcanic outgassing, which accompanied the subsequent thermal history of the terrestrial planets, in response to radiogenic heating and density redistribution in the mantle continuing from the hottest earliest stage which was fueled by the potential energy of accretion. The outflow of low-melting-point basaltic material from the mantle was accompanied by the de-gassing of volatiles.

In fact, the formation of planets and their atmospheres followed a more complex scenario. Various classes of models are based on common ideas concerning the dust and gas from which larger-sized bodies (planetesimals) subsequently accumulated, consolidating through collisions into protoplanets, but the role of gas in the protoplanetary nebula during the final stage of formation differs among models. Of the two fundamental ideas, one assumes the presence of a gigantic gaseous envelope, later lost, within which the final accretion of material occurred (Cameron, 1963, 1978; Cameron et al., 1980). The other model proposes dissipation of the surrounding protoplanetary gas before formation was completed. In the second case, the versions are heterogeneous accretion (Lewis, 1974; Turekian and Clark, 1975; Anders and Owen, 1977) and implantation of solar wind onto dust particles and larger fragments (Wetherill, 1981; Donahue et al., 1981; McElroy and Prather, 1981). Each model is called upon to explain not only the accretion process but also the observed ratios of volatiles in the current atmospheres of the planets.

The hypothesis that the composition of original material from which the planets formed corresponded to cosmic elemental abundances, is likewise oversimplified. According to current notions, the chemical elements that entered into the composition of the protoplanetary nebula were formed during nucleosynthesis reactions much earlier than the formation of the solar system—more than 10 billion years ago (Burbidge and Burbidge, 1982). Additional elements may have been implanted into the original material of the nebula by a supernova explosion near the future solar system. Data from isotopic cosmochemistry suggests that supernova remnants and primitive material of the solar system were identical (Schramm, 1978; Lee, 1979; Lavrukhina, 1983); this evidence supports current cosmogenic models.

As one moves farther from the Sun, a definite fractionation of elements occurs owing to their condensation in an order inversely proportional to their volatility, and consequently a certain change in the initial composition of the constituent material takes place with heliocentric distance. This change was evidently caused by the thermodynamic state in that region of the protoplanetary cloud where the corresponding mineralogical associations formed. Their formation during the overall fractionation of metal and silicate components, in this case, was controlled by chemical equilibrium conditions and, at the same time, was linked with the selective accumulation of condensates as they emerged. Electromagnetic and corpuscular radiation may also have exerted a considerable influence; this radiation swept away a portion of the original nebular material that did not accumulate—primarily volatiles not absorbed by the original particles. This process is usually associated with young stars passing through the T Tauri evolutionary phase, when a star loses mass at a colossal rate—approximately one solar mass in 10^6 years.

The planets were apparently formed during subsequent accumulation, including the accretion of parent, high-temperature condensates (as an internal core), onto which metallic condensates and silicates precipitated from later condensation stages. During the final stage the lowest temperature condensates settled (see Makalkin, 1980; J. Smith, 1982; Rudnik and Sobotovich, 1985; Voytkevich, 1988). A fundamental redistribution of the constituent material occurred 10^6–10^8 years after the beginning of accretion (which may have had a homogeneous or non-homogeneous nature, depending on the presence of gas in the protoplanetary cloud or after its dissipation). This redistribution had a bidirectional trend as a function of density, which led to the formation of the basic planetary shells—an iron core and silicate-oxide mantle (Fig. 9.1).

The fundamental differentiation process, culminating in the formation of the core and accompanied by large-scale release of gases from the interior, apparently was completed on Earth only several hundred million years after its formation (4.55 billion years ago). Through this process, and the late stages of accretion, the bulk of the atmosphere and hydrosphere was formed. The formation of an atmosphere on Venus can probably be attributed to this same period. The two planets may differ, however, in their de-gassing mechanisms, which are regulated by their thermal history. For example, differences may have occurred owing to concentrations of volatiles in the mantle material, which affected heterogeneous migration of melts and solutions as it melted and the core formed. The main reason for the diverging paths of subsequent evolution was evidently the different surface

Fig. 9.1. The sequence of fundamental events in the history of matter, up to the formation of Earth and neighboring planets (according to Voytkevich, 1988).

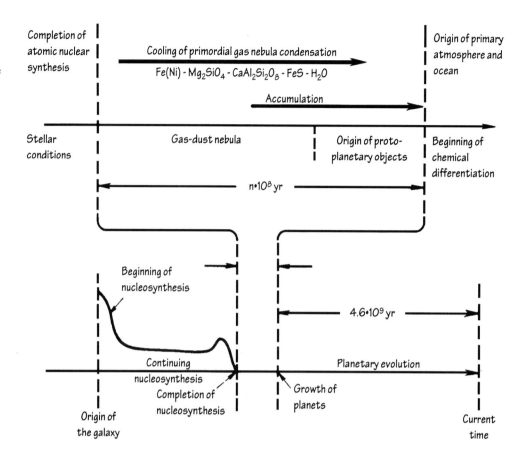

conditions and processes of lithospheric-atmospheric interaction. We shall now examine the scenario that most likely occurred on Venus.

A Simple Evolutionary Model

One must admit that Earth was lucky, because if it had formed only 15–20 million km closer to the Sun (a quarter of the distance between the orbits of Venus and Earth), our favorable climatic conditions would hardly have developed. Things would be worse if we were to move Earth all the way to the orbit of Venus. The following simple estimates, which are in close agreement with more rigorous theoretical models, convince us of this (see Rasool and de Bergh, 1970; Pollack, 1971; Kuzmin and Marov, 1974; Kasting and Pollack, 1983; Kasting et al., 1984; Kasting, 1988; Kasting et al., 1993).

If we assume that the original albedo of Earth was determined entirely by the surface and corresponded to the lunar albedo (~ 0.07), then, given the current luminosity of the Sun, its effective temperature would be 275 K. This temperature value is overstated, because it does not take into account the Sun's slowly increasing luminosity (by 25–30%) after its transition to a main-sequence star, 4.6 billion years ago (Newman and Rood, 1977). One should also take into account the increase in the Earth's albedo and the increasing greenhouse effect, as the atmosphere started to form. In this case an interesting paradox arises: at the initial luminosity level the mean temperature of the terrestrial surface turns out to be lower than the freezing point of sea water (Ringwood, 1961). But this conclusion contradicts modern geological and paleontological data, according to which primitive organisms arose on Earth not less than ~3.2 billion years ago and oceans formed even earlier. The most ancient stromatolites—stratified formations in limestone and dolomite masses—formed as a result of the activity of blue-green algae colonies during this period.

One can eliminate the contradiction by assuming that in the early Precambrian the terrestrial atmosphere contained, in addition to carbon dioxide gas and water, a relatively small amount of ammonia (on the order of several ten thousandths of a percent). Under this assumption, the temperature could rise above the freezing point of water over the course of the first 2 billion years of Earth's existence, owing to the strong greenhouse effect created by ammonia (Sagan and Mullen, 1972). Ammonia, however, rapidly dissociates, in response to solar ultraviolet radiation, into nitrogen and hydrogen, which do not create a greenhouse effect. Attempts have been made to obtain the same result by assuming that the ancient atmosphere of Earth (and of Mars) contained a significantly greater amount of CO_2 than it does now, equivalent to a surface pressure on the order of 60 atm (Pollack and Black, 1979; Owen et al., 1979; Cess et al., 1980).

Similar considerations hold for our neighboring planets. One should keep in mind, however, possible differences in the composition of the protoplanetary material during the condensation stage of solar gas into mineral particles of various sizes and composition. This difference may further affect the differentiation of the original mantle (in particular, the degree of melting of low-melting-point silicate fractions enriched in volatiles), the formation of basaltic magma, and the accumulation of the lithophile and atmophile elements forced out onto the surface. In addition, one cannot exclude the possibility that in that part of the protoplanetary nebula where Venus was formed, the water content of planetesimals was much lower, because it was too hot for hydrated minerals to remain stable (Lewis, 1972). The probability of this is small, however, if one takes into account the exchange of planetesimals between the accretion zones of Earth and Venus, as follows from Wetherill's model (1975, 1981). Also, the existence of hydrated silicates in nebular condensates, as a major source of water for the terrestrial planets, has been called into question because chemical kinetic modeling has led to the conclusion that the necessary hydration reactions would be kinetically inhibited at the appropriate temperatures (Lewis and Prinn, 1984). Thus one may be forced to invoke mixing of lower temperature condensates from further out in the solar system as a source of a (perhaps large) portion of the water and other volatiles for the terrestrial planets. Such a source would not discriminate greatly between Venus and Earth.

For Venus, with the same initial albedo as assumed for the young Earth, the equilibrium temperature turns out to be no less than 325 K; up to a pressure of 0.2 atm this is higher than the boiling point of water. Thus in order to retain water, Venus must have possessed an original atmosphere almost two orders of magnitude more dense than that of Earth. It would be more reasonable to assume that carbon dioxide, along with water vapor, gradually accumulated in the atmosphere. This in turn contributed to a further increase in the surface temperature owing to the greenhouse effect and to the transport of even greater amounts of CO_2 and H_2O into the atmosphere due to an equilibrium state, characterized by relationships between mineral phases and volatiles at the surface, the most important of which is the carbonate-silicate interaction in the surface layer of the planet's crust.

In other words, a positive feedback loop may have occurred. Such a development of events on Venus has been called a runaway greenhouse effect. In this case, given additional heating from the late stages of accretion, the surface temperature for a certain period may have exceeded the current value. But a more interesting question is, To what extent is the reverse situation likely? That is, because of less de-gassing or for some other reason, could the temperature have been much more moderate sometime in Venus' early history? To answer this question, it is necessary to examine the nature of processes responsible for lithospheric-atmospheric interaction (see Chapter 5). We shall return to this question at the end of this chapter.

The equilibrium between the partial pressure of carbon dioxide and carbonate concentrations in the crust is one of the most characteristic manifestations of the chemical interaction between a planet's atmosphere and lithosphere, as discussed by Urey (1951, 1952). In carbonate reactions with silicic acid, the most widespread of which on Earth are calcites and magnesites (dolomites), carbon dioxide gas is released, and calcium and magnesium silicates are formed. An example is reaction (5.52). On Earth this cyclical process (Fig. 9.2) is activated in the presence of water: water currents on the surface wash away Ca and Mg cations from silicate rock, which subsequently interact with the carbonate anions of sea water. Sea organisms play an important role in this process. Carbonates accumulate in sedimentary rock, are remelted at depth during metamorphism, and again release CO_2 into the atmosphere. The complete cycle takes several hundred thousand (~ 0.5 million) years. The ratio between the carbonate content in the crust and CO_2 in the atmosphere is often referred to as wollastonite equilibrium. These reactions (Fig. 9.3) are reversible.

It is possible that the CO_2 abundance in Venus' atmosphere is controlled by these and similar reactions in regions on the planet's surface that contain sufficient calcium and magnesium. In this case, if CO_2 is locked in sedimentary rock on Earth, then under Venusian conditions it is essentially completely released into the atmosphere. In fact, the amount of carbon bound in Earth's sediments is estimated to be the equivalent of 53 bars of CO_2, which is comparable to the CO_2 content in Venus' atmosphere (Ronov and Yaroshevsky, 1967; Vinogradov and Volkov, 1971; Hunten, 1993).

In reality the situation is more complicated and, for such a heterogeneous natural system, includes a large number of minerals and mineral associations found in equilibrium with the atmosphere, in addition to atmospheric gases (primarily SO_2) that affect the stability of carbonates (Lewis, 1970; Lewis and Prinn, 1984; Fegley and Treiman, 1992). In Chapter 5 three cycles, responsible for the global budget of sulfur components, were studied in detail; one of these (the slow geological cycle) includes a sequence of transformations with the participation of pyrite (FeS_2), water, carbon dioxide, sulfur compounds, calcite ($CaCO_3$), anhydrite ($CaSO_4$) (which is formed as a result of calcite reaction with sulfur dioxide), and FeO. Recall that Barsukov et al. (1980) and Lewis and Kreimendahl (1980)

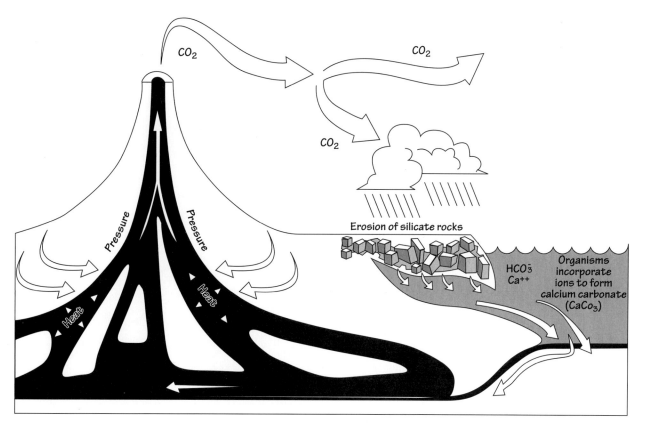

Fig. 9.2. The carbonate-silicate geochemical cycle.

also considered a buffering reaction of pyrite-calcite-anhydrite, inhibiting the occurrence of free oxygen in the atmosphere and simultaneously contributing to the release of CO_2 (see reaction [5.53]). Such an approach, which includes a complex calculation of the phase composition of the system, open with respect to H_2O, CO_2, CO, SO_2, HCl, HF, and other volatile components of Venus' troposphere, provides an idea of the maximum possible changes in the initial rock during chemical weathering. In other words, we arrive at a model of equilibrium mineral association of the soil, in which diffusion exchange processes of volatiles and petrogenic components are completed. A series of thermochemical reactions, with the participation of the above-named components (as well as halogens), along with the carbonate-silicate cycle, can in principle maintain the system in equilibrium. Thus, the abundance of carbon dioxide in Venus' atmosphere, from a geochemical point of view, is perhaps understandable. However, a model by Bullock and Grinspoon (1996), which includes the radiative properties of geochemically buffered gases, has pointed toward possible instabilities in the Venusian climate system.

The situation with water is considerably more complicated to explain. Assuming geochemical similarity in the evolutionary processes of planetary interiors and the degassing of volatiles, the amount of water outgassed on Venus would have to correspond to the volume of the terrestrial hydrosphere, which is approximately 1370 million km^3, or more than $1.37 \cdot 10^{24}$ g (Vinogradov, 1967). Moreover, on the surface of Venus water is not retained, because the temperature there is higher than the critical value, 647 K. This assertion remains valid even for aqueous solutions (brines), for which the critical temperature is

Fig. 9.3. An equilibrium diagram for the carbonate-silicate cycle under Venus' surface conditions (without accounting for the presence of SO_2). The main reactions are indicated in the diagram. The dashed line shows the change in CO_2 pressure in the near-surface troposphere; the zero level is at $R_{\venus} = 6051.5$ km.

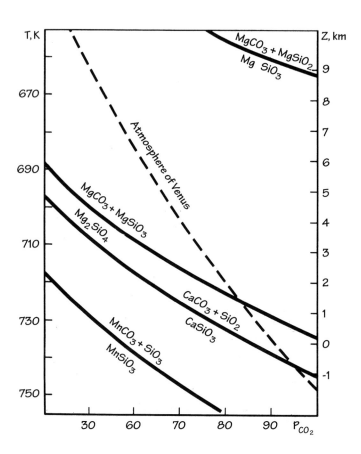

usually somewhat higher (675–700 K). As far as the atmosphere is concerned, if we accept a mean relative water vapor content of 0.005%, the amount of water turns out to be $3.5 \cdot 10^{19}$ g. This exceeds the water content in Earth's atmosphere ($1.3 \cdot 10^{19}$ g) but is almost five orders of magnitude less than the water reserves in the hydrosphere (Fig. 9.4).

One should keep two additional considerations in mind. First, by analogy with Earth, a certain amount of water could have been retained in the Venusian crust—in the form of both chemically bound (constitutional and crystallized) mineral water and free water, evidently found in a vapor state. The water content in the terrestrial crust, according to various sources, is estimated to be between 4–5% and 30–50% of the mass of the hydrosphere, about 25% of which is from bound water. As applied to Venus, the lower of the indicated limits (4–5%) seems more likely. In particular, data from soil analysis on Aphrodite Terra (see Chapter 3) indicate the possibility of rock formation from a moist melt (given 1–1.5% H_2O in the original material). Second, water may be in Venus' mantle. According to ideas by Vinogradov (1967), only a minor fraction of the volatiles contained or produced in Earth's mantle was de-gassed into the atmosphere and hydrosphere over the entire geological history of our planet. In particular, the volume of the hydrosphere, according to his estimates, does not exceed 7.5% of the overall water reserves in the mantle. For Venus the estimates are considerably less optimistic. For example, according to Walker (1975, 1977) and Holland (1978), de-gassing of its mantle was complete, in accordance with completely de-gassed C and N. Nevertheless, if Venus' interior contains a gaseous

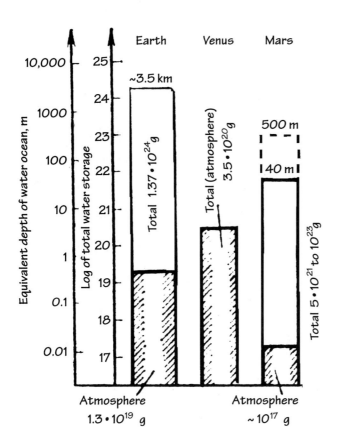

Fig. 9.4. A comparison of the water content (including atmospheric water) on planets of the terrestrial group.

component comparable to Earth's, and possibly two times higher with respect to C and N (Donahue and Pollack, 1983), then Venus may have the potential to "acquire" a hydrosphere, if its climate changes.

The question of water and the possible mechanisms for its loss will be given special consideration, taking into account the exceptional importance of this problem for understanding planetary evolution as a whole, not only on Venus. For now, though, we shall direct our attention to a most interesting source of information—the concentration of noble gases in the atmosphere, which limits the range of speculation concerning processes that occurred in the past.

Limitations Imposed by the Concentration of Noble Gases

To clarify the evolutionary path of the atmosphere and ancient climate of Venus, mass spectrometer measurements of the concentration of minor admixtures, primarily noble gases, and the ratios of fundamental isotopes in the planet's atmosphere are of extreme importance (see Tables 5.5 and 5.6). Noble gases, which do not chemically interact with atmospheric components or surface rocks, are ideal witnesses of events that occurred on the planet over its geological history, or even in an earlier epoch if we are speaking of primordial isotopes. By comparing the measured concentrations of inert gases with their absolute and relative concentrations in Earth's atmosphere and the gas fraction of meteorites, one can judge the degree of original fractionation of volatiles and the degree of de-gassing that occurred on the planet.

On Venus the ratio of the radiogenic argon isotope ^{40}Ar to the primordial isotopes ^{36}Ar and ^{38}Ar is approximately equal to unity, whereas on Earth this ratio is nearly three hundred times greater, and on Mars three thousand times greater. At the same time, the absolute ^{40}Ar content on Venus is about six times less than on Earth. In other words, compared to the terrestrial and, especially, Martian atmospheres, the atmosphere of Venus is highly enriched in the primordial isotopes of argon: although ^{40}Ar is six times less abundant than on Earth, ^{36}Ar is almost a hundred times more abundant. The amount of neon is also nearly an order of magnitude greater in the Venusian atmosphere than in that of Earth, although the isotopic ratios for both planets do not differ greatly. A similar tendency is maintained with respect to krypton and xenon; their absolute concentration on Venus is approximately an order of magnitude higher, although one should keep in mind the somewhat greater errors in measuring these gases.

In Fig. 9.5, the absolute concentrations of the primordial isotopes of noble gases for the three planets are plotted with respect to their solar concentration and compared to the primitive material of meteorites — Ⓒ1 class carbonaceous chondrites. All three planets are highly depleted in noble gases. The deficit is especially great on Mars and less on Venus, whereas Earth occupies a nearly intermediate position. In this case, the lighter the element, the higher the degree of depletion, compared to heavy elements. Venus is closest of all to carbonaceous chondrites, especially in Ne and Ar. Moreover, the concentrations of noble gases on Venus display considerable differences from the so-called planetary ratio characteristic of Earth and Mars: whereas the $^{36}Ar/^{20}Ne$ ratio is approximately the same for Earth and Venus, the $^{36}Ar/^{84}Kr$ ratio on Venus is more reminiscent of the solar ratio than the terrestrial (Hunten et al., 1988).

If we take the ratios of the measured volume concentrations of primordial argon in the atmosphere per unit mass of the body (for Venus, Earth, and Mars they are equal to $5 \cdot 10^{-6}, 2 \cdot 10^{-8}$, and $0.5 \cdot 10^{-10}$ cm^3/g, respectively) and plot them on a logarithmic scale as a function of distance from the Sun, then these ratios turn out to be distributed on a straight line. Note that certain types of carbonaceous chondrites from Ⓒ and Ⓗ classes fit this curve (Ⓒ3 has approximately 10^{-6} cm^3/g, and higher temperature Ⓗ chondrites have, on average, $2 \cdot 10^{-8}$ cm^3/g). Hence it follows that Venus' atmosphere is highly over-enriched in the primordial isotopes of argon compared to the neighboring planets (calculated per gram of constituent material). A similar situation is observed with respect to neon (Hunten et al., 1988).

Thus we run up against a persistent trend, and some curious anomalies (such as the increasing depletion of light, inert gases with increasing distance from the Sun, or the nearly solar ratio of heavy inert gases on Venus). Some of these results may be accidental; stochastic impacts of large, volatile, rich objects during, or shortly after, accretion may have played an as-yet-undetermined role (Owen et al., 1992). Yet, an explanation may also be found in the condensation sequence, mentioned earlier, for the formation of planetary and meteoritic material at various distances from the Sun, and in associated mechanisms dur-

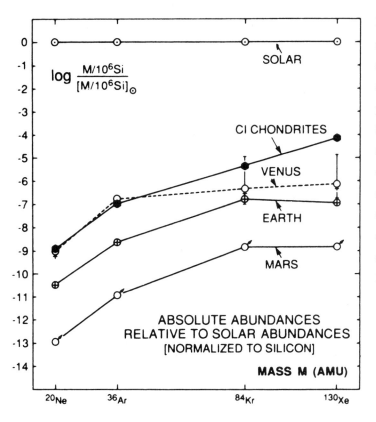

Fig. 9.5. Depletion in the absolute concentrations of inert gases in the atmospheres of the terrestrial planets (normalized to 10^6 Si atoms on the planet) and in (CI) carbonaceous chondrites, compared to corresponding concentrations in the Sun, with the same normalization (according to Hunten et al., 1988).

ing the accumulation and subsequent evolutionary stages. Thus the question of atmospheric and climatic evolution is linked with the general cosmogonic problem.

One could suppose that four fundamental mechanisms were at work with various degrees of effectiveness: heterogeneous accretion, unequal degrees of de-gassing, fractionation of the original protoplanetary cloud, and stochastic accretion of large, volatile-rich impactors. The heterogeneous accretion model has been considered in detail by Lewis (1974), Turekian and Clark (1975), and Anders and Owen (1977), based on analysis of the concentrations of volatiles in terrestrial, planetary, and meteoritic material. The fundamental idea of this model is that after formation of the main planetary mass, meteorites, consisting of the last low-temperature condensates containing the main mass of volatiles (primitive meteorites), as well as comets consisting of more than half water ice, fell onto the body during the final stage. Differences in the positions of the terrestrial planets thus could not have played a significant role in the formation of volatile reserves. The degree of subsequent planetary de-gassing, including the surface layer of resultant meteoritic material, could have had a stronger effect.

The idea of heterogeneous accretion is contrary to ideas based on the cold condensation (or successive accumulation) model. This model predicts that with increasing distance from the Sun, the volatile content must increase; that is, there must be proportionally more volatiles on Mars than on Venus and Earth. In fact, the picture is quite different. The atmosphere of these planets were thus more likely formed from a relatively thin covering (veneer), which accumulated over a certain period of time and basically depended on the mass of the planet.

Among the late condensates we have primarily carbonaceous chondrites, enriched in hydrated silicates, gases, and even organic materials. In their hydrogen content, the group Ⓗ chondrites are quite similar to group Ⓒ chondrites (especially Ⓒ3). Assuming identical factors for the release of primordial isotopes of argon and hydrogen on Earth and Venus during de-gassing, one could conclude that comparable amounts of water were forced out onto the surfaces of both planets over geological time. Such a possibility is not contradicted by commonly accepted ideas concerning the effectiveness of cometary and meteoritic bombardment of planets in the inner regions of the solar system, which continues to the modern era but was especially intense during the early evolutionary stages of the solar system. According to Wasserburg et al. (1977), for example, lunar bombardment by meteorites during the first 700 million years exceeded the current value by many orders of magnitude.

These considerations, nevertheless, do not explain the unusually high Ar, Kr, and Xe concentrations in Venus' atmosphere (for example, the ^{36}Ar content per unit mass of the planet exceeds by several times the analogous concentration even for type Ⓒ3 carbonaceous chondrites, which are enriched in this isotope). In order to overcome this difficulty and, to some extent, rescue the idea of heterogeneous accretion, Hunten et al. (1988) put forward the idea of an icy planetesimal, with a solar composition of noble gases, falling onto Venus. This is a modification of the idea of a cometary bombardment source, with the difference being that the required size of such a planetesimal would have to have been at least an order of magnitude larger than the nucleus of Halley's comet. If such a body had not formed at the very distant periphery of the solar system (at $T \geq 35$ K), it could not have retained Ne; this would explain the deficit of Ne with respect to other noble gases on Venus.

The discovered regularity in the distribution of noble gases on the planet could, to a certain degree, be caused by a third mechanism—the difference in the initial ratios of elements as the planets formed. In this case one would have to assume that during accumulation Venus captured from the pre-planetary cloud (of solar but non-meteoritic composition; see Fig. 9.5) more of the parent argon, neon, krypton, and xenon isotopes, and their related groups of other elements, than Earth, whereas Mars captured still less. Could such a sharp fractionation of parent material have occurred in a relatively small region of the solar system, within less than 0.8 AU? And why for Ne is the solar ratio not maintained but is instead almost two orders of magnitude less, so that the ^{20}Ne/^{36}Ar ratio corresponds to the terrestrial value? What role could solar wind irradiation of the original grains of material from which the planets close to the Sun accumulated, as proposed by Wetherill (1981), have played? And if noble gases of solar composition were actually implanted in this original material, to what extent was this source effective for accumulating Ar, Kr, and Xe? In addition, was subsequent heating sufficient so that only the lighter Ne dissipated, or did another mechanism work in parallel? And, finally, why did irradiation not leave similar traces of implantation in the parent material of Earth? Unfortunately, this third mechanism does not provide an answer to these questions.

We will now turn to the radiogenic isotope ^{40}Ar, to consider the second proposed mechanism for explaining the difference in the evolutionary paths. In contrast to parent isotopes, which carry certain cosmogonic information, this isotope, a decay product of potassium (with a half-life of 1.5 billion years), to a greater extent serves as a tracer of the intrinsic thermal evolution of the planet. The fact that there is considerably less ^{40}Ar in Venus' atmosphere than in Earth's suggests that the degree of de-gassing on our neighboring planet was substantially less (0.25–0.5, according to estimates by Hart et al., 1979). At the same time, evidence suggests that at an early stage (probably in the first billion years) immense geological processes were taking place on both planets (see Chapter 3) and, evidently, nearly comparable amounts of carbon dioxide gas, nitrogen, and possibly water were de-gassed. In order to eliminate this contradiction, Donahue and Pollack (1983), and Morrison and Owen (1988), assumed that the initial period of vigorous geological activity on Venus was replaced by a prolonged period (3–4 billion years) of relatively weak activity; this impeded the de-gassing process (which continued without interruption on Earth) and, consequently, the transport of ^{40}Ar into the atmosphere. They found confirmation of this hypothesis in the observable morphology of the planet's surface, which they considered to be static forms, rather than a manifestation of current tectonic activity. As far as internal heat is concerned, which in particular was generated through the decay of ^{40}K, they imagined that its removal from the interior could be facilitated through scattered volcanism and other regions of higher-than-normal heat emission, associated with hot spots on Venus' surface.

These ideas seemed quite convincing before the Magellan mission, especially taking into account the very similar K/U ratios for the surface material of Earth and Venus (see Table 3.2). Magellan, however, has revealed a tectonically active planet, although the rate of tectonic activity over the past billion years has not been conclusively determined, and the record of earlier tectonic activity has apparently been lost. One interpretation of the Venusian cratering record is that tectonic and volcanic activity declined dramatically about one billion years ago, as the interior cooled (see Chapters 3 and 4).

The Problem of Water

The key question in the evolution of Venus and its atmosphere remains: Was there water on Venus and, if so, in what amounts did it exist and how was it lost? The argument presented in the previous section seems to carry the most weight; both planets probably received approximately the same amount of volatiles (as a consequence of heterogeneous accretion during the final phase of accumulation or later). The opposite point of view, that Venus formed as a "dry" planet, seems less likely. The idea of a "parent ocean" on Venus, which remains attractive, at the same time supports the decisive role of the heliocentric position of the inner planets in their climatic evolution. In any case, from this point of view it seems more natural to explain why Venus lost such a huge mass of water, comparable to the volume of the terrestrial hydrosphere, whereas Mars "preserved" a much more moderate amount of water in ice form in its cryosphere.

By what means could such a loss occur? Donahue and Pollack (1983) analyzed four basic mechanisms in detail, each of which runs up against another complex problem, which we will consider separately: how the huge amount of hydrogen formed is lost from the atmosphere. The first mechanism, according to the simple evolutionary model presented earlier, proposes that the water removed from the interior was not retained on the surface; rather, it accumulated in the atmosphere as vapor, which was subjected to intense photodissociation by solar ultraviolet radiation. The hydrogen that was formed dissipated into space, whereas the oxygen bonded with surface rocks (Walker, 1975). In the second mechanism, it is assumed that the expelled water reacted with FeO, contained in the surface rock, and the hydrogen then dissipated into space (Watson et al., 1982). The third mechanism assumes that reactions of water vapor with reduced gases, mainly CO, released in parallel, were the primary loss channel, resulting in the formation of carbon dioxide gas, which accumulated in the atmosphere, and hydrogen, which was not retained by the planet (Pollack and Yung, 1980). Finally, according to a fourth mechanism, cyclical transformations of water in Venus' interior were possible, with partial thermal dissociation, as a result of which oxygen reacted with magma, while the hydrogen migrated into the atmosphere and from there dissipated into space. Each of these mechanisms has its merits and limitations. And each may contribute to the explanation of the loss of Venus' ocean.

That view that Venus possessed a thick hydrosphere has been supported by analysis of the atmospheric concentration of hydrogen and its heavy isotope deuterium, with a mass two times greater than H. This analysis, carried out on mass spectrometer measurements from the Pioneer-Venus Large probe by Donahue et al. (1982), led to an important conclusion: the ratio of deuterium (D) to hydrogen (H) in Venus' atmosphere is $(1.6 \pm 0.2) \cdot 10^{-2}$; this proved to be two orders of magnitude greater than in Earth's atmosphere. This observation was more recently confirmed through ground-based near-infrared spectroscopy by de Bergh et al. (1991). Such a high deuterium concentration could be explained by the fractionation of these isotopes with non-thermal escape of hydrogen from the atmosphere, where it presumably accumulated through the vaporization of water removed from the interior (the "parent ocean") and subsequent dissociation of the water vapor by ultraviolet radiation. In this case, theoretical estimates suggest that as long as the relative hydrogen content in the atmosphere exceeds approximately 2% (Watson et al., 1982), a hydrodynamic escape mechanism, under which fractionation of hydrogen and deuterium does not occur, is at work.

The idea behind a hydrodynamic mechanism, studied in detail by Hunten (1982) and Chamberlain and Hunten (1987), is that at a high dissipation rate for H atoms, their flow may carry away heavier elements along with them. In other words, the gases effectively interact with one another, generating blowoff-assisted escape; this process enables us to explain not only the deuterium concentration in Venus' atmosphere but other similar phenomena—in particular the loss of sulfur atoms on Jupiter's satellite Io, which may be carried off by the flow of atmospheric oxygen (Hunten, 1985; Hunten et al., 1987; Zahnle and Kasting, 1986). Until now, the role of this mechanism, including during the accumulation stage of the planets, has possibly been underestimated. If direct capture of gas from the protoplanetary nebula could have occurred to form the original atmosphere of the accreting planet, then the atmosphere would primarily have been composed of hydrogen, whose subsequent hydrodynamic escape, in principle, would have been able to ensure the observed fractionation of inert gases (Hunten et al., 1988).

Other non-thermal mechanisms of escape are also possible; these mechanisms are associated not with hydrodynamics but with aeronomy. To explain the 40% concentration of the heavy nitrogen isotope ^{15}N in the Martian atmosphere, McElroy (1972) and McElroy et al. (1977) proposed dissociation of the nitrogen molecule by impacts of energetic electrons, formed through photoionization. The kinetic energy acquired by one of the atoms may prove sufficient for escape. By analogy, McElroy et al. (1982) provided an interpretation of the fact that Venus' atmosphere is enriched in deuterium, based on the idea of dissociative recombination of O_2^+ with the formation of fast O atoms, which in themselves are not capable of escaping the atmosphere, although they may approximately double their velocity upon collision with H atoms. In turn, Kumar et al. (1985) considered charge exchange between hydrogen atoms and hot protons of the solar wind to be dominant in separating D and H. Both processes operate effectively in Venus' present atmosphere, although the escape flux is still uncertain, probably by a factor of several (Gurwell and Yung, 1993; Hodges, 1993; Donahue and Hartle, 1992).

If we proceed from the assumption that over the history of Venus only thermal and aeronomic mechanisms operated to dissipate hydrogen, then the lower limit for the amount of water driven off turns out to be about half a percent of the volume of Earth's hydrosphere; this precisely corresponds to the measured D/H ratio. It is hardly likely, however, that the actual volume of the Venusian hydrosphere matched this lower limit; taking into account the geochemical considerations presented earlier, and a possible hydrodynamic escape mechanism, it is much more likely that the volume was closer to that found on Earth.

Meanwhile, Grinspoon (1987) and Grinspoon and Lewis (1988) proposed a different interpretation for the measured D/H ratio. In their model the D/H ratio does not necessarily reflect a declining water abundance but rather results from a steady state balance between hydrogen sources and loss with selective escape of the lighter isotope. They pointed out that, with current estimates of the non-thermal hydrogen escape flux, the lifetime of water is on the order of 100 million years. Because this is much less than the age of Venus, it is more likely that the current escape flux is maintained by a recent hydrogen source rather than the vestiges of a primordial ocean. They suggested that a likely source of hydrogen is from the impact of cometary material. They then succeeded in producing the observed deuterium concentration in Venus' atmosphere, given specific assumptions on the degree of fractionation of hydrogen isotopes upon dissipation from the atmosphere. More recently Grinspoon (1993) proposed that, in light of Magellan findings, volcanic outgassing may maintain the current water abundance in steady state with non-thermal escape. This model successfully produces the observed fractionation if the deuterium fraction of the source H on Venus is enhanced over the terrestrial value by a factor of 10. Alternatively, the fractionation observed could have resulted from escape of water outgassed during a global resurfacing episode within the last 1 billion years (Grinspoon, 1993).

Although we do not have definitive evidence revealing whether early Venus was a dry or wet planet, the classical idea of a large initial abundance of water (more precisely, water vapor) seems more arguable, although it is likely that water on Venus is currently in a steady state, which is reflected in the observed D/H ratio. The question whether significant water reserves accumulated in the atmosphere, promoting the creation of a more energetic greenhouse effect, or whether the water loss occurred more or less uniformly, still remains open. But even if water vapor in the atmosphere did not accumulate and its loss was uniform, then one must assume that the flow of hydrogen molecules from the atmosphere required to evacuate the expelled amount of water would reach an enormous value—nearly $7 \cdot 10^{10}$ $cm^{-2} \cdot sec^{-1}$. This is approximately 3–4 orders of magnitude higher than the current hydrogen dissipation rates from the atmospheres of Earth and Venus, and larger values seem unrealistic. At a loss rate of 10^7 $cm^{-2} \cdot sec^{-1}$, an equivalent layer of water of only ~9 m could have been lost over Venus' entire history (Donahue and Pollack, 1983).

Is a hydrodynamic escape mechanism capable of coping with such a flow? It is difficult to answer this question clearly. Similarly, the sweeping by the solar wind during its intense phase could have been quite effective but probably occupied a quite limited time period. In principle, we can circumvent this difficulty by postulating a thermal (or, as it is still called, Jeans) dissipation mechanism, if we keep in mind that given a large water content in the atmosphere a very small moist adiabatic temperature gradient is established. As a result, temperature falls off very slowly with altitude, and a cold trap, suppressing the escape of hydrogen, is shifted from the level of the tropopause (as on Earth) to much higher. The conditions for water vapor condensation are achieved here at a higher saturated vapor pressure, which means at a higher X_{H_2O} value. In turn, at high altitudes (~100 km) under intense ultraviolet radiation and for high X_{H_2O}, extremely efficient photodissociation of water vapor will occur; this makes a large flow of escaping hydrogen possible. It is difficult, however, to base the possibility of losing all the hydrogen on this mechanism, if we take into account the drop in the dissipation rate with a decrease in X_{H_2O} and, accordingly, a lowering of the altitude of the cold trap and its temperature. The diffusion rate through the stratomesosphere to the level of the exobase, which controls the escape of H atoms and is likewise subject to changes, may impose an additional limitation.

If the first of these numerous mechanisms for water loss (photodissociation) is valid, then, in addition to hydrogen loss, it is necessary to understand which processes were responsible for binding an immense mass of liberated oxygen upon H_2O breakdown. Atmospheric constituents, even taking into account cloud chemistry, could hardly have played a decisive role here. It is unlikely that all excess oxygen went to oxidize carbon, on the strength of the more realistic assumption about the primary origin of atmospheric CO_2 through de-gassing of the interior. As far as dissipation of oxygen from the atmosphere is concerned, an exospheric temperature of several thousand degrees Kelvin would be required; this is many times greater than the current value (see Chapter 8) and would hardly be possible under conditions of radiative cooling by carbon dioxide molecules. It is more likely that the oxygen combined with surface rocks (and partially with gases that migrated into the atmosphere at high temperature), but this would force us to infer significantly greater tectonic activity on Venus, compared to Earth, required for the efficient delivery of fresh, unoxidized (unweathered) material from depth to the surface.

In addition to oxidizing processes at depth, one must also assume efficient surface erosion. In this case, the oxygen combination rate would have to have been approximately two orders of magnitude higher than currently on Earth's surface ($\sim 3 \cdot 10^{14}$ g/year). On Earth, however, a significant fraction of this goes to oxidizing organic material in sedimentary rock. One can remove (or at least reduce) this limitation by assuming that at high temperature the atmosphere on Venus reacts energetically with the surface, and that in the epoch attributed to the loss of water, chemical equilibrium with the surface rock was not yet achieved (Walker, 1975). Given the rapid kinetics of the processes, however, the situation would have stabilized quickly, and therefore one can assume that conditions subsequently approached a state of equilibrium. This is apparently preserved in the modern epoch, and it may be that deviation from equilibrium can be created only by volcanic exhalations.

In other words, the cyclical process of atmospheric-surface rock interaction (rock cycle) during certain periods on Venus, given an abundance of atmospheric oxygen, could have been similar to the terrestrial process, but at a very different temperature and, of course, lacking the biogenic processes that destroy equilibrium. In this case the question arises, How long could free oxygen, by virtue of its own chemical activity, have remained in a free state after depletion of water reserves? Earth's atmosphere is, in this respect, an exception, because oxygen here is constantly reproduced through biogenic means owing to the breakdown of carbon dioxide by plants. Recall, by the way, that nitrogen in the terrestrial atmosphere is partially linked with the presence of organic life; its cyclical exchange with the atmosphere is maintained by the chemical activity of live organisms (bacteria). In the atmosphere of Venus, nitrogen is entirely the product of de-gassing, and is present in greater abundance than on Earth (see Chapter 5). Its high concentration, in contrast to the essential lack of oxygen and given the impossibility of denitrification, can evidently be explained by the absence of channels of chemical loss due to the high inertness of N_2 molecules. This is suggested by the high upper limits for detecting nitrogen compounds in Venus' atmosphere (see Table 5.5). In the final analysis, the lack of biogenic processes on Venus (processes that definitely contribute to the equilibrium concentration of CO_2 on Earth, along with carbonate-silicate interaction) is interrelated with the extremely low water content, which is the key component for the appearance and development of life.

More than 45 years ago Fesenkov (1951) gave a correct interpretation based on limited observational data; in an atmosphere of optimism concerning life on our neighboring planet, he wrote: "On Venus oxygen has not been detected. True, on this planet the surface is always covered by a layer of clouds, but it is not subject to doubt that there is also no oxygen below this layer, if there is none above. There is also no water vapor on this planet, at least in perceptible amounts. Carbon dioxide exists in amounts several hundred times greater than on the Earth. Consequently, one can imagine that for certain reasons life on Venus, possibly as a result of the relatively high temperature unsuitable for stable protein compounds, could not have developed, and carbon dioxide, released from the interior of the planet, was not broken down. The abundance of carbon dioxide on Venus suggests that the decomposing activity of planet life could not have developed there" (p. 22).

The question of how to eliminate oxygen in the event of its appearance through H_2O photodissociation is perhaps no less critical than the question of how to eliminate hydrogen. This makes other water loss mechanisms, mentioned earlier, somewhat more attractive, because they are less subject to this limitation. In the second mechanism, in order for FeO oxidation to proceed vigorously, in addition to the intense delivery of fresh magmatic material to the surface necessary in this case, one should also assume an extremely high surface temperature, reaching the solidus temperature for basalts. Then active absorption of water occurs with the release of hydrogen, regulated by the equilibrium ratio H_2/H_2O, which in turn depends on the degree of ion oxidation in the surface rock (Donahue and Pollack, 1983). Therefore, one is obliged to assume a higher FeO/Fe_2O_3 ratio for the original igneous rock of Venus than for terrestrial rock. As far as the high surface temperature is concerned, in principle it can be ensured through the greenhouse effect, given the assumed initial high water vapor content in the atmosphere. By decreasing its content by approximately an order of magnitude from the original value (possibly equivalent to the H_2O abundance in Earth's oceans), the temperature will noticeably drop, which will become an impediment to the continued efficiency of the FeO reaction with H_2O.

A similar mechanism for the evacuation of water, based on thermal dissociation of water vapor reacting over free iron contained in the surface rock, was earlier proposed by Sorokhtin (1974). This reaction ($3Fe + 4H_2O \rightarrow Fe_3O_4 + 4H_2$) also takes place at high temperatures, along with the serpentization reaction and the binding of de-gassed CO_2 and H_2O. The essential absence of oxygen in Venus' atmosphere is explained by the higher Fe/FeO ratio in Venusian material owing to the greater degree of thermal fractionation in the protoplanetary cloud. However, the necessary assumption concerning the retention of free iron in the mantle, which continues to combine actively with oxygen released as the core grows, seems highly improbable. As far as the crust is concerned, one should not talk about free iron but about the degree of oxidation of the surface layer, and in particular about the ratio of bivalent to trivalent iron Fe^{2+}/Fe^{3+}.

The situation with the remaining mechanisms is perhaps even less certain. As has already been mentioned, an outgassing source rich in CO_2 seems most likely. Nevertheless, the oxidation reaction of CO to CO_2 with the assistance of water, given high concentrations of de-gassed CO and H_2O, could have been quite efficient. But a situation where the necessary balance of de-gassing products is ensured, leading finally to the current extremely low mixing ratios of both gases, appears quite artificial (Pollack and Yung, 1980; Donahue and Pollack, 1983). Moreover, it is necessary to assume that the CO/CO_2 ratio in de-gassed volatiles on Venus was approximately two orders of magnitude higher than on Earth; it is difficult to find a reason for this. Similar considerations could be suggested concerning the loss of water through cyclical processes in the crust and upper mantle—combining in the sedimentary rock and thermally dissociating as they sink. The intensity of the thermal dissociation reaction is characterized here by the H_2/H_2O ratio. These processes are quite efficient owing to global plate tectonics on Earth and, thus, the question reduces to this: Were they more effective on the early Venus, and must the higher H_2/H_2O ratio in de-gassing products from the interior, compared to terrestrial conditions (Donahue and Pollack, 1983), have been responsible for this?

Was Venus Different?

Any discussion of the collection of available data and existing views concerning the origin and evolution of Venus and its atmosphere would be incomplete if we did not touch upon one more hypothesis, proposed to answer the question that is the heading for this final section. Kasting et al. (1988) considered the problem of climatic evolution on all terrestrial planets from a point of view differing from traditional ideas and, based on model calculations, came up with some interesting results.

The fundamental idea, developed by Walker et al. (1981), is that upon atmospheric-lithospheric interaction controlled by the carbonate-silicate cycle, there may develop a negative feedback loop, stabilizing the climate system, instead of a positive feedback loop, which amplifies the greenhouse effect. So, under Earth conditions, raising the temperature (for example, by an increase in the Sun's luminosity) would lead to higher than normal evaporation of the oceans, which would intensify rainfall and the flushing of cations from silicate rock (that is, "weathering"; see Fig. 9.2). As a result, a greater amount of atmospheric CO_2 would be combined in carbonates, and the greenhouse effect would weaken. By contrast, upon lowering the temperature, less moisture would enter the atmosphere and weathering of silicate rock would decrease. At the same time, through continuing carbonate metamorphism, CO_2 would gradually accumulate in the atmosphere, the greenhouse effect would be amplified, and the temperature would rise, that is, the temperature would return to its initial value. In postulating the given cyclical mechanism, Kasting et al. suggest that Earth always possessed a moderate climate.

It was probably a different situation on Mars, where a moderate climate existed for only the first billion years; evidence of this is seen in dried riverbeds with approximately the same dating. Apparently Mars' small size—and not its 1.5 times greater distance from the Sun—produced the decisive effect on the trend of its further climatic evolution. With the smaller reserves of radiogenic elements (internal heat sources) in this relatively small planet, carbonate metamorphism was disrupted, and with it the release of free CO_2 into the atmosphere. This, in addition to greater vulnerability to loss of atmosphere from large impacts on this small world (Melosh and Vickery, 1989), led to the current state of Mars. It is thus possible that large carbonate deposits are to be found in the Martian crust.

As far as Venus is concerned, its climatic situation over a certain period of time could also have been favorable, if from the very beginning the prevailing mechanism was not a "classical" runaway greenhouse effect but rather a much more quiescent mechanism with a negative feedback relation, which Kasting et al. (1988) called a moist greenhouse effect (Fig. 9.6). According to their calculations, such a situation could have been realized on Venus if, given a solar radiation flux approximately 1.1 times greater than that delivered to Earth (taking into account differences in the planets' positions and changes in the Sun's lu-

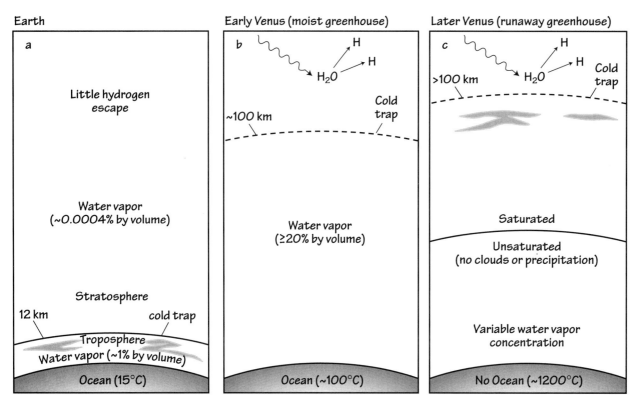

Fig. 9.6. A possible scenario for the loss of water on Venus upon transition from a moist greenhouse to a runaway greenhouse effect (according to Kasting et al., 1988). The critical relative water vapor content in the atmosphere is 20%; upon exceeding this threshold value, a negative inverse relation, which preserves a moderate temperature at Venus' surface by means of the carbonate-silicate cycle, alternates to a positive relation, the cold trap is shifted upward, water vapor undergoes intense photodissociation, and hydrogen is dissipated into space. At left, for comparison, a stable situation with a negative inverse relation, which regulates and maintains the moderate climate on Earth, is shown.

minosity), the water vapor content in the atmosphere did not exceed 20%. In this case, effective binding of carbon dioxide gas in carbonates occurred through weathering of silicate rock and rainfall, which inhibited its accumulation in the atmosphere and the increase in the greenhouse effect. In other words, the situation is reminiscent of that considered for Earth with an enhanced near-surface temperature (a temperature of 70° C corresponds to a 20% water vapor content in the atmosphere with a pressure of 1 bar).

How long could such conditions be maintained on Venus? Evidently up until the moment when the H_2O content in the atmosphere exceeded the threshold of $\sim$20%, owing to a further increase in the Sun's luminosity; this increased the release of heat of condensation and raised the altitude of the cold trap. Starting from this moment, the scenario almost completely resembles that realized in a runaway greenhouse effect, and over several million years Venus could have lost its hydrosphere owing to the escape of hydrogen. Thus, up until a certain moment, its distance from the Sun may not have prevented extremely moderate climatic conditions on Venus. The decisive event in the drastic change leading to the current situation was the loss of a negative feedback loop in the carbonate-silicate cycle owing to a decrease, and then a cessation, of water precipitation from the atmosphere, with a subsequent breakdown of carbonates, accumulation of CO_2 in the atmosphere, and a temperature increase.

If we recall that, because of the highly reflective clouds, Venus currently receives even less solar energy than Earth, then it is easy to imagine a neighboring planet quite unlike the present one. The transition from a moist to a runaway greenhouse effect might not have happened if the increase in the planet's albedo had occurred earlier and had compensated for the increase in energy from the Sun. In this case Venus would most likely have possessed an extremely moist carbon dioxide atmosphere with a surface pressure in the range of several bars and a temperature of several tens of degrees Celsius. Such a planet, in principle, would have been suitable for organic life.

The actual timing of the transition from a moist greenhouse Venus with liquid water oceans to a runaway greenhouse Venus with an extremely hot and dry atmosphere is difficult to estimate. It is certainly possible that our neighboring world possessed liquid oceans for 2 billion years or more. This has interesting implications for considerations of life elsewhere in our solar system. The prerequisites for life are often taken to be liquid water, organic material, an energy source, shielding from solar ultraviolet radiation, and some measure of environmental stability. Life on present-day Mars and Venus is precluded by the requirement for liquid water, unless we postulate extremely acid-resistant strains in the Venusian clouds or life in subsurface thermal springs on Mars. Early in solar-system history, however, all three large terrestrial planets may have possessed the necessary conditions for biology. Mars clearly shows evidence of liquid water in the form of runoff channels on the most ancient, highly cratered terrain. On Venus, such direct evidence seems to have been lost, owing to a more vigorous and extended geological history. Nevertheless, a habitable epoch on Venus that was much more long lived than this ancient epoch on Mars is entirely consistent with our current understanding.

Conclusion

....

Owing to continued focused efforts manifested in the launch of twenty space probes, a broad program of ground-based observations, and the active interest of numerous groups of qualified investigators—experimentalists and theorists—the level of our knowledge of Venus in the late 1990s is quite high. At the same time, because of the sporadic nature of individual results and the limited nature (in space and time) of the measurements conducted, we are still far from understanding many natural phenomena on the planet, and we lack clear laws governing their formation and development as one of the evolutionary branches of bodies in the solar system.

At present, much is known to us. We have measurements of conditions on Venus' surface and altitude profiles of atmospheric parameters that enable us to judge the nature and depth of temperature and density variations. We know the basic chemical composition of the atmosphere, including most minor contaminants (or their upper detection limit), and we know the concentrations and isotopic ratios of inert gases. For several regions we know how solar radiation is attenuated with altitude in the atmosphere and clouds, and how its spectral composition varies, how much energy is absorbed at the level of the cloud layer, and how much reaches the surface. On the whole, the structure, vertical extent, and micro-physical properties of the clouds have been ascertained, and the sulfuric acid composition of the fundamental mode of polydisperse particles has been identified, although concerning the composition of the second and third modes the information is not definite. Data on atmospheric dynamics have been obtained, including wind velocity profiles with respect to altitude, which make it possible to put together some general concepts concerning the global circulation system, whose characteristic feature is its four-day superrotation. Essentially the entire surface has been imaged with hundred-meter resolution. Relief details and types of geological structures have been identified with characteristic manifestations of volcanic and tectonic activity. In several places the elemental composition of the surface rock was measured or its nature was determined from gamma-ray spectral signatures caused by natural radioactivity; the surface was found to be composed of various types of basalts. This finding confirmed the idea that differentiation of the constituent material occurred on the planet. The absence of an intrinsic magnetic field on Venus has been established, and information on the structure and chemical composition of the neutral upper atmosphere and ionosphere—and the nature of its interaction with the solar wind plasma (in the absence of a magnetic field)—has been obtained. Based on experimental data and model calculations, the most likely geochemical cycles through which one could obtain the necessary equilibrium conditions have been analyzed, and models have been developed that provide an approach for understanding the nature of heat exchange in Venus' atmosphere.

This collection of data is, unfortunately, insufficient to answer many key questions about Venus. First of all, we do not understand why Venus, which is so close to Earth and of a similar size, would form with a very slow retrograde rotation, devoid of satellites, and with no magnetic field, whereas Earth has high rotational momentum, a satellite, and a strong magnetic field. Were these differences linked to stochastic events in the late stages of planetary formation? What role did the Moon, with its excess rotational momentum, play in Earth's evolution? To what extent was differentiation of Venus' constituent material similar to Earth's? Little is yet known about Venus' internal structure; the models that have been developed are largely based on analogy with Earth, taking into account certain limitations imposed by the surface properties and gravitational multiple moments of Venus. Discussions concerning geologic processes are quite speculative; no direct information is available on the mineral composition of the rock or the concentration of isotopes, which are the basis for geochronological analysis. Hence the origin of the rocks making up the surface structures, their ages, and other properties of the crust are unknown. No direct evidence reduces the uncertainty in such details of atmospheric-lithospheric interaction as the paths and rates for the exchange of sulfur and carbon compounds with the participation of halogens, primarily chlorine and other minor admixtures. Likewise, large uncertainties exist in conditions for thermochemical equilibrium at the surface and photochemical equilibrium in the clouds and above-cloud atmosphere, including certain chemical and gas-phase reactions, and catalytic processes. For now, only the principal paths in the chains of transformation are understood.

Existing data are insufficient for explaining Venus' thermal regime, including the relative contribution of atmospheric components and clouds to radiative transfer, as well as peculiar features of planetary circulation. Despite the huge progress made from studying the concentration of inert gases and certain isotopes (mainly, the ratio of deuterium to hydrogen), little can be said about the history of water on Venus. Whether the planet was born humid or dry, whether a primordial ocean existed, and if so, how it was lost, the rate of hydrogen dissipation, and the degree of surface rock oxidation are all unclear.

And finally, although the current climate is, in a general sense, understood, no sound basis exists for understanding the paleoclimate; such understanding is necessary to answer at least the following fundamental questions:

- Were the dramatically different current natural conditions on Earth and Venus predetermined by the nature of material fractionation in the proto-planetary nebula, or were they determined by laws governing subsequent evolutionary stages controlled by their different distances from the Sun?
- Was accretion on Venus homogeneous or heterogeneous, and what was the role of subsequent impacts on the surface by meteorites and comets containing low-temperature condensates?
- How, when, and in what amount were volatiles de-gassed into the Venusian atmosphere? What kind of exchange existed between surface volatiles and the atmosphere, and what were the main channels for their transformation?
- When did the existing climate appear on Venus? Was it stable over a time comparable to the age of the solar system, or did Venus experience more "favorable" periods when the atmosphere cooled and, consequently, its thickness and chemical composition changed? (For example, at an early stage,

given low solar luminosity, a moist greenhouse effect might have been possible, later to be replaced by a runaway greenhouse effect; or Venus could have experienced something similar to the epochs known on Earth as the great ice ages.)

- What relationship do the tectonic processes observed today on Venus have to the terrestrial system of global plate tectonics?
- Finally, is outgassing continuing, and if so, how intensely? What is the balance between outgassing products and gas loss from Venus' atmosphere during various epochs?

The list of problems awaiting solution could be continued. Aside from direct study of Venus, finding the answers to these questions is most important for further progress in comparative planetology and planetary cosmogony. Although the 1990s are a period of relative calm in the study of our neighboring planet compared with the past three turbulent decades, the analysis of the large amount of data recently returned by Magellan, and incorporation of these results into general models of planetary evolution, will for years to come occupy scientists who study Venus. There are no other actual projects on the horizon. The Russian space program, as of the end of the 1980s, was reoriented toward the investigation of Mars, and then declined, and a return to research on Venus, including the sending of long-lived (possibly mobile) stations, is several decades off, most likely. Japanese scientists have expressed great interest in Venus and have considered several original projects. The prospects of these projects, and the time frame for their implementation, are uncertain, however. Several missions to Venus, including a new orbiter and a probe of the atmospheric composition, have been proposed for the new class of small Discovery missions in the United States, but these missions are years off at the earliest. Evidently we face a period of many years for in-depth interpretation of Magellan data and of the rest of our extensive cumulative information on Venus. Comprehensive analysis of this information, incorporating our developing knowledge concerning other celestial bodies and the solar system as a whole, will take place based on a comparative planetological approach.

In conclusion, let us again emphasize that the cardinal and, in many respects, intriguing problem of Venus' climate has an intimate bearing on the problem of predicting the future climate of our home planet. This problem boils down to a search for the reasons behind the striking differences in the climates of Earth, Venus, and Mars, which, on the scale of the solar system, are located at insignificant distances from one another. The conditions on Venus and its path of development serve as a warning to humanity. Acute ecological problems already face us today. We need to combine our efforts to avoid the undesirable consequences of industrial development. The example of Venus clearly shows that the goal of planetary research is not limited to, and is not exhausted by, the accumulation of knowledge about how our neighboring worlds were constructed and how they function. By contrast, planetary research is one area in the fundamental sciences that will be intimately involved in solving many technical-economic and social problems on the path of the further development of civilization.

Chapter 1: Venus as a Planet of the Solar System

1. An inverse logarithmic scale on which a decrease of five magnitudes (referred to as stellar magnitudes) represents an increase in brightness by a factor of 100.

2. "Phase" defines the proportion of the disk that is illuminated. It is characterized by a phase angle Q, which is the angle between the Sun, the center of the planet, and the observer.

3. The boundary between the illuminated (day) and unilluminated (night) hemispheres.

4. Note that e, characterizing the real oblateness of the planet, differs from the dynamical compression $\alpha = (a - b)/a$ where a and b are semimajor and semiminor axes of the rotational ellipsoid that approximates the rotating planet in a hydrostatic equilibrium. α characterizes the mass distribution inside the planet and, hence, its gravitational potential. e and α are in close coincidence for Jupiter and Saturn, but they are quite different for the Moon and inner planets, in particular for Venus.

5. The improved value for the rotational period from the Magellan data is 243.0184 $\pm$ 0.0001 days (Davies et al., 1992).

6. The integral albedo of the planet, A (Bond albedo), is determined by corresponding values of monochromatic (depending on the wavelength λ) albedo and monochromatic flux of solar radiation

$$A = \int_0^\infty A(\lambda)S(\lambda)d\lambda / \int_0^\infty S(\lambda)d\lambda,$$

where $S(\lambda)$ is the monochromatic flux of solar radiation.

Chapter 2: The History of Investigations

1. The Planck radiation formula is $B_v(T) = 2hv^3/c^2[exp(hv/kT) - 1]^{-1}$, where h is Planck's constant, k is the Boltzmann constant, v is the frequency of the radiation, and c is the speed of light.

2. The single scattering albedo, ω (also known as the probability of quantum survival), is the ratio of the scattering coefficient σ to the extinction coefficient κ, which represents the sum of the scattering σ and absorption $\varkappa$ coefficients, that is, $\omega = \sigma/(\sigma + \varkappa)$. In a purely scattering atmosphere ($\varkappa = 0$), we have $\omega = 1$; in a purely absorbing atmosphere ($\varkappa = \infty$), we have $\omega = 0$.

3. The asymmetry parameter, g, characterizes how the scattered radiation is distributed in space, that is, as a function of phase angle Q; therefore, it is also called the phase function. ω and g represent the fundamental optical characteristics of an atmosphere.

4. By the effective radar reflection cross sections, S_e, we mean the planet's geometric cross sectional area, πR^2, multiplied by the surface coefficient of reflection, k_s, which depends on the dielectric constant of the surface rock, and then multiplied by the coefficient of directionality, q, which characterizes the scattering of reflected radiation by surface irregularities. The quantity S_e is determined directly from radar measurements and is usually expressed in fractions of the geometrical area of the planet's cross section, $\sigma_e = S_e/\pi r^2$.

5. R. Jastrow and S. I. Rasool, eds., *The Venus Atmosphere* (New York: Gordon and Breach, 1969).

6. *Soviet Space Research* 14, nos. 5, 6 (1976).

7. *Soviet Space Research* 17, no. 5 (1979); *J. Geophys. Res.* 85, no. A13 (1980). See also Hunten et al., 1983.

8. *Soviet Space Research* 21, no. 2 (1983).

Chapter 3: The Surface

1. The current upgrading of the Arecibo telescope and radar system will significantly improve the sensitivity, allowing resolution close to 1 km over large areas over $330°$ longitude, the longitude of the sub-Earth point at close approach (Campbell, 1993).

2. A summary of Venera results and pre-Magellan ideas about the geology of Venus may be found in Barsukov et al., 1992. For more in-depth discussion of Magellan results, the reader is referred to the *Journal of Geophysical Research* 97, nos. E8 and E10 (1992). Magellan gravity results are discussed in *Icarus* 112, no. 1 (1994).

3. The International Astronomical Union (IAU) decreed that all morphological features on Venus' surface would be given female names, including mythological names, historical names, and names commemorating the distinguished women from cultures around the world who contributed significantly to science, art, literature, and so on. Maxwell Montes on Ishtar Terra is the only admitted exception, for historical reasons.

Chapter 4: The Internal Structure and Thermal History

1. Along with these values, the non-standard coefficients C_{22} and S_{22}, computed directly from measurements, are also usually used.

2. Such a transition occurs under conditions of very high pressures, when the external atomic shells are "crushed" and free electrons appear that ensure high conductivity analogous to metals. The density of metallic hydrogen can be roughly estimated, assuming that the distance between protons is on the order of the Bohr radius, $a_0 = \hbar^2/me^2 = 0.529 \cdot 10^{-8}$cm (here m and e are the mass and charge of the electron $\hbar = h/2\pi$, where h is Planck's constant). Because the mass of the proton is $m_p = 1.67 \cdot 10^{-24}$ g, we obtain $\rho \sim m_p/a_0^3 \sim 10$ g/cm^3. Actually, this estimate turns out to be almost an order of magnitude too high; more rigorous, although not very reliable, calculations indicate that at a pressure of 2.6 Mbars, metallic hydrogen is found in a state of thermodynamic equilibrium with molecular hydrogen, and its density is equal to 1.16 g/cm^3 (Ginzburg, 1985). For transition of helium into a metallic state, a pressure of approximately 90 Mbars is required, which is not reached inside Jupiter or Saturn.

3. By their incidence (frequency of impact), stony meteorites are dominant—about 96% in relative number (including differentiated stones)—whereas the impact frequency of iron meteorites does not exceed 3%, and that of stony-iron meteorites does not exceed 1%. Among the stony meteorites, which are composed primarily of lithophile elements forming stable natural compounds with oxygen (sodium, calcium, magnesium, aluminum, silicon, and, of course, oxygen), chondrites head the list (they received this designation because of the spherical particles contained in their structure, on the order of 1 mm in size—chondrules, consisting of silicate minerals). Other meteorites of this class (achondrites) are nearly an order of magnitude less abundant, and by their texture (erosion of the boundaries between the chondrules and embedded material) one can assume that during their formation they underwent significantly more intense heating than the chondrites.

4. This factor is a measure of quality (resonance properties) of the oscillatory system and serves to characterize a rate of its deviation from ideal elasticity. It is usually expressed in terms of the dissipation function, $1/Q$. The Q factor also characterizes selective features of an oscillation sys-

tem and shows by what factor the amplitude of established forced oscillations in resonance is higher than the amplitude of forced oscillations at frequencies that do not coincide with the natural frequency of the system. The physical nature of dissipation in planetary bodies is obviously connected with the forces of viscous friction, and with breaking the ideal structure of crystal lattices (impurities, dislocations, disorder, and so on).

Chapter 5: The Lower and Middle Atmosphere

1. Electrostatic discharge, resulting from the accumulation of static charge on the surface of each probe during its descent through the atmosphere, is identified as the most likely cause; this rendered the measurement equipment inoperative (see Seiff et al., 1980).

2. Strictly speaking, aside from refraction, deviation in the propagation velocity of the radio waves in the atmosphere, from the speed of light in free space, also affects the phase shift. The magnitude of this effect, however, is negligibly small and may be ignored.

3. By "zonal winds," we mean a condition where the pressure gradient in the meridional direction creates a horizontal component of the centripetal force, counterbalancing the velocity of the zonal flow on a planetary scale (see Chapter 7).

4. In this and subsequent reactions, the dot over the component denotes an intermediate (excited) state of the radical (component).

5. The working principle of this type of electrical sensor is based on the relationship between the water vapor pressure, above the surface of a saturated LiCl salt solution, and temperature.

6. $S_{n(g)}$ denotes sulfur allotropes in the gaseous phase.

7. In the following reaction, ν is the collision rate of atoms; $O(^1D)$ and $O(^3P)$ denote oxygen atoms in excited and ground states, respectively.

Chapter 7: The Thermal Regime and Dynamics of the Atmosphere

1. By overall net flux we mean the algebraic sum of the descending and ascending fluxes of scattered solar radiation. Similar terminology is subsequently used in analyzing thermal radiation flux in the atmosphere. In comparing the overall solar and thermal fluxes, one should keep in mind that they have opposite signs. Hence thermal net flux $S = W\uparrow - W\downarrow$ (see Fig. 7.1).

2. Wave number ν is defined as the ratio of 1 cm (expressed in μm) to λ (in μm): $\nu(\text{cm}^{-1}) = (10^4 \mu\text{m/cm})/\lambda(\mu\text{m})$. Hence, for $\lambda = 1 \mu\text{m}$, $\nu = 10,000 \text{ cm}^{-1}$; for $\lambda = 20 \mu\text{m}$, $\nu = 500 \text{ cm}^{-1}$, and so on.

3. The angular momentum of the atmosphere at a certain level Z is written as $Q = \rho a U_x \cos \varphi$, where U_x is the horizontal wind velocity at altitude Z. It is evident that with an increase in φ, Q drops; nevertheless, mass transport to the poles occurs owing to meridional transport. The change in Q with altitude is determined by the ρ and U_x profiles.

Chapter 8: The Upper Atmosphere

1. This routinely measured value is used as an index of solar activity, which is rather well correlated with the shortwave (EUV) solar radiation responsible for the energy input into the planetary upper atmosphere; it is more convenient than the Wolf number, which is proportional to the number of sunspots (see, for example, Marov and Kolesnichenko, 1987, for more detail).

2. We should note that the almost constant conditions near the homopause, at $Z = 100$ km, adopted in the VIRA model, are a simplification. In fact, as is obvious from the experimental data summarized by Keating et al. (1985) and Seiff et al. (1985), temperature variations at 100 km may

reach 15–20 K. It is doubtful, however, that these changes would significantly affect the exospheric temperature variations.

3. In addition to direct photodissociation of CO_2, the formation of CO and O is caused by dissociative recombination of CO^+.

4. In quantum mechanics and spectroscopy A, B, C, . . . denote successive electron states (as opposed to the ground state X); in turn, each state is characterized by the projection of the electron orbital moment onto the molecular axis (or by the quantum number Λ), the overall spin of the electrons s, and properties of symmetry. States with $\Lambda = 0, 1, 2, \ldots$ are denoted by the Greek letters Σ, Π, Δ, ... The projection of the spin onto the axis may take on 2s + 1 values; each term is split in accordance with this quantity. The multiplicity of the term is indicated at the upper left, for example $^2\Sigma$, $^3\Pi$ (s = 1/2 and s = 1, respectively). The corresponding interaction between the molecular rotation and electron motion causes the terms to split in two, because the projection of the electron orbital moment changes sign upon reflection in the plane passing through the molecular axis. In the $\Lambda = 0$ state, reflection does not change the electron energy, and the wave function is simply multiplied by $+1$ or -1. This property of symmetry for Σ terms is indicated in the upper right (Σ^+ or Σ^-). Finally, if the molecule consists of identical atoms, yet another property of symmetry emerges, which conveys the condition that energy is invariant with respect to simultaneous sign changes in the coordinates of all electrons and nuclei. Here, the wave function is multiplied by $+1$ or -1; this is denoted by the g or u indices to the lower right, for example, Σ_g, Σ_u, Π_u. As a rule, the ground state of diatomic molecules has complete symmetry, and the ground term is $^1\Sigma_g^+$. The O_2 molecule, with ground term $^3\Sigma_g^-$, and the NO molecule, with ground term $^2\Pi$, are exceptions. As an example we shall give the first excited (metastable) state of N_2: $A^3\Sigma_u^+$. It is useful to remember that in the case of ionized molecules, one can add primes to the letters A, B, C, . . . (or a, b, c, . . .). Dipole transitions between various electron states (or, in other words, dipole transitions with emission or absorption of light) obey so-called selection rules. They depend on the type of bond between the electron orbital motions, their spin, and the rotation of the molecule. The most important of these are $\Delta\Lambda = 0 \pm 1$; the multiplicity 2s + 1 does not change; the $\Sigma^+ \rightleftarrows \Sigma^-$ and g → g; u → u transitions are forbidden (see Hertzberg, 1949; Landau and Lifschitz, 1964; Zeldovich and Raizer, 1966).

5. As is obvious from the dimensions, the coefficient of turbulent diffusion serves as the characteristic equilibration rate in a physical system, because the equalization time t_r is written in the form $t_r = \text{const}L^2/D$, where L is the characteristic size of the region and D is the coefficient that characterizes the physical process (diffusion, thermal conductivity, or kinematic viscosity).

The early phases of Venus space exploration with the Venera and Mariner spacecraft are summarized in the monograph *Physics of the Planet Venus* (Kuzmin and Marov, 1974), which was translated into English by Leo Kanner Associates for NASA (NASA-TT-F-16226). Marov's book *Planets of the Solar System* (Marov, 1986) discusses the comparative planetology approach, with specific focus on Venus and Mars. More detailed analyses of pre-Magellan Venus studies are available in *Venus* (Hunten et al., 1983), *The Planet Venus* (Barsukov and Volkov, 1989), and *Venus Geology, Geochemistry, and Geophysics: Research Results from the Soviet Union* (Barsukov et al., 1992). A more recent book specifically focused on the Magellan imaging and geologic data is *Venus, the Geological Story* (Cattermole, 1994). A popular history of Venus, from mythology to Magellan results, is *Venus Revealed* (Grinspoon, 1997). An up-to-date collection of scientific chapters on Venus is being published in *Venus II* (Bougher et al., 1998).

In the following list of references, Russian titles have been translated into English.

Adams, W. S., and T. Dunham (1932). Absorption bands in the infra-red spectrum of Venus. Publ. Astron. Soc. Pacif. 44: 243–247.

Adel, A., and V. M. Slipher (1934). Concerning the carbon dioxide content of the atmosphere of the planet Venus. Phys. Rev. 46: 240.

Akim, E. L., Z. P. Vlasova, and I. V. Chuyko (1978). Determination of dynamical oblateness of Venus based on trajectory measurements of the first artificial satellites Venera 9 and Venera 10. Doklady Akad. Nauk USSR 240 (no. 3): 556–561.

Akimoto, S., Y. Matsui, and Y. Syono (1976). High pressure crystal-chemistry of orthosilicates and the formation of the mantle transition zone. In *The Physics and Chemistry of Minerals and Rocks*, ed. R. G. J. Strens. London: John Wiley.

Aleksandrov, Yu. N., M. B. Vasilev, A. S. Vyschlov, V. M. Dubrovin, A. L. Zaitsev, M. A. Kolosov, A. A. Krymov, G. I. Makovoz, G. M. Petrov, N. A. Savich, V. Z. Samovol, L. N. Samoznaev, A. I. Siborenko, A. F. Khasyanov, and D. Ya. Shtern (1976). Preliminary results of two-frequency radioscopy of the daytime ionosphere of Venus from Venera 9 and Venera 10. Kosmich. Issled. 14: 816–823.

Allen, D. A., and J. W. Crawford (1984). Cloud structure on the dark side of Venus. Nature 307: 222–224.

Anders, E., and T. Owen (1977). Origin and abundances of volatiles. Science 198: 453–465.

Anderson, J. D., G. E. Pease, L. Efron, and R. C. Tausworthe (1968). Determination of the mass of Venus and other astronomical constants from the radio tracking of Mariner V. Astron. J. 43 (no. 2): 2–12.

Anderson, R. C., J. G. Pipes, A. L. Broadfoot, and L. Wallace (1969). Spectra of Venus and Jupiter from 1800 to 3200 Å. J. Atmos. Sci. 26: 874–878.

Andreeva, N. E., V. P. Volkov, Yu. I. Sidorov, and I. L. Khodakovsky (1985). On the concentration of minor components in the near-surface Venus atmosphere: Physico-chemical implications. Lunar Planet Sci. 15.

Andreichikov, B. M. (1978). Distribution of water vapor in the upper troposphere of Venus based on the data of Venera 4, 5, and 6 probes. Geokhimiya 1: 11–15.

Apt, J., and J. Leung (1982). Thermal periodicities in the Venus atmosphere. Icarus 49: 423–427.

Arkani-Hamed, J., and M. N. Toksoz (1984). Thermal evolution of Venus. Phys. Earth Planet. Inter. 34: 232–250.

Arkani-Hamed, J., G. G. Schaber, and R. G. Strom (1993). Constraints on the thermal evolution of Venus inferred from Magellan data. J. Geophys. Res. 98: 5309–5315.

Arrehenius, G. (1923). *The Life Path of the Planets*. Moscow: Gosizdat.

Arvidson, R. E., R. A. Brackett, M. K. Shepard, N. R. Izenberg, and B. Fegley, Jr. (1994). Microwave signatures and surface properties of Ovda Regio and surroundings, Venus. Icarus 112: 171–186.

Arvidson, R. E., R. Greeley, M. C. Malin, R. S. Saunders, N. Izenberg, J. J. Plaut, E. R. Stofan, and M. K. Shepard (1992). Surface modification of Venus as inferred from Magellan observations of plains. J. Geophys. Res. 97: 13303–13318.

Ash, M. E., R. P. Ingalls, G. H. Pettengill, I. I. Shapiro, W. B. Smith, M. A. Slade, D. B. Campbell, R. B. Dyce, R. Jurgens, and T. W. Thompson (1968). The case of radar radius of Venus. J. Atmos. Sci. 25 (no. 4): 560–563.

Asimow, Paul D., and J. A. Wood (1992). Fluid outflows from Venus impact craters: Analysis from Magellan data. J. Geophys. Res. 97: 13643–13666.

Aumann, H. H., J. V. Martonchin, and J. S. Orton (1982). Airborne spectroscopy and spacecraft radiometry of Venus in the far infrared. Icarus 49: 227.

Avduevsky, V. S., N. F. Borodin, V. V. Kuznetsov, A. I. Lifshits, M. Ya. Marov, V. V. Mikhnevich, and M. K. Rozhdestvensky (1968a). Temperature, pressure and density of Venus' atmosphere according to "Venera 4" measurements. Doklady Akad. Nauk USSR 179 (no. 2): 310.

Avduevsky, V. S., M. Ya. Marov, and M. K. Rozhdestvensky (1968b). Model of the planet Venus based on results of the measurements made by the Soviet automatic interplanetary station Venera 4. J. Atmos. Sci. 25 (no. 4): 537–545. Reprinted in *The Venus Atmosphere* (1968), ed. R. Jastrow and S. I. Rasool (New York: Gordon and Breach).

Avduevsky, V. S., M. Ya. Marov, and M. K. Rozhdestvensky (1970a). A tentative model of the atmosphere of the planet Venus based on the results of measurements of Venera 5 and 6. J. Atmos. Sci. 27 (no. 4): 561–568.

Avduevsky, V. S., M. Ya. Marov, A. I. Noykina, V. I. Polezhaev, and F. S. Zavelevich (1970b). Heat transfer in the Venus atmosphere. J. Atmos. Sci. 27 (no. 4): 569–579.

Avduevsky, V. S., F. S. Zavelevich, M. Ya. Marov, A. I. Noykina, and V. I. Polezhaev (1971). Numerical simulation of radiative-convective heat exchange in the atmosphere of Venus. Kosmich. Issled. 9 (no. 2): 280–291.

Avduevsky, V. S., M. Ya. Marov, and M. K. Rozhdestvensky (1972a). Model of Venus' atmosphere from direct measurements. In *Physics of the Moon and Planets*. Moscow: Nauka.

Avduevsky, V. S., F. S. Zavelevich, M. Ya. Marov, A. I. Noykina, and V. I. Polezhaev (1972b). Thermal regime and convective motions in the lower layers of Venus' atmosphere. In *Physics of the Moon and Planets*. Moscow: Nauka.

Avduevsky, V. S., M. Ya. Marov, B. E. Moshkin, and A. P. Ekonomov (1973a). The results of direct measurements of illumination in the atmosphere of Venus from Venera 8 spacecraft. Dokl. Acad. Nauk USSR 210 (nos. 4–6): 799–802.

Avduevsky, V. S., M. Ya. Marov, B. E. Moshkin, and A. P. Ekonomov (1973b). Venera 8: Measurements of solar illumination through the atmosphere of Venus. J. Atmos. Sci. 30 (no. 6): 1215–1218.

Avduevsky, V. S., and M. Ya. Marov (1976). Study of Venus with the Venera spacecraft. In *Aeromechanics and Gas Dynamics*, ed. V. V. Struminsky. Moscow: Nauka.

Avduevsky, V. S., N. F. Borodin, V. P. Burtsev, Ya. V. Malkov, M. Ya. Marov, S. F. Morozov, M. K. Rozhdestvensky, R. S. Romaov, S. S. Sokolov, V. G. Fokin, Z. P. Cheremukhina, and V. I. Shkirina (1976a). Automatic station Venera 9 and Venera 10: Functioning of descent vehicles and measurement of atmospheric parameters. Kosmich. Issled. 14: 577.

Avduevsky, V. S., Yu. M. Golovin, F. S. Zavelevich, V. Ya. Likhushin, M. Ya. Marov, D. A. Melnikov, Ya. I. Merson, B. E. Moshkin, K. A. Razin, L. I. Chernoshcekov, and A. P. Ekonomov (1976b). Preliminary results of the investigation of the light regime in the atmosphere and on the surface of Venus. Kosmich. Issled. 14: 735–741.

Avduevsky, V. S., S. L. Vishnevetsky, I. A. Golov, Yu. Ya. Karpeisky, A. D. Lavrov, V. Ya. Kukushkin, M. Ya. Marov, D. A. Melnikov, N. I. Pomogin, N. N. Pronina, K. A. Razin, and V. G. Fokin (1976c). Measurement of wind velocity on the surface of Venus during the operation of stations Venera 9 and Venera 10. Kosmich. Issled. 14: 622–625.

Avduevsky, V. S., N. F. Borodin, V. N. Vasilev, A. G. Godnev, V. P. Karyagin, V. A. Koveryanov, V. M. Kovtunenko, R. S. Kremnev, V. M. Pavlova, M. K. Rozhdestvensky, V. I. Serbin, K. G. Sukhanov, G. R. Uspensky, and Z. P. Cheremukhina (1979). Parameters of Venus atmosphere at Venera 11 and 12 landing sites. Kosmich. Issled. 17: 539–544.

Avduevsky, V. S., M. Ya. Marov, Yu. N. Kulikov, V. P. Shari, A. Ya. Gorbachevsky, G. R. Uspensky, and Z. P. Cheremukhina (1983a). Structure and parameters of the Venus atmosphere according to Venera probe data. In *Venus*, ed. D. M. Hunten, L. Colin, T. M. Donahue, and V. I. Moroz. Tucson: University of Arizona Press.

Avduevsky, V. S., A. G. Godnev, V. V. Semenchenko, G. R. Uspensky, and Z. P. Cheremukhina (1983b). Characteristics of the stratosphere of Venus from measurements of the overloads during the braking of the Venera 13 and Venera 14 spacecraft. Kosmich. Issled. 21 (no. 2): 149.

Ayers, G. P., R. W. Gillett, and J. L. Gros (1980). On the vapor pressure of sulfuric acid. Geophys. Res. Lett. 7: 433–436.

Baker, V. R., G. Komatsu, T. J. Parker, V. C. Gulick, J. S. Kargel, and J. S. Lewis (1992). Channels and valleys on Venus: Preliminary analysis of Magellan data. J. Geophys. Res. 97: 13421–13444.

Banerdt, W. B., and C. G. Sammis (1992). Small-scale fracture patterns on the volcanic plains of Venus. J. Geophys. Res. 97: 16149–16166.

Barabaschev, N. P. (1928). Photometric studies of the brightness distribution over Venus' disk. Kharkov Astron. Obs. Pub. no. 2: 3–11.

Barabaschev, N. P., and A. T. Chikirda (1950). Photographic spectrophotometry of Venus, Mars, Jupiter and Saturn. Proceed. Kharkov. Obs. 1 (no. 9): 19–24.

———. (1952). Concerning change in the light index of Venus. Proceed. Kharkov Obs. 2 (no. 10): 5–7.

Barabaschev, N. P., and D. E. Semeykin (1935). Etudes photométriques de Vénus. Kharkov Astron. Obs. Pub. no. 5: 29–37.

Barker, E. S. (1975a). Comparison of simultaneous CO_2 and H_2O observations of Venus. J. Atmos. Sci. 32: 1071.

———. (1975b). Observations of Venus water vapor over the disk of Venus: The 1972–74 data using the H_2O lines at 8197 A and 8176 A. Icarus 25: 268.

———. (1979). Detection of SO_2 in the UV spectrum of Venus. Geophys. Res. Lett. 6: 117–120.

Barmin, I. V., and A. A. Shevchenko (1983). Soil collection device for the "Venera 13–14" spacecraft. Kosmich. Issled. 21 (no. 2): 171–175.

Barrett, A. H. (1961). Microwave absorption and emission in the atmosphere of Venus. Astrophys. J. 133: 281.

Barsukov, V. L., and V. P. Volkov, eds. (1989). *The Planet Venus*. Moscow: Nauka.

Barsukov, V. L., I. L. Khodakovsky, V. P. Volkov, and R. P. Florensky (1980). The geochemical model of the troposphere and lithosphere of Venus based on new data. Space Res. 20: 197.

Barsukov, V. L., V. P. Volkov, and I. L. Khodakovsky (1982). The crust of Venus: Theoretical models of chemical and mineral composition. Proc. 13th Lunar Planet. Sci. Conf., J. Geophys. Res. 87: A 3–9.

Barsukov, V. L., A. T. Basilevsky, V. P. Volkov, and V. N. Zharkov (1983). Venus geology, geochemistry and geophysics: Research results from the USSR. In *Venus*, ed. D. M. Hunten, L. Colin, T. M. Donahue and V. I. Moroz. Tucson: University of Arizona Press.

Barsukov, V. L., Yu. A. Surkov, and L. P. Moskaleva (1984). New data on the composition, structure and properties of Venus rock obtained by Venera 13 and Venera 14. J. Geophys. Res. 91 (no. 4): B393–B402.

Barsukov, V. L., A. T. Basilevsky, R. O. Kuzmin, M. S. Markov, V. P. Kryuchkov, O. V. Nikolaeva, A. A. Pronin, A. L. Sukhanov, I. M. Chernaya, G. A. Burba, N. N. Bobina, and V. P. Shashkina (1985). The major types of structures in the northern Venus hemisphere. Astron. Vestn. (Solar System Research) 19: 3–14.

Barsukov, V. L., A. T. Basilevsky, G. A. Burba, N. N. Bobina, V. P. Kryuchkov, R. O. Kuzmin, O. V. Nikolaeva, A. A. Pronin, L. B. Ronca, I. M. Chernaya, V. P. Sashkina, A. V. Garanin, E. R. Kushky, M. S. Markov, A. L. Sukhanov, V. A. Kotelnikov, O. N. Rzhiga, G. M. Petrov, Yu. N. Alexandrov, A. I. Sidorenko, A. F. Bogomolov, G. I. Skrytnik, M. Yu. Bergman, L. V. Kudrin, I. M. Bokshtein, M. A. Kronrad, P. A. Chechiya, Yu. S. Tyuflin, S. A. Kadnichansky, and E. L. Akim (1986a). The geology and geomorphology of the Venus surface as revealed by the radar images obtained by Veneras 15 and 16. J. Geophys. Res. 91: D378–D398.

Barsukov, V. L., Yu. A. Surkov, L. Dmitriev, and I. L. Khodakovsky (1986b). Geochemical study of Venus by landers of Vega-1 and Vega-2 probe. Geochimiya: 275–289.

Barsukov, V. L., S. P. Borunov, V. P. Volkov, V. A. Dorofeyeva, M. Yu. Zolotov, S. V. Porotkin, Yu. V. Semenov, Yu. I. Sidorov, I. L. Khodakovsky, and A. I. Shapkin (1986c). Estimation of mineral composition of soil at the landing sites of Venera 13, 14 and Vega-2 landers according to thermodynamical calculations. Dokl. Akad. Nauk USSR 287: 415–417.

Barsukov, V. L., A. T. Basilevsky, V. P. Volkov, and V. N. Zharkov, eds. (1992). *Venus Geology, Geochemistry, and Geophysics: Research Results from the Soviet Union*. Tucson: University of Arizona Press.

Barth, C. A., A. I. F. Stewart, S. W. Bougher, D. M. Hunten, S. J. Bauer, and A. F. Nagy (1992). Aeronomy of the current Martian atmosphere. In *Mars*, ed. H. H. Kieffer, B. M. Jakosky, C. W. Snyder, and M. S. Matthews. Tucson: University of Arizona Press.

Basharinov, A. E., Yu. N. Vetukhnovskaya, V. N. Galaktionov, I. B. Drozdovskaya, S. T. Egorov, M. A. Kolosov, V. D. Krotikov, N. N. Krupenio, A. D. Kuzmin, V. A. Ladygin, L. I. Malofeev, E. P. Omelchenko, V. S. Troitsky, N. Ya. Shapirovskaya, A. M. Shutko, and O. B. Schuko (1973). Results of radio occultation of the planet Mars from "Mars-3" data. Kosmich. Issled. 11: 803–808.

Basharinov, A. E., Yu. N. Vetukhnovskaya, V. N. Galaktionov, S. T. Egorov, M. A. Kolosov, N. N. Krupenio, A. D. Kuzmin, V. A. Ladygin, L. I. Malofeev, V. S. Troitsky, N. Ya. Shapirovskaya, and A. M. Shutko (1976). Radio astronomical measurements from the "Mars-5" spacecraft. Kosmich. Issled. 14: 73–81.

Basilevsky, A. T. (1981). On some peculiarities of the structure of impact craters on planets and satellites of the solar system. Dokl. Acad. Nauk USSR 258: 323–325.

———. (1986). Structure of central and eastern regions of Ishtar Terra and some problems of Venusian tectonics. Geotektonika 20: 282–288.

———. (1989). The planet next door. Sky and Telescope 77: 360–368.

Basilevsky, A. T., and J. W. Head (1988). The geology of Venus. Ann. Rev. Earth Planet. Sci. 16: 295–317.

Basilevsky, A. T., and C. M. Weitz (1992). Venera 9, 10 and 13 landing sites as seen by Magellan. Lunar Planet. Sci. 23: 67–68.

Basilevsky, A. T., B. A. Ivanov, V. P. Kryuchkov, R. O. Kuzmin, A. A. Pronin, and I. M. Chernaya (1985). Impact craters on Venus based on radar images from Venera 15/16. Dokl. Akad. Nauk USSR 282: 671–674.

Basilevsky, A. T., A. A. Pronin, L. B. Ronca, V. P. Kryuchkov, and A. L. Sukhanov (1986). Styles of tectonic deformations on Venus: Analysis of Venera 15 and 16 data. J. Geophys. Res. 91: D 399–D 411.

Basilevsky, A. T., B. A. Ivanov, G. A. Burba, I. M. Chernaya, V. P. Kryuchkov, O. V. Nikolaeva, D. B. Campbell, and L. B. Ronca (1987). Impact craters on Venus: A continuation of the analysis of data from the Venera 15 and 16 spacecraft. J. Geophys. Res. 92: 12869–12901.

Basilevsky, A. T., O. V. Nikolaeva, and C. M. Weitz (1992). Geology of the Venera 8 landing site region from Magellan data: Morphological and geochemical considerations. J. Geophys. Res. 97 (no. E10): 16315–16335.

Bauer, S. J., L. M. Brace, H. A. Taylor, T. K. Breus, A. J. Kliore, W. S. Knudsen, A. F. Nagy, C. T. Russell, and N. A. Savich (1985). The Venus ionosphere. In *Venus International Reference Atmosphere*, ed. A. J. Kliore, V. I. Moroz, and G. M. Keating. Adv. Space Res. 5 (no. 11): 233–267.

Becker, R. H., and R. O. Pepin (1984). The case for a Martian origin of shergotites: Nitrogen and noble gases in EETA 79001. Earth Planet. Sci. Lett. 69: 225–242.

Beletsky, V. V. (1975). Resonance phenomena in the rotational motions of artificial and natural bodies. Preprint, M. V. Keldysh Institute App. Math. Acad. Nauk USSR (no. 10).

Beletsky, V. V., and S. I. Trushin (1974). Resonances in the rotation of bodies and the generalized Cassini Laws. In *The Mechanics of Solid Bodies*, no. 6. Kiev: Naukova Dumka.

Beletsky, V. V., and A. A. Khentov (1995). Resonance rotations of celestial bodies. Nizhegorodsky Gumanitarniy Center, Nizhniy Novgorod.

Belopolsky, A. A. (1900). Ein Versuch die Rotationsgeschwindigkeit des Venus Aequators auf spectrographische Wege zu bestimmen. Astron. Nachr. 152 (no. 3641): 236–276.

———. (1903). Preliminary results of rotational studies of Venus near the axis. Izv. Imp. Akad. Nauk 18: 28–29.

———. (1911). *Uber die Rotation von Venus*. St. Petersburg: Obs. Centr. Poulkovo.

Belton, M. J. S. (1968). Theory of the curve of growth and phase effects in a cloudy atmosphere: Application to Venus. J. Atmos. Sci. 25: 596–609.

Belton, M. J. S., and D. M. Hunten (1966). Water vapor in the atmosphere of Venus. Astrophys. J. 146: 307–308.

———. (1968). A search for O_2 on Mars and Venus: A possible detection of oxygen in the atmosphere of Mars. Astrophys. J. 153: 963–974.

Belton, M. J. S., D. M. Hunten, and R. M. Goody (1968). Quantitative spectroscopy of Venus in the region 8,000–11,000 Å. In *The Atmospheres of Venus and Mars*, ed. J. C. Brandt and M. B. McElroy. New York: Gordon and Breach.

Belton, M. J. S., G. Smith, D. A. Elliott, K. Klaasen, and G. E. Danielson (1976a). Space-time relationships in UV markings on Venus. J. Atmos. Sci. 33: 1383–1393.

Belton, M. J. S., G. Smith, G. Schubert, and A. D. Del Genio (1976b). Cloud patterns, waves and convection in the Venus atmosphere. J. Atmos. Sci. 33: 1394–1417.

Berttaux, J. L., E. Chassefiere, and V. G. Kurt (1985). Venus EUV measurements of hydrogen and helium from Venera 11 and Venera 12. Adv. Space Res. 5 (no. 9): 119–124.

Bézard, B., C. de Bergh, D. Crisp, and J.-P. Maillard (1990). The deep atmosphere of Venus revealed by high-resolution night-side spectra. Nature 345: 508–511.

Bézard, B., C. de Bergh, J.-P. Maillard, D. Crisp, J. Pollack, and D. H. Grinspoon (1991). High-resolution spectroscopy of Venus' nightside in the 2.3, 1.7, and 1.1–1.3 micron windows. Bull. Am. Astron. Soc. 23: 1192.

Bézard, B., C. de Bergh, B. Fegley, J.-P. Maillard, D. Crisp, T. Owen, J. B. Pollack, and D. Grinspoon (1993). The abundance of sulfur dioxide below the clouds of Venus. Geophys. Res. Lett. 20: 1587–1590.

Billoti, F., and J. Suppe (1992). Wrinkle ridges and topography on Venus. Geol. Soc. Am. Abstr. with Programs 24: A 195 (abstract).

Bills, B. G. (1992). Venus satellite orbital decay, ephemeral ring formation and subsequent crater production. GRL 19: 1025–1028.

Bindschadler, D. L., and J. W. Head (1988). Diffuse scattering of radar on the surface of Venus: Origin and implications for the distribution of soils. Earth, Moon and Planets 42: 133–149.

———. (1989). Characterization of Venera 15/16 geologic units from Pioneer Venus reflectivity and roughness data. Icarus 77: 3–20.

Bindschadler, D. L., G. Schubert, and W. M. Kaula (1992a). Coldspots and hotspots: Global tectonics and mantle dynamics of Venus. J. Geophys. Res. 97: 13495–13532.

Bindschadler, D. L., A. deCharon, K. K. Beratan, S. E. Smrekar, and J. W. Head (1992b). Magellan observations of Alpha Regio: Implications for formation of complex ridged terrains on Venus. J. Geophys. Res. 97: 13563–13578.

Birnbaum, G., W. Ho, and A. Rosenberg (1971). Farintrated collision induced absorption in CO_2. 2. Pressure dependence in the gas phase and absorption with liquid. J. Chem. Phys. 55: 1039–1045.

Bogard, D. D., L. E. Nyquist, and P. H. Johnson (1984). Noble gas contents of shergotites and implications for the Martian origin of SNC meteorites. Geochim. Cosmochim. Acta 48: 1723–1740.

Borucki, W. J., J. W. Dyer, and J. R. Phillips (1991). PVO search for Venusian lightning. J. Geophys. Res. 96: 1033–1043.

Bottema, M., W. Plummer, and J. Strong (1964). Water vapor in the atmosphere of Venus. Astrophys. J. 139: 1021–1022.

———. (1965). A quantitative measurement of water vapor in the atmosphere of Venus. Ann. Astrophys. 28 (no. 1): 225–230.

Bougher, S. W., and R. G. Roble (1991). Comparative terrestrial planet thermospheres, 1: Solar cycle variation of global mean temperatures. J. Geophys. Res. 96: 11045–11055.

Bougher, S. W., R. E. Dickinson, E. C. Ridley, and R. G. Roble. (1988a). Venus mesosphere and thermosphere, 3: Three dimension general circulation with coupled dynamics and composition. Icarus 73: 545–573.

Bougher, S. W., R. E. Dickinson, R. G. Roble, and E. C. Ridley (1988b). Mars thermospheric general circulation model: Calculations for the arrival of Phobos at Mars. Geophys. Res. Lett. 15: 1511–1514.

Bougher, S. W., J.-C. Gerard, A. I. F. Stewart, and C. G. Fesen (1990). The Venus nitric oxide night air glow: Model calculations based on the Venus thermospheric general circulation model. J. Geophys. Res. 95: 6271–6284.

Bougher, S., D. Hunten, and R. J. Phillips, ed. (1998). *Venus II*. Tucson: University of Arizona Press.

Boyer, C. (1960). Observations à Brazzaville. Astronomie 74 (no. 9): 376–378.

Boyer, C., and H. Camichel (1961). Observation photographique de la planète Vénus. Ann. Astrophys. 24: 531–535.

Boyer, C., and P. Guerin (1969). Etude de la rotation rétrograde, en 4 jours, de la couche extérieure buageuse de Vénus. Icarus 11: 338–355.

Brace, L. H., and A. J. Kliore (1991). Space Sci. Rev. 55: 81.

Brace, L. H., H. A. Taylor, T. I. Gombosi, A. J. Kliore, W. C. Knudsen, and A. F. Nagy (1983). The ionosphere of Venus: Observations and their interpretations. In *Venus*, ed. D. M. Hunten, L. Colin, T. M. Donahue, and V. I. Moroz. Tucson: University of Arizona Press.

Brackett, R. A., B. Fegley, Jr., and R. E. Arvidson (1995). Volatile transport on Venus and implications for surface geochemistry and geology. J. Geophys. Res. 100: 1553–1564.

Branon, J. F., J. L. Fox, and H. S. Porter (1993). Evidence for day-to-night ion transport at low solar activity in the Venus pre-dawn ionosphere. Geophysical Res. Lett. 20 (no. 23): 2739–2742.

Bridge, H. S., A. J. Lazarus, C. W. Snyder, E. J. Smith, L. Davis, P. H. Coleman, and D. E. Jones (1967). Plasma and magnetic fields near Venus. Science 158: 1669–1673.

Bridge, H. S., A. J. Lazarus, J. D. Scudder, K. W. Ogilvie, R. E. Hartle, J. R. Asbridge, S. J. Bame, and G. L. Siscoe (1974). Observations at Venus encounter by plasma science experiments on Mariner 10. Science 183: 1293–1296.

Broadfoot, A. L., S. Kumar, M. J. S. Belton, and M. B. McElroy (1974). Ultraviolet observations of Venus from Mariner 10: Preliminary results. Science 183: 1315–1318.

Bullen, R. E. (1936). The variation of density and the ellipticities of strata of equal density within the Earth. Mon. Not. R. Astr. Soc. Geophys. (suppl. 3): 395.

Bullock, M. A., D. H. Grinspoon, and J. W. Head (1993). Venus resurfacing rates: Constraints provided by 3-D Monte Carlo simulations. Geophys. Res. Lett. 20: 2147–2152.

Bullock, M. A., and D. H. Grinspoon (1996). The stability of climate on Venus. J. Geophys. Res. 101, 7521–7529.

Burbridge, E. M., and G. R. Burbidge (1982). Nucleosynthesis in galaxies. In *Essays in Nuclear Astrophysics*, ed. C. A. Barnes, D. D. Clayton, and D. N. Schramm. Cambridge: Cambridge University Press.

Burch, D. E., and D. A. Gryvnak (1971). Absorption of infrared radiant energy by CO_2 and H_2O: Absorption by CO_2 between 1100 and 1835 cm^{-1} (9.1–5.5 μm). J. Opt. Soc. Amer. 61 (no. 4): 499–503.

Bursa, M., and Z. Sima (1985). Dynamic and figure parameters of Venus and Mars. Adv. Space Res. 5 (no. 8): 43–46.

Cameron, A. G. W. (1963). The origin of the atmospheres of Venus and the Earth. Icarus 2 (no. 3): 249.

———. (1978). Physics of the primitive solar nebula and of giant gaseous protoplanets. In *Protostars and Planets*, ed. T. Gehrels. Tucson: University of Arizona Press.

———. (1982). The abundance of chemical elements and nuclides in the solar system. In *Essays in Nuclear Astrophysics*, ed. C. A. Barnes, D. D. Clayton, and D. N. Schramm. Cambridge: Cambridge University Press.

Cameron, A. G. W., and J. Pollack (1976). On the origin of the solar system and of Jupiter and its satellites. In *Jupiter*, ed. T. Gehrels. Tucson: University of Arizona Press.

Cameron, A. G. W., W. M. De Campli, and P. H. Bodenheimer (1980). Numerical experiments with giant gaseous protoplanets embedded in the primitive solar nebula. Lunar Planet Sci. 11: 122–124.

Campbell, B. A. (1994). Merging Magellan emissivity and SAR data for analysis of Venus surface dielectric properties. Icarus 112: 187–203.

Campbell, D. B., and B. A. Burns (1980). Earth-based radar imagery of Venus. J. Geophys. Res. 85: 8271–8281.

Campbell, B. A., and D. B. Campbell (1992). Analysis of volcanic surface morphology on Venus from comparison of Arecibo, Magellan and terrestrial airborne radar data. J. Geophys. Res. 97: 16299–16315.

Campbell, D. B., R. B. Dyce, R. P. Ingalls, G. H. Pettengill, and I. I. Shapiro (1972). Venus: Topography revealed by radar data. Science 175: 514–516.

Campbell, D. B., R. B. Dyce, and G. H. Pettengill (1976). New radar image of Venus. Science 193: 1123–1124.

Campbell, D. B., B. A. Burns, and V. Boriakoff (1979). Venus: Further evidence of impact cratering and tectonic activity from radar observations. Science 204: 1424–1427.

Campbell, D. B., J. W. Head, J. K. Harmon, and A. A. Hine (1983). Venus: Identification of banded terrain in the mountains of Ishtar Terra. Science 221: 644–647.

Campbell, D. B., J. W. Head, J. K. Harmon, and A. A. Hine (1984). Venus: Volcanism and rift formation in Beta Region. Science 226: 167–170.

Campbell, D. B., J. W. Head, A. A. Hine, J. K. Harmon, D. A. Senske, and P. A. Fisher (1989). Styles of volcanism on Venus: New Arecibo high resolution radar data. Science 246: 373–377.

Campbell, D. B., N. J. S. Stacy, W. I. Newman, R. E. Arvidson, E. M. Jones, G. S. Musser, A. Y. Roper, and C. Schaller (1992). Magellan observations of extended impact crater related features on the surface of Venus. J. Geophys. Res. 97: 16249–16278.

Carlson, R. W., K. H. Baines, T. Encrenaz, F. W. Taylor, P. Drossart, L. W. Kamp, J. B. Pollack, E. Lellouch, A. D. Collard, S. B. Calcutt, D. H. Grinspoon, P. R. Weissman, W. D. Smythe, A. C. Ocampo, G. E. Danielson, F. P. Fanale, T. V. Johnson, H. H. Kieffer, D. L. Matson, T. B. McCord, and L. A. Soderblom (1991). Galileo infrared imaging spectrometer measurements at Venus. Science 253: 1541–1548.

Carlson, R. W., L. W. Kamp, K. Baines, J. Pollack, D. Grinspoon, T. Encrenaz, P. Drossart, and F. Taylor (1993). Distinct Venus cloud types as observed by the Galileo near infrared mapping spectrometer. Planet. Space Sci. 41: 477–486.

Carpenter, R. L. (1966). Study of Venus by CW-radar: Results of the 164 conjunction. Astron. J. 69: 2–11.

———. (1970). A radar determination of the rotation of Venus. Astron. J. 75 (no. 1): 61–66.

Cassen, P., and A. Summers (1983). Models of the formation of the solar nebula. Icarus 53 (no. 16): 26–40.

Cattermole, P. J. (1994). *Venus, the Geological Story*. Baltimore: Johns Hopkins University Press.

Cess, R. D., V. Ramanathan, and T. Owen (1980). The Martian paleoclimate and enhanced atmospheric carbon dioxide. Icarus 41: 159–165.

Chamberlain, J. W. (1965). The atmosphere of Venus near her cloud tops. Astrophys. J. 141: 1184–1205.

Chamberlain, J. W., and Hunten, D. M. (1987). *Theory of Planetary Atmospheres*. 2nd ed. New York: Academic Press.

Chamberlain, J. W., and Kuiper, G. P. (1956). Rotational temperature and phase variation of the carbon dioxide bands of Venus. Astrophys. J. 124 (no. 2): 399–405.

Chandrasekhar, S. (1950). *Radiative Transfer*. Oxford: Clarendon Press.

Chase, S. C., L. D. Kaplan, and G. Neugebauer (1963). The Mariner 2 infrared radiometer experiment. J. Geophys. Res. 68 (no. 22): 6157–6169.

Chase, S. C., E. D. Miner, and D. Morrison (1974). Preliminary infrared radiometry of Venus from Mariner 10. Science 183 (no. 4131): 1291–1292.

Chen, J. H., and G. J. Wasserburg (1986). Formation ages and evolution of Shergotty and its parent planet from U-Th-Pb systematics. Geochim. Cosmochim. Acta 50: 955–968.

Chub, E. V., and O. L. Yakovlev (1980). Temperature and zonal circulation of Venus based on the data of radio probe experiments. Kosmich. Issled. 18: 331–336.

Cimino, J. B., C. Elachi, A. J. Kliore, D. J. McCleese, and I. R. Patel (1980). Polar cloud structure as derived from the Pioneer Venus orbiter. J. Geophys. Res. 85: 8082–8088.

Clancy, R. T., and D. O. Muhleman (1985). Diurnal variations in the Venus mesosphere from CO microwave spectra. Icarus 64: 157–182.

Cloutier, P. A., and R. E. Daniel (1973). Ionospheric currents induced by solar wind interaction with planetary atmospheres. Planet. Space Sci. 21: 463–474.

Cloutier, P. A., T. F. Tascioue, R. E. Daniel, Jr., H. A. Taylor, and R. S. Wolff (1983). Physics of the interaction of the solar wind with the ionosphere of Venus: Flow/field models. In: *Venus*, ed. D. M. Hunten, L. Colin, T. M. Donahue, and V. I. Moroz. Tucson: University of Arizona Press.

Cloutier, P. A., L. Kramer, and H. A. Taylor, Jr. (1993). Observations of the nightside Venus ionosphere: Final encounter of the Pioneer Venus orbiter mass spectrometer. Geophysical Res. Lett. 20 (no. 23): 2731–2734.

Coblentz, W. W., and C. O. Lampland (1924). Radiometric measurements on Venus. Publ. Astron. Soc. Pacif. 36: 274–276.

———. (1925). Measurements of planetary radiation. Lowell Obs. Bull. 85 (no. 3): 91–134.

Coffeen, D. L. (1968). Optical polarization of Venus. J. Atmos. Sci. 25: 643–648.

Coffeen, D. L., and T. Gehrels (1969). Wavelength dependence of polarization, 16: Atmosphere of Venus. Astron. J. 74 (no. 3): 446–460.

Coffeen, D. L., and J. E. Hansen (1974). Polarization studies of planetary atmosphere. In: *Planets, Stars, and Nebulae Studied with Photopolarimetry*, ed. T. Gehrels. Tucson: University of Arizona Press.

Colin, L., and D. M. Hunten (1977). Pioneer-Venus experiment descriptions. Space Sci. Rev. 20: 451–525.

Colombo, G., and I. Shapiro (1965). The rotation of the planet Mercury. Astrophys. J. 145: 298.

Connes, J., and P. Connes (1966). Near-infrared planetary spectra by Fourier spectroscopy, 1: Instruments and results. J. Opt. Soc. Amer. 56: 896–910.

Connes, J., P. Connes, W. S. Benedict, and L. D. Kaplan (1967). Traces of HCl and HF in the atmosphere of Venus. Astrophys. J. 147: 1230–1237.

Connes, P., J. Connes, L. D. Kaplan, and W. S. Benedict (1968). Carbon monoxide in the Venus atmosphere. Astrophys. J. 152: 731–747.

Conway, R. R., R. P. McCoy, C. A. Barth, and A. L. Lane (1979). TUE detection of sulfur dioxide in the atmosphere of Venus. Geophys. Res. Lett. 6: 629–631.

Counselman, C. C. III, S. A. Gourevitch, R. W. King, and G. B. Loriot (1980). Zonal and meridional circulation of the lower atmosphere of Venus determined by radio interferometry. J. Geophys. Res. 85: 8026–8030.

Courtin, R., P. Lena, M. de Muizon, D. Rouan, C. Nicollier, and J. Wijnbergen (1979). Far infrared photometry of planets Saturn and Venus. Icarus 38: 411–419.

Covey, C., E. J. Pitcher, and J. P. Brown (1986). General circulation model simulations of superrotation in slowly rotating atmospheres: Implications for Venus. Icarus 69: 202–220.

Craig, R. A., R. T. Reynolds, and B. Ragent (1983). Sulfur trioxide in the lower atmosphere of Venus? Icarus 53 (no. 1): 1–9.

Cravens, T. E., L. H. Brace, H. A. Taylor, C. T. Russell, W. C. Knudsen, K. L. Miller, A. Barnes, J. D. Michalov, F. L. Scarf, S. J. Quenon, and A. F. Nagy (1982). Disappearing ionospheres on the nightside of Venus. Icarus 51: 271–282.

Crisp, D., D. A. Allen, D. H. Grinspoon, and J. B. Pollack (1991). The dark side of Venus: Near infrared images and spectra from the Anglo-Australian Observatory. Science 253: 1263–1266.

Cruikshank, D. P. (1967). Sulfur compounds in the atmosphere of Venus, 2: Upper limits for the abundance of COS and H_2S. Comm. Lunar Planet. Lab. 6: 199–200.

Cruikshank, D. P., and G. P. Kuiper (1967). Sulfur compounds in the atmosphere of Venus, 1: An upper limit for the abundance of SO_2. Comm. Lunar Planet. Lab. 6: 195–197.

Cruikshank, D. P., and G. T. Sill (1967). The infrared spectrum of carbon suboxide, 2: Region 2–15 microns. Comm. Lunar Planet. Lab. 6: 204–205.

Crumpler, L. S., and J. W. Head (1987). Bilateral topographic symmetry across Aphrodite Terra, Venus. J. Geophys. Res. Lett. 14: 607.

Crumpler, L. S., J. W. Head, and J. K. Harmon (1987). Regional linear cross-strike discontinuities in western Aphrodite Terra, Venus. Geophys. Res. Lett. 14 (no. 6): 607–610.

Crumpler, L. S., J. W. Head, and J. C. Aubele (1993). Relation of major volcanic center concentration on Venus to global tectonic patterns. Science 261: 591–595.

Danjon, M. A. (1949). Photométrie et colorimétrie des planètes Mercure et Vénus. Bull. Astron. de Paris 14 (no. 4): 315–345.

Davies, M. E., T. R. Colvin, P. G. Rogers, P. W. Chodas, W. L. Sjogren, E. L. Akim, V. A. Stepanyantz, Z. P. Vlasova, and A. I. Zakhorov (1992). The rotation period, direction of the north pole, and geodetic control network of Venus. J. Geophys. Res. 97: 13141–13152.

De Bergh, C., B. Bézard, T. Owen, D. Crisp, J.-P. Maillard, and B. L. Lutz (1991). Deuterium on Venus: Observations from Earth. Science 251: 547–549.

De Bergh, C., B. Bézard, D. Crisp, J.-P. Maillard, T. Owen, J. Pollack, and D. Grinspoon (1995). Water in the deep atmosphere of Venus from high resolution spectra of the night side. Adv. Space Res. 15: 479–488.

Deirmendjian, D. (1964). A water cloud interpretation of Venus' microwave continuum. Icarus 3 (no. 2): 109–120.

———. (1968). Introductory comments of Venus' lower atmosphere. In *The Atmospheres of Venus and Mars*, ed. J. C. Brandt and M. B. McElroy. New York: Gordon and Breach.

———. (1969). *Electromagnetic Scattering on Spherical Polydispersions*. New York: Elsevier.

Del Genio, A. D., and W. B. Rossow (1982). Temporal variability of ultraviolet cloud features in the Venus stratosphere. Icarus 51: 391–415.

Del Genio, A. D., J. T. Lunetta, and J. H. Lee (1986). Evidence for long-term cyclic behavior in Venus equatorial top dynamics. Bull. Amer. Astron. Soc. 18: 793.

Deming, D., F. Espenak, and D. Jonnings (1983). Observations of the 10 μm natural laser emission from the mesospheres of Mars and Venus. Icarus 55 (no. 3): 347–355.

Dickinson, R. E. (1972). Infrared radiative heating and cooling in the Venusian mesosphere, 1: Global mean radiative equilibrium. J. Atmos. Sci. 29: 1531–1556.

Dickinson, R. E. (1976). Venus mesosphere and thermosphere temperature structure, 1: Global mean radiative and conductive equilibrium. Icarus 27: 479–493.

Dickinson, R. E., and S. W. Bougher (1986). Venus mesophere and thermosphere, 1: Heat budget and thermal structure. J. Geophys. Res. 91: 70–80.

Dickinson, R. E., and E. C. Ridley (1972). Numerical solution for the composition of the thermosphere in the presence of a steady subsolar-to-antisolar circulation with application to Venus. J. Atmos. Sci. 29: 1557–1570.

———. (1975). A numerical model for the dynamics and composition of the Venusian thermosphere. J. Atmos. Sci. 32: 1219–1231.

———. (1977). Venus mesosphere and thermosphere temperature structure, 2: Day night variations. Icarus 30: 163–178.

Dickinson, R. E., E. C. Ridley, and R. G. Roble (1981). A three-dimensional general circulation model of the thermosphere. J. Geophys. Res. 86: 1499–1512.

———. (1984). Thermospheric general circulation with coupled dynamics and composition. J. Atmos. Sci. 43: 205–219.

Diner, D. J., and J. A. Westphal (1978). Phase coverage of Venus during the 1975 apparition. Icarus 36: 119–126.

Diner, D. J., J. A. Westphal, and F. P. Schloerb (1976). Infrared imaging of Venus: 8–14 micrometers. Icarus 27 (no. 2): 191–195.

Dobrovolskis, A. R., and A. P. Ingersoll. (1980). Atmospheric tides and the rotation of Venus, 1: Tidal theory and balance of torques. Icarus 41: 1–17.

Dolginov, Sh. Sh., Ye. G. Yeroshenko, and L. N. Zhuzgov (1968). The study of the magnetic field from the interplanetary station Venera-4. Kosmich. Issled. 6: 561–575.

Dolginov, Sh. Sh., Ye. G. Yeroshenko, and L. Davis (1969). On the nature of the magnetic field near Venus. Kosmich. Issled. 7: 747–752.

Dolginov, Sh. Sh., L. N. Zhuzgov, V. A. Sharova, and V. B. Buzin (1978). Magnetic field and magnetosphere of the planet Venus. Kosmich. Issled. 16: 657–687.

Dollfus, A. (1953). Observations visuelles et photographiques des planètes Mercure et Vénus. L'astronomie 67 (no. 2): 61–75.

———. (1955). Etude visuelle et photographique de l'atmosphère de Vénus. L'astronomie 69 (no. 11): 413–425.

———. (1956). Polarisation de la lumière renvoyée par les corps solides et les nuages naturales. Ann. Astron. 19 (no. 2): 83–113.

———. (1957). Etude des planètes par la polarization de leur lumière. Suppl. Ann. d'Astrophys. 4: 117.

———. (1964). L'eau sur Vénus et Mars. Astronomie 78: 41–56.

———. (1968). Détermination optique du diamètre de la planète Vénus. C. R. Acad. Sci. Paris 267, ser. B: 1304.

———. (1972). New optical measurements of planetary diameters, part 2: Planet Venus. Icarus 17 (no. 1): 104–115.

———. (1975). Venus: Evolution of the upper atmospheric clouds. J. Atmos. Sci. 32: 1060–1070.

Donahue, T. M., and R. E. Hartle (1992). Solar cycle variations in H+ and D+ densities in the Venus ionosphere: Implications for escape. Geophys. Res. Lett. 19: 2449–2452.

Donahue, T. M., and R. R. Hodges (1992). The past and present water budget on Venus. J. Geophys. Res. 97: 6083–6091.

Donahue, T. M., and J. B. Pollack (1983). Origin and evolution of the atmosphere of Venus. In *Venus*, ed. D. M. Hunten, L. Colin, T. M. Donahue, and V. I. Moroz. Tucson: University of Arizona Press.

Donahue, T. M., J. H. Hoffman, and R. R. Hodges (1981). Krypton and Xenon in the atmosphere of Venus. Geophys. Res. Lett. 8: 513–516.

Donahue, T. M., J. H. Hoffman, R. R. Hodges, and A. J. Watson (1982). Venus was wet: A measurement of the ratio of deuterium to hydrogen. Science 216: 630.

Donahue, T. M., D. H. Grinspoon, R. E. Hartle, and R. R. Hodges (1998). Ion neutral escape of hydrogen and deuterium: Evolution of water. In *Venus II*, ed. S. Bougher, D. Hunten, and R. J. Phillips. Tucson: University of Arizona Press.

Dorofeeva, V. A., N. E. Andreeva, V. P. Volkov, and I. L. Khodakovsky (1981). Elementary sulfur in the troposphere of Venus. Geochimiya 11: 1638–1651.

Drake, F. D. (1962). 10cm observations of Venus near superior conjunction. Nature 195: 894.

Drossart, P., B. Bézard, Th. Encrenaz, E. Lellouch, M. Roos, F. W. Taylor, A. D. Collard, J. Pollack,

D. H. Grinspoon, R. W. Carlson, K. Baines, and L. W. Kamp (1993). Search for spatial variations of the H_2O abundance in the lower atmosphere of Venus from NIMS-Galileo. Planet.
Space Sci. 41: 495–504.

Dunham, T. (1948). Spectroscopic observations of the planets at Mount Wilson Observatory. In
The Atmospheres of the Earth and Planets, ed. G. P. Kuiper. Chicago: University of Chicago
Press.

Dunne, J. A. (1974). Mariner 10 Venus encounter. Science 183: 1289–1291.

Dziewonski, A. M., and D. L. Anderson (1981). Preliminary reference Earth model. Phys. Earth
Planet. Inter. 25: 297–356.

Dziewonski, A. M., A. L. Hales, and E. R. Lapwood (1975). Parametrically simple Earth models
consistent with geophysical data. Phys. Earth Planet. Inter. 10 (no. 1): 12–48.

Eddy, J. (1977). Integral flux of solar energy. In *The Solar Output and Its Variation*, ed. O. R.
White. Boulder: Colorado Associated University Press.

Ekonomov, A. P., Yu. M. Golovin, and B. E. Moshkin (1980). Visible radiation near the surface of
Venus: Results and their interpretation. Icarus 41: 65–75.

Ekonomov, A. P., B. E. Moshkin, V. I. Moroz, Yu. M. Golovin, V. I. Gnedych, and A. V. Grigoriev
(1983). Experiment on UV photometry on the descent modules Venera 13 and Venera 14.
Kosmich. Issled. 21 (no. 2): 254–268.

Eneev, T. M., and N. N. Kozlov (1981a). Model of the accumulation process for forming planetary systems, 1: Numerical experiments. Astron. Vestnik (Solar System Research) 15 (no. 2):
80–94.

———. (1981b). Model of the accumulation process for forming planetary systems, 2: Rotation of
planets and relationship of the model with the theory of gravitational instability. Astron. Vestnik (Solar System Research) 15 (no. 3): 131–141.

Ertel, D., V. I. Moroz, V. M. Linkin, R. S. Kremnev, I. Nopirakovsky, X. Bekker-Ros, V. Bervald, B.
Deler, A. V. Dyachkov, X. Drisher, L. V. Zasova, I. A. Zelenov, V. V. Kerzhanovich, A. N.
Lipatov, I. A. Matsigorin, V. Skrbek, M. Ulikh, E. A. Ustinov, G. Sellberg, C. E. Renin,
K. Shefer, D. Shpenkukh, V. Shtadthaus, A. A. Shurukov, X. Shtudemund, R. Shuster, and
X. Yan (1984). Venera 15–16: Initial results of infrared spectrometry experiments. Pisma As-
tron. Zh. 10 (no. 4): 243–252.

Esposito, L. W. (1980). Ultraviolet contrasts and the absorbers near the Venus cloud tops. J. Geophys. Res. 85: 8151–8157.

———. (1981). Absorber seen near the Venus cloud tops from Pioneer-Venus. Adv. Space Res. 1:
163–166.

———. (1984). Sulfur dioxide: Episodic injection shows evidence for active Venus volcanism. Science 223: 1072–1074.

Esposito, L. W., and L. D. Travis (1982). Polarization studies of the Venus UV contrasts: Cloud
height and haze variability. Icarus 51: 374–390.

Esposito, L. W., J. R. Winick, and A. T. F. Stewart (1979). Sulfur dioxide in the Venus atmosphere:
Distribution and implications. Geophys. Res. Lett. 6: 601.

Esposito, L. W., R. G. Knollenberg, M. Ya. Marov, O. B. Toon, and R. P. Turco (1983). The clouds
and hazes of Venus. In *Venus*, ed. D. M. Hunten, L. Colin, T. M. Donahue, and V. I. Moroz.
Tucson: University of Arizona Press.

Esposito, L. W., M. Copley, R. Eckert, L. Gates, A. I. Stewart, and H. Worden (1988). Sulfur dioxide at the Venus cloud top, 1978–1986. J. Geophys. Res. 93 (no. D5): 5267–5276.

Evans, J. V., and G. N. Taylor (1959). Radioecho observations of Venus. Nature 184 (no. 4696):
1358–1359.

Ezersky, V. I. (1957). Photographic photometry of Venus. Charkov Astron. Obs. Pub. 12: 73–165.

Fegley, B., Jr., and A. H. Treiman (1992). Chemistry of atmosphere-surface interactions on Venus and Mars. In *Venus and Mars: Atmospheres, Ionospheres, and Solar Wind Interactions*, ed. J. G. Luhmann, M. Tatrallyay, and R. O. Pepin. AGU Monograph 66.

Fegley, B., Jr., A. H. Treiman, and V. L. Sharpton (1993). Venus surface mineralogy: Observational and theoretical constraints. Lunar and Planet. Sci. 22: 3–19.

Feldman, P. D., H. W. Moos, J. T. Clarke, and A. L. Lane (1979). Identifications of the UV night-glow from Venus. Nature 279: 221–222.

Ferriso, C. C., C. B. Ludwig, and A. L. Thompson (1966). Empirically determined infrared absorption coefficients of H_2O from 300 to 3000°K. J. QSRT 6 (no. 3): 241.

Fesenkov, V. G. (1951). Evolution problems of the Earth and planet. Izvestiya Akad. Nauk KazSSR, ser. Astronomiya i fizika 5: 19–30.

Fink, U., H. P. Larson, G. P. Kuiper, and R. F. Poppen (1972). Water vapor in the atmosphere of Venus. Icarus 17: 617–631.

Fjeldbo, G., and V. R. Eshleman (1969). Atmosphere of Venus as studied with the Mariner 5 dual radio-frequency occultation experiment. Radio Sci. 4 (no. 10): 879–897.

Fjeldbo, G., A. J. Kliore, and V. R. Eshleman (1971). The neutral atmosphere of Venus as studied with the Mariner V radio occultation experiments. Astron. J. 76 (no. 2): 123–140.

Florensky, C. P., V. P. Volkov, and O. V. Nikolaeva (1978). A geochemical model of the Venus troposphere. Icarus 33: 537.

Florensky, C. P., A. T. Basilevsky, A. A. Pronin, and G. A. Burba (1979). Results of geological-morphological analysis of Venusian panoramas. In *First Panoramas of the Surface of Venus*, ed. M. V. Keldysh. Moscow: Nauka, 107–127.

Florensky, C. P., A. T. Basilevsky, V. P. Kryuchkov, R. O. Kuzmin, M. K. Naraeva, O. V. Nikolaeva, A. A. Pronin, A. S. Selivanov, Yu. S. Tyuflin, and E. M. Chernaya (1983a). Geological-morphological analysis of the "Venera 13–14" panoramas. Kosmich. Issled. 21: 340–350.

Florensky, C. P., A. T. Basilevsky, V. P. Kryuchkov, R. O. Kuzmin, O. V. Nikolaeva, A. A. Pronin, E. M. Chernaya, Yu. S. Tyuflin, A. S. Selivanov, M. K. Naraeva, and L. B. Ronca (1983b). Venera 13 and 14: Sedimentary rocks on Venus. Science 221: 57–59.

Ford, P. G., and G. H. Pettengill (1992). Venus topography and kilometer-scale slopes. J. Geophys. Res. 97: 13103–13114.

Fox, J. L. (1992). The chemistry of the nightside ionosphere of Venus. Planet. Space Sci. 40: 1663.

Fox, J. L., and S. W. Bougher (1991). Structure, luminosity and dynamics of the Venus thermosphere. Space Sci. Rev. 55: 357–489.

Fox, J. L., and A. Dalgarno (1981). Ionization, luminosity, and heating of the upper atmosphere of Venus. J. Geophys. Res. A 86 (no. 9): 629–639.

Galtsev, A. P., and M. A. Odishariya (1970). CO_2 absorption bands: 7.2, 7.8, 9.4, 10.4 and 15 μm at higher than normal pressure. Izvestiya Akad. Nauk USSR, Fizika atmosphery i okeana 6 (no. 9): 881–888.

Galtsev, A. P., and V. M. Osipov (1971). Calculation of CO_2 IR transmission bands at high temperatures and pressures: A system of bands at 9.4–10.4μm. Izvestiya Akad. Nauk USSR, Fizika atmosphery i okeana 7 (no. 8): 857–870.

Garvin, J. B., J. W. Head, M. T. Zuber, and P. Helfenstein (1984). Venus: The nature of the surface from Venera panoramas. J. Geophys. Res. 89: 3381–3399.

Garvin, J. B., J. W. Head, G. H. Pettengill, and S. H. Zisk (1985). Venus global radar reflectivity and correlations with elevation. J. Geophys. Res. 90: 6859–6871.

Gehrels, T., and R. E. Samuelson (1961). Polarization-phase relations for Venus. Astrophys. J. 134 (no. 3): 1022–1024.

Gelman, B. G., V. G. Zolotukhin, N. I. Lamonov, B. V. Levchuk, A. N. Lipatov, L. M. Mukhin, D. F. Nenarokov, V. A. Rotin, and B. P. Okhotnikov (1979). An analysis of the chemical composition of the atmosphere of Venus on an AMS of the Venera 12 using a gas chromotograph. Kosmich. Issled. 17: 708.

Gelman, B. G., Yu. V. Drozdov, and V. V. Melnikov (1986). Chemical analysis of aerosol of the Venus clouds by the method of reactive gas chromatography on descent module of the Vega spacecraft. Pisma Astron. Zh. 12 (no. 2): 106–109.

Gierasch, P. J. (1970). The four-day rotation in the stratosphere of Venus: A study of radiative driving. Icarus 13: 25–33.

———. (1975). Meridional circulation and the maintenance of the Venus atmospheric rotation. J. Atmos. Sci. 32: 1038–1044.

Gierasch, P. J., R. Goody, and P. Stone (1970). The energy balance of planetary atmospheres. Geophys. Fluid Dyn. 1: 1–18.

Gierasch, P. J., A. P. Ingersoll, and R. T. Williams (1973). Radiative instability of a cloudy planetary atmosphere. Icarus 19: 473–481.

Ginzburg, V. L. (1985). *Concerning Physics and Astrophysics*, 2nd ed. Moscow: Nauka.

Gold, T., and S. Soter (1969). Atmospheric tides and the resonant rotation of Venus. Icarus 11: 356–366.

———. (1971). Atmospheric tides and the 4-day circulation on Venus. Icarus 14: 16–20.

———. (1979). Theory of the Earth-synchronous rotation of Venus. Nature 277: 280–281.

Goldreich, P., and S. J. Peale (1967). Spin-orbit coupling in the solar system, 2: The resonant rotation of Venus. Astron. J. 72: 662–668.

Goldsmith, D., and T. Owen (1992). *The Search for Life in the Universe*, 2nd ed. Reading, Mass.: Addison-Wesley.

Goldstein, R. M. (1964). Venus characteristics by Earth-based radar. Astron. J. 69 (no. 1): 12–18.

Goldstein, R. M., and R. L. Carpenter (1963). Rotation of Venus: Period estimated from radar measurements. Science 139 (no. 3558): 910.

Goldstein, R. M., and H. C. Rumsey (1972). A radar image of Venus. Icarus 17: 699–703.

Goldstein, R. M., R. R. Green, and H. C. Rumsey (1976). Venus radar images. J. Geophys. Res. 81: 4807–4817.

———. (1978). Venus radar brightness and altitude images. Icarus 36: 334–352.

Golitsyn, G. S. (1970). A similarity approach to the general circulation of planetary atmospheres. Icarus 13: 1–24.

———. (1973). *Introduction to the Dynamics of Planetary Atmospheres*. Leningrad: Gidrometeoizdat.

———. (1982). Problems of the Venus atmospheric dynamics. NASA Tech. Mem. 76816.

Golovin, Yu. M., and E. A. Ustinov (1982). Subcloud aerosol of the atmosphere of Venus. Kosmich. Issled. 20 (no. 1): 104–110.

Good, J. C., and F. P. Schloerb (1983). Limits on Venus: SO_2 abundance profile from interferometric observations at 3.4 mm wavelength. Icarus 53: 538.

Goody, R. M. (1964). *Atmospheric Radiation*. London: Oxford University Press.

———. (1969). Motion of planetary atmospheres. Ann. Rev. Astron. Astrophys. 7: 303–352.

Goody, R. M., and A. R. Robinson (1966). A discussion of the deep circulation of the atmosphere of Venus. Astrophys. J. 146: 339–355.

———. (1970). The deep circulation. Report presented at the 4th Arizona Conference on Planetary Atmospheres, Tucson, 2–4 March.

Gordiets, B. F., and Yu. N. Kulikov (1985). On the role of turbulence in energetics of the night thermosphere of Venus. Astron Vestnik (Solar System Research) 19 (no. 4): 289–295.

Grebowsky, J. M., R. E. Hartle, J. Kar, P. A. Cloutier, H. A. Taylor, Jr., and L. H. Brace (1993). Ion measurements during Pioneer Venus reentry: Implications for solar cycle variation of ion composition and dynamics. Geophys. Res. Lett. 20 (no. 23): 2735–2738.

Greeley, R., R. E. Arvidson, C. Elachi, M. A. Geringer, J. J. Plaut, R. S. Saunders, G. Schubert, E. R. Stofan, E. J. P. Thouvenot, S. D. Wall, and C. M. Weitz (1992). Aeolian features on Venus: Preliminary Magellan results. J. Geophys. Res. 97: 13319–13346.

Grimm, R. E., and R. J. Phillips (1991). Gravity anomalies, compensation mechanisms, and the geodynamics of western Ishtar Terra, Venus. J. Geophys. Res. 96: 8305–8324.

———. (1992). Anatomy of a Venusian hot spot: Geology, gravity, and mantle dynamics of Eistla Region. J. Geophys. Res. 97: 16035–16054.

Gringauz, K. I., M. I. Verigin, T. K. Breus, and T. Gombosi (1979). The interaction of electrons in the optical umbra of Venus with the planetary atmosphere: The origin of the nightside ionosphere. J. Geophys. Res. 84: 2123–2127.

Grinspoon, D. H. (1987). Was Venus wet? Deuterium reconsidered. Science 238: 1702–1704.

———. (1993). Implications of the high deuterium-to-hydrogen ratio for the sources of water in Venus' atmosphere. Nature 363: 428–431.

———. (1997). *Venus Revealed*. Reading, Mass.: Addison-Wesley.

Grinspoon, D. H., and J. S. Lewis (1988). Cometary water on Venus: Implications of stochastic impacts. Icarus 74: 21–35.

Grinspoon, D. H., J. B. Pollack, B. R. Sitton, R. W. Carlson, L. W. Kamp, K. H. Baines, T. Encrenaz, and F. W. Taylor (1993). Probing Venus' cloud structure with Galileo NIMS. Planet. Space Sci. 41: 515–542.

Guest, J. E., M. H. Bulmer, J. Aubele, K. Beratan, R. Greeley, J. W. Head, G. Michaels, C. Weitz, and C. Wiles (1992). Small volcanic edifices and volcanism in the plains of Venus. J. Geophys. Res. 97: 15949–15966.

Gurwell, M. A., and Y. L. Yung (1993). Fractionation of hydrogen and deuterium on Venus due to collisional ejection. Planet. Space Sci. 41: 410.

Gutenberg, B. (1963). *Physics of the Earth's Interior*. Moscow: Inostrannaya Literatura.

Halley, E. (1686). An historical account for the trade winds and monsoons. . . . Phil. Trans. Roy. Soc. 16: 153–158.

Haltiner, G. J., and F. L. Martin (1957). *Dynamical and Physical Meteorology*. New York: McGraw-Hill.

Hansen, J. E. (1972). Information contained in the intensity and polarization of scattered sunlight. Inst. Space Studies Preprint, Sept., New York.

Hansen, J. E., and A. Arking (1971). Clouds of Venus: Evidence for their nature. Science 171: 669–672.

Hansen, J. E., and J. W. Hovenier (1974). Interpretation of the polarization of Venus. J. Atmos. Sci. 31: 1137–1160.

Hansen, J. E., and S. Matsushima (1967). The atmosphere and surface temperature of Venus: A dust insulation model. Astrophys. J. 150: 1139.

Hansen, J. E., and L. D. Travis (1974). Light scattering in planetary atmosphere. Space Sci. Rev. 16: 527–610.

Hart, R., J. Dymond, and L. Hogan (1979). Preferential formation of the atmosphere-sialic crust system from the upper mantle. Nature 278: 156–159.

Hartle, R. E., and H. A. Taylor (1983). Identification of deuterium ions in the ionosphere of Venus. Geophys. Res. Lett. 10 (no. 10): 965–968.

Hartman, W. K., and S. M. Vail (1986). Giant impactors: Plausible sizes and populations. In *Origin of the Moon*, ed. W. K. Hartman, R. J. Phillips, and G. J. Taylor. Houston: Lunar and Planetary Institute.

Head, J. W., and L. S. Crumpler (1987). Evidence for divergent planet-boundary characteristics and crustal spreading on Venus. Science 238: 1380–1385.

———. (1990). Venus geology and tectonics: Hotspot and crustal spreading models and questions for the Magellan mission. Nature 346: 525–533.

Head, J. W., A. R. Petterfreund, J. B. Garvin, and S. H. Zisk (1985). Surface characteristics of Venus derived from Pioneer-Venus altimetry, roughness and reflectivity measurements. J. Geophys. Res. 90: 6873–6885.

Head, J. W., L. S. Crumpler, J. C. Aubele, J. E. Guest, and R. S. Saunders (1992). Venus volcanism: Classification of volcanic features and structures, associations, and global distribution from Magellan data. J. Geophys. Res. 97: 13153–13198.

Hedin, A. E., H. B. Nieman, W. T. Kasprzak, and A. Seiff (1983). Global empirical model of the Venus thermosphere. J. Geophys Res. 88: 73.

Herrick, R. R., and R. J. Phillips (1992). Geological correlations with the interior density structure of Venus. J. Geophys. Res. 97: 16017–16034.

———. (1994). Effects of the Venusian atmosphere on incoming meteroids and the impact crater population. Icarus 112: 253–281.

Herzberg, G. (1949). *Spectra and Structure of Diatomic Molecules*. Moscow: Inostrannaya Literatura.

———. (1952). Laboratory absorption spectra obtained with long path. In *The Atmospheres of Earth and Planets*, 2nd ed., ed. G. P. Kuiper. Chicago: University of Chicago Press.

Heyden, F. J., C. C. Kiess, and H. K. Kiess (1959). Spectrum of Venus in the violet and near ultraviolet. Science 130 (no. 3383): 1195.

Hide, R., N. T. Birch, L. V. Morrison, D. J. Shea, and A. A. White (1980). Atmospheric angular momentum fluctuations and changes in the length of the day. Nature 286: 114–117.

Hizen, B., and M. Tarp (1974). *The Flow of the Ocean*. New York: Lamont-Doherty Geological Observatory, Columbia University.

Ho, C. M., R. S. Strangeway, and C. T. Russell (1993). Evidence for four Langmuir oscillations and a low density cavity in the Venus magnetotail. Geophys. Res. Lett. 20 (no. 23): 2775–2778.

Ho, W., I. A. Kaufman, and P. Thaddeus (1996). Laboratory measurement of microwave absorption in model of atmosphere of Venus. J. Geophys. Res. 71: 5091.

Hodges, R. R., Jr. (1993). Isotopic fractionation of hydrogen in planetary exosphere due to ionosphere-exosphere coupling: Implications for Venus. J. Geophys. Res. 98 (no. E6): 10833–10838.

Hoffman, J. H., R. R. Hodges, M. B. McElroy, T. M. Donahue, and M. Kolpin (1979). Composition and structure of the Venus atmosphere: Results from Pioneer-Venus. Science 205: 49–52.

Hoffman, J. H., R. R. Hodges, Jr., T. M. Donahue, and M. B. McElroy (1980a). Composition of the Venus lower atmosphere from the Pioneer-Venus mass spectrometer. J. Geophys. Res. 85 (no. A13): 7882.

Hoffman, J. H., V. I. Oyama, and U. von Zahr (1980b). Measurements of the Venus lower atmosphere composition: A comparison of results. J. Geophys. Res. 85: 7871–7881.

Holland, H. D. (1978). *The Chemistry of the Atmosphere and Oceans*. New York: Wiley.

Holton, J. R. (1979). *An Introduction to Dynamic Meteorology*, 2nd ed. New York: Academic Press.

Hou, A. Y., and R. Goody (1985). Diagnostic requirements for the superrotation on Venus. J. Atmos. Sci. 42: 413–432.

Howard, H. T., G. L. Tyler, G. Fjeldbo, A. J. Kliore, G. S. Levy, D. L. Brunn, R. Dickinson, R. E. Edelson, W. L. Martin, R. B. Postal, B. Seidel, T. T. Sesplaukis, D. L. Shirley, S. T. Stelzried,

D. N. Sweetnam, A. E. Zygielbaum, P. B. Esposito, J. D. Anderson, I. I. Shapiro, and R. D. Reasenberg (1974). Venus mass, gravity field, atmosphere and ionosphere as measured by Mariner 10 dual-frequency radio system. Science 183: 1297–1301.

Howard, I. N., D. E. Burche, and D. Williams (1956). Infrared transmission on synthetic atmospheres, 1–4. J. Opt. Soc. Am. 46: 176–190.

Hsu, N. C., G. M. Keating, and W. T. Willcockson (1992). First Magellan measurements of the Venus thermosphere. EOS Trans. AGU 73 (no. 43): 332.

Huggins, W. (1863). The Tulse Hill spectroscope. Astr. and Astrophys. 12.

Hunten, D. M. (1971). Composition and structure of planetary atmospheres. Space Sci. Rev. 12: 539–599.

———. (1982). Thermal and nonthermal escape mechanisms for terrestrial bodies. Planet. Space Sci. 30: 773–783.

———. (1985). Blowoff of an atmosphere and possible application to Io. Geophys. Res. Lett. 12: 271–273.

———. (1993). Atmospheric evolution of the terrestrial planets. Science 259: 915–920.

Hunten, D. M., L. Colin, T. M. Donahue, and V. I. Moroz, eds. (1983). *Venus.* Tucson: University of Arizona Press.

Hunten, D. M., R. O. Pepin, and J. C. G. Walker (1987). Mass fractionation in hydrodynamic escape. Icarus 69: 532–549.

Hunten, D. M., R. O. Pepin, and T. C. Owen (1988). Planetary atmosphere. In *Meteorites and the Early Solar System*, ed. J. F. Kerridge and M. S. Matthews. Tucson: University of Arizona Press.

Hunten, D. M., T. M. Donahue, J. C. G. Walker, and J. F. Kasting (1989). Escape of atmospheres and loss of water. In *Origin and Evolution of Planetary and Satellite Atmospheres*, ed. J. B. Pollack, S. K. Atreya, and M. S. Matthews. Tucson: University of Arizona Press.

Intriligator, D. S., R. E. Hartle, H. Perez-de-Tejada, and G. L. Siscoe (1993). Initial PVO evidence of electron depletion signatures downstream of Venus. Geophys. Res. Lett. 20 (no. 23): 2779–2782.

Irvine, W. M. (1968). Monochromatic phase curves and albedos for Venus. J. Atmos. Sci. 25 (no. 4): 610–616.

Istomin, V. G., K. V. Grechnev, V. A. Kochnev, and L. N. Ozerov (1979). Composition of low atmosphere of Venus according to mass-spectrometer data. Kosmich. Issled. 17: 703–707.

Istomin, V. G., K. V. Grechnev, and V. A. Kochnev (1983). Venera 13 and Venera 14: Mass spectrometry of the atmosphere. Kosmich. Issled. 21: 410.

Ivanov, B. A. (1992). Venera 13 and 14 landing sites: Geology from Magellan data. Proc. Lunar Planet. Sci. 23: 579–580.

Ivanov, B. A., and A. T. Basilevsky (1987). Comparison of crater retention age on Earth and Venus. Astron. Vestnik (Solar System Research) 21: 136–143.

Ivanov, B. A., A. T. Basilevsky, V. P. Kryuchkov, and I. M. Chernaya (1986). Impact craters of Venus: Analysis of Venera 15 and 16 data. J. Geophys. Res. 91: D 414–430.

Ivanov, B. A., I. V. Nemchinov, V. A. Svetsov, A. A. Provalov, V. M. Khazins, and R. J. Phillips (1992). Impact cratering on Venus: Physical and mechanical models. J. Geophys. Res. 97: 16167–16182.

Ivanov-Kholodny, G. S., M. A. Kolosov, N. A. Savich, Yu. N. Aleksandrov, M. B. Vasilev, A. S. Vyshlov, V. M. Dubrovin, A. L. Zaytsev, A. V. Michailov, G. M. Petrov, V. A. Samovol, L. N. Samoznayev, I. A. Sidorenko, and F. A. Hasyznov (1979). Daytime ionosphere of Venus as studied with Venera 9 and 10 dual-frequency radio occultation experiments. Icarus 39: 209–213.

Izakov, M. N. (1986). Formation of solid matter of the pre-planetary nebula and the composition of chondrites. Astron. Vestnik (Solar System Research) 20 (no. 1): 35–49.

Izakov, M. N., and M. Ya. Marov (1989). Venus' upper atmosphere. In *The Planet Venus*, ed. V. L. Barsukov and V. P. Volkov. Moscow: Nauka.

Izakov, M. N., O. P. Krasitsky, and A. V. Pavlov (1981). Model of ion composition of the dayside Venus ionosphere. Kosmich. Issled. 29 (no. 5): 733–748.

Janes, D. M., S. W. Squyres, D. L. Bindschadler, G. Baer, G. Schubert, V. L. Sharpton, and E. R. Stofan (1992). Geophysical models for the formation and evolution of coronae on Venus. J. Geophys. Res. 97: 16055–16069.

Janssen, M. A., and M. J. Klein (1981). Constraints on the composition of the Venus atmosphere from microwave measurements near 1.35 cm wavelength. Icarus 46: 58–69.

Jastrow, R., and S. I. Rasool (1963). Radiative transfer in the atmosphere of Venus. In *Space Research 3*, ed. W. Priester Amsterdam: North-Holland.

Jeffreys, H. (1937). The density distribution of the inner planets. Mon. Not. R. Astr. Soc. Geophys. Suppl. 4: 62.

John, C. E., and S. B. Nicholson (1922). The physical constituents of the atmosphere of Venus. Phys. Rev. 19: 444.

Johnson, C. L., and D. T. Sandwell (1992). Joints in Venusian lava flows. J. Geophys. Res. 97: 13601–13610.

Johnson, T. V., C. M. Yeates, R. Young, and J. Dunne (1991). The Galileo Venus encounter. Science 253: 1516–1518.

Jurgens, R. F. (1970). Some preliminary results of the 70 cm radar studies of Venus. Radio Sci. 5 (no. 2): 435–442.

Jurgens, R. F., R. M. Goldstein, H. R. Rumsey, and R. R. Green (1980). Images of Venus by three-station radar interferometry: 1977 results. J. Geophys. Res. 85: 8282–8294.

Kalinin, V. A., and N. A. Sergeeva (1979). The internal structure of Mars and Venus. Izv. Akad. Nauk USSR, ser. Fizika Zemli 10: 3.

Kalnay de Rivas, E. (1973). Numerical models of the circulation of the atmosphere of Venus. J. Atmos. Sci. 31: 763–779.

———. (1975). Further numerical calculations of the circulation of the atmosphere of Venus. J. Atmos. Sci. 32: 1017–1024.

Kar, J. (1996). Recent advances in planetary ionospheres. Space Sci. Rev. 77: 193–266.

Kargel, J. S., R. Kirk, B. Fegley, Jr., and A. H. Treiman (1994). Carbonate-sulfate volcanism on Venus? Icarus 112: 219–252.

Kasprzak, W. T., H. B. Niemann, A. E. Hedin, S. W. Bougher, and D. M. Hunten (1993a). Neutral composition measurements by the Pioneer Venus neutral mass spectrometer during orbiter re-entry. Geophys. Res. Lett. 20 (no. 23): 2747–2750.

Kasprzak, W. T., H. B. Niemann, A. E. Hedin, and S. W. Bougher (1993b). Wave-like perturbations observed at low altitudes by the Pioneer Venus orbiter neutral mass spectrometer during orbiter entry. Geophys. Res. Lett. 20 (no. 23): 2755–2758.

Kasting, J. F. (1988). Runaway and moist greenhouse atmospheres and the evolution of Earth and Venus. Icarus 74: 472–494.

Kasting, J. F., and J. B. Pollack (1983). Loss of water from Venus, 1: Hydrodynamic escape of hydrogen. Icarus 53: 479–508.

Kasting, J. F., J. B. Pollack, and T. P. Ackerman (1984). Response of Earth's atmosphere to increases in solar flux and implications for loss of water from Venus. Icarus 57: 335–355.

Kasting, J. F., O. B. Toon, and J. B. Pollack (1988). How climate evolved on the terrestrial planets. Scientific American 256: 90–97.

Kasting, J. F., D. P. Whitmire, and R. T. Reynolds (1993). Habitable zones around main sequence stars. Icarus 101: 108–128.

Kaula, W. M. (1968). *An Introduction to Planetary Physics: The Terrestrial Planets*. New York: John Wiley.

———. (1979). Thermal evolution of Earth and Moon growing by planetesimal impacts. J. Geophys. Res. 84: 999–1008.

Kaula, W. M., D. L. Bindschadler, R. E. Grimm, V. L. Hansen, K. M. Roberts, and S. E. Smrekar (1992). Styles of deformation in Ishtar Terra and their implications. J. Geophys. Res. 97: 16085–16120.

Kawabata, K. D., D. L. Coffeen, J. E. Hansen, W. A. Lane, M. Sato, and L. D. Travis (1980). Cloud and haze properties from Pioneer-Venus polarimetry. J. Geophys. Res. 85: 8129–8140.

Keating, G. M., and N. C. Hsu (1993). The Venus atmospheric response to solar cycle variations. Geophys. Res. Lett. 20 (no. 23): 2751–2754.

Keating, G. M., R. H. Tolson, and E. W. Hinson (1979). Venus thermosphere and exosphere: First satellite drag measurements of an extraterrestrial atmosphere. Science 203: 772–774.

Keating, G. M., J. Y. Nicholson, and L. R. Lake (1980). Venus upper atmosphere structure. J. Geophys. Res. 85: 7941–7956.

Keating, G. M., J. L. Bertaux, S. W. Bougher, T. E. Cravens, R. E. Dickinson, A. E. Hedin, V. A. Krasnopolsky, A. F. Nagy, J. Y. Nicholson, III, L. J. Paxton, and U. von Zahn (1985). Models of Venus' neutral upper atmosphere: Structure and composition. In *The Venus International Reference Atmosphere*, ed. A. J. Kliore, V. I. Moroz, and G. M. Keating. Adv. Space Res. 5 (no. 11): 117–172.

Keldysh, M. V. (1977). Venus exploration with the Venera 9 and Venera 10 spacecraft. Icarus 30: 605–625.

Keldysh, M. V., ed. (1979). *First Panoramas of the Surface of Venus*. Moscow: Nauka.

Keldysh, M. V., and M. Ya. Marov (1981). *Space Research*. Moscow: Nauka.

Kemurdzhian, A. L., V. V. Gromov, and V. V. Shvarev (1978). Investigation of the physico-chemical properties of extraterrestrial grounds. In *The Achievements of Soviet Science in the Exploration of Outer Space*. Moscow: Nauka.

Kemurdzhian, A. L., P. N. Brodsky, V. V. Gromov, V. P. Grushin, I. E. Kiselev, G. V. Kozlov, A. V. Mitskevich, V. G. Perminov, P. S. Sologub, A. D. Stepanov, A. V. Turobinsky, V. N. Turchaninov, and E. N. Yudkin (1983). Preliminary results of determination of physico-mechanical properties of Venus soil from the Soviet automatic stations Venera 13 and Venera 14. Kosmich. Issled. 21 (no. 3): 323–330.

Kerzhanovich, V. V., and S. S. Limaye (1985). Circulation of the atmosphere from the surface to 100 km. In *The Venus International Reference Atmosphere*, ed. A. J. Kliore, V. I. Moroz, and G. M. Keating. Adv. Space Res. 5 (no. 11): 59–83.

Kerzhanovich, V. V., and M. Ya. Marov (1974). Circulation and dustiness of the Venus atmosphere from wind velocity measurements of interplanetary station Venera-8. Dokl. Akad. Nauk USSR 215: 554–557.

———. (1977). On the wind velocity measurements from Venera spacecraft data. Icarus 30 (no. 2): 320–325.

———. (1983). The atmospheric dynamics of Venus according to Doppler measurements by the Venera entry probes. In *Venus*, ed. D. M. Hunten, L. Colin, T. M. Donahue, and V. I. Moroz. Tucson: University of Arizona Press.

Kerzhanovich, V. V., M. Ya. Marov, and M. K. Rozhdestvensky (1972). Data on dynamics of the subcloud Venus atmosphere from Venera space probes measurements. Icarus 17: 659–674.

Kerzhanovich, V. V., F. I. Kozlov, A. S. Selivanov, Yu. S. Tyuflin, and V. I. Khizhnichenko (1979a).

Structure of Venusian cloud layer according to Venera 9 television pictures. Kosmich. Issled. 17: 57–66.

Kerzhanovich, V. V., Yu. F. Makarov, M. Ya. Marov, M. K. Rozhdestvensky, and V. P. Sorokin (1979b). Venera 11 and Venera 12: Preliminary estimates for the wind speed and turbulence in the atmosphere of Venus. Moon Planets 23: 261–270.

Kerzhanovich, V. V., Yu. F. Makarov, M. Ya. Marov, E. P. Molotov, M. K. Rozhdestvensky, V. P. Sorokin, N. M. Antsibor, V. D. Kustodiev, and V. I. Pouchkov (1979c). An estimate of the wind velocity and turbulence in the atmosphere of Venus on the basis of reciprocal Doppler measurements by Venera 11 and Venera 12 spacecraft. Kosmich. Issled. 17: 569–575.

Kerzhanovich, V. V., N. M. Antsibor, V. D. Kustodiev, Yu. F. Makarov, I. A. Matsygorin, E. P. Molotov, V. P. Sorokin, K. G. Sukhanov, V. F. Tikhonov, V. P. Karyagin, and B. I. Motzulev (1982). Estimates of wind velocity from Doppler measurements of Venera 13 and Venera 14 space probes: First results. Pisma Astron. Zh. 8: 414–418.

Kerzhanovich, V. V., M. Ya. Marov, and V. I. Moroz (1983). Proposals for Venus International Reference Atmosphere: Dynamics and structure below 100 km. Preprint, Moscow: Space Research Institute, USSR Academy of Sciences.

Khodakovsky, I. L. (1982). Atmosphere-surface interactions on Venus and implications for atmospheric evolution. Planet. Space Sci. 30: 803.

Khodakovsky, I. L., V. P. Volkov, Yu. I. Sidorov, and M. V. Borisov (1978). Geochemical model of the Venus troposphere and crust from new data. Geokhimiya 12: 1748–1758.

Khodakovsky, I. L., V. P. Volkov, Yu. I. Sidorov, and M. V. Borisov (1979). Venus: Preliminary prediction of the mineral composition of surface rocks. Icarus 39: 352–363.

King, E. S. (1923). Revised magnitudes and color-indices of the planets. Harv. Obs. Ann. 85 (no. 4): 63–71.

———. (1929). Photovisual magnitudes of stars and planets. Harv. Obs. Ann. 81 (no. 4): 201–215.

King, J. I. F. (1964). A quantitative greenhouse model of the Venus atmosphere. AIAA Bull. 1 (no. 1): 23.

Kirk, R. L., L. A. Soderblom, and E. M. Lee (1992). Enhanced visualization for interpretation of Magellan radar data: Supplement to the Magellan special issue. J. Geophys. Res. 97: 16371–16380.

Kislyakov, A. G., A. D. Kuzmin, and A. E. Solomonovich (1962). Radio emission from Venus at the 4 mm wavelength. Astron. Zhur. 39 (no. 3): 410–417.

Kliore, A. J., and L. F. Mullen (1989). The long-term behavior of the main peak of the dayside ionosphere of Venus during solar cycle 21 and its implications on the effect of the solar cycle upon the electron temperatures in the main peak region. J. Geophys. Res. 94: 13339.

Kliore, A. J., and I. R. Patel (1980). The vertical structure of the atmosphere of Venus from Pioneer-Venus orbiter radio occultations. J. Geophys. Res. 85 (no. A13): 7957.

———. (1982). Thermal structure of the atmosphere of Venus from Pioneer-Venus radio occultations. Icarus 52 (no. 2): 320.

Kliore, A. J., G. S. Levy, D. L. Cain, G. Fjeldbo, and S. I. Rasool (1967). Atmosphere and ionosphere of Venus from the Mariner 5 S-Band radio occultation measurement. Science 158: 1683–1688.

Kliore, A. J., D. L. Cain, G. S. Levy, G. Fjeldbo, and S. I. Rasool (1969). Structure of the atmosphere of Venus derived from Mariner 5 S-Band measurements. In *Space Res. IX*. Amsterdam: North-Holland.

Kliore, A. J., C. Elachi, I. R. Patel, and J. B. Cimino (1979a). Liquid content of the lower clouds of Venus as determined from Mariner 10 radio occultation. Icarus 37: 51–72.

Kliore, A. J., I. R. Patel, A. F. Nagy, T. E. Cravens, and T. I. Gombosi (1979b). Initial observations of the nightside ionosphere of Venus from Pioneer-Venus orbiter radio occultations. Science 205: 99–102.

Kliore, A. J., V. I. Moroz, and G. M. Keating, eds. (1985). *Venus International Reference Atmosphere.* Oxford: Pergamon Press.

Klose, K. B., J. A. Wood, and A. Hashimoto (1992). Mineral equilibria and the high radar reflectivity of Venus mountaintops. J. Geophys. Res. 97: 16353–16370.

Knollenberg, R. G. (1984). A re-examination of the evidence for large, solid particles in the clouds of Venus. Icarus 57 (no. 2): 161–183.

Knollenberg, R. G., and D. M. Hunten (1980). Microphysics of the clouds of Venus: Results of the Pioneer Venus particle size spectrometer experiment. J. Geophys. Res. 85: 8039–8058.

Knollenberg, R. G., L. Travis, M. Tomasko, P. Smith, B. Kagent, L. Esposito, D. McCleese, J. Martonchik, and R. Beer (1980). The clouds of Venus: A synthesis report. J. Geophys. Res. 85: 8059–8081.

Knuckles, C. F., M. K. Sinton, and W. M. Sinton (1961). UBV photometry of Venus. Lowell Obs. Bull. 5: 153–156.

Knudsen, W. C., K. Spenner, K. L. Miller, and V. Novak (1980). Transport of O^+ ions across the Venus terminator and implications. J. Geophys. Res. 85: 7803–7810.

Knudsen, W. C., A. J. Kliore, and R. C. Whitten (1987). Solar cycle changes in the ionization source of the nightside Venus ionosphere. J. Geophys. Res. 92: 13391–13398.

Kondratyev, K. Ya. (1969). *Radiation in the Atmosphere.* New York: Academic Press.

Kondratyev, K. Ya., and G. E. Hunt (1982). *Weather and Climate on Planets.* New York: Pergamon Press.

Kondratyev, K. Ya., O. A. Avaste, M. P. Fedorova, and K. E. Yakushevskaya (1967). *Radiation Field of the Earth as a Planet.* Leningrad: Gidrometeoizdat.

Konopliv, A. S., and W. L. Sjogren (1994). Venus spherical harmonic gravity model to degree and order 60. Icarus 112: 42–54.

Korolkov, D. V., Yu. N. Pariysky, Yu. N. Timofeeva, and S. E. Khaikin (1963). Higher resolution radio astronomy observations of Venus. Dokl. AN USSR 149: 65–67.

Kotelnikov, V. A., V. M. Dubrovin, M. D. Kislik, E. B. Korenberg, V. P. Minashin, V. A. Morozov, N. I. Nikitsky, G. M. Petrov, O. N. Rzhiga, and A. M. Shakhovskoy (1962). Radar observations of Venus. Dokl. AN USSR 145 (no. 5): 1035–1038.

Kotelnikov, V. A., M. D. Dubinskii, M. D. Kislik, and D. M. Tsvetkov (1963). Refinement of the astronomical unit from radar results of the planet Venus in 1961. Iskusstv. Sputn. Zemli 17: 101–106.

Kotelnikov, V. A., Yu. N. Aleksandrov, L. V. Apraksin, V. M. Dubrovin, M. D. Kislik, V. I. Kuznetsov, G. M. Petrov, O. N. Rzhiga, A. V. Frantsesson, and A. M. Shakhovskoy (1965). Radar observations of Venus in the Soviet Union in 1964. Dokl. AN USSR 163 (no. 1): 50–53.

Koval, I. K. (1958). Absolute photometry of Venus in the ultraviolet and infrared. Astron. Zhur. 35 (no. 5): 792–796.

Kozlovskaya, S. V. (1982). Internal structure of Venus and the iron content of the terrestrial planets. Astron. Vestnik (Solar System Research) 16 (no. 1): 3.

Kozyrev, N. A. (1954a). On the light of the night sky of Venus. Izv. Krymskoy Astrophys. Obs. 12: 169–176.

———. (1954b). Molecular absorption of light in violet spectrum of Venus. Izv. Krymskoy Astrophys. Obs. 12: 177–184.

Krasnopolsky, V. A. (1980). Venera 9 and 10: Spectroscopy of scattered radiation in overcloud atmosphere. Kosmich. Issled. 18: 899.

———. (1983). Venus spectroscopy in the 3000–8000 Å region by Veneras 9 and 10. In *Venus,* ed.

D. M. Hunten, L. Colin, T. M. Donahue, and V. I. Moroz. Tucson: University of Arizona Press.

———. (1986). *Photochemistry of the Atmospheres of Mars and Venus.* New York: Springer Verlag.

Krasnopolsky, V. A., and V. A. Parshev (1979). On the chemical composition of the Venus troposphere and the cloud layer based on Venera 11, 12, and Pioneer-Venus measurements. Kosmich. Issled. 17: 763.

———. (1981a). Photochemistry of Venus atmosphere at heights above 50 km: Initial data for calculation. Kosmich. Issled. 19 (no. 1): 87–103.

———. (1981b). Photochemistry of Venus atmosphere at heights above 50 km: Results of calculation. Kosmich. Issled. 19 (no. 2): 261–278.

———. (1983). Photochemistry of the Venus atmosphere. In *Venus*, ed. D. M. Hunten, L. Colin, T. M. Donahue, and V. I. Moroz. Tucson: University of Arizona Press.

Kraus, J. C. (1956). Rotation period of the planet Venus as determined by radio observations. Nature 178 (no. 4535): 687–688.

———. (1960). Apparent radio radiation at 11-m wavelength from Venus. Nature 186 (no. 4723): 462.

Kreslavsky, M. A., A. T. Basilevsky, and Yu. G. Shkuratov (1988). Prognosis of the distribution of the tessera terrain on Venus using Pioneer-Venus and Venera 15/16 data. Astron. Vestn. (Solar System Research) 22 (no. 4): 277–286.

Kropotkin, P. N. (1989). Genesis of ring structures on the Moon, Earth and other planets. Izv. Akad. Nauk USSR, series Geologiya (no. 7): 3–14.

Krotikov, V. D. (1962). Certain electrical characteristics of terrestrial rock and their comparison with characteristics of the lunar surface layer. Izvestiya vuzov, Radiofizika 5 (no. 6): 1057–1061.

Ksanfomality, L. V. (1979). Lightning in Venus' cloud layer. Kosmich. Issled. 12: 747–762.

———. (1980). Venera 9 and 10 thermal radiometry. Icarus 41: 36–64.

———. (1983a). Search for microseisms on Venus. Kosmich. Issled. 21: 355–360.

———. (1983b). Electrical activity of the Venus atmosphere, 1: Measurements from descending probes. Kosmich. Issled. 21 (no. 2): 279–296.

———. (1985). *The Planet Venus.* Moscow: Nauka.

Ksanfomality, L. V., N. M. Vasilchikov, N. V. Goroshkova, E. V. Petrova, A. P. Suvorov, and V. K. Khondyrev (1982). Low-frequency electromagnetic field in the atmosphere of Venus from Venera 13 and Venera 14 measurements. Pisma Astron. Zh. 8: 424–428.

Kucinskas, A. B., D. L. Turcotte, J. Huang, and P. G. Ford (1992). Fractal analysis of Venus topography in Tinatin Planitia and Ovda Regio. J. Geophys. Res. 97: 13635–13642.

Kuiper, G. P. (1947). Infrared spectra of planets. Astrophys. J. 106: 252.

———. (1954). Determination of the pole of rotation of Venus. Astrophys. J. 120: 603–605.

———. (1962). Infrared spectra of stars and planets, 1: Photometry of the infrared spectrum of Venus, 1–2.5 microns. Comm. Lunar and Planet. Lab. 1 (nos. 14–16): 83–117.

———. (1969). On the nature of the Venus clouds. In *Planetary Atmospheres*, IAU Symposium no. 40, ed. C. Sagan, T. C. Owen, and H. J. Smith. Dordrecht: Reidel.

———. (1971). On the nature of the Venus clouds. In *Planetary Atmospheres*, IAU Symposium no. 40, ed. C. Sagan, T. Owen, and H. J. Smith. Dordrecht: Reidel.

Kuiper, G. P., and F. F. Forbes (1967). High altitude spectra from NASA CV 990 Jet, 1: Venus, 1–2.5 microns, resolution 20 per cm. Comm. Lunar Planet. Lab. 6: 177–189.

Kuiper, G. P., F. F. Forbes, D. L. Steinmetz, and R. T. Mitchell (1969). High altitude spectra from NASA CV 990 Jet, 2: Water vapor on Venus. Comm. Lunar Planet. Lab. 6: 209–228.

Kumar, S., and A. L. Broadfoot (1975). Helium 584 Å airglow emission from Venus: Mariner 10 observations. Geophys. Res. Lett. 2: 357–360.

Kumar, S., D. M. Hunten, and A. L. Broadfoot (1978). Non-thermal hydrogen in the Venus exosphere: The ionospheric source and the hydrogen budget. Planet. Space Sci. 26: 1063.

Kumar, S., D. M. Hunten, and H. A. Taylor, Jr. (1981). H_2 abundance in the atmosphere of Venus. Geophys. Res. Lett. 8: 237–239.

Kumar, S., D. M. Hunten, and J. B. Pollack (1985). Nonthermal escape of hydrogen and deuterium from Venus and implications for loss of water. Icarus 55: 369–389.

Kunde, V. G., R. A. Hanel, and L. Herath (1977). High spectral resolution ground-based observations of Venus in the 450–1250 cm^{-1} region. Icarus 32: 210–224.

Kurt, V. G., E. K. Sheffer, and S. B. Dostovalov (1972). Study of ultraviolet radiation by the "Venera-4" spacecraft. In: *Physics of the Moon and Planets*. Moscow: Nauka.

Kuzmin, A. D. (1965). Measurements of the brightness temperature on the illuminated side of Venus at a wavelength of 10.6 cm. Astron. Zh. 42 (no. 6): 1281–1286.

———. (1967). Radio physical research on Venus. Moscow: VINITI.

———. (1983). Radio astronomical studies of Venus. In *Venus*, ed. D. M. Hunten, L. Colin, T. M. Donahue, and V. I. Moroz. Tucson: University of Arizona Press.

Kuzmin, A. D., and B. J. Clark (1965). Polarization measurements and the brightness temperature distribution of Venus at a wavelength of 10.6 cm. Astron. Zh. 43 (no. 3): 595–617.

Kuzmin, A. D., and M. Ya. Marov (1974). *Physics of the Planet Venus*. Moscow: Nauka. Trans. Leo Kanner Associates, NASA-TT-F-16226, Washington, D.C.: NASA.

Kuzmin, A. D., and A. E. Solomonovich (1960). Radio emissions of Venus in the 8-mm wavelength range. Astron. Zh. 37 (no. 2): 297–300.

———. (1961). Results of observations of the radio emission of Venus in 1961. Astron. J. 38 (no. 6): 1115–1117.

———. (1962). Observations of radio emissions of Venus and Jupiter in the 8-mm wavelength range. Astron. Zh. 39 (no. 4): 660–668.

Landau, L. D., and E. M. Lifschitz (1964). *Statistical Physics*, 2nd ed. Moscow: Nauka.

Landerer, J. (1890). Comptes rendus. Acad. Sci. Paris 110: 210–212.

Lavrukhina, A. K. (1983). Special features of physical-mechanical processes in the evolution of space material. Meteoritika 42: 3–22.

Leberl, F. W., K. E. Maurice, J. K. Thomas, C. E. Leff, and S. D. Wall (1992a). Images and topographic relief at the North Pole of Venus. J. Geophys Res. 97: 13667–13674.

Leberl, F. W., J. K. Thomas, and K. E. Maurice (1992b). Initial results from the Magellan stereo experiment. J. Geophys. Res. 97: 13675–13689.

Lee, T. (1979). New isotopic clues to solar system formation. Rev. Geophys. Space Phys. 17: 1591–1611.

Leovy, C. B. (1973). Rotation of the upper atmosphere of Venus. J. Atmos. Sci. 30: 1218–1220.

———. (1987). Zonal winds near Venus' cloud top level: An analytic model of the equatorial wind speed. Icarus 69: 193–201.

Leovy, C. B., and J. B. Pollack (1973). A first look at atmospheric dynamics and temperature variations on Titan. Icarus 19 (no. 2): 195.

Levin, B. Yu. (1964). *Origin of the Earth and Planets*, 4th ed. Moscow: Nauka.

Levin, B. Yu., and S. V. Maeva (1975). Riddles of the origin and thermal history of the Moon. In *Cosmochemistry of the Moon and Planets*, ed. A. P. Vinogradov. Moscow: Nauka.

Lewis, J. S. (1968). Composition and structure of the clouds of Venus. Astrophys. J. 152: L79–L83.

———. (1969). Geochemistry of volatile elements on Venus. Icarus 11: 367–385.

———. (1970). Venus: Atmospheric and lithospheric composition. Earth Planet. Sci. Lett. 10: 73–80.

———. (1971). The atmosphere, clouds, and surface of Venus. Amer. Sci. 59: 557–566.

———. (1972). Metal/silicate fractionation in the solar system. Earth Planet. Sci. Lett. 15: 286–290.

———. (1973). Chemistry of the planets. Ann. Rev. Phys. Chem. 24: 339.

———. (1974). The temperature gradient in the solar nebula. Science 186: 440–443.

Lewis, J. S., and B. Fegley (1982). Venus: Halide cloud condensation and volatile element inventories. Science 216 (no. 4551): 1223–1224.

Lewis, J. S., and D. H. Grinspoon (1990). Vertical distribution of water in the atmosphere of Venus: A simple thermochemical explanation. Science 249: 1273–1275.

Lewis, J. S., and F. A. Kreimendahl (1980). Oxidation state of the atmosphere and crust of Venus from Pioneer-Venus observations. Icarus 42: 330–337.

Lewis, J. S., and R. G. Prinn (1984). *Planets and Their Atmospheres: Origin and Evolution.* London: Academic Press.

Limaye, S. S. (1984). Morphology and movements of polarization features on Venus: Pioneer OCPP observations. Icarus 57: 362–385.

———. (1987). Atmospheric dynamics on Venus and Mars. Adv. Space Res. 7 (no. 12): 39–53.

Limaye, S. S., and V. E. Suomi (1977). A normalized view of Venus. J. Atmos. Sci. 34: 205–215.

———. (1981). Cloud motions on Venus: Global structure and organization. J. Atmos. Sci. 38: 1220–1235.

Limaye, S. S., C. J. Grund, and S. P. Burre (1982). Zonal mean circulation at the cloud level on Venus: Spring and Fall 1979 OCPP observations. Icarus 51: 416–439.

Limaye, S. S., C. Grassotti, and M. J. Kueremeyer (1988). Venus: Cloud level circulation during 1982 as determined from Pioneer cloud photopolarization images. Icarus 73: 193–211.

Link, F., and L. Neuzil (1957). Sur la couleur de Vénus. Mem. Soc. Roy. Sci. Liège 18: 156–159.

Linkin, V. M., J. Blamont, S. I. Devyatkin, A. V. Dyachkov, S. P. Ignatova, V. V. Kerzhanovich, A. N. Lipatov, C. Malique, B. I. Stadnyuk, Ya. V. Sanotskiy, P. G. Stolarchuk, A. V. Terterashvili, A. A. Shurupov, G. A. Frank, and L. I. Hlyustova (1986a). Thermal structure of the Venus atmosphere from Vega 2 lander measurements. Kosmich. Issled. 25 (no. 5): 659–672.

Linkin, V. M., J. Blamont, A. N. Lipatov, A. A. Shurupov, C. Malique, S. P. Ignatova, G. A. Frank, L. I. Hlyustova, A. V. Terterashvili, A. Seiff, V. V. Kerzhanovich, B. Ragent, R. Young, E. Elson, R. Preston, A. Ingersoll, and D. Crisp (1986b). Thermal structure of the Venus atmosphere in the middle cloud layer. Pisma Astron. Zh. 12 (no. 1): 36.

Liu, L. (1977). The system enstatite-pyrope at high pressure and temperatures and the minerology of the Earth's mantle. Earth Planet. Sci. Lett. 36: 237–245.

Logan, L. M., G. R. Hunt, D. A. Long, and J. P. Dybwad (1974). Absolute infrared radiance measurements of Venus and Jupiter. Air Force Cambridge Res. Lab. TR74-0573.

Lomonosov, M. V. (1955). *The Complete Works* 5. Moscow: Acad. Nauk USSR.

Lukashevich, N. L., and V. P. Shari (1982). Certain features of the phase matrix for single scattering of a semidispersive system of spherical particles. Preprint, M. V. Keldysh Inst. Appl. Math. Akad. Nauk USSR (no. 15).

Lyot, D. (1929). Recherches sur la polarisation de la lumière des planètes et de quelques substances terrèstres. Ann. Obs. Paris (sect. Meudon 8): 161.

Magee, R. K., J. E. Guest, J. W. Head, and M. G. Lancaster (1992). Mylitta Fluctus, Venus: Rift-related, centralized volcanism and the emplacement of large-volume flow units. J. Geophys. Res. 97: 15991–16016.

Makhalkin, A. B. (1980). Possibility of formation of an originality in homogeneous Earth. Phys. Earth Planet. Inter. 22 (nos. 3–4): 302–312.

Malin, M. C. (1992). Mass movements on Venus: Preliminary results from Magellan cycle 1 observations. J. Geophys. Res. 97: 16337–16352.

Markov, M. S., Ya. B. Smirnov, and L. F. Dobrzhinetskaya (1989). Tectonics on Venus and the early Precambrian. Earth, Moon, and Planets 45: 101–113.

Marov, M. Ya. (1971). Model of Venus' atmosphere. Dokl. Akad. Nauk USSR 196 (no. 1): 67–70.

———. (1972). Venus: A perspective at the beginning of planetary exploration. Icarus 16: 415–461.

———. (1978). Results of Venus missions. Ann. Rev. Astron. Astrophys. 16: 141–170.

———. (1979a). The atmosphere of Venus: Venera data. Fund. Cosmic Phys. 5: 1–46.

———. (1979b). Research on Venus' atmosphere. Astron. Vestnik 13 (no. 1): 3–23.

———. (1982). Highlights of Venera missions of Venus. Planet. Rep. 2 (no. 6): 14–17.

———. (1986). *Planets of the Solar System*, 2nd ed. Moscow: Nauka.

———. (1994). The inner planets. In *Space Biology and Medicine*, vol. 1, 2nd ed., ed. J. Ramel and V. Kotelnikov. NASA–Russian Acad. of Sci. Moscow: Nauka.

Marov, M. Ya., ed. (1992). *Atlas of the Terrestrial Planets and Their Satellites*. Moscow: Institute for Geodesy, Aeromapping and Cartography.

Marov, M. Ya., and A. V. Kolesnichenko (1987). *Introduction to Planetary Aeronomy*. Moscow: Nauka.

Marov, M. Ya., and O. L. Ryabov (1972). A model atmosphere of Venus. Preprint, M. V. Keldysh Inst. Appl. Math. Akad. Nauk USSR (no. 39).

Marov, M. Ya., and V. P. Shari (1973). Transfer of longwave radiation in Venus' lower atmosphere. Preprint, M. V. Keldysh Inst. Appl. Math., Akad. Nauk USSR (no. 23).

Marov, M. Ya., V. S. Avduevsky, M. K. Rozhdestvensky, H. F. Borodin, and V. V. Kerzhanovich (1971). Preliminary results of research on Venus' atmosphere from the "Venera-7" spacecraft. Kosmich. Issled. 9 (no. 4): 570–579.

Marov, M. Ya., V. S. Avduevsky, V. V. Kerzhanovich, M. K. Rozhdestvensky, N. F. Borodin, and O. L. Ryabov (1973a). Measurements of temperature, pressure and wind velocity in Venus' atmosphere from the "Venera-8" spacecraft. Dokl. Acad. Nauk USSR 210 (no. 3): 559–562.

———. (1973b). Venera 8: Measurements of temperature, pressure and wind velocity of the illuminated side of Venus. J. Atmos. Sci. 30: 1210–1214.

Marov, M. Ya., V. S. Avduevsky, N. F. Borodin, A. P. Ekonomov, V. V. Kerzhanovich, V. P. Lysov, B. Ye. Moshkin, and M. K. Rozhdestvensky (1973c). Preliminary results on the Venus atmosphere from the Venera 8 descent module. Icarus 20: 407–421.

Marov, M. Ya., B. V. Byvshev, Yu. P. Baranov, I. S. Kuznetsov, V. N. Lebedev, V. E. Lystsev, A. V. Maksimov, G. K. Popandopulo, V. R. Razdolin, V. A. Sandimirov, and A. M. Frolov (1976a). Nephehemetric investigations on the stations Venera 9 and Venera 10. Kosmich. Issled. 14 (no. 5): 729–734.

Marov, M. Ya., V. N. Lebedev, and V. E. Lystsev (1976b). Some preliminary estimates of aerosol component in the Venus atmosphere. Pisma Astron. Zh. 2: 251–256.

Marov, M. Ya., B. V. Byvshev, Yu. P. Baranov, V. N. Lebedev, V. E. Lystsev, A. V. Maximov, K. N. Manuylov, and A. M. Frolov (1979). Aerosol component of atmosphere of Venus from data of Venera 11. Kosmich. Issled. 17: 743–746.

Marov, M. Ya., V. E. Lystsev, V. N. Lebedev, N. L. Lukashevich, and V. P. Shari (1980). The structure and microphysical properties of the Venus clouds: Venera 9, 10 and 11 data. Icarus 44: 608–639.

Marov, M. Ya., B. V. Byvshev, Yu. P. Baranov, V. N. Lebedev, N. L. Lukashevich, A. V. Maksimov, K. K. Manuylov, A. M. Frolov, and V. P. Shari (1983). Investigation of structure of the Venus clouds using nephelometers aboard Venera 13 and Venera 14 landers. Kosmich. Issled. 21 (no. 2): 269–278.

Marov, M. Ya., A. P. Galtsev, and V. P. Shari (1984). H_2O profile in Venus' lower atmosphere from measurements of the effective radiation flux. Kosmich. Issled. 22 (no. 2): 267–272.

———. (1985). Transfer of thermal radiation and the water content of Venus' atmosphere. Astron. Vestnik 19 (no. 1): 15–41.

Marov, M. Ya., V. P. Volkov, Yu. A. Surkov, and M. L. Rivkin (1989a). Lower atmosphere. In *The Planet Venus*, ed. V. L. Barsukov and V. P. Volkov. Moscow: Nauka.

Marov, M. Ya., A. P. Galtsev, and V. P. Shari (1989b). Thermal conditions of Venus' atmosphere. In *The Planet Venus*, ed. V. L. Barsukov and V. P. Volkov. Moscow: Nauka.

———. (1989c). Estimation of the influence of sulfur dioxide on heat transfer in the Venus atmosphere. In *Proceedings of the 19th Lunar and Planetary Conference*. Houston: Lunar and Planetary Institute.

Marov, M. Ya., V. I. Shematovich, and D. V. Bisikalo (1996). Nonequilibrium aeronomic processes: A kinetic approach to the mathematical modeling. Space Sci. Rev. 76 (nos. 1–2): 1–204.

Martynov, D. Ya. (1962). On the radius of Venus. Astron. Zh. 39 (no. 4): 653–659.

Martynov, D. Ya., and M. M. Pospergelis (1961). Notes on photometric analysis of the structure of Venus' atmosphere. Astron. Zh. 38: 558–561.

Massey, S. T., D. M. Hunten, and D. R. Sonell (1983). Day and night models of the Venus thermosphere. J. Geophys. Res. 88 (no. 5): 3955–3969.

Masursky, H. E., E. Eliason, P. G. Ford, G. E. McGill, G. H. Pettengill, G. G. Schaber, and G. Scubert (1980). Pioneer-Venus radar results: Geology from images and altimetry. J. Geophys. Res. 85: 8232–8260.

Mayer, C. H., R. M. Sloanaker, and T. P. McCullough (1957). Radiation from Venus at 3.15 cm wavelength. Astron. J. 62 (no. 1): 26–27.

Mayer, C. H., T. P. McCullough, and R. M. Sloanaker (1958). Observations of Venus at 3.15 cm wavelength. Astrophys. J. 127: 1–10.

———. (1960). Observations of Venus at 10.2 cm wavelength. Astron. J. 65: 349–350.

———. (1962). 3.15 cm observations of Venus in 1961. Mem. Soc. Roy. Sci. Liège 7: 357–363.

Mayr, H. G., I. Harris, W. T. Kasprzak, M. Dube, and F. Varosi (1988). Gravity waves in the upper atmosphere of Venus. J. Geophys. Res. 93: 11247–11262.

McElroy, M. B. (1972). Mars: An evolving atmosphere. Science 175: 443–445.

McElroy, M. B., and M. J. Prather (1981). Noble gases in the terrestrial planets: Clues to evolution. Nature 293: 535–539.

McElroy, M. B., N. D. Sze, and Y. L. Yung (1973). Photochemistry of the Venus atmosphere. J. Atmos. Sci. 30: 1437–1447.

McElroy, M. B., T. Y. Kong, and Y. L. Yung (1977). Photochemistry and evolution of Mars' atmosphere: A Viking perspective. J. Geophys. Res. 82: 4379–4388.

McElroy, M. B., M. J. Prather, and J. M. Rodriguez (1982). Escape of hydrogen from Venus. Science 215: 1614–1615.

McGill, G. T., J. L. Warner, M. C. Malin, R. E. Arvidson, E. Eliason, S. Nozette, and R. D. Reasenberg (1983). Topography, surface properties, and tectonic evolution. In *Venus*, ed. D. M. Hunten, L. Colin, T. M. Donahue, and V. I. Moroz. Tucson: University of Arizona Press.

McKenzie, D., P. G. Ford, F. Liu, and G. H. Pettengill (1992a). Pancakelike domes on Venus. J. Geophys. Res. 97: 15967–15976.

McKenzie, D., J. M. McKenzie, and R. S. Saunders (1992b). Dike emplacement on Venus and on Earth. J. Geophys. Res. 97: 15977–15990.

McKenzie, D., P. G. Ford, C. Johnson, B. Parsons, D. Sandwell, R. S. Saunders, and S. C. Solomon (1992c). Features on Venus generated by plate boundary processes. J. Geophys. Res. 97: 13533–13544.

McSween, H. Y., Jr. (1984). SNC meteorites: Are they Martian rocks? Geology 12: 3–6.

Melosh, H. J., and A. M. Vickery (1989). Impact erosion of the primordial Martian atmosphere. Nature 338: 487–489.

Menzel, D., and F. Whipple (1954). The case for H_2O clouds on Venus. Astron. J. 59: 329–330.

———. (1955). The case for H_2O clouds on Venus. Publ. Astron. Soc. Pacif. 67 (no. 396): 161–168.

Miller, K. L., W. C. Knudsen, and K. Spenner (1984). The dayside Venus ionosphere, 1: Pioneer-Venus retarding potential analyzer experimental observations. Icarus 57: 386–409.

Monin, A. S. (1977). *History of the Earth*. Leningrad: Nauka.

Moore, H. J., J. J. Plaut, P. M. Schenk, and J. W. Head (1992). An unusual volcano on Venus. J. Geophys. Res. 97: 13479–13494.

Moore, P. (1961). *The Planet Venus*, 3rd ed. London: Faber and Faber.

Morgan, J. W., and E. Anders (1980). Chemical composition of Earth, Venus, and Mercury. Proc. Nat. Acad. Sci. USA 77: 6973–6977.

Morgan, P., and R. J. Phillips (1983). Hot spot heat transfer and its application to Venus. J. Geophys. Res. 88: 8305–8317.

Moroz, V. I. (1963). Infrared spectrum of Venus (1–2.5). Astron. Zh. 40 (no. 1): 144–153.

———. (1964). New observations of Venus' infrared spectrum (1.2–3.8). Astron. Zh. 41 (no. 4): 711–719.

———. (1968). The CO_2 bands and some optical properties of the atmosphere of Venus. Astron. Zh. 11: 653–661.

———. (1981). The atmosphere of Venus. Space Sci. Rev. 29: 3–127.

Moroz, V. I., N. A. Parfentev, and N. F. Sanko (1979). Spectrophotometric experiment in the Venera 11 and 12 modules, 2: Analysis of Venera spectra by layer-addition method. Kosmich. Issled. 17: 727–742.

Moroz, V. I., Yu. M. Golovin, B. E. Moshkin, and A. P. Ekonomov (1981). Spectrophotometric experiment on the Venera 11 and 12 descent modules, 3: Results of the photometric measurements. Kosmich. Issled. 19: 599–612.

Moroz, V. I., A. P. Ekonomov, Yu. M. Golovin, B. E. Moshkin, and N. F. Sanko (1983a). Solar radiation scattered in the Venus atmosphere: The Venera 11 and 12 data. Icarus 53: 509–537.

Moroz, V. I., B. E. Moshkin, A. P. Ekonomov, A. V. Grigoryev, V. I. Gnedykh, and Yu. M. Golovin (1983b). Spectrophotometric experiment on board the Venera 13 and Venera 14 probes. Preliminary analysis of H_2O absorption bands in spectra. Kosmich. Issled. 21: 246–253.

Moroz, V. I., A. P. Ekonomov, B. E. Moshkin, H. E. Revercomb, L. A. Sromovsky, J. T. Shofield, D. Spankuch, F. W. Taylor, and M. G. Tomasko (1985a). Solar and thermal radiation in the Venus atmosphere. In *Venus International Reference Atmosphere*, ed. A. J. Kliore, V. I. Moroz, and G. M. Keating. Adv. Space Res. 5 (no. 11): 197–232.

Moroz, V. I., B. Deler, E. A. Ustinov, K. Shefer, L. V. Zasova, D. Shpenkukh, A. V. Dyachkov, R. Dyubua, V. M. Linkin, D. Ertel, B. B. Kerzhanovich, I. Nopirakovsky, I. A. Matsigorin, X. Bekker-Ros, A. A. Shurukov, V. Shtadthaus, and A. N. Lipatov (1985b). Infrared experiment aboard the Venera 15 and Venera 16 probes. Preliminary results of spectral analysis in the region of H_2O and SO_2 absorption bands. Kosmich. Issled. 23: 236–247.

Morrison, D., and T. Owen (1988). *The Planetary System*. New York: Addison-Wesley.

Mottinger, N. A., W. L. Sjorgenm, and B. G. Bills (1985). Venus gravity: An harmonic analysis and geophysical implications. J. Geophys. Res. 90: C739–C756.

Mueller, R. F. (1963). Chemistry and petrology of Venus: Preliminary deductions. Science 141: 1046–1047.

———. (1964). A chemical model for the lower atmosphere of Venus. Icarus 3 (no. 4): 285–298.

Muhleman, D. O. (1961). Early results of the 1961 LPL Venus radar experiment. Astron. J. 66 (no. 67): 292.

Mukhin, L. M., B. G. Gelman, N. I. Lamonov, V. V. Melnikov, D. F. Nenarokov, B. P., Okhotnikov, V. A. Rotin, and V. A. Khokhlov (1983). Gas chromotographical analysis of the chemical composition of the atmosphere of Venus by Venera 13 and 14. Kosmich. Issled. 21: 225–230.

Müller, G. (1893). Helligkeitsbestimmungen der grossen Planeten und einiger Asteroiden. Publ. Observ. Potsdam 8 (no. 30): 197–389.

Muller, R. (1926). Photometrische tagesbeobachtungen des planeten Venus von G. Muller. Astron. Nachr. 227 (no. 5429): 65–72.

Murray, B. C., R. L. Wildey, and J. A. Westphal (1963). Infrared photometric mapping of Venus through the 8- to 14-micron atmospheric window. J. Geophys. Res. 68: 4813–4818.

Murray, B. C., J. S. Belton, G. E. Danielson, M. E. Davies, D. Gault, B. Hapke, B. O'Leary, R. G. Strom, V. Suomi, and N. Trask (1974). Venus: Atmospheric motion and structure from Mariner 10 pictures. Science 183: 1307–1315.

Mustel, E. R. (1960). *Stellar Atmosphere*. Moscow: Physmatgiz.

Nagy, A. F., T. E. Cravens, R. H. Chen, H. A. Taylor, Jr., L. H. Brace, and H. C. Brinton (1979). Comparison of calculated and measured ion densities on the dayside of Venus. Science 205: 107–109.

Nagy, A. F., T. E. Cravens, and T. I. Gombosi (1983). Basic theory and model calculations of the Venus ionosphere. In *Venus*, ed. D. M. Hunten, L. Colin, T. M. Donahue, and V. I. Moroz. Tucson: University of Arizona Press.

Namiki, N., and S. C. Solomon (1994). Impact crater densities on volcanoes and coronae on Venus: Implications for volcanic resurfacing. Science 165: 929–933.

Ness, N. F., K. W. Behannon, R. P. Lepping, Y. C. Whang, and K. H. Schatten (1974). Magnetic field observations near Venus: Preliminary results from Mariner 10. Science 183: 1301–1304.

Newbauer, J. A. (1959). Radar contact with Venus. Astronautics 4 (no. 5): 50–124.

Newman, M. J., and R. T. Rood (1977). Implications of solar evolution for the Earth's early atmosphere. Science 198: 1035–1037.

Newman, M. J., G. Shubert, A. J. Kliore, and I. R. Patel (1984). Zonal winds in the middle atmosphere of Venus from Pioneer Venus radio occultation data. J. Atmos. Sci. 41 (no. 12): 1901–1913.

Niemann, H. B., W. T. Kasprzak, A. E. Hedin, D. M. Hunten, and N. W. Spencer (1980). Mass spectrometric measurements of the neutral gas composition of the thermosphere and exosphere of Venus. J. Geophys. Res. 85: 7817–7827.

Nikolaeva, O. V., L. B. Ronka, and A. T. Basilevsky (1986). Circular features on the plains of Venus as evidence of its geologic history. Geochimiya 5: 579–589.

Nozette, S., and J. S. Lewis (1982). Venus: Chemical weathering of igneous rocks and buffering of atmospheric composition. Science 216: 181–183.

Ohring, G., and J. Mariano (1964). The effect of cloudiness on a greenhouse model of the Venus atmosphere. J. Geophys. Res. 69: 165.

O'Nions, R. K., N. M. Ivensen, and P. J. Hamilton (1979). Geochemical modeling of mantle differentiation and crustal growth. J. Geophys. Res. 84: 6091.

Opik, E. J. (1961). The aeolosphere and atmosphere of Venus. J. Geophys. Res. 66: 2807.

Osipov, V. M., and A. P. Galtsev (1971). Calculation of CO_2 infrared bands transmission under high temperatures and pressures. Izvestiya Acad. Nauk USSR, Physics of Atmosphere and Ocean 7 (no. 8): 857–870.

Ostriker, J. P. (1963). Radiative transfer in a finite gray atmosphere. Astrophys. J. 138 (no. 1): 281.

Owen, T. (1967). Water vapor on Venus: A dissent and clarification. Astrophys. J. 150: L121–L123.

———. (1968). A search for minor constituents in the atmosphere of Venus. J. Atmos. Sci. 25: 583.

Owen, T., and C. Sagan (1972). Minor constituents in planetary atmospheres: Ultraviolet spectroscopy from the Orbiting Astronomical Observatory. Icarus 16: 557–568.

Owen, T., R. D. Cess, and V. Ramanathan (1979). Early Earth: An enhanced carbon dioxide greenhouse to compensate for reduced solar luminosity. Nature 277: 640–642.

Owen, T., A. Bar-Nun, and I. Kleinfeld (1992). Possible cometary origin of heavy noble gases in the atmospheres of Venus, Earth and Mars. Nature 358: 43–46.

Oyama, V. I., G. C. Carle, F. Woeller, and J. B. Pollack (1979). Venus' lower atmosphere composition: Analysis by gas chromotography. Science 203: 802–804.

Oyama, V. I., G. C. Carle, F. Woeller, J. B. Pollack, R. T. Reynolds, and R. A. Craig (1980). Pioneer Venus gas chromotography of the lower atmosphere of Venus. J. Geophys. Res. 85: 7891–7902.

Parisot, J. P., and G. Moreels (1984). Photochemistry of fluorine in the mesosphere of Venus. Icarus 59 (no. 1): 69–73.

Parmentier, E. M., and P. C. Hess (1992). Chemical differentiation of a convecting planetary interior: Consequences for a one plate planet such as Venus. Geophys. Res. Lett. 19: 2015–2018.

Parshin, I. A. (1948). Research on the planet Venus in 1948. Vestnik LGU (no. 2): 144–146.

Pavri, B., J. W. Head, K. B. Klose, and L. Wilson (1992). Steep-sided domes in Venus: Characteristics, geologic setting, and eruption conditions from Magellan data. J. Geophys. Res. 97: 13445–13478.

Paxton, L. J. (1985). Pioneer Venus orbiter ultraviolet spectrometer limb observations. J. Geophys. Res. A 90 (no. 6): 5089–5096.

Paxton, L. J., and A. I. F. Stewart (1982). The "hot" carbon corona of Venus. EOS Trans. Amer. Geophys. Union 63 (no. 18): 367.

Perez-de-Tejada, H. (1995). Plasma boundary in planetary ionosheaths. Space Sci. Rev. 72: 655–675.

Petryakov, I. V., B. M. Andreychikov, and B. N. Korchuganov (1981). Utilization of FP filter for investigation of the Venus clouds. Dokl. Akad. Nauk USSR 258 (no. 1): 57–59.

Pettengill, G. H. (1995). Venus Highlands: What is that mystery material? Fall 1995 Meeting of the American Geophysical Union, Abstract.

Pettengill, G. H., and R. Price (1961). Radar echoes from Venus and a new determination of the solar parallax. Planet and Space Sci. 5 (no. 1): 71–74.

Pettengill, G. H., H. W. Briscol, J. V. Evans, E. Gehrels, G. M. Hyde, L. G. Kraft, R. Price, and W. B. Smith (1962). A radar investigation of Venus. Astron. J. 67 (no. 4): 181–199.

Pettengill, G. H., R. B. Dyce, and A. Sancher (1964). Radar measurements on the Arecibo 1000-foot antenna. Report at the 45th Annual Assembly of the American Geophysical Union, Washington, D.C.

Pettengill, G. H., P. G. Ford, W. E. Brown, W. M. Kaula, H. Masursky, and G. E. McGill (1979). Venus: Preliminary topographic and surface imaging results from the Pioneer Orbiter. Science 205: 91–93.

Pettengill, G. H., E. Eliason, P. G. Ford, G. B. Loriot, H. Masursky, and G. E. McGill (1980). Pioneer Venus radar results: Altimetry and surface properties. J. Geophys. Res. 85 (no. A13): 8261–8270.

Pettengill, G. H., P. G. Ford, and S. Nozette (1982). Venus: Global surface radar reflectivity. Science 217: 640–642.

Pettengill, G. H., P. G. Ford, and B. D. Chapman (1988). Venus surface electromagnetic properties. J. Geophys. Res. 93 (no. B12): 881–892.

Pettengill, G. H., P. G. Ford, and R. J. Wilt (1992). Venus surface radiothermal emission as observed by Magellan. J. Geophys. Res. 97: 13091–13102.

Pettit, E. S., and S. B. Nicholson (1924). Radiation from dark hemisphere of Venus. Publ. Astron. Soc. Pacif. 36: 227–228.

———. (1955). Temperatures of the bright and dark sides of Venus. Publ. Astron. Soc. Pacif. 67: 293–303.

Phillips, R. J. (1994). Estimating lithospheric properties at Atla Regio, Venus. Icarus 112: 147–170.

Phillips, R. J., and V. L. Hansen. (1994). Tectonic and magmatic evolution of Venus. Ann. Rev. Earth Planet. Sci. 22: 597–654.

Phillips, R. J., and M. C. Malin (1983). The interior of Venus and tectonic implications. In *Venus*, ed. D. M. Hunten, L. Colin, T. M. Donahue, and V. I. Moroz. Tucson: University of Arizona Press.

Phillips, R. J., W. M. Kaula, G. E. McGill, and M. C. Malin (1981). Tectonics and evolution of Venus. Science 212: 879–887.

Phillips, R. J., R. F. Raubertas, R. E. Arvidson, I. C. Sarkar, R. R. Herrick, N. Izenberg, and R. E. Grimm (1992). Impact craters and Venus resurfacing history. J. Geophys. Res. 97: 15923–15948.

Pieters, C. M., J. M. Head, W. Patterson, S. Pratt, J. Garvin, V. L. Barsukov, A. T. Basilevsky, I. L. Khodakovsky, A. S. Selivanov, A. S. Panfilov, Yu. M. Gektin, and Y. M. Naraeva (1986). The color of the surface of Venus. Science 234: 1379–1383.

Plass, G. N., and V. R. Stull (1963). Carbon dioxide absorption for path lengths applicable to the atmosphere of Venus. J. Geophys. Res. 68 (no. 5): 1355–1363.

Plass, G. N., and V. R. Wyatt (1962). Carbon dioxide absorption for path length applicable to the atmosphere of Venus. Aeronutronic Publication no. U 1844.

Plaut, J. J., and R. E. Arvidson (1992). Comparison of Goldstone and Magellan radar data in the equatorial plains of Venus. J. Geophys. Res. 97: 16279–16292.

Pollack, J. B. (1969a). A non-gray CO_2-H_2O greenhouse model of Venus. Icarus 10: 314–341.

———. (1969b). Temperature, structure of non-gray planetary atmospheres. Icarus 10: 301–313.

———. (1971). A non-gray calculation of the runaway greenhouse: Implications for Venus' past and present. Icarus 14: 295–306.

———. (1979). Climatic change on the terrestrial planets. Icarus 31: 479–553.

Pollack, J. B., and D. C. Black (1979). Implications of the gas compositional measurements of Pioneer Venus for the origin of planetary atmospheres. Science 205: 56–59.

———. (1982). Noble gases in planetary atmospheres: Implications for the origin and evolution of atmospheres. Icarus 51: 169–198.

Pollack, J. B., and J. Cuzzi (1979). Scattering by non-spherical particles of size comparable to the wavelength: A new semi-empirical theory and its application to tropospheric aerosols. J. Atmos. Sci. 37: 868–881.

Pollack, J. B., and D. Morrison (1970). Venus: Determination of atmospheric parameters from microwave spectrum. Icarus 12: 376–390.

Pollack, J. B., and C. Sagan (1965). The infrared limb darkening of Venus. J. Geophys. Res. 10: 4403–4426.

———. (1967). A critical test of the electrical discharge model of the Venus microwave emission. Astrophys. J. 150: 699.

Pollack, J. B., and R. Young (1975). Calculations of radiative and dynamical state of the Venus atmosphere. J. Atmos. Sci. 32: 1025–1037.

Pollack, J. B., and Y. L. Yung (1980). Origin and evolution of planetary atmospheres. Ann. Rev. Earth Planet. Sci. 8: 425–487.

Pollack, J. B., D. W. Strecker, F. C. Witteborn, E. Erickson, and B. J. Baldwin (1978). Properties of the clouds of Venus as inferred from airborne observations of its near-infrared reflectivity spectrum. Icarus 34: 28–45.

Pollack, J. B., B. Ragent, R. Boese, M. Tomasko, J. Blamont, R. G. Kuollenberg, L. W. Esposito, L. Travis, and D. Wiedman (1980a). Distribution and source of the UV absorption in Venus' atmosphere. J. Geophys. Res. 85: 8141–8150.

Pollack, J. B., O. B. Toon, and R. Boese (1980b). Greenhouse models of Venus' high surface temperature, as constrained by Pioneer Venus measurements. J. Geophys. Res. 85: 8223–8231.

Pollack, J. B., J. B. Dalton, D. H. Grinspoon, R. B. Wattson, R. Freedman, D. Crisp, D. Allen, B. Bézard, C. de Bergh, L. P. Giver, Q. Ma, and R. Tipping (1993). Near-infrared light from Venus' nightside: A spectroscopic analysis. Icarus 103: 1–42.

Price, M., and J. Suppe (1994). Mean age of rifting and volcanism on Venus deduced from impact crater densities. Nature 372: 756–759.

Prinn, R. G. (1971). Photochemistry of HCl and other minor constituents in the atmosphere of Venus. J. Atmos. Sci. 28: 1058–1068.

———. (1973). Venus: Composition and structure of visible clouds. Science 182: 1132–1133.

———. (1975). Venus: Chemical and dynamical processes in the stratosphere and mesosphere. J. Atmos. Sci. 32: 1237–1247.

———. (1979). On the possible roles of gaseous sulfur and sulfanes in the atmosphere of Venus. J. Geophys. Res. Lett. 6: 807–810.

Prokofiev, V. K. (1964). On the presence of oxygen in the atmosphere of Venus, 2. Izv. Krymskoy Astrophys. Obs. 31: 276–280.

———. (1965). On the presence of oxygen in the atmosphere of Venus, 3. Izv. Krymskoy Astrophys. Obs. 34: 243–251.

Prokofiev, V. K., and N. N. Petrova (1963). On the abundance of oxygen in the Venus atmosphere. Izv. Krymskoy Astrophys. Obs. 29: 3–14.

Pronin, A. A. (1986). The structure of Lakshmi Planum, an indication of horizontal astenosphere flows on Venus. Geotectonika 20: 271–281.

Ragent, B., and J. Blamont (1980). Structure of the clouds of Venus: Results of the Pioneer Venus nephelometer experiment. J. Geophys. Res. 85: 8089–8105.

Ragent, B., L. W. Esposito, M. G. Tomasko, M. Ya. Marov, V. P. Shari, and V. N. Lebedev (1985). Particulate matter in the Venus atmosphere. In *The Venus International Reference Atmosphere*, ed. A. J. Kliore, V. I. Moroz, and G. M. Keating. Adv. Space Res. 5 (no. 11): 85–116.

Rasool, S. I., and C. de Bergh (1970). The runaway greenhouse and the accumulation of CO_2 in the Venus atmosphere. Nature 226 (no. 5250): 1037–1039.

Revercomb, H. E., L. A. Sromovsky, and V. E. Suomi (1982). Reassessment of net radiation measurements in the atmosphere of Venus. Icarus 52 (no. 1): 279–300.

———. (1985). Net thermal radiation measurements in the atmosphere of Venus. Icarus 61 (no. 3): 521–538.

Richardson, R. S. (1955). Observations of Venus made at Mount Wilson in the winter of 1954–1955. Publ. Astron. Soc. Pacif. 67 (no. 398): 304–314.

———. (1958). Spectroscopic observations of Venus' rotation made at Mount Wilson in 1958. Publ. Astron. Soc. Pacif. 70 (no. 414): 251–260.

Riehl, H. (1978). *Introduction to the Atmosphere*, 3rd ed. New York: McGraw-Hill.

Ringwood, A. E. (1961). Changes in solar luminosity and some possible terrestrial consequences. Geochim. Cosmochim. Acta 21: 295–296.

Ringwood, A. E., and D. L. Anderson (1977). Earth and Venus: A comparative study. Icarus 30 (no. 2): 243–253.

Roberts, R. E., J. E. A. Selby, and L. M. Biberman (1976). Infrared continuum absorption by atmospheric water vapor in the 8–12 μm window. Appl. Opt. 15 (no. 9): 2085–2090.

Rodriguez, J. M., M. J. Prather, and M. B. McElroy (1984). Hydrogen on Venus: Exospheric distribution and escape. Planet. Space Sci. 32: 1235–1355.

Rogers, A. E., and R. P. Ingalls (1970). Radar mapping of Venus with interferometric resolution of the range-Doppler ambiguity. Radio Sci. 5 (no. 2): 425–433.

Rogers, A. E., R. P. Ingalls, and L. P. Rainville (1972). The topography of a swath around the equator of the planet Venus from the wavelength dependence of the radar cross section. Astron. J. 77 (no. 1): 100–103.

Ronov, A. B., and A. A. Yaroshevsky (1967). Chemical structure of the Earth's crust. Geokhimiya 11: 1285–1310.

Roos, M., P. Drossart, Th. Encrenaz, E. Lellouch, B. Bézard, R. W. Carolson, K. Baines, L. W. Kamp, F. W. Taylor, A. D. Collard, S. B. Calcutt, J. B. Pollack, and D. H. Grinspoon (1993). The upper clouds of Venus: Determination of the scale height from NIMS-Galileo infrared data. Planet. Space Sci. 41: 505–514.

Ross, F. E. (1928). Photographs of Venus. Astrophys. J. 68: 57–92.

Ross, J. E., and L. H. Aller (1976). The chemical composition of the Sun. Science 1991 (no. 4233).

Rosse, W. P. (1878). Preliminary note on some measurements of the polarization of the light coming from the Moon and from the planet Venus. Sci. Proc. Roy. Dublin Soc. 1: 19–20.

Rossow, W. B. (1978). Cloud microphysics: Analysis of the clouds of Earth, Venus, Mars and Jupiter. Icarus 36: 1–5.

———. (1982). A general circulation model of a Venus-like atmosphere. J. Atmos. Sci. 40: 273–302.

Rossow, W. B., and C. Sagan (1975). Microwave boundary conditions on the atmosphere and clouds of Venus. J. Atmos. Sci. 32: 1164–1176.

Rossow, W. B., A. D. Del Genio, S. S. Limaye, L. D. Travis, and P. H. Stone (1980). Cloud morphology and motions from Pioneer Venus images. J. Geophys. Res. 85: 8107–8128.

Rothman, L. S., R. R. Gamache, A. Goldman, L. R. Brown, R. A. Tooth, H. M. Pickett, R. L. Poynter, J. M. Fland, C. Camy-Peyret, A. Barbe, N. Husson, C. P. Rinsland, and M. A. H. Smith (1987). HITRAN Database: 1986 edition. Appl. Opt. 26: 4058–4097.

Rubey, W. W. (1951). Geological history of sea water. Bull. Geol. Soc. Amer. 62 (no. 9): 1111.

Rudnik, V. A., and E. V. Sobotovich (1985). *Early History of the Earth*. Moscow: Nedra.

Russel, H. N. (1916). The stellar magnitudes of the Sun, Moon, and planets. Astrophys. J. 43 (no. 2): 103–129.

Russell, C. T. (1976). The magnetic moment of Venus, Venera-4 measurements reinterpreted. Geophys. Res. Lett. 3: 125.

Russell, C. T., and O. L. Vaisberg (1983). The interaction of the solar wind with Venus. In *Venus*, ed. D. M. Hunten, L. Colin, T. M. Donahue, and V. I. Moroz. Tucson: University of Arizona Press.

Russell, C. T., R. C. Elphic, and J. A. Slavin (1979a). Initial Pioneer Venus magnetic field results: Dayside observations. Science 203: 745.

———. (1979b). Initial Pioneer Venus magnetic field results: Nightside observations. Science 205: 114.

———. (1980). Limits on the possible intrinsic magnetic field of Venus. J. Geophys. Res. 85: 8319–8332.

Russell, C. T., R. J. Strangeway, J. G. Luhmann, and L. H. Brace (1993). The magnetic state of the lower ionosphere during Pioneer Venus entry phase. Geophys. Res. Lett. 20 (no. 23): 2723–2726.

Ruzmaikina, T. V., and S. V. Maeva (1986). Study of the formation process of the proto-planetary disk. Astron. Vestnik (Solar System Research) 20 (no. 3): 212–227.

Rzhiga, O. N. (1988). *New Epoch in the Exploration of Venus (Radar Imaging from Venera 15 and 16 Spacecraft)*. Moscow: Znanie.

Safronov, V. S. (1969). *Evolution of the Pre-Planetary Cloud and Formation of the Earth and Planets*. Moscow: Nauka.

Safronov, V. S., and A. V. Vityazev (1983). *Origin of the Solar System*. Series Astronomiya 24. Moscow: VINITI.

Sagan, C. (1960a). The radiation balance of Venus. JPL Tech. Rept.: 32–34.

———. (1960b). The surface temperature of Venus. Astron. J. 65: 352–353.

———. (1962). Structure of the lower atmosphere of Venus. Icarus 1: 151–169.

———. (1967). Atmospheres, origins of Earth and planets. In *International Dictionary of Geophysics*. Oxford: Pergamon Press.

———. (1969). Microwave radiation from Venus: Thermal versus non-thermal models. Comments on Astrophys. and Space Phys. 1: 94–100.

Sagan, C., and G. Mullen (1972). Earth and Mars: Evolution of atmospheres and surface temperature. Science 177: 52–56.

Sagan, C., and J. B. Pollack (1969). On the structure of the Venus atmosphere. Icarus 10: 274–281.

Sagdeev, R. Z., R. S. Kremnev, V. M. Linkin, J. Blamont, R. Preston, and A. S. Selivanov (1986). Vega Venus balloon experiment. Pisma Astron Zh. 12 (no. 1): 10–17.

Samuelson, R. E. (1967). Greenhouse effect in semi-infinite scattering atmospheres: Application to Venus. Astrophys. J. 147: 782–798.

———. (1968). The particulate medium in the atmosphere of Venus J. Atmos. Sci. 25 (no. 4): 634–643.

Sandwell, D. T., and G. Schubert (1992). Flexural ridges, trenches, and outer rises around coronae on Venus. J. Geophys. Res. 97: 16069–16084.

Sanko, N. F. (1980). Gaseous sulfur in the atmosphere of Venus. Kosmich. Issled. 18: 600.

Saunders, R. S. (1992). Foreword. J. Geophys. Res. 97: 15921–15922.

Saunders, R. S., A. J. Spear, P. C. Allin, R. S. Austin, A. L. Berman, R. C. Chandlee, J. Clark, A. V. deCharon, E. M. De Jong, D. G. Griffith, J. M. Gunn, S. Hensley, W. T. K. Johnson, C. E. Kirby, K. S. Leung, D. T. Lyons, G. A. Michaels, J. Miller, R. B. Morris, A. D. Morrison, R. G. Piereson, J. F. Scott, S. J. Shaffer, J. P. Slonski, E. R. Stofan, T. W. Thompson, and S. D. Wall (1992). Magellan mission summary. J. Geophys. Res. 97: 13067–13090.

Savich, N. A., V. E. Andreev, A. S. Vyshlov, A. L. Gavrik, V. P. Konovalov, N. V. Laptev, V. A. Marmulev, A. P. Masterton, A. S. Nabatov, Yu. N. Orlov, I. K. Osmolovsky, L. N. Salyuznaev, and D. Ya. Shtern (1986a). Daytime ionosphere of Venus from radio occultation data from "Venera 15–16" in 1983. Radiotekhnika i elektronika 31 (no. 11): 2113–2120.

———. (1986b). Nighttime ionosphere of Venus from radio occultation data from "Venera 15–16." Radiotekhnika i elektronika 31 (no. 3): 433–439.

Scarf, F. L., and C. T. Russell (1983). Measurements of Venus lightning from the Pioneer orbiter. Geophys. Res. Lett. 10: 1192–1195.

Scarf, F. L., W. W. L. Taylor, C. T. Russell, and L. H. Brace (1980). Pioneer Venus plasma waves observations: The solar wind-Venus interaction. J. Geophys. Res. 85: 7599–7612.

Schaber, G. G., R. G. Strom, H. J. Moore, L. A. Soderblom, R. L. Kirk, D. J. Chadwick, D. D. Dawson, L. R. Gaddis, J. M. Boyce, and J. Russell (1992). Geology and distribution of impact craters on Venus: What are they telling us? J. Geophys. Res. 97: 13257–13302.

Schloerb, F. P., S. E. Robinson, and W. M. Irvine (1980). Observation of CO in the stratosphere of Venus via its $J = 0$–1 rotation transition. Icarus 43: 121–127.

Schmidt, O. Yu. (1957). *Origin of the Earth and Planets*, 4th ed. Moscow: Nauka.

Schofield, J. T., and D. J. Diner (1983). Rotation of Venus' polar dipole. Nature 305: 116.

Schofield, J. T., and F. M. Taylor (1982). Net thermal emission from the Venus atmosphere. Icarus 52: 245.

———. (1983). Measurements of the mean, retrieved solar fixed temperature and cloud structures of the middle atmosphere of Venus. Q. J. R. Met. Soc. 109: 57–80.

Schofield, J. T., F. W. Taylor, and D. J. McCleese (1982). The global distribution of water vapor in the middle atmosphere of Venus. Icarus 52 (no. 2): 263–278.

Schramm, D. N. (1978). Supernovae and solar system formation. In *Protostars and Planets*, ed. T. Gehrels. Tucson: University of Arizona Press.

———. (1982). r-Process and nuclear cosmochronology. In *Essays in Nuclear Astrophysics*, ed. C. A. Barner, D. D. Clayton, and D. N. Schramm. Cambridge: Cambridge University Press.

Schubert, G. (1979). Subsolidus convection in the mantles of terrestrial planets. Ann. Rev. Earth Planet. Sci. 7: 289.

———. (1983). General circulation and the dynamical state of the Venus atmosphere. In *Venus*, ed. D. M. Hunten, L. Colin, T. M. Donahue, and V. I. Moroz. Tucson: University of Arizona Press.

Schubert, G., and J. Whitehead (1969). Moving flame experiment with liquid mercury: Possible implications for the Venus atmosphere. Science 163: 71–72.

Schubert, G., and R. E. Young (1970). The 4-day Venus circulation driven by periodic thermal forcing. J. Atmos. Sci. 27: 523–528.

Schubert, G., C. Covey, A. Del Genio, L. S. Elson, G. Keating, A. Seiff, R. E. Young, J. Apt, C. C. Counselman, A. J. Kliore, S. S. Limae, H. E. Revercomb, L. A. Sromovsky, V. E. Suomi, F. W. Taylor, R. Woo, and U. von Zahn (1980). Structure and circulation of the Venus atmosphere. J. Geophys. Res. 85 (no. A13): 8007–8025.

Schubert, G., M. N. Ross, D. J. Stevenson, and T. Spohn (1988). Mercury's thermal history and the generation of its magnetic field. In *Mercury*, ed. F. Vilas, C. R. Chapman, and M. S. Matthews. Tucson: University of Arizona Press.

Schubert, G., S. C. Solomon, D. L. Turcotte, M. J. Drake, and N. H. Sleep (1992). Origin and thermal evolution of Mars. In *Mars*, ed. H. H. Kieffer, B. M. Jakosky, C. W. Snyder, and M. S. Matthews. Tucson: University of Arizona Press.

Schultz, P. H. (1992). Atmospheric effects on Ejecta emplacement and crater formation on Venus from Magellan. J. Geophys. Res. 97: 16183–16248.

Sclater, J. G., C. Jaupart, and D. Galson (1980). The heat flow through oceanic and continental crust and the heat loss of the Earth. Rev. Geophys. Space Phys. 18: 269–311.

Secchi, A. (1864). Sur les raies atmosphériques des planètes. Rendus Acad. Sci. Paris 59: 182; 60: 543.

Seiff, A. (1982). Dynamical implications of observed thermal contrasts in Venus' upper atmosphere. Icarus 51: 574–592.

———. (1983). Thermal structure of the atmosphere of Venus. In *Venus*, ed. D. M. Hunten, L. Colin, T. M. Donahue, and V. I. Moroz. Tucson: University of Arizona Press.

Seiff, A., and D. B. Kirk (1982). Structure of the Venus mesosphere and lower thermosphere from measurements during entry of the Pioneer Venus probes. Icarus 49: 49–70.

Seiff, A., D. B. Kirk, R. E. Young, R. C. Blanchard, J. T. Findlay, G. M. Kelly, and S. C. Sommer (1980). Measurements of thermal structure and thermal contrasts in the atmosphere of Venus and related dynamical observations: Results from the four Pioneer Venus probes. J. Geophys. Res. 85 (no. A13): 7903–7933.

Seiff, A., J. T. Schofield, A. J. Kliore, F. W. Taylor, K. E. Revercomb, L. A. Gromovsky, V. V. Kerzhanovich, V. I. Moroz, and M. Ya. Marov (1985). Models of the structure of the atmosphere of Venus from the surface to 100 kilometers altitude. In *The Venus International Reference Atmosphere*, ed. A. J. Kliore, V. I. Moroz, and G. M. Keating. Adv. Space Res. 5 (no. 11): 3–58.

Selivanov, A. S., V. P. Chemodanov, M. K. Naraeva, et al. (1976). Television experiment on the surface of Venus. Kosmich. Issled. 14 (no. 5): 674–677.

Selivanov, A. S., Yu. M. Gektin, M. A. Gerasimov, B. I. Nosov, M. K. Naraeva, A. S. Panfilov, A. S. Titov, A. C. Fokin, and V. P. Chemodanov (1983a). Continuation of television studies of Venus' surface from landers. Kosmich. Issled. 21: 176–182.

Selivanov, A. S., N. A. Avatkova, I. M. Bokshtein, et al. (1983b). First color panoramas of Venus' surface, transmitted by the "Venera 13–14" spacecraft. Kosmich. Issled. 21: 183–189.

Senske, D. A. (1990). Geology of the Venus equatorial regions from Pioneer Venus radar imaging. Earth, Moon, and Planets 50–51: 305–327.

Senske, D. A., G. G. Schaber, and E. R. Stofan (1992). Regional topographic rises on Venus: Geology of Western Eistla Regio and comparison to Beta Regio and Atla Regio. J. Geophys. Res. 97: 13395–13420.

Shapiro, I. I., D. Campbell, and W. M. De Campli (1979). Nonresonance rotation of Venus? Astrophys. J. Lett. 230: 123–126.

Shari, V. P. (1976). Radiative thermal fluxes in the lower atmosphere of Venus. Kosmich. Issled. 14 (no. 1): 97–110.

Shari, V. P., and N. L. Lukashevich (1977). Characteristics of non-polarized light scattering for gamma distribution of spherical particles. Preprint, M. V. Keldysh Inst. Appl. Math., USSR Acad. Nauk (no. 105).

Sharonov, V. V. (1951). Photometric comparison of crepuscular phenomena on the Earth and Venus. Astron. Zh. 28 (no. 5): 382–387.

———. (1953a). Refraction and crepuscular phenomena in Venus' atmosphere. Vestn. Leningr. Univ. 8: 51–80.

———. (1953b). Colorimetric observations of Venus and Jupiter. Astron. Tsirkulyar 138: 7.

———. (1958). *Nature of the Planets*. Moscow: Fizmatgiz.

———. (1963). Visual colorimetric comparison of the planets with the Sun. Astron. Zh. 30 (no. 5): 532–539.

———. (1965). *The Planet Venus*. Moscow: Nauka.

Shepard, M. K., R. E. Arvidson, B. Fegley, Jr., and R. A. Brackett (1994). A ferroelectric model for the low emissivity of highlands on Venus. Geophys. Res. Lett. 21: 469–472.

Shimazaki, T., R. C. Whitten, and H. T. Woodward (1984). The dayside Venus ionosphere. Icarus 60 (no. 3): 654–674.

Sill, G. T. (1972). Sulfuric acid in the Venus clouds. Comm. Lunar Planet. Lab. 9: 191–198.

———. (1975). The composition of the ultraviolet dark markings on Venus. J. Atmos. Sci. 32: 1201–1204.

———. (1983). The clouds of Venus: Sulfuric acid by the lead chamber process. Icarus 53 (no. 1): 10–17.

Simino, J. B. (1982). The composition and vertical structure of the lower cloud deck on Venus. Icarus 51 (no. 2): 334–357.

Simonenko, A. K. (1985). *Asteroids or Thorny Path of Research*. Moscow: Nauka.

Sinton, W. M. (1963a). Infrared observations of Venus. Mem. Soc. Roy. Sci. Liège 7 (no. 1): 300–313; 364–368.

———. (1963b). Recent infrared spectra of Mars and Venus. J. Quant. Spectrosc. and Radiat. Transfer *3* (no. 4): 551–558.

Sinton, W. M., and J. Strong (1960). Radiometric observations of Venus. Astrophys. J. 131 (no. 2): 470–490.

Sjorgen, W. L., R. J. Phillips, P. W. Birkeland, and R. N. Wimberly (1980). Gravity anomalies on Venus. J. Geophys. Res. 85: 8295–8302.

Sjorgen, W. L., B. G. Bills, P. W. Birkeland, N. A. Nottinger, S. J. Ritke, and R. Phillips (1983). Venus gravity anomalies and their correlations with topography. J. Geophys. Res. 88: 1119–1128.

Slipher, V. M. (1909). The spectra of the major planets. Lowell Obs. Bull. no. 42.

———. (1933). Spectrographic studies of the planets. Monthly Notices of the RAS 93: 657.

Smith, J. V. (1982). Heterogeneous growth of meteorites and planets. J. Geology 90 (no. 1): 2–48.

Smith, W. B., R. P. Ingalls, I. I. Shapiro, and M. E. Ash (1970). Surface-height variations on Venus and Mercury. Radio Sci. 5 (no. 2): 411–423.

Smrekar, S. E. (1994). Evidence for active hotspots on Venus from analysis of Magellan gravity data. Icarus 112: 2–26.

Smrekar, S. E., and S. C. Solomon (1992). Gravitational spreading of high terrain in Ishtar Terra, Venus. J. Geophys. Res. 97: 16121–16148.

Sobolev, V. V. (1944). On the optical properties of Venus' atmosphere. Astron. Zh. 21 (no. 5): 241–244.

———. (1964). Study of Venus' atmosphere, 1. Astron. Zh. 41: 97–103.

———. (1968). Study of Venus' atmosphere, 2. Astron. Zh. 45 (no. 1): 169–176.

———. (1972). *Light Scattering in Planetary Atmospheres*. Moscow: Nauka.

Solomon, S. C. (1979). Formation, history, and energetics of cores in the terrestrial planets. Phys. Earth Planet. Int. 19: 168–182.

———. (1993). A tectonic resurfacing model for Venus (abstract). Lunar Planet Sci. 24: 1331–1332.

Solomon, S. C., and J. W. Head (1982). Mechanisms for lithospheric heat transport on Venus: Implications for tectonic style and volcanism. J. Geophys. Res. 87: 9239–9246.

Solomon, S. C., S. E. Smrekar, D. L. Bindschadler, R. E. Grimm, W. M. Kaula, G. E. McGill, R. J. Phillips, R. S. Saunders, G. Schubert, S. W. Squyres, and E. R. Stofan (1992). Venus tectonics: An overview of Magellan observations. J. Geophys. Res. 97: 13199–13256.

Sorokhtin, O. G. (1974). *Global Evolution of the Earth*. Moscow: Nauka.

Spinrad, H. (1962a). On the continuous spectrum of Venus and Jupiter. Publ. Astron. Soc. Pacif. 74 (no. 437): 156–158.

———. (1962b). A search for water vapor and trace constituents in the Venus atmosphere. Icarus 1 (no. 3): 266–270.

———. (1962c). Spectroscopic temperature and pressure measurements in the Venus atmosphere. Publ. Astron. Soc. Pacif. 74 (no. 438): 187–201.

Spinrad, H., and E. H. Richardson (1965). An upper limit to the molecular oxygen content of the Venus atmosphere. Astrophys J. 141: 282–286.

Spinrad, H., and S. J. Shawl (1966). Water vapor on Venus—a confirmation. Astrophys. J. 146: L328–L329.

Spreiter, J. R. (1976). Solar wind interactions with the planets Mercury, Venus and Mars. Proceedings of a seminar held in November 1975, Moscow. NASA SP-397: 135.

Squyres, S. W., D. M. Janes, G. Baer, D. L. Bindschadler, G. Schubert, V. L. Sharpton, and E. R. Stofan (1992a). The morphology and evolution of coronae on Venus. J. Geophys. Res. 97: 13611–13634.

Squyres, S. W., D. G. Janokowski, M. Simons, S. C. Solomon, B. H. Hager, and G. E. McGill

(1992b). Plains tectonism on Venus: The deformation belts of Lavinia Planitia. J. Geophys. Res. 97: 13579–13600.

Sromovsky, L. A., H. E. Revercomb, and V. E. Suomi (1985). Temperature structure of the lower atmosphere of Venus: New results derived from Pioneer Venus entry probe measurements. Icarus 2: 458.

Staelin, D. H., and A. H. Barrett (1966). Spectral observations of Venus near 1-centimeter wavelength. Astrophys. J. 144 (no. 1): 352.

Starodubtseva, O. M. (1989). Variations of polirametric properties of Venus based on ground-based observations in 1980. Astron. Vestnik (Solar System Research) 23 (no. 3): 233–242.

Stevenson, D. J., T. Spohn, and G. Schubert (1983). Magnetism and thermal evolution of the terrestrial planets. Icarus 54 (no. 3): 466.

Stewart, A. I. F. (1972). Mariner 6 and 7 ultraviolet spectrometer experiment: Implications of CO_2, CO, and O airglow. J. Geophys. Res. 77 (no. 1): 55–68.

Stewart, A. I. F., and C. A. Barth (1979). Ultraviolet night airglow of Venus. Science 205: 59–62.

Stewart, A. I. F., D. E. Anderson, L. W. Esposito, and C. A. Barth (1979). Ultraviolet spectroscopy of Venus: Initial results from the Pioneer Venus orbiter. Science 203: 777–779.

Stofan, E. R., S. W. Squyres, V. L. Sharpton, G. Schubert, D. M. Janes, G. Baer, and D. L. Bindschadler (1992). Global distribution and characteristics of coronae and related features on Venus: Implications for origin and relation to mantle process. J. Geophys. Res. 97: 13347–13378.

Stone, P. H. (1968). Some properties of Hadley regimes on rotating and non-rotating planets. J. Atmos. Sci. 25 (no. 4): 644–657.

———. (1975). The dynamics of the atmosphere of Venus. J. Atmos. Sci. 32: 1005–1016.

Strangeway, R. J. (1991). Plasma waves at Venus. Space Sci. Rev. 55: 275–316.

———. (1993). The Pioneer Venus orbiter entry phase. Geophys. Res. Lett. 20 (no. 23): 2715–2717.

Strangeway, R. J., C. T. Russell, C. M. Ho, and L. H. Brace (1993). Plasma waves observed at low altitudes in the tenuous Venus nightside ionosphere. Geophys. Res. Lett. 20 (no. 23): 2767–2770.

Strom, R. G., G. G. Schaber, and D. D. Dawson (1994). The global resurfacing of Venus. J. Geophys. Res. 99: 10899–10926.

Strong, J., M. D. Ross, and C. B. Moore (1960). J. Geophys. Res. 65: 2526.

Sukhanov, A. L. (1986). Parquet: Regions of a real plastic dislocation. Geotectonika 20: 294–305.

Sukhanov, A. L., and A. A. Pronin (1988). Spreading features on Venus. Lunar Planet. Sci. 19: 1147–1148.

Suomi, V. E. (1974). Cloud motions on Venus. In *The Atmosphere of Venus*, ed J. Hansen. New York: Goddard Institute of Space Studies.

Suomi, V. E., and S. S. Limaye (1978). Venus: Further evidence of vortex circulation. Science 201: 1009–1011.

Suomi, V. E., L. A. Sromovsky, and H. Revercomb (1980). Net radiation in the atmosphere of Venus: Measurements and interpretations. J. Geophys. Res. 85: 8200–8218.

Suppe, J., and C. Connors (1992). Critical taper wedge mechanics of fold-and-thrust belts on Venus: Initial results from Magellan. J. Geophys. Res. 97.

Surkov, Yu. A., and L. P. Moskaleva (1990). Cosmochemical research of bodies in the solar system. In *The Third Space Decade*, ed. B. V. Rauschenbach. Moscow: Nauka.

Surkov, Yu. A., B. M. Andreichikov, and O. M. Kalinkina (1973). On the ammonia content of Venus' atmosphere from "Venera-8" data. Doklady Akad. Nauk USSR 213 (no. 2): 296–298.

Surkov, Yu. A., F. F. Kirnozov, V. K. Khristianov, B. N. Korchuganov, V. N. Glazer, and V. F. Ivanov

(1976a). Density of surface rock on Venus from data obtained by the Venera 10 automatic interplanetary station. Kosmich. Issled. 14: 697–703.

Surkov, Yu. A., F. F. Kirnozov, V. N. Glazov, A. G. Dunchenko, and L. P. Tatsil (1976b). The content of natural radioactive elements in Venusian rock as determined by Venera 9 and Venera 10. Kosmich. Issled. 14: 704–709.

Surkov, Yu. A., B. M. Andreichikov, and O. M. Kalinkina (1977). Gas-analysis equipment of the automatic interplanetary stations Venera 4, 5, 6, and 8. Space. Sci. Instrum. 3: 301–310.

Surkov, Yu. A., V. F. Ivanova, A. N. Pudov, B. I. Verkin, N. N. Bagrov, and A. P. Pilipenko (1978). Mass spectral study of the chemical compositions of the Venus atmosphere by the unmanned space probes Venera 9 and Venera 10. Geochimiya 4: 506.

Surkov, Yu. A., F. F. Kirnozov, V. I. Guryanov, V. N. Glazov, A. G. Dunchenko, V. Kurochkin, E. Raspotny, L. Kharitonova, and L. P. Tatsil (1981). Venus cloud aerosol investigation on the Venera 12 space probe (preliminary data). Geochimiya 7: 3–9.

Surkov, Yu. A., F. F. Kirnozov, and V. N. Glazov (1982). New data on aerosol of the Venus clouds (Preliminary results of investigation on Venera 14 spacecraft). Pisma Astron. Zh. 8 (no. 11): 700–704.

Surkov, Yu. A., V. F. Ivanov, A. N. Pudov, V. A. Pavlenko, N. A. Davydov, and D. N. Sheinin (1983). The water vapor content of Venus' atmosphere from "Venera 13–14" data. Kosmich. Issled. 21 (no. 2): 231–235.

Surkov, Yu. A., V. L. Barsukov, L. P. Moskalyova, V. P. Kharyukova, and A. L. Kemurdzhian (1984). New data on the composition, structure, and properties of Venus rock obtained by Venera 13 and Venera 14. J. Geophys. Res. 89: B393–B402.

Surkov, Yu. A., F. F. Kirnozov, V. N. Glazov, A. G. Dunchenko, and L. P. Tatsil (1987a). Uranium, thorium, and potassium in the Venusian rocks at the landing sites of Vega 1 and 2. J. Geophys. Res. 92: E537–E540.

Surkov, Yu. A., L. P. Moskaleva, V. P. Charynkova, A. I. Dudin, G. G. Smirnov, S. E. Zaytseva, A. N. Tichomirov, and O. S. Manvelyan (1987b). Elementary composition of Venus rocks in the Northeast region of Aphrodite Terra (from Vega 2 lander data). Kosmich. Issled. 25 (no. 5): 751–761.

Surkov, Yu. A., O. P. Scheglov, M. L. Rivkin, D. N. Sheinin, and N. A. Davydov (1987c). Water vapor distribution in the middle and lower atmosphere of Venus. Kosmich. Issled. 25 (no. 5): 678–694.

Surkov, Yu. A., V. F. Ivanova, A. N. Pudov, E. P. Sheretov, B. I. Kolotilin, M. P. Safomov, P. Toma, G. Israel, J. Lespanol, D. Imbo, A. Ozer, and D. Caramel (1987d). Determination of chemical composition of aerosols in the Venus clouds by mass spectral device "Malachit" from "Vega 1" space station. Kosmich. Issled. 25 (no. 5): 744–752.

Sytinskaya, N. N. (1948). *Absolute Photometry of Extended Celestial Objects*. Leningrad: Leningrad State University.

Sze, N. D., and M. B. McElroy (1975). Some problems in Venus aeronomy. Planet. Space Sci. 23: 763–786.

Taylor, F. W. (1975). Interpretation of Mariner 10 infrared observations of Venus. J. Atmos. Sci. 32: 1101–1106.

Taylor, F. W., D. J. Diner, L. S. Elson, M. Manner, D. J. McCleese, J. V. Martonchik, P. E. Reichley, J. T. Houghton, J. Delderfield, J. T. Schofield, S. E. Bradley, and A. P. Ingersoll (1979). Infrared remote sounding of the middle atmosphere of Venus from the Pioneer orbiter. Science 205: 65–67.

Taylor, F. W., R. Beer, M. T. Chanine, D. J. Diner, L. S. Elson, R. D. Haskins, D. J. McCleese, J. V. Martonchik, P. E. Reichley, S. P. Bradley, J. Delderfield, J. T. Schofield, C. Farmer, L. Froidevaux, J. Leung, M. T. Coffey, and J. C. Gille (1980). Structure and meteorology of the middle atmosphere of Venus: Infrared remote sensing from the Pioneer orbiter. J. Geophys. Res. 85 (no. A13): 7963–8006.

Taylor, F. W., D. M. Hunten, and L. V. Ksanfomality (1983). The thermal balance of the middle and upper atmosphere of Venus. In *Venus*, ed. D. M. Hunten, L. Colin, T. M. Donahue, and V. I. Moroz. Tucson: University of Arizona Press.

Taylor, F. W., J. T. Schofield, and P. J. Valdes (1985). Temperature structure and dynamics of the middle atmosphere of Venus. Adv. Space Res. 5 (no. 9): 5–7.

Taylor, H. A., Jr., and P. A. Cloutier (1992). Non-evidence of lightning and associated volcanism at Venus. Space Sci. Rev. 61: 387–391.

Taylor, H. A., Jr., H. C. Brinton, S. J. Bauer, R. E. Hartle, P. A. Cloutier, and R. E. Daniell (1980). Global observations of the composition and dynamics of the ionosphere of Venus: Implications for the solar wind interaction. J. Geophys. Res. 85: 7765–7777.

Taylor, H.A., Jr., R. E. Hartle, H. B. Niemann, L. H. Brace, R. E. Daniell, Jr., S. J. Bauer, and A. J. Kliore (1982). Observed composition of the ionosphere of Venus. Icarus 51 (no. 2): 283–285.

Taylor, W. W. L., F. L. Scarf, C. T. Russell, and L. M. Brace (1979). Absorption of whistler mode waves in the ionosphere of Venus. Science 205: 112–114.

Theis, R. F., and L. B. Brace (1993). Solar cycle variations of electron density and temperature in the Venusian nightside ionosphere. Geophys. Res. Lett. 20 (no. 23): 2719–2722.

Thompson, R. (1970). Venus general circulation is a merry-up-round. J. Atmos. Sci. 27: 1107–1116.

Thomson, J. H., J. E. Ponsonby, G. N. Taylor, and R. S. Roger (1961). A new determination of the solar parallax by means of radar echoes from Venus. Nature 190 (no. 4775): 519–520.

Tikhonov, A. N., and V. Ya. Arsenin (1979). Methods of solution of incorrect problems. Moscow: Physmatgiz, Nauka.

Titov, D. V. (1983). On the possibility of aerosol formation in the chemical reaction of SO_2 with NH_3 in the Venus atmosphere conditions. Kosmich. Issled. 21 (no. 3): 401–409.

Toksoz, M. N., and D. H. Johnston (1975). Evolution of the Moon and terrestrial planets. In *Cosmochemistry of the Moon and Planets*, ed. A. P. Vinogradov. Moscow: Nauka.

Toksoz, M. N., A. T. Hsui, and D. H. Johnston (1978). Thermal evolution of terrestrial planets. Moon and Planets 18: 281–320.

Tomasko, M. G. (1983). The thermal balance of the lower atmosphere of Venus. In *Venus*, ed. D. M. Hunten, L. Colin, T. M. Donahue, and V. I. Moroz. Tucson: University of Arizona Press.

Tomasko, M. G., L. R. Doose, P. H. Smith, and A. Odell (1980a). Measurements of the flux of sunlight in the atmosphere of Venus. J. Geophys. Res. 85: 8167–8186.

Tomasko, M. G., P. H. Smith, V. E. Suomi, L. A. Sromovsky, H. E. Revercomb, F. W. Taylor, D. J. Martonchik, A. Seiff, R. Boese, J. B. Pollack, A. P. Ingersoll, G. Schubert, and C. C. Covey (1980b). The thermal balance of Venus in light of the Pioneer Venus mission. J. Geophys. Res. 85: 8187–8199.

Toon, O. B., and R. P. Turco (1982). The ultraviolet absorber on Venus: Amorphous sulphur. Icarus 51: 358–373.

Toon, O. B., B. Ragent, D. Colburn, J. Blamont, and C. Cot (1982). Large solid particles in the clouds of Venus: Do they exist? Icarus 57: 143–160.

Traub, W. A., and N. P. Carleton (1973). A search for H_2 and O_2 on Venus (abstract). Bull. Amer. Astron. Soc. 5: 299–300.

Travis, L. D. (1975). On the origin of ultraviolet contrasts on Venus. J. Atmos. Sci. 32: 1190–1200.

———. (1978). Nature of the atmospheric dynamics on Venus from power spectrum analysis of Mariner 10 images. J. Atmos. Sci. 35: 1584–1595.

Travis, L. D., D. L. Coffeen, J. E. Hansen, K. Kawabata, A. A. Lacis, W. A. Lane, S. S. Limaye, and H. M. Stone (1979a). Orbiter cloud photopolarimeter investigation. Science 203: 781–785.

Travis, L. D., D. L. Coffeen, A. D. Del Genio, J. E. Hansen, K. Kawabata, A. A. Lacis, W. A. Lane, S. S. Limaye, W. B. Rossow, and H. M. Stone (1979b). Cloud images from the Pioneer Venus orbiter. Science 205: 74–76.

Tryka, K. A., and D. O. Muhleman (1992). Reflection and emission properties on Venus: Alpha Regio. J. Geophys. Res. 97: 13379–13394.

Tsesevich, V. P. (1955). On the direction of rotation axes of Venus. Astron. Circular no. 158: 15–18.

Turco, R. P., R. C. Whitten, and O. B. Toon (1982). Stratospheric aerosols: Observation and theory. Rev. Geophys. Space Phys. 20 (no. 2): 233–279.

Turcotte, D. L. (1993). An episodic hypothesis for Venusian tectonics. J. Geophys. Res. 98: 17061–17068.

Turekian, K. K., and S. P. Clark (1975). The non-homogeneous accumulation model for terrestrial planet formation. J. Atmos. Sci. 32: 1257–1261.

Tyler, G. L., R. A. Simpson, M. J. Maurer, and E. Holmann (1992). Scattering properties of the Venusian surface: Preliminary results from Magellan. J. Geophys. Res. 97: 13115–13140.

Ulrich, R. K. (1982). S-Process. In *Essays in Nuclear Astrophysics*, ed. C. A. Barner, D. D. Clayton, and D. N. Schramm. Cambridge: Cambridge University Press.

Urey, H. C. (1951). The origin and development of the Earth and the terrestrial planets. Geochim. et Cosmochim. Acta 1 (no. 4): 209.

———. (1952). *The Planets, Their Origin and Development*. New Haven: Yale University Press.

Van de Hulst, H. C. (1948). Scattering in a planetary atmosphere. Astrophys. J. 107: 220.

Varanasi, P. (1971). Line width and intensities H_2O-CO_2 mixtures. J. Quant. Spectrosc. and Radiat. Transfer 11 (no. 1): 223–230.

Vasin, V. G. (1978). Numerical experiments with a moving heat source in an annular layer of compressible gas. Preprint, M. V. Keldysh Inst. Appl. Math. Akad. Nauk USSR (no. 93).

———. (1987). Numerical experiments with the "Four-day circulation model of Venus' atmosphere," in Venus' atmosphere. In *Proceedings of M. V. Keldysh Inst. Appl. Math. Akad. Nauk USSR*, ed. M. Ya. Marov, Moscow.

Vasin, V. G., and M. Ya. Marov (1977). On the problem of circulation mechanism on Venus. Preprint, M. V. Keldysh Inst. Appl. Math., Acad. Nauk USSR (no. 6).

Vaucouleurs, G. de (1964). Geometric and photometric parameters of the terrestrial planets. Icarus 3 (no. 3): 187.

Vaucouleurs, G. de, and D. H. Menzel (1960). Results of the occultation of Regulus by Venus, July 7, 1959. Nature 188 (no. 4744): 28–33.

Victor, W. K., and R. Stevens (1961). Exploration of Venus by radar. Science 134 (no. 3471): 46–48.

Vinogradov, A. P. (1971). *Introduction to Geochemistry of the Ocean*. Moscow: Nauka.

———. (1975). Differentiation of lunar material. In *Cosmochemistry of the Moon and Planets*, ed. A. P. Vinogradov. Moscow: Nauka.

Vinogradov, A. P., and V. P. Volkov (1971). On vollastonite equilibrium as mechanism determining the composition of Venus' atmosphere. Geochimiya 7: 755.

Vinogradov, A. P., Yu. A. Surkov, G. M. Chernov, F. F. Kirnozov, and G. V. Nazarkina (1966). Preliminary results of gamma radiation measurements of the lunar surface from the "Luna 10" spacecraft. Kosmich. Issled. 4 (no. 6): 874–879.

Vinogradov, A. P., Yu. A. Surkov, and C. P. Florensky (1968). The chemical composition of Venus' atmosphere based on the data of the interplanetary station Venera 4. J. Atmos. Sci. 25: 535–536.

Vinogradov, A. P., Yu. A. Surkov, B. M. Andreichikov, O. M. Kalinkina, and I. M. Grechischeva (1970). Chemical composition of the Venus atmosphere. Kosmich. Issled. 8 (no. 4): 578–580.

Vinogradov, A. P., Yu. A. Surkov, F. F. Kirnozov, and V. N. Glazov (1973). The contents of uranium, thorium, and potassium rocks of Venus as measured by Venera-8. Icarus 20: 253–259.

Vinogradov, A. P., Yu. A. Surkov, L. P. Moskaleva, and F. F. Kirnozov (1975). Intensity measurements of the spectral gamma radiation composition of Mars from the "Mars-5" spacecraft. Dokl. Akad. Nauk USSR 223: 1336–1339.

VIRA-85. *See* Kliore et al., 1985.

Vogel, H. C. (1874). Untersuchungen uber die spectra der Planeten, gecronte preischrift von der Konig. Leipzig: Gesellsch. in Kopenhagen.

Voitkevich, G. V. (1988). *Fundamentals of the Theory of the Earth's Origin.* Moscow: Nedra.

Volkov, V. P. (1983). *Chemistry of the Atmosphere and Surface of Venus.* Moscow: Nauka.

———. (1987). Scheme of sulfur rotation in the outer shells of planet Venus. Astron. Vestnik (Solar System Research) 21 (no. 2): 165–169.

Volkov, V. P., and I. L. Khodakovsky (1984). Physical-chemical modeling of the mineral composition of Venusian surface rock. Geokhimiya i Kosmokhimiya 11: 32–38.

Volkov, V. P., I. L. Khodakovsky, V. A. Dorofeeva, and V. L. Barsukov (1979). Basic physico-chemical factors controlling chemical composition of the Venus clouds. Geochimiya 12: 1759–1766.

———. (1982). On possible condensates of the main cloud layer of the planet Venus. Geochimiya 1: 3–22.

Volkov, V. P., M. Yu. Zolotov, and I. L. Khodakovsky (1986). Lithospheric-atmospheric interaction on Venus. In *Chemistry and Physics of Terrestrial Planets*, ed. S. K. Saxena. New York: Springer-Verlag.

Volkov, V. P., M. Ya. Marov, V. N. Lebedev, Yu. I. Sidorov, and V. P. Shari (1989). Clouds. In *The Planet Venus*, ed. V. L. Barsukov and V. P. Volkov. Moscow: Nauka.

von Zahn, U., and V. I. Moroz (1985). Composition of the Venus atmosphere below 100 km altitude. In *The Venus International Reference Atmosphere*, ed. A. J. Kliore, V. I. Moroz, and G. M. Keating. Adv. Space Res. 5 (no. 11): 173–196.

von Zahn, U., D. Krankowsky, K. Mauersberger, A. O. Nier, and D. M. Hunten (1979). Venus thermosphere: In situ composition measurements, the temperature profile, and the hemopause altitude. Science 203: 768–770.

von Zahn, U., K. H. Fricke, D. M. Hunten, D. Krankowsky, K. Mauersberger, and A. O. Nier (1980). The upper atmosphere of Venus during morning conditions. J. Geophys. Res. 85: 7829–7840.

von Zahn, U., S. Kumar, H. Niemann, and R. Prinn (1983). Composition of the Venus atmosphere. In *Venus*, ed. D. M. Hunten, L. Colin, T. M. Donahue, and V. I. Moroz. Tucson: University of Arizona Press.

Vorder Bruegge, R. W., and J. W. Head (1989). Fortuna Tessera, Venus: Evidence of horizontal convergence and crustal thickening. Geophys. Res. Lett. 16: 699–702.

Walker, J. C. G. (1975). Evolution of the atmosphere of Venus. J. Atmos. Sci. 32: 1248–1255.

———. (1977). *Evolution of the Atmosphere.* New York: Macmillan.

Walker, R. G., and C. Sagan (1966). The ionospheric model of the Venus microwave emission: An obituary. Icarus 5: 105.

Walker, J. C. G., P. B. Hays, and J. F. Kasting (1981). A negative feedback mechanism for the long-term stabilization of Earth's surface temperature. J. Geophys. Res. 86: 9776–9782.

Waltersheld, R. L., G. Schubert, M. Newman, and A. J. Kliore (1985). J. Atmos. Sci. 42: 1982–1990.

Ward, W. R., and W. M. De Campli (1979). Comments on the Venus rotation pole. Astrophys. J. 230: L117–L121.

Ward, D. B., G. E. Gull, and M. Harwitt (1977). Far infrared spectrum observations of Venus, Mars, and Jupiter. Icarus 30: 295–300.

Wasserburg, G. J., and D. J. De Paolo (1979). Models of Earth structure inferred from neodymium and strontium isotopic abundances. Proc. Nat. Acad. Sci. USA 75: 3594.

Wasserburg, G. J., and D. A. Popanastassioo (1982). Some short-lived nuclides in the early solar system—the connection with interstellar medium. In *Essays in Nuclear Astrophysics*, ed. C. A. Barner, D. D. Clayton, and D. N. Schramm. Cambridge: Cambridge University Press.

Wasserburg, G. J., G. J. F. MacDonald, F. Hoyle, and W. A. Fowler (1964). Relative contribution of uranium, thorium, and potassium to heat production in the Earth. Science 143: 465–467.

Wasserburg, G. J., D. A. Popanastassioo, F. Tera, and J. C. Huneke (1977). Outline of a lunar chronology. In *The Moon: A New Appraisal*. London: Royal Society.

Watson, A. J., T. M. Donahue, D. H. Stedman, R. G. Knollenberg, B. Ragent, and J. Blamont (1979). Oxides of nitrogen and the clouds of Venus. Geophys. Res. Lett. 6: 743–746.

Watson, A. J., T. M. Donahue, and J. C. G. Walker (1982). The dynamics of a rapidly escaping atmosphere: Applications to the evolution of Earth and Venus. Icarus 48: 150–166.

Wattson, R. B., and L. S. Rothman (1986). Determination of vibrational energy levels and parallel band intensities of CO_2 by direct numerical diagonalization. J. Mol. Spectrosc. 119: 83–100.

———. (1992). Direct numerical diagonalization: Wave of the future. J. Quant. Spectrosc. Rad. Transfer 48: 763–780.

Wegener, A. (1924). *The Origin of Continents and Oceans*. London: Methuen.

Weitz, C. M., and A. T. Basilevsky (1993). Magellan observations of the Venera and Vega landing site regions. J. Geophys. Res. 98: 17069–17097.

Wetherill, G. W. (1975). Late heavy bombardment of the Moon and terrestrial planets. Proc. Lunar Planet. Sci. Conf. 4: 1539–1559.

———. (1981). Solar wind origin of ^{36}Ar on Venus. Icarus 46: 70–80.

———. (1985). Giant impacts during the growth of the terrestrial planets. Science 228: 877–879.

Whitecomb, S. E., R. M. Hildebrand, J. Keene, R. F. Stiening, and D. A. Harper (1979). Submillimeter brightness temperatures of Venus, Jupiter, Uranus and Neptune. Icarus 38: 75–80.

Wildt, R. (1940). Note on the surface temperature of Venus. Astrophys. J. 91: 266–268.

———. (1966). The greenhouse effect in a gray planetary atmosphere. Icarus 5: 24.

Williams, B. G., N. A. Mottinger, and N. D. Panagiotacopulos (1983). Venus gravity field: Pioneer Venus orbiter navigation results. Icarus 56 (no. 3): 578–589.

Wilson, W. J., M. J. Klein, R. K. Kakar, S. Gulkis, E. T. Olsen, and P. T. P. Ho (1981). Venus, 1: Carbon monoxide distribution and molecular-line searches. Icarus 45: 624–627.

Winick, J. R., and A. I. Stewart (1980). Photochemistry of SO_2 in Venus' upper cloud layers. J. Geophys. Res. 85: 7849–7860.

Woo, R., J. W. Armstrong, and A. Ishimaru (1980). Radio occultation measurements of turbulence in the Venus atmosphere by Pioneer Venus. J. Geophys. Res. 85: 8031–8038.

Woo, R., J. W. Armstrong, and A. J. Kliore (1982). Small-scale turbulence in the atmosphere of Venus. Icarus 52: 335–345.

Wood, J. A. (1963). Physics and chemistry of meteorites. In *The Solar System 4: The Moon, Meteorites and Comets*, ed. B. M. Middlehurst and G. P. Kuiper. Chicago: University of Chicago Press.

Wood, J. A. (1975). Overview of types of lunar rock and a comparison of the lunar and terrestrial crust. In *Cosmochemistry of the Moon and Planets*, ed. A. P. Vinogradov. Moscow: Nauka.

Yakovlev, O. I. (1974). Propagation of radiowaves in the solar system. Moscow: Radio.

Yakovlev, O. I., and S. S. Matyugov (1982). Neutral atmosphere of Venus as measured by radio occultation by Venera 9 and Venera 10. Manuscript.

Yakovlev, O. I., A. I. Efimov, S. S. Matyugov, T. S. Timofeeva, E. V. Chub, and G. D. Yakovleva (1978). Radioscopy of the nighttime atmosphere of Venus by probes Venera 9 and Venera 10. Kosmich. Issled. 16: 88–93.

Young, A. T. (1973). Are the clouds of Venus sulfuric acid? Icarus 18: 564–582.

———. (1975). The clouds of Venus. J. Atmos. Sci. 32: 1125–1132.

———. (1977). An improved Venus cloud model. Icarus 32: 1–26.

———. (1979). Chemistry and thermodynamics of sulfur on Venus. Geophys. Res. Lett. 6: 49–50.

———. (1983). Venus cloud microphysics. Icarus 56: 568–577.

Young, A. T., and L. D. G. Young (1973). Comments on the composition of the Venus cloud tops in light of recent spectroscopic data. Astrophys. J. 179: 39–43.

Young, L. D. G. (1972). High resolution spectra of Venus: A review. Icarus 17: 632–658.

———. (1974). Infrared spectra of Venus. In *Exploration of Planetary Systems*, ed. A. Woszczyk and A. Iwaniszwska. Dordrecht: Reidel.

Young, R. E., G. Schubert, and K. E. Torrance (1972). Non-linear motions induced by moving thermal waves. J. Fluid Mech. 54: 163–187.

Yung, Y. L., and W. D. De More (1982). Photochemistry of the stratosphere of Venus: Implications for atmospheric evolution. Icarus 51: 199.

Zahnle, K. J. (1992). Airburst origin of dark shadows on Venus. J. Geophys. Res. 97: 10243–10255.

Zahnle, K. J., and J. F. Kasting (1985). Mass fractionation during transonic hydrodynamic escape and applications for loss of water from Venus and Mars. Icarus 68: 462–480.

Zahnle, K. J., J. F. Kasting, and J. B. Pollack (1988). Evolution of a steam atmosphere during Earth's accretion. Icarus 74: 62–97.

Zasova, L. V., V. A. Krasnopolsky, and V. I. Moroz (1981). Vertical distribution of SO_2 in upper cloud layer of Venus and origin of UV absorption. Adv. Space Res. 1: 13–16.

Zasova, L. V., D. Shpenkukh, V. I. Moroz, K. Shefer, E. A. Ustinov, V. Deler, V. M. Linkin, D. Ertel, A. V. Dyachkov, X. Bekker-Ros, I. A. Matsygorin, I. Nopirovsky, V. V. Kerzhanovich, R. Dyubua, A. N. Lipatov, V. Shtadthaus, and A. L. Shurupov (1985). Infrared experiment on Venera 15 and Venera 16 spacecraft, 3: Some conclusions on the structure of clouds based on the spectra analysis. Kosmich. Issled. 23 (no. 2): 221–235.

Zeldovich, Ya. B., and Yu. P. Raizer (1966). *Physics of Shock Waves and High-Temperature Hydrodynamic Phenomena*. Moscow: Nauka.

Zhang, M. H. G., J. G. Luhmann, and A. J. Kliore (1990). An observational study of the nightside ionospheres of Mars and Venus with radio occultation methods. J. Geophys. Res. 95: 17095–17102.

Zharkov, V. N. (1983). *Interior Structure of the Earth and Planets*. Moscow: Nauka.

———. (1985). Models of Venus' internal structure. In *The Planet Venus*, by L. V. Ksanfomality. Moscow: Nauka.

Zharkov, V. N., and V. P. Trubitsyn (1980). *Physics of Planetary Interiors*. Moscow: Nauka.

Zharkov, V. N., and I. Ya. Zasursky (1981). Distribution of shear stresses in the silicate shell of Venus. Astron. Vestnik 15 (no. 1): 11–16.

———. (1982). Physical model of Venus. Astron. Vestnik 16 (no. 1): 18–26.

Zharkov, V. N., S. V. Kozlovskaya, and I. Ya. Zasursky (1981). Interior structure and comparative analysis of the terrestrial planets. Adv. Space Res. 1 (no. 7): 117.

Zöllner, J. C. F. (1865). Photometrische Untersuchungen mit besonderer Rücksicht auf die physiche Beschaffenheit der Himmelskörper. Leipzig.

Zolotov, M. Yu. (1985). Sulfur-containing gases in the Venus atmosphere and stability of carbonates. Lunar Planet. Sci. 16: 942–943.

Zolotov, M. Yu., and I. L. Khodakovsky (1985). Composition of volcanic gases on Venus. Lunar and Planet. Sci. 16, pt. 2: 946–947.

Zotkin, I. T., and A. N. Chigorin (1953). Observations of Venus in 1949 and 1950. Bull. VAGO, no. 12: 3–9.

Zuev, V. E. (1966). *Transmittance of the Atmosphere for Visible and Infrared Rays*. Moscow: Radio.

Note: Page numbers in italics refer to figures.